AF518361

Green Technology for Environment Management

THE EDITORS

Dr. Anil Kumar Biswal awarded Ph. D. in Botany from Utkal university. He has been imparting teaching at both undergraduate and post graduate level for last 32 years. At present, he is serving as Professor and Head, Department of Botany, Maharaj Sriram Chandra Bhanjdeo University. To his credit,he has supervised twenty M.Phil. students and four students have been awarded Ph.D degree under his guidance. He was a receipient of "Dr. Gopinath Panigrahi commemoration award for significant contribution in Taxonomy" for the year 2016 by Orissa Botanical Botanical Society. He is life members of several learned societies namely Orissa Botanical Society, National Envirnmental Science Academy and Indian Association for Angiosperm Taxonomy. He has been conferred with Fellow of the Association for Plant Taxonomy, India (FAPT) for recognition in the of field Plant Taxonomy. He has authored five books namely Forest tribes of Orissa Vol.3: The Juang; Similipal Biosphere Reserve:Floral Diversity-Vol.I ; Coastal Biodiversity of undivided Balasore District; Pteridophytic Flora of Similipal and Facets of Similipal. He has made an humble contribution to the floristic studies of Odisha. He has added Exacum tenue (Gentianaceae)as new record for India, Coralodiscus (Gesneriaceae) as a new generic record and Nymphoides parviflora (Menyanthaceae) as new taxon for Eastern Ghats. He has been actively involved in creating awareness among students and general public for conservation and sustainable utilization of Biodiversity .

Dr. H.N. Thatoi, is a Professor, Department of Biotechnology, North Orissa University, Odisha, India and is also presently working as Director of Centre for Similipal Studies, Director, HRDC, Maharaj Sriram Chandra Bhanjdeo University, North Orissa University. He has also served as the Reader and Head, Department of Biotechnology, Biju Patnaik University and Technology, Odisha, Asst. Prof. (Microbiology), Basic Science College, OUAT, Bhubaneswar and Sr. Scientist, M.S. Swaminathan Research Foundation, Chennai. Prof. Thatoi obtained his M. Sc, M. Phil. and Ph. D. from Utkal University, Odisha, and has more than 18 years of teaching and 25 years of research experience. He has contributed immensely to the field of microbiology and biotechnology throughout his research and teaching career. His research activities are basically based on exploration microbial diversity, bioremediation, biofuel production, mangrove biology, bioactivity studies in medicinal plants and mushrooms, ethnopharmacology and in silico drug design and modeling etc. Prof. Thatoi has handled many research projects funded by the State and Central Government organisations including DST, Govt. of Odisha, UGC-DAE, Govt. of India, Department of Forest, Govt. of Odisha, etc. So far, 19 students have obtained Ph.D.s under his guidance and some 7 students continuing. Besides, several M.Sc., M.Tech and B.Tech students have received his guidance for their dissertation works. Prof. Thatoi has published more than 250 research papers in various nationally and internationally reputed journals, book chapters and proceedings of seminars. Prof. Thatoi has also authored 20 reference books and Seminar Proceedings by various notable publishers like, Elsevier publication, Springer publication, Taylor & Francis publication (CRC & Apple Publications), Studium Press LLC USA, IK International, Narosa publishers, Biotech Books, APH Publication, North Orissa University Publication etc. He has also authored two text books on Microbiology and Immunology, published by India Tech Publication, New Delhi, and Practical Biotechnology: Principles and Protocols by IK International Publication, New Delhi for Graduation and Post-Graduation students in the field of Life Sciences and Biotechnology.

Green Technology for Environment Management

– Editors –

Dr. Anil Kumar Biswal

Dr. H.N. Thatoi

2023

Daya Publishing House®

A Division of

Astral International Pvt. Ltd.

New Delhi – 110 002

ISBN 978-93-5461-379-1

Publisher's Note:

Every possible effort has been made to ensure that the information contained in this book is accurate at the time of going to press, and the publisher and author cannot accept responsibility for any errors or omissions, however caused. No responsibility for loss or damage occasioned to any person acting, or refraining from action, as a result of the material in this publication can be accepted by the editor, the publisher or the author. The Publisher is not associated with any product or vendor mentioned in the book. The contents of this work are intended to further general scientific research, understanding and discussion only. Readers should consult with a specialist where appropriate.

Every effort has been made to trace the owners of copyright material used in this book, if any. The author and the publisher will be grateful for any omission brought to their notice for Acknowledgment in the future editions of the book.

Published by : **Daya Publishing House®**
A Division of
Astral International Pvt. Ltd.
– ISO 9001:2015 Certified Company –
4736/23, Ansari Road, Darya Ganj
New Delhi-110 002
Ph. 011-43549197, 8130496929
e-mail: info@astralint.com
Website: www.astralint.com

Digitally Printed at : **Replika Press Pvt. Ltd.**

Preface

Our Earth is considered as a natural ecosystem which can support all life forms on it. The current rate of resource utilisation and consumption pattern are depleting the planet's finite resources and damaging life-supporting ecosystems. The human activities lead to enable land use changes of great magnitude and to alter global carbon and nitrogen cycle. A large number of toxic substances are increasingly found in air, water, soil, and flora and fauna. Humans in their quest to derive food, feed and other valuable resources modified the earth's landscape. Rapid industrialization, urbanization, land use changes and unprecedented natural resource use driven by rising human population lead to several environmental crisis. The environment on account of human's action has been experiencing imbalances and ecological catastrophe. In recent years, the environmental issues like global climate change, biodiversity loss, the rapid depletion of natural resources, degradation of global commons, stratospheric ozone depletion have been restricting the safe operating space and transgressing the planetary boundaries endangering the existence of human societies. The global environmental problems if not scientifically managed at this time, it may end up in the civilizational collapse. The global environmental challenges can be managed by adopting sustainable green technologies which is based on the principles of environmental sustainability with social and ecological sustainability. Green technologies also known as sustainable technology is the application of environmental friendly technologies aimed to safeguard Earth's life support system, minimize the environmental degradation, mitigate climate change through low carbon emissions, conserve the natural resources for posterity and strive to maintain environmental health, which can bring human well-being and curb the negative impacts of human involvement. Green technology offers us the best hope to reverse the effects of climate change and pollution. The book epitomizes the potential of green technologies for environmental management and sustainable development. It will cater to the needs of environmentalists, microbiologists, agriculturalists and those who are interested in environmental stewardship and sustainability paradigms. The book covers some of the topics discussed in the seminar Green technologies for environmental management conducted by Department of Botany, Maharaj Sriram Chandra Bhanjadeo University (Erstwhile North Orissa University), Baripada, Odisha, along with invited articles from various experts working in the field of biodiversity conservation and livelihood management. The book contains 35 papers comprising biodiversity issues and livelihood managements in two different sections. The book is extremely important for students, researchers, policymakers to the environment in general and plant science in particular.

Editors

Contents

Chapter 1

India's Biodiversity Hotspot: Eastern Himalaya-Characteristics and Diversity

Sudhansu Sekhar Dash

Botanical Survey of India,
CGO Complex, 3 MSO Building, Kolkata, West Bengal
e-mail: ssdash2002@gmail.com

ABSTRACT

Eastern Himalayan region, in the broad sense is one of the regions having richest numbers of species and varied fragile complex eco-systems in the North Temperate Zone. The region is also amongst the world's 36 Hotspots of biodiversity designated by Conservation International. This unique region, with an area of 92000 sq. km and with elevations ranging from 100 meters to 8575 meters, harbours approximately 7,517 species of vascular plants, belonging to 193families and 1304 genera constituting 26 percent of flora and more than 10 percent of endemics of the country. Subtropical floral elements (31.58 per cent) have the maximum representation followed by tropical elements (29.77 per cent), temperate (24.27 per cent) and alpine (7.79 per cent). Orchidaceae is the most dominant family followed by Leguminosae and Poaceae. The genus Rhododendron is the most dominant in the region with more than 97 per cent of Indian species belonging to it. Moreover, 466 species are exclusively endemic to the region. High mountain ranges and deep valleys of Vadse hills of Kurung Kumey, Anini and Dalai valley of Dibang valley, Lali Ani of Papum Pare district, Zero plateau of Arunachal Pradesh; Lhonak valley of Sikkim are the noteworthy high endemic areas. The flora of eastern Himalaya is exceptionally interesting not only for richness in extremely diversifications, but also for its phyto-geographic affinities. Many species found in the state have significantly contributed to the economic development and commercial interest of the inhabitants of this region. During the period of expeditions we have been able to collect a huge number of specimens of vascular plants from nearly all parts of eastern Himalaya. An account of the observations on the characteristics and diversity of the plants, based on the survey and exploration, are presented here.

Keywords: *Eastern Himalaya, Arunachal Pradesh, Distribution, Floristic diversity, Himalayan flora, Phyto-geographical affinities.*

Introduction

Study Area: Location Topography and Climate

The eastern Himalaya encompassing a geographical area of more than 92000 sq. km is well known for its distinct and compact physiographic diversity that has produced almost all possible type of ecosystems and vegetation ranging from tropical evergreen forests to alpine grasslands. The region is one of the youngest, fragile and complex eco-systems in the world.

The mountain ranges with beautiful landscapes play important role in controlling local atmospheric circulation, the hydrological cycle and water resources which provide a wide range of ecosystem services to the locality. The region finds place amongst the world's 36 biodiversity hotspots of designated by Conservation International (Ref). The precipitation of the region fluctuates tremendously due varied topography, climate, hydrology and elevations ranging between 100 meters to 8575 meters. The region receives more than 300 cm of rainfall annually. The average temperature ranges between 5°C to 35°C at foothills and much below freezing point at higher reaches. The region also receives vast amounts of water from major water bodies such as Teesta, Brahmaputra, Lohit, Subansiri, Kurung and Kamengthat originates from Tibetan plateau.

Past Studies

To recognize the biological significance of the Himalayas, several scientific studies on different aspects have already been conducted by various organizations since British era which dated back to early part of the 17th century. Among the pioneers visited the Mishimee hills in Arunachal Pradesh were two Europeans namely Lieutenant R.Wilcox and Captain Bedford (1825-1828) during their Assam survey of geographical discoveries. W.Griffith made first botanical collections from the state during 1836, while first floristic account of the state was given by Robinson (1841). Hooker (1854, 1906) provided very detailed account on the vegetation and flora of the region. The floristic diversity of the region has also attracted a number of European botanists and explorers since 18th Century (Buchanan-Hamilton, 1820; Roxburgh, 1820-24; Griffith, 1847; Hooker, 1854,1872-1897; Hooker and Thomson, 1855; Clarke 1889, Burkill, 1924-1925, 1965; Kingdom Ward, 1929,1960; *etc.*). Extensive explorations were resulted many classical floristic accounts of which *Flora of Mishmee Hills* (Griffith, 1847); *Botany of Abor Expedition* (I.H. Burkill, 1924-25);

Botanical Expedition in the Mishmi Hills (Kingdom Ward, 1929-1931); A *Sketch of the vegetation of Aka hills* (Bor, 1938), Flora of *Lohit Valley* (Kingdom Ward, 1953) and *The flora of Aka hills* (K.P.Biswas, 1941) *etc.*, were noteworthy.

Establishment of regional centers at Shillong in (1955), Itanagar in (1977) and Sikkim (1979) have accelerated the survey work in this region in the post re-organization of BSI. Extensive explorations were carried out to study the floristic account of Arunachal Pradesh (Panigrahi and Naik, 1961; Panigrahi, 1965, 1966; Rao and Joseph, 1965; Panigrahi and Joseph, 1966; Sastry, 1966; Rao and Deori, 1980; Rao and Murti, 1990; Rao, 1994) *etc.* A contribution to the flora of Namdapha, Arunachal Pradesh (Chauhan *et al.*, 1996); Materials for the flora of Arunachal Pradesh, vol. I, II, III published by Botanical Survey of India enumerates the plants of Arunachal Pradesh. Orchid flora of Arunachal Pradesh (Chowdhery, 1998), Flora of Lower Subansiri(Pal, 2013), Flora of Kurung-Kumey district (Dash and Singh, 2017), among other major contributions made by the scientists of the Arunachal Pradesh unit of the survey besides numerous reports of new species. Well known botanically curious plants such as *Sapria himalayana* (Rafflesiaceae), *Balanophora dioica, Rhopalocnemis phalloides* (Balanophoraceae), *Boschniakia himalaica* (Orobonchaceae), *Galeola falconeri* (Orchidaceae), *Monotropa uniflora* (Monotropaceae) *etc.*, and many commercially useful medicinal plants like species of *Aconitum, Aquilaria, Cinnamomum, Coptis, Curcuma, Gentiana, Picrorhiza, Sinopodo phyllum, Rubia, Taxus wallichiana, etc.* were also reported to grow abundantly in the this region.

Vegetation

Though broadly the vegetation of eastern Himalaya can be categorized as Tropical, Subtropical. Temperate, Subalpine and Alpine depending upon the altitude, yet numerous vegetation sub-types, each with their characteristic floristic association are found in the region. Climatic condition and altitude play a decisive role in determining species composition, dominance, community structure and pattern of flowering time across the landscapes. On the east and south east, the vegetation comprises of mixed broadleaved deciduous and evergreen forests with such characteristic plants of south east Asian-Malaysian elements. A number of taxa such as *Dipeterocarpus retusus, Acrocarpus fraxinifolius, Actinidia callosa, Baccaurea ramiflora, Bauhinia purpurea, Bischofia javanica, Calamus* sp., *etc.* are the typical, all of which occur in South Myanmar and Malayan peninsula, many ranged to Indo-China. While in the central Arunachal Pradesh the vegetation is subtropical and dominated by Chinese and Sino-Japanese floristic elements such as *Acer, Aesculus* (buckeye), several genera within laurels (Lauraceae), *Meliosma, Evodia, Betula* (birch), *Quercus* (oak), *Lithocarpus, Fagus*(beech), *Elaeocarpus* and *Ailanthus* (Dash and Singh 2017). *Terminalia myriocarpa, etc.* Forest of these types of vegetation can be characterized by their close canopy, dense underground cover and luxuriant growth of the epiphytes. The middle canopy is formed by the medium sized trees, *viz. Albizia procera, Archidendron clyperaria, Callicarpa macrophylla,*

Castanopsis lanceifolia, Dysoxylum spp. (Dash and Singh 2018). A subtropical pure stand of palm forest is typical features of Arunachal Pradesh, where the three species of *Arenga* are mostly dominated.

The temperate vegetation occurs between 1800m and 3500m. On the basis of species composition and altitudes, temperate forests are categorized into broadleaved temperate forest, mixed temperate coniferous forest. This type of forest are characterized by mostly Laurels, *Quercus, Acer, Castanopsis, Magnolia,* mixed with different species of *Rhododendrons*, in different proportions. Between 2700m to 3500m, *Abies* (Fir), *Picea* (Spruce), *Betula* and other boreal plants intermix to form a typical east Himalayan mixed temperate coniferous forests. Full grown, well-formed oak and fir trees reach an elevation of around 3,800 meters in some places and intermix with alpine meadows, scree slopes, and *Rhododendron* thickets. These type of forests are characterized by the mixed population of *Tsuga-Pinus-Taxus* series of conifers with *Rhododendron* species in West Kameng district, pure stands of *Cupressus* in Upper Siang of Arunachal Pradesh, *Abies- Taxus*in West Siang or *Picea-Larix-Abies* in Kurungkumey, Dombeyang valley of North Sikkim and Tawang districts. The important Rhododendron species that are found mixed are *Rhododendron arboreum* subsp. *cinnamomeum, R. arizelum, R. campylogynum, R. cerasinum, R. dalhousiạe, R. falconeri, R. faucium, etc.* The extreme northern parts of the state is dominated with the vegetation comprises stunted bushy growth of *Juniperus* spp., *Rhododendron lepidotum, R. anthopogon, Salix zangica, Sorbus microphylla, Berberis angulosa, B. macrosepala, Rosa sericea etc.* The herbaceous species that occur in northern ridges are *Bistorta vivipara, Diapensia* spp., *Koenigia islandica, Oxyriạ digyna, Pinguicula alpina, Sagina saginoides, Thalictrum alpinum.* Herbaceous vegetation reaches up to 5500 meters, although few plant specimens have been collected above 4,500 meters except from the Tawang district and some parts of Sikkim.

Above 3,500 meters vegetation types generally dominated by a highly diverse flora of herbaceous plants and shrubs (Dash, 2012). The alpine meadows are usually intermixed with conifer forests of primarily fir, spruce and junipers in protected ravines and gorges. *Rhododendron* thickets are a conspicuous feature of the landscape, especially in West Siang, Kurung-Kumey and Tawang districts. The rich diversity of *Rhododendron* are more prominent on places that are protected from direct sunlight, a phenomenon also noted on south-facing slopes in northern Myanmar (Burma) by Kingdon Ward (1937).

Secondary Forests

Due to some of the irrevocable transformations of the forest lands such as landslides, expansion of infrastructure, forest fire, unplanned shifting cultivation the soil profile of freshly disturbed forest lands are extremely vulnerable to various invasion. These abandoned plots ultimately revert to scrubby jungles or secondary forests, providing an easy spread for single dominant species to take over (Dash, 2012). Depending on the extant of disturbances, the secondary forest is seen up to

2000m elevation in Arunachal Pradesh particularly in Papum Pare, Kurung-Kumey, Changlang and Lohit districts. The upper canopy formed with one or two species, while the lower canopy is shrubby and mixed type. The common species that are dominant in upper canopy are *Abroma augusta, Callicarpa arborea, Osbeckia nutans, Pseudodissochaeta assamica, Saurauia armata, etc.* while the shrubby species are *Aralia foliosa, Verbascum thapsus, Boehmeria penduliflora, Debregeasia longifolia, Girardinia diversifolia, Glochidion hirsutum, Maesa indica, Melastoma malabathircum, Oxyspora paniculata etc.*

Floral Diversity

The vascular plant diversity is truly impressive, with as many as a one third of India's vascular flora of 18,666 species growing just in the eastern Himalaya. The region is estimated to have at least 7517 flowering plants which cater just more than 42 per cent of Indian flora and about 68 per cent of Himalayan flora. The dicotyledons are represented by 73 per cent of total species present and monocotyledon represented by 27 per cent. About 71 per cent families and 32 per cent of genera of angiosperms of India are represented in Arunachal Pradesh and Sikkim (Singh and Dash, 2018), of which 10 per cent flora is endemic (Singh *et al.,* 2015). There are 768 Orchids, 430 legumes, 110 taxa of *Rhododendron*, 89 taxa of *Bamboo*, 85 species of *Impatiens*, 42 species of *Begonia*, 34 species of *Citrus*, 29 species of *Dioscorea*, 18 species of *Rattan*, 653 Fern species, and 29 Gymnosperms found in the state (Dash and Singh, 2017; Hajra and Mudgal, 1997; Rao, 2007; Mao, 2010; Dash, 2009; Dash, 2010; Dash and Mao, 2011; Taj and al., 2009, 2009a; Taj, 2014, Singh *et al.,* 2019). Well known botanically curious plants and many commercially useful medicinal plants *viz.* members of *Aconitum, Aquilaria, Cinnamomum, Coptis, Curcuma, Gentiana, Picrorhiza, Sinopodo phyllum, Rubia, Taxus wallichiana, etc.* were also reported to grow abundantly in the state (Dash and Singh, 2018)

An analysis of the first ten dominating families shows that, the family Orchidaceae is most dominant family in eastern Himalaya followed by Leguminosae, Asteraceae and Poaceae. The ratio of dicot to monocot species as 2.8:1 whereas, genera to species as 3.4:1. Monotypic families are also show great species diversity. Out of 42 families that are monotypic, families like Balsaminaceae, Begoniaceae are represented 85 species of *Impatiens* and 42 species of *Begonia* respectively, while the monotypic monocots families like Dioscoreaceae represented by 26 species of *Dioscorea* while Smilacaceae is represented by 23 species of *Smilax* respectively.

Some of the high altitude members of genus *Arenaria, Bistorta, Draba, Rhodiola, Saussurea, etc.*, are highly adaptive for their survival. Members of certain groups like *Orchids, Bamboos, Rhododendrons, Ficus, Hedychiums, Impatiens, etc.* exhibit remarkable diversity in Arunachal Pradesh. Rhododendrons is the most dominant genera in the state followed by *Bulbophyllum* (69), *Impatiens* (85), *Ficus* (59), Dendrobium (52), Primula (47) *etc.*

Endemic Species

Eastern Himalaya is richer in endemics than the Western Himalaya. No families are endemic to this region; however all the endemics are manifested at species, subspecies or variety level. The altitudinal gradient of the mountain ranges is one of the decisive factors in shaping the spatial patterns of the species diversity and its uniqueness. Western Himalaya contributes 297 endemic taxa while Eastern Himalaya contributes 466 endemic taxa. Western and Eastern Himalaya contributes 7 per cent and 11 per cent of total endemic taxa, respectively. Arunachal Pradesh and Sikkim, known for their high biological diversity contribute 183 taxa and 160 taxa respectively. It is also interesting to note that, endemic diversity greatly differ along the altitudinal gradient. Among the Eastern Himalayan endemic angiosperms, family Orchidaceae with 77 taxa is the most dominant, followed by the family Ericaceae (29 taxa), Lauraceae (27 taxa) and Balsaminaceae (22 taxa). Several genera show relatively higher richness in endemism in Eastern Himalaya, *Impatiens* (22 species), *Piper* (17 species), *Rhododendron* (10 species) and *Agapetes* (10 species) are dominant genera. As the altitudinal gradient increases, more endemic species are encountered in the Himalayas. Analysis of data on endemic taxa in Himalaya revealed that, maximum representation is in Alpine, followed by temperate, and subtropical regions.

Unique Ecosystem Diversity

The vast geographic area with its diverse climatic conditions and topography has given rise to various types of ecosystems in the eastern Himalaya. The diversity of ecosystems not only a great source of huge germplasm, but also great economic services to the local inhabitants of this region.

High Altitudes Alpine Ecosystems

Occurrence of pristine alpine grasslands ecosystems in eastern Himalaya though uncommon, some of these unique biodiversity rich landscapes occurred between snow and tree lines particularly in Vadse hills of Kurung-Kumey, Yarlang valley of West Siang and Anini of Dihang valley. The maximum diversity of plants is observed during July–August and the above ground biomass is low during December. Herbs are the main growth form of vegetation in alpine zones with occasional intermixing of alpine woody elements (shrubs). Some of the common shrubs are *Ephedra gerardiana, Juniperus communis, J. recurva, Rhododendron anthopogon, R. lepidotum, Ribes glaciale*. The alpine meadows are dominated by the herbaceous species of *Anaphalis, Anemone, Artemisia, Aster, Astragalus, Bupleurum, Caltha, Corydalis, Gentiana, Geranium, Poa, Polygonum, Potentilla, Primula, Rhodiola, Saussurea, Saxifraga, etc.* Besides these, the grasses and sedge are found to be dominant the steep alpine slope.

High altitude wetlands are also an important component of these ecosystems due to not only their role in maintaining ecological balance, carbon sink, but also considered sacred. The species that are found in these wetlands are mainly of Tibetan affinity. Members from Ranunculaceae which are moderately represented; *Draba, Stellaria. Arenaria, Primula,* are well represented by the gregarious patches. Legumes are mainly represented by the species of *Astragalus* and *Oxytropis.* Tufted deep rooted perennial hers are represented by many species of *Potentilla, Sibbaldia, Spiraea* and *Cotoneaster, Saxifraga* in upper reaches. Compositae are well distributed with more abundance of species of *Saussurea* and *Cremanthodium.*

In contrast, the diversity of Llonakh Valley of Sikkim is of rare kind and forms a unique alpine community. The species occur in this high altitude are predominantly of Tibetan affinity. Species from Ranunculaceae moderately represented. Among the few shrubby plants of the higher regions are the species of *Berberis* which persist far into the upper Llonakh valley where they are much dwarfed and generally prostrate. *Corydalis* is well represented and rich in species. The Cruciferae particularly species from *Draba, Lychnis, Stellaria*and *Arenaria* are well represented, while members from Caryophyllariaceae such Arenarias are represented by the tufted form in Llonakh. Legumes are moderately represented by the species of *Astragalus* and *Oxytropis.* Members of Rosaceae are conspicuous features of the vegetation with many species of *Potentilla, Sibbaldia, Spiraea* and *Cotoneaster. Saxifraga* is another dominant genus in upper reaches. Umbelliferae are represented by prostrate *Cortia. Pleurospermum* and *Trachydium* grow on open wind-swept areas of Llonakh. With the exception of *Galium,* Rubiaceae are absent from the higher altitudes. Compositae are well distributed with abundance of species of *Saussurea*in the valley. *Anaphalis* and *Artemisia* are common towards the dry Tibetan region. *Pedicularis* is another dominant genus of this valley; their bright colours are a conspicuous feature in sparse Llonakh vegetation. Prostrate species of *Salix* are prevalent right up to the melting snow. Species of *Habenaria* and *Cypripedium* sparingly represent the Orchidaceae. Juncaceae, Cyperaceae and Graminae are less prevalent. *Ephedra* and dwarf Junipers ascend to over 5000 m altitude.

A Great Source of High Germplasm

The biodiversity of eastern Himalaya have been a the source of wild genetic resources of *Pyrus, Prunus, Sorbus, Rubus, Ribes, Musa, Hedychium, Oryza, Momordica, Cucumis, Solanum, Piper, Amomum, Alpinia, Curcuma, Zingiber* and *Saccharum.* Large quantities of medicinal and aromatic plants havealso been commercially exploited from eastern Himalaya. Some of the species which we have collected are *Aconitum heterophyllum, Colchicum luteum, Coptisteeta, Curcuma zedoria, Dactylorhiza latifolia, Dioscorea deltoides, Ephedragerardiana, Gentiana kurroa, Nardostachys jatamansi, Podophyllum hexandrum.* This region also harbours a number of wild ornamentals with potential for cultivation which need to be mention here that *Primula, Rhododendron, Balsum, Gingers,* and whole family *Orchidaceae.* Bamboo and Rattan diversity in the state, particularly in subtropical zones is remarkable and it contributes to the daily needs of the millions of people.

Biodiversity in an Economic Context

Many species have been contributing significantly to the daily need of the local inhabitants for sustenance. Plants are may be used as fruits(*Rubus niveus, Malus sikkimensis, Prunus cornuta, Mangifera sylvatica, etc.*) or leafy vegetables (*Houttuynia cordata, Persicaria chinensis, Phytolacca acinosa, etc.*). The floral stalks of *Musa* spp., the underground rhizome or tubers of *Dioscorea bulbifera, Codonopsis ovata, Apios cornea, etc.*) are cooked.

Similarly some of the important medicinal species that used in traditional systems and also have high commercial values are *Aconitum ferox, Saussurea gossypiphora, Gentiana kurroo, Picrorhiza scrophuliflora, Ephedra gerardiniana, Rhododendron anthopogon, Alpinia nigra, Neopicrorhiza scruphulariiflora, Ephedra gerardiniana, Rubia cordifolia, etc.* Some of the common dyes yielding plants in Arunachal are *Berberis asiatica, Buteamonosperma, Carthamus tinctorius, Curcuma domestica, Hedychium spicatum, Indigo feratinctoria, Juglans regia, Lagerstroemia parviflora, Mallotus philippensis, Rhododendron arboretum, Rubia cordifolia, Rumexha status, Woodfordia fructicosa etc.*

The Nature of Challenge Ahead

The isolation of remote landscapes of the state, topographic relief and varied climatic condition of eastern Himalaya have resulted a unique kind of diverse flora. The conservation of the eastern Himalaya's species rich and distinctive biota is critically important for heritage, ethical, intrinsic and utilitarian value and the roles that it plays in the maintenances of ecosystem services. Habitat destruction, heavy grazing, clearing of forest land for agriculture, fragmentation of habitats, invasion of weeds are some of the major issues need to address. Shift in abundance, changes in distribution of species, depletion of genetic diversity, habitat destruction of endemic species are the different major results of such changes.

Conservation Measures

Inventory and correct identity of each biodiversity components is essential prerequisite knowledge to understand the ecosystem services and it's quantitative scientifically management. Being a signatory of Convention on Biological Diversity (CBD), India is committed to deliver outcomes of biodiversity research and documentation of all plant resources of the country. Botanical Survey of India, the leading organization of our country playing the central role in documenting, compiling and updating of all information on Plant diversity. To conserve and scientifically manage the rich diversity of flowering plants of the state a well-established and intensively monitored protected area network is in force. Presently, the region has two biosphere research, 4 national parks and 19 wildlife sanctuaries. These protected areas, located in different locations, are part of the cultural heritage and are also play an important role in conservation of biodiversity reinforced with the traditional practices (Dash, 2017).

REFERENCES

Bor, N.L. 1938. A sketch of the vegetation of Aka Hills, Assam a synecological study. India. For. Rec. (n.s) Bot.1(4): 103-222.

Burkill, I. H. 1924. The Botany of Abor expeditions. *Rec. Bot. Surv. India* 10(1): 1 - 154.

Burkill, I.H. 1925. The Botany of Abor expeditions. *Rec. Bot. Surv. India* 10(2): 155-420.

Chauhan A.S. 1996. A contribution to the flora of Namdapha, Arunachal Pradesh. Botanical Survey of India.Kolkata. 478pp.

Chowdhery, H. J., G. S. Giri, G. D. Pal and A. Pramanik. 1996. Materials for the flora of Arunachal Pradesh. Vol. 1.Ranunculaceae-Dipsacaceae. In: Hajra, P. K., D. M. Verma and G. S. Giri (eds.), Botanical Survey of India, Kolkota, India.

Clarke, C. B. 1898. Subareas of British India illustrated by the detailed distribution of the Cyparaceae in that Empire, J. Kinn.Soc. London 34: 1 - 146.

Dash, S. S. 2009.The genus *Desmodium*Desv.and its allied genera in Arunachal Pradesh. *Bull. Arunachal Forest Research* 25(1 and 2): 18 - 33.

Dash, S. S. 2010.Native Species of *Begonia* L. (Begoniaceae) in Arunachal Pradesh-Diversity and Distribution. *Bulletin of Arunachal Forest Research* 26 (1 and 2): 29 - 43. 2010 (2011).

Dash, S. S. 2012.Floristic composition of early recovery vegetation of traditional shifting cultivation areas of Kurungkumey district of Arunachal Pradesh. Indian Forester. 138(9): 767 - 775.

Dash, S. S. and A. A. Mao. 2011. A new species of *Codonopsis* Wall. (Campanulaceae) from Arunachal Pradesh, India, *Nelumbo* 53: 17 - 20.

Dash, S. S. and A. A. Mao. 2011. Additions to the flowering plants of India from Arunachal Pradesh, *Nelumbo* 53: 228 - 235.

Dash, S. S. and M. Sanjappa. 2011. Two new species and a new distributional record of *Archidendron* F. Muell. (Leguminosae: Mimosoideae) from India. *Nelumbo* 53: 7 - 16.

Dash, S.S and Singh, P. 2011. Trees of Sikkim. In Arrawatia, M.L. and Tambe, Sandeep (eds.). *Biodiversity of Sikkim Exploring and Conserving a Global Hotspot.* Information and Public Relations Department, Government of Sikkim, Gangtok. pp. 89 - 124.

Dash, S.S and P. Singh. 2017. Flora of Kurung-Kumey District, Arunachal Pradesh. Botanical Survey of India. Kolkata Pp 794.

Griffith, W. 1847. Journals of travels in Assam, Burma, Bhootan, Afghanistan and the neighboring countries

Hajra, P. K. and V. Mudgal 1997.Diversity in Hotspots - An Overview.Botanical Survey of India, Calcutta.pp. 1 – 12.

Hooker, J. D. 1906. A sketch of the flora of British India. London.

Hooker, J. D. and T. Thomson. 1855. Himalayan Journal, Reeve and Col, London

Kingdon-Ward, F. 1929.Botanical exploration in the Mishmi Hills.*Himalay.Journ.* 1: 51 – 59.

Kingdon-Ward, F. 1929-30. Botanical Exploration: Mishmi Hills, Assam. *Proc. Linn. Soc. London*: 60 – 64.

Kingdon-Ward, F. 1930.The Forests of the North East Frontier of India.*Empire For. Journ.* 9: 11 – 31.

Kingdon-Ward, F. 1930.The Seinghku and Delei Valleys, North-East Frontier of India.*Geogr.Journ.*75: 412 – 435.

Kingdon-Ward, F. 1953.The Lohit valley in 1950. Proc. Linn. Soc. London 164: 2 – 8.

Mao, A.A. 2010. The genus Rhododendron in north-east India. *BotanicaOrientalis* 7: 26 34.

Pal,G. D 2013. Flora of Lower Subansiri, Arunachal Pradesh, Botanical Survey of India, Kolkata, pp 578

Panigrahi, G. 1965. Studies in the Monocot flora of Assam and NEFA (Families: Burmanniaceae to Dioscoreaceae). *Proc. Nation. Acad. Sci. India* 36B: 353 – 368.

Panigrahi, G. and J. Joseph. 1966. A botanical tour to Tirap Frontier Division, NEFA (India). *Bull. Bot. Surv. India* 8: 142 – 157.

Panigrahi, G. and S. K. Kar. 1967. The family Compositae in Assam and NEFA. *Bull Bot. Surv. India* 8(3-4): 228 – 236.

Panigrahi, G. and V. N. Naik. 1961. A Botanical tour to Subansiri Frontier Division (NEFA). *Bull. Bot. Surv. India* 3(3 and 4): 361–388.

Rao, A. N. 2007. Orchid flora of north-east India – An up-to-date analysis.*Bulletin of Arunachal Forest Research* 23: 6 – 38.

Rao, A. N., O. Apang, A. A. Mao and S. S. Dash. 2010. Two new orchid records from Arunachal Pradesh, *India. J. Ochid. Soc. India* 24(1 and 2): 97-100.

Rao, A.S. 1974. The vegetation and phytogeopgraphy of Assam - Burma.*In Ecology and Biogeography in India*. (M.S. Mani ed.) 204 - 246. The Hague: MonographiaeBiologicae. 23

Rao, R. R. 1994. Biodiversity in India (Floristic Aspects).Bishen Singh Mahendra Pal Singh, Dehradun. Pp 315.

Rao, R. R. and N. C. Deori. 1980. A preliminary census of the orchids of Kameng district, Arunachal Pradesh. *Indian J. Forest.* 3(3): 255 – 269.

Rao, R. R. and S. K. Murti. 1990. North-east India, a major centre for plant diversity in India. *Indian J. Forest.*13: 214 - 222.

Rao, R. S. 1972. Botanical Survey of India works in Arunachal. Arunachal Pradesh Bulletin (Independence Silver Jubilee issue) 46 - 76.

Rao, R. S. and J. Joseph. 1965. Observation of the flora of Siang frontier division of NEFA. *Bull. Bot. Surv. India* 7: 138–161.

Robinson, W. 1941. A descriptive account of Assam: with a sketch of the local geography and concise history of the tea plant of Assam. Calcutta, Ostell and Legate.XV, 421 p.

Sahni, K.C. 1969. A contribution to the flora of Kameng and Subansiri district, NEFA. *Indian For.* 95(5): 330–352.

Sastry, A. R. K. 1968.*Alniphyllum*Mats, and *Huodendron*Rehder two additional generic records to the Indian Styraceae. *Bull. Bot. Surv. India* 9(1-4): 297 - 298.

Singh, P. and S. S. Dash, 2018. In Das, A.P. and S. Bera (eds). Plant Diversity in Indian Himalayan Regions: An Overview in Plant Diversity in the Himalaya Hotspot Region. Vol. I. 1-43. Bishen Singh Mahendra Pal Singh, Dehradun

Singh, P. S.S.Dash, B.K.Sinha. 2019. Plants of Indian Himalayan Region. (In press) Botanical Survey of India.Kolkata.

Singh, P.; Kartigeyan, K.; Laxminarasingham, P. and Dash, S.S. 2015.*Endemic Vascular plants of India*.Botanical Survey of India, Kolkata.

Taj, R. K., G. Tayeng, T. Ronrang and A. Arunachalum. 2009. Four new records of bamboos to India from Arunachal Pradesh. *Bulletin of Arunachal Forest Research* 25(1 and 2): 14 - 17.

Taj, R.K. 2014.Status of Bambo and its Productivity in Arunachal Pradesh.Proceeding of National Seminar on Bomboo Productivity in forest and non-forest areas. Forest Research Institute, Derha Dun. Pp. 53-85.

Taj, R.K., Tayeng, G., Thungdok, D.K., Gab, R., Norbu, T. and Sinha, G.N. 2009a.Bamboo Flowering and its monitoring in Arunachal Pradesh.SFRI information Bulletin no.31.

Chapter 2

Assessment of Genomic Relationship among the Species of *Cordia* (Boragicaceae) using RAPD and ISSR Markers with a Note on Taxonomy of *C. dichotoma* - *C. myxa* Complex

Subrat Kumar Kar, Pankajini Bal and Pratap Chandra Panda*

Taxonomy and Conservation Division, Regional Plant Resource Centre (RPRC), Bhubaneswar – 751 015, Odisha
**e-mail: pankajini.bal@gmail.com*

ABSTRACT

The inter and intra-specific genetic diversity and genomic relationship among 4 wild and cultivated species of Cordia (Boraginaceae) namely, Cordia dichotoma, Cordia myxa, Cordia lutea and Cordia sebestena have been evaluated using RAPD and ISSR molecular markers. Twelve RAPD primers and 15 ISSR primers were used for assessment of genetic variability of 13 accessions of above 4 species and high level of polymorphism was observed. All accessions of C. dichotoma and C. myxa formed clear clusters justifying their taxonomic identity as distinct species. The closeness of both the species from molecular point of view appear justified as they belong to Cordia sect. Myxa as per the ITS1 sequence data of Gottschling et al. (2005) and also based on previous morphological and bio-geographical studies. However, two unrelated species such as Cordia lutea and C. sebestena came in a cluster, which was not in accordance with the infra-generic classification based on molecular analysis and morphological data.

***Keywords**: Cordia, RAPD, ISSR, nNTSYS-PC.*

Introduction

The family Boraginaceae is comprised of about 2,600 species under 121 genera (Mabberley, 2017) but The Plant List (theplantlist.org/1.1) records the occurrence of 155 genera and 2,686 species in the world. Of these, the genus *Cordia* is represented by about 250 species of trees and shrubs, distributed mostly in tropics of North and South America and poorly represented in Africa and Asia. In India, 13 species of the genus are reported to occur in the wild (Thirupathi *et al.,* 2008).

The species of the genus *Cordia* are especially included in studies of natural products of plant origin due to their diversity of chemical, biological and pharmacological properties (Matias *et al.,* 2010 a and b). Different parts of the plant such as fruits, leaves, stem bark, seeds and roots of most species of the genus *Cordia*, especially *Cordia dichotoma, C. myxa, C. obliqua, C. verbenacea, C. martinicensis, C. salicifolia, C. spinescens, C. latifolia, C. ulmifolia* have long been used in traditional medicine for the treatment of various illnesses of human beings. Several ethno-pharmacological applications as anti-microbial, anti-inflammatory, anthelmintic, analgesic and diuretic agents and for treating digestive, respiratory, uro-genital, cardiac, vascular and blood disorders have been reported (Bhattacharya and Saha, 2013; Costa *et al.,* 2008; Matias *et al.,* 2010 a and b; Medeiros *et al.,* 2007; Paulino*et al.,* 2011; Scheeren *et al.,*2002; Sertié *et al.,* 1990).

In the last few decades, scientific studies of *Cordia* species have intensified; demonstrating the great interest in phytochemical, biological and pharmacological research (Matias *et al.,* 2015) and wide classes of compounds including flavanol glycosides, pyrrolizidine alkaloids, triterpenes, cardiaquinone, sphingolipids, and cerebrosides have been isolated and characterized for possible applications. However, the inter and intra-specific genetic variability and application of molecular markers for solving taxonomic ambiguities among and within certain species of the genus *Cordia* have not been studied so far except few exotic species.

Derero *et al.* (2011) estimated the genetic diversity among 22 populations of *Cordia africana*, a declining forest tree species in Ethiopia using AFLPs and chloroplast microsatellite markers and found high level of genetic variability within and among the populations investigated. They concluded that efficient gene flow via pollen and seed is likely to be the main factor contributing to the maintenance of genetic diversity in natural and disturbed conditions. In a similar study by Latifi *et al.* (2018), the genetic variations among 7 populations of *Cordia myxa* collected from southern parts of Iran were analyzed with the application of CDDP (Conserved DNA-Derived Polymorphism) molecular markers. Maximum genetic variability was obtained for Shoosh population and intra-population diversity was found to be higher than inter-population diversity. The existence of high genetic diversity among the accessions of *Cordia dichotoma* from Nanded District in Maharashtra has also been reported (Nandedkar and Mulani, 2016). This diversity assessment was based on study of morphology and RAPD molecular markers.

In view of availability of scanty literature on assessment of genetic diversity of the genus *Cordia* and in consideration of the long-standing taxonomic and nomenclatural ambiguity of the *C. dichotoma- C. obliqua- C. myxa* complex, the genetic diversity and genomic relationship of 4 species of *Cordia* namely, *C. dichotoma, C. myxa, C. sebestena* and *C. lutea* were studied using RAPD and ISSR markers.

Materials and Methods

Plant Materials

Four different species of *Cordia* were collected from Bhubaneswar and adjoining regions in Odisha (Figures 2.1a-f). The morphological differences between *Cordia myxa* and *Cordia dictotoma* are shown in Figure 2.2 a-f. Fresh and young leaves were taken for DNA isolation. Some samples were preserved for future use at -85°C temperature. The herbarium specimens were prepared and deposited in the Herbarium of Regional Plant Resource Centre (RPRC), Bhubaneswar, Odisha, India to serve as voucher specimens.

The details of four species and accessions of *Cordia* collected for the study are given in Table 2.1.

Table 2.1: Details of Four Species and Accessions of *Cordia* Collected for the Study

Sl.No.	Name of the Species	No. of Accessions	Localities of Collection
1	*Cordia dichotoma* G. Forst.	1	RPRC, Bhubaneswar, Odisha
2	*Cordia dichotoma* G. Forst.	1	CRPF square, Bhubaneswar, Odisha
3	*Cordia dichotoma* G. Forst.	1	Old town, Bhubaneswar, Odisha
4	*Cordia lutea* Lam.	1	RPRC, Bhubaneswar, Odisha
5	*Cordia sebestena* Linn.	1	RPRC, Bhubaneswar, Odisha
6	*Cordia myxa* Linn.	4	Tulsipur, Cuttack, Odisha
7	*Cordia myxa* Linn.	4	Khandagiri, Bhubaneswar, Odisha

Genomic DNA Isolation and Quantification

Tender leaves were collected from mature plants and stored at -80 °C prior to use. Total genomic DNA was extracted by using the protocol described by Doyle and Doyle (1990) with required modifications. The quality and concentration of DNA was examined by ethidium bromide stained agarose gel electrophoresis and spectrophotometric analysis. The quantification was done in comparison with the

Figure 2.1a-f: (a) *Cordia lutea,* (b) *C. sebestena,* (c) *C. dichotoma,* (d) *C. myxa,* (e) Fruits of *C. myxa,* (f) Fruits of *C. dichotoma.*

Figure 2.2a-f: Comparative Morphological Features of *Cordia myxa* and *C. dichotoma* - (a) Fruiting branches, (b) Leaves, (c-d) Fruits, (e) T. S. of fruits, (f) Seeds.

known standard (λ-DNA). After quantification, the DNA was diluted with $T_{10}E_1$ buffer to a working concentration of 25 ng/μl of PCR analysis.

Random Amplified Polymorphic DNA (RAPD) Analysis

Eighteen random decamer oligonucleotide Operon primers from A, C, D and N series were used for RAPD analysis. Out of these, ten primers responded well and gave very good amplification. The RAPD analysis was performed as per the methodology described by Williams *et al.* (1990). Each amplification reaction mixture of 25 ml volume contained 2.5 ml of 10X assay buffer (100 mM Tris-Cl, pH 8.3, 500 mM KCl, 15 mM $MgCl_2$ and 0.1 per cent gelatin), 200 mM of each dNTPs (dATP, dCTP, dGTP and dTTP) (MBI Fermantas, Lithuania) 15 ng of primer, 0.5 unit of Taq DNA polymerase (Bangalore Genei Pvt. Ltd, Bangalore, India) and 25 ng of template DNA. The amplification reaction was carried out in Gene Amp PCR System 9700 (Applied Biosystems, Germany). The amplification was performed in three steps PCR. Initial denaturation of the template DNA was carried out at 94 °C for 5 min for one cycle. The second step was carried out for 42 cycles and each cycle consisted of three temperature steps *i.e.* one min at 92 °C for denaturation of template, one min at 37 °C for primer annealing followed by two min at 72 °C for primer extension. The final step consisted of only one cycle *i.e.* 7 min at 72 °C for complete polymerization. The soak temperature was 4 °C. After the completion of the PCR 2.5 ml of 6X loading dye (MBI Fermentas, Lithuania) was added and amplification products were detected using 1.5 per cent agarose gel stained with 0.5 μg/ml ethidium bromide at 65V for 2 hours.

Inter Simple Sequence Repeat (ISSR) Analysis

Inter Simple Sequence Repeats were used for PCR amplification. Twenty anchored and non-anchored microsatellites were used as primers. These simple sequence repeats were synthesized and procured from Genei (Bangalore Genei Pvt. Ltd, Bangalore, India). Out of twenty ISSR primers, eleven primers showed reproducibility. The ISSR analysis was performed as per the methodology given by Zietkiewicz*et al.* (1994). Each amplification reaction mixture of 25 ml contained 20ng of template DNA, 2.5ml of 10X assay buffer (100mM Tris-HCl pH 8.3, 0.5M KCl and 0.01 per cent gelatin), 1.5mM $MgCl_2$, 200mm each of dNTPs, 44ng of primer and 0.5U Taq DNA polymerase. The amplification was carried out in a Thermal Cycler. The first cycle consisted of denaturation of template DNA at 94°C for 5 min, primer annealing at specific temperature for particular primer (as indicated in the Table 2.2) for 1 min and primer extension at 72 °C for 2 min. In the subsequent 42 cycles, the period of denaturation was reduced to 1 min while the primer annealing and primer extension time was the same as in the first cycle. The last cycle consisted of only primer extension at 72°C for 7 min. the amplified products were resolved in 2 per cent agarose gel stained with ethidium bromide.

Table 2.2: Details of RAPD Analysis in *Cordia* spp.

Sl. No.	*Primer*	*Nucleotide Sequence*	*No. of Bands Present*	*No. of Polymorphic Bands*	*No. of Monomorphic Bands*	*No. of Unique Band*	*Resolving Power*	*Primer Index*	*Percentage (per cent)*
1	**LKA-46**	5'TCCGATGCTG3'	7	6	1	3	2.77	2.07	85.71
2	**RAPD-NS**	5'GGGTTGTGGG3'	8	6	2	1	3.69	2.45	75
3	**OPA-2**	5'TGCCGAGCTG3'	7	7	0	2	7.08	2.79	100
4	**OPA-5**	5'$(AG)_4$TCTTG3'	6	6	0	1	3.85	2.60	100
5	**OPA-10**	5'GTGATCGCAG3'	8	8	0	0	11.38	2.56	100
6	**OPC-2**	5'GTGAGGCGTC3'	9	8	1	0	6.62	3.96	88.89
7	**OPC-5**	5'GATGACCGCC3'	7	5	2	1	3.69	1.61	71.43

Data Analysis and Construction of Phylogenetic Tree

The presence/absence of bands in RAPD/ISSR analysis was recorded in binary (0, 1) form. All the bands (polymorphic and monomorphic) were taken into account for calculation of similarity with a view to avoid over-/underestimation of the distance (Gherardi*et al.*, 1998). Jaccard's coefficient of similarity (Jaccard, 1908) was measured and a dendrogram based on similarity coefficients generated by the un-weighted pair group method using arithmetic averages (UPGMA) (Sneath and Sokal, 1973) and SHAN clustering. The statistical analysis was done using the computer package nNTSYS-PC (Rohlf, 1997). Resolving power (Rp) of the RAPD primer was calculated according to Prevost and Wilkinson (1999): Rp = ΣIB, where IB (band informativeness) = 1 D [2/(0.5 D *P*)], *P* being the proportion of the 5 species containing the band.

Results

In this investigation, the DNA was isolated from 13 accessions of four different species of *Cordia* namely, *Cordia dictotoma*, *Cordia myxa*, *Cordia sebastena* and *Cordia lutea* collected from various parts of Odisha and their genetic diversity was assessed using RAPD and ISSR markers.

RAPD Analysis

Twelve (12) random decamer oligonucleotide primers were used for fingerprinting of 13 accessions of *Cordia* species and out of these the 7 primers namely (LKA 46, RAPD-NS, OPA 2, OPA 5, OPA 10, OPC 2, OPC 5) responded and produced reproducible bands. The details of the primers producing scorable bands, their nucleotide sequence, and number of amplicons, number of polymorphic/ monomorphic bands, and number of unique bands, resolving power, primer index and percentage of polymorphism are presented in Table 2.2. A total of 52 bands were amplified, out of which 46 bands were polymorphic in nature, 8 bands were unique and 6 were monomorphic bands. The highest number of fragments (9) was amplified by the primer OPC 02 and lowest (5) by the primer OPA 05. The Resolving Power (RP) was highest in OPA 10 (11.38) and lowest in LKA-46, whereas Primer Index was highest (3.96) and lowest (1.61) for OPC-2 and OPC-5 respectively. The lowest RP (2.77) was observed in case of LKA 46 and the lowest PI (1.61) in OPC 05. The RAPD banding pattern is shown in Figure 2.3.

The dendrogram constructed by SHAN clustering using Jaccard's similarity coefficient showed a distinct cluster of *Cordia sebestena* and *Cordia lutea* with one accession from each segregated from the rest 11 accessions at similarity level of 58 per cent (Figure 2.4). Four genotypes of *C. myxa* from Khandagiri hills, Bhubaneswar andall the three accessions of *C. dichotoma* formed one cluster sharing a node at about 74 per cent genetic similarity with the clade containing 4 accessions of *C. myxa* from Cuttack. Subsequently, the 4 genotypes of *C. myxa* of Cuttack got separated in group of two at 73 per cent similarity level. The maximum similarity of 94.9 per cent between two accession of *C. myxa* (CMK1 and CMK2) from Khandagiri,

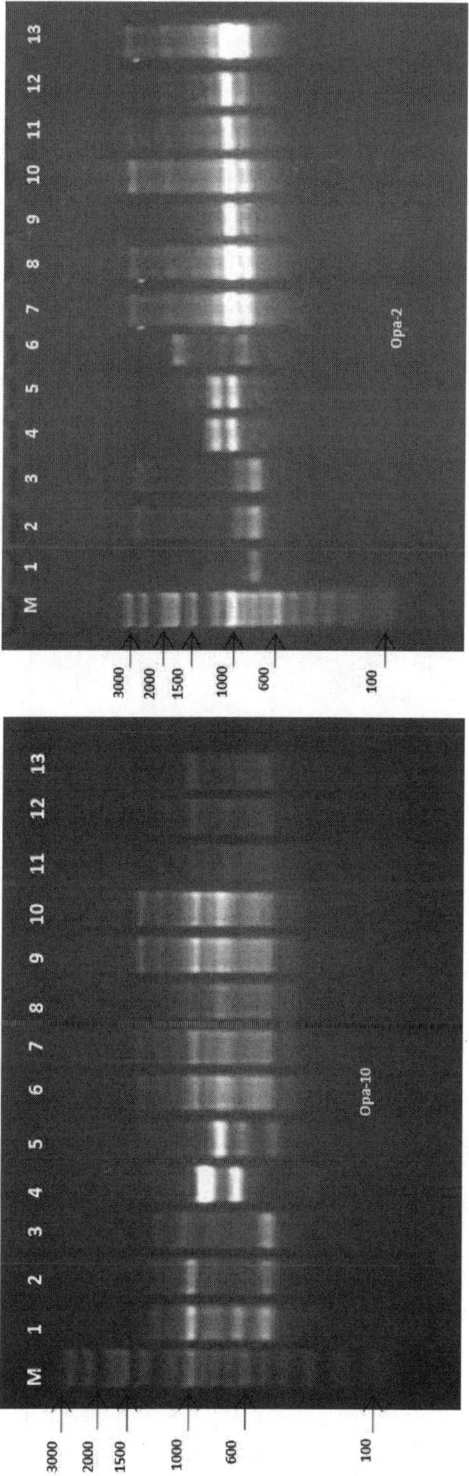

Figure 2.3: Details of RAPD Banding Pattern in *Cordia* spp.

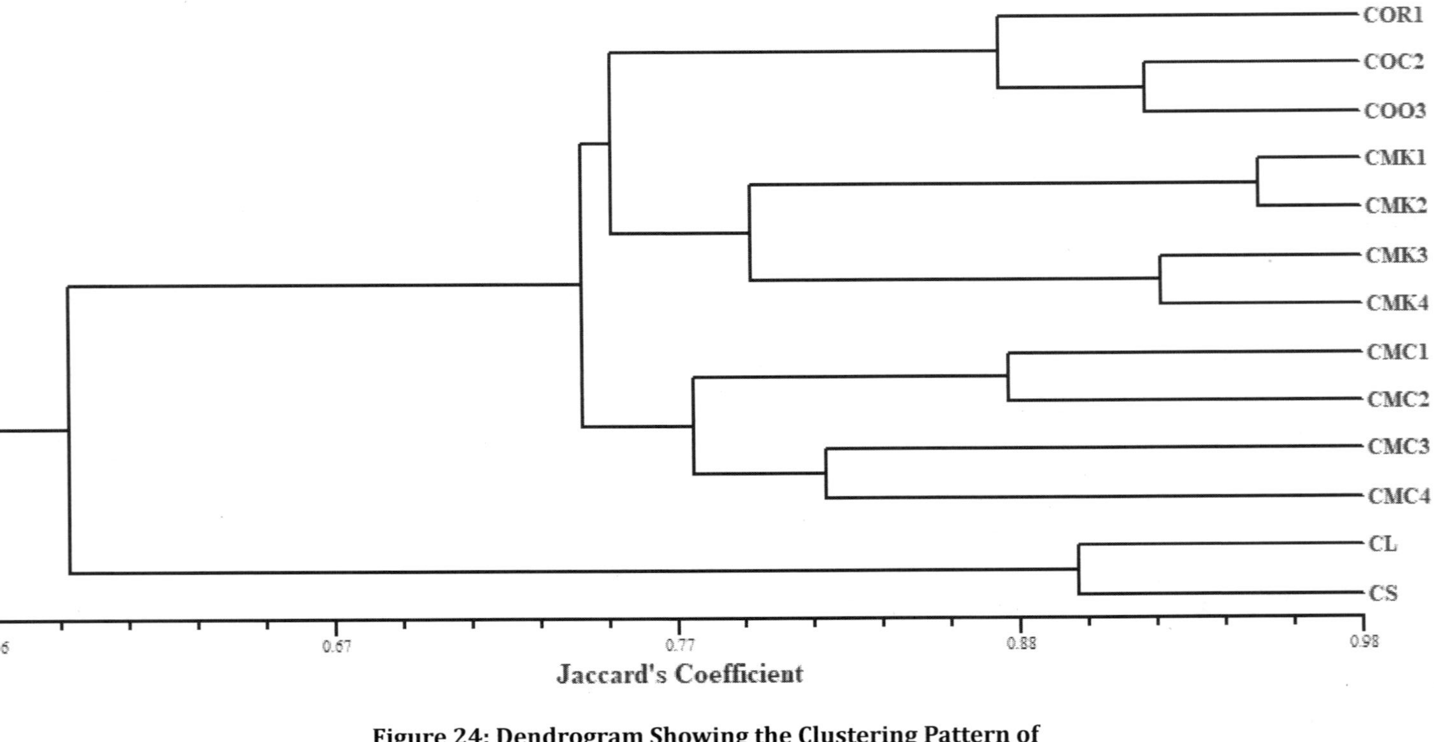

Figure 24: Dendrogram Showing the Clustering Pattern of 13 Samples of 4 Species of *Cordia* as Revealed from RAPD Markers.

Table 2.3: Jaccard's Table Revealed from RAPD Markers Showing the Similarity among and within the 13 Samples of 4 *Cordia* Species

	COR1	*COC2*	*COO3*	*CL*	*CS*	*CMC1*	*CMC2*	*CMC3*	*CMC4*	*CMK1*	*CMK2*	*CMK3*	*CMK4*
COR1	1.000												
COC2	0.853	1.000											
COO3	0.886	0.914	1.000										
CL	0.600	0.583	0.579	1.000									
CS	0.595	0.579	0.575	0.893	1.000								
CMC1	0.707	0.732	0.762	0.600	0.595	1.000							
CMC2	0.667	0.737	0.725	0.553	**0.550**	0.872	1.000						
CMC3	0.806	0.784	0.816	0.595	0.590	0.825	0.789	1.000					
CMC4	0.692	0.718	0.750	0.579	0.615	0.762	0.725	0.816	1.000				
CMK1	0.769	0.795	0.780	0.575	0.571	0.791	0.714	0.800	0.780	1.000			
CMK2	0.725	0.750	0.738	0.575	0.571	0.791	0.714	0.756	0.738	**0.949**	1.000		
CMK3	0.711	0.784	0.725	0.595	0.590	0.738	0.659	0.700	0.683	0.800	0.756	1.000	
CMK4	0.700	0.769	0.756	0.590	0.585	0.810	0.732	0.732	0.714	0.829	0.786	0.919	1.000

Bhubaneswar and minimum (55 per cent) between *C. sebestena* (CS) and *C. myxa* from Cuttack (CMC2) was observed (Table 2.3).

ISSR Analysis

Out of 15 primers used, thirteen ISSR primers yielded 89 bands. Of these, 3 were monomorphic and rest (84) was polymorphic in nature including 7 unique bands. The details of the ISSR primers and band details are presented in Table 2.4 and Figure 2.5. Maximum of 11 no. of polymorphic bands were amplified with primer PCP-12 and minimum polymorphism (2 bands) with the primer ISSR-10. Highest Resolving Power and Primer Index were calculated as 10.15 and 6.20 from ISSR-17 and PCP-12 respectively. Similarly, lowest values of Resolving Power (1.85) in PCP-8 and Primer Index (0.15) in ISSR-10 were recorded.

The dendrogram constructed on the basis of SHAN clustering using Jaccard's similarity showed similar trend and relationship as observed in case of RAPD analysis. A distinct and separate cluster of *Cordia sebestena* and *Cordia lutea* was observed having 29 per cent genetic similarities (Figure 2.6). *C. myxa* and *C. dictotoma* formed distinct clusters along with their respective accessions. All three accessions of *C. dictotoma* came in a group and had more than 95 per cent similarity among them. Each of the 4 accessions of *C. myxa* collected from Khandagiri hills and Cuttack formed district clades based on their locality of distribution sharing a node at 84 per cent similarity level. As per the Jaccard's similarity coefficient, 100 per cent similarity was recorded between two accessions of Cuttack (CMC1 and CMC4) and least similarity was 26.9 per cent between CMK1-CL, CMK3-CL and CMK4-CS (Table 2.5).

Combined Molecular Marker Analyses (RAPD + ISSR)

When the data obtained from RAPD and ISSR analyses were considered in combination, a slightly different pattern of relationship emerged in the dendrogram. The dendrogram constructed on the basis of both RAPD and ISSR data revealed segregation of the *Cordia lutea* and *C. sebestena* in the first place from the large cluster of C. *dictotoma* and *C. myxa* with a similarity of more than 39 per cent (Figure 2.7). All the 3 accessions of *C. dictotoma* and 8 accessions of *C. myxa* formed clear clusters according to their species affiliation. *C. myxa* genotypes also got segregated to two clusters of 4 each based on their locality of collection. As per Jaccard's similarity analysis, two accessions of *C. myxa* from Khandagiri (CMK1-CMK2) had highest similarity (95.5 per cent) between them, whereas minimum (37.3 per cent) was noted between *C. myxa* accession of Khandagiri and *C. lutea* (CMK1-CL). Eight accessions of *C. myxa* formed two very distinct groups of 4 each in the dendrogram justifying their geographical area of occurrence. Within *C. myxa* genotypes, accessions CM5, CM6, CM7 and CM8 were found to be genetically close; CM5 and CM6 being the closest of all with over 80 per cent level of similarity.

The jaccard's similarity analysis revealed that all the samples were genetically related to each other having an average similarity of 0.629. The values showed that

Table 2.4: Details of ISSR Analysis in *Cordia* spp.

Sl.No.	Primer	Nucleotide Sequence	No. of Bands Present	No. of Polymorphic Bands	No. of Monomorphic Bands	No. of Unique Band	Resolving Power	Primer Index	Percentage (Per cent)
1	**PCP-1**	$(GAC)_5$	8	8	0	1	4.62	3.15	100
2	**PCP-2**	$(AGG)_5$	9	9	0	0	6.92	2.05	100
3	**PCP-4**	$T(GA)_4$	6	6	0	1	3.54	2.27	100
4	**PCP-7**	$(GGA)_4$	8	8	0	0	5.38	2.71	100
5	**PCP-8**	$(GTG)_5$	3	1	0	0	1.85	0.15	33.33
6	**PCP-9**	$(GACAC)_4$	7	7	0	1	7.08	2.06	100
7	**PCP-10**	$(AG)_8$	5	4	1	0	7.54	1.14	80
8	**PCP-11**	$(GA)_8G$	4	4	0	1	4.92	1.07	100
9	**PCP-12**	$(GACA)_8G$	11	11	0	1	8.62	6.20	100
10	**ISSR-2**	$(AG)_8G$	10	9	1	0	7.54	4.37	90
11	**ISSR-10**	$(GACA)_4G$	2	1	1	0	1.85	0.15	50
12	**ISSR-17**	$G(CT)_8$	8	8	0	1	10.15	2.34	100
13	**ISSR-22**	$(CTGT)_4C$	8	8	0	1	6.77	3.72	100

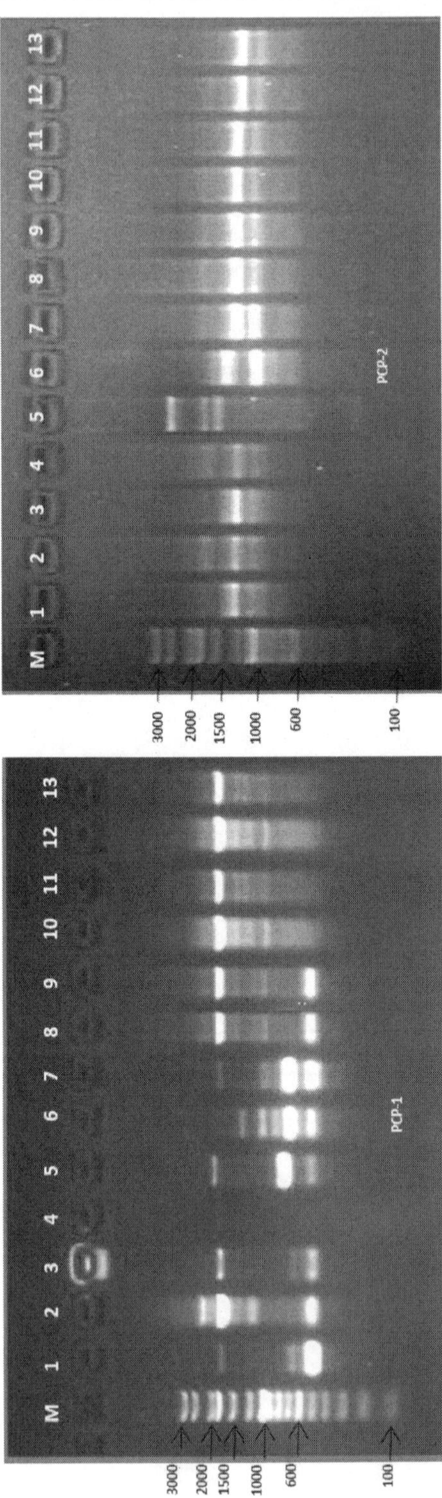

Figure 2.5: Details of ISSR Banding Pattern in *Cordia* spp.

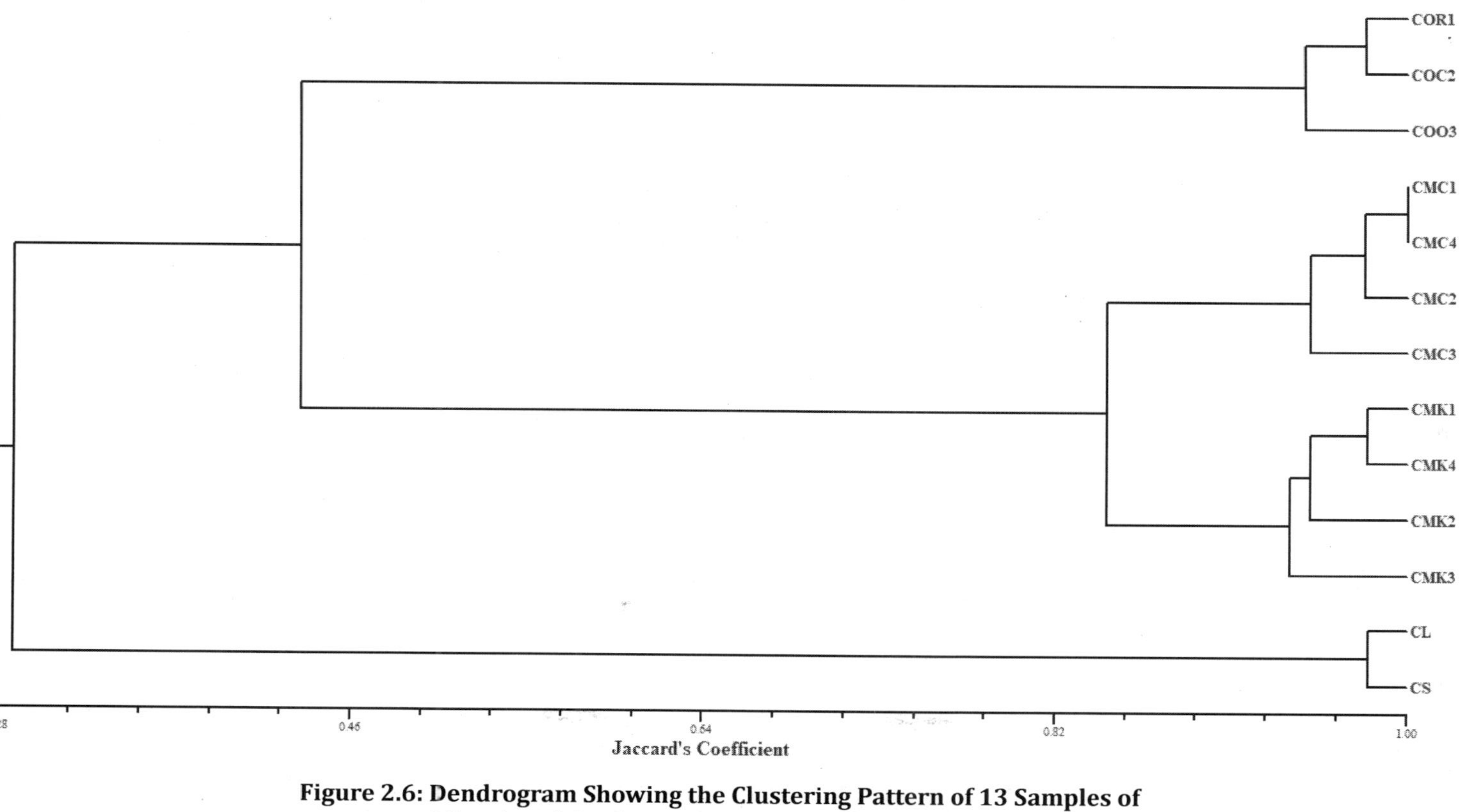

Figure 2.6: Dendrogram Showing the Clustering Pattern of 13 Samples of 4 Species of *Cordia* as Revealed from ISSR Markers.

Table 2.5: Jaccard's Table Revealed from ISSR Markers Showing the Similarity among and within the 13 Samples of 4 *Cordia* Species

	COR1	*COC2*	*COO3*	*CL*	*CS*	*CMC1*	*CMC2*	*CMC3*	*CMC4*	*CMK1*	*CMK2*	*CMK3*	*CMK4*
COR1	1.000												
COC2	0.979	1.000											
COO3	0.938	0.958	1.000										
CL	0.311	0.307	0.307	1.000									
CS	0.315	0.311	0.311	0.980	1.000								
CMC1	0.444	0.438	0.460	0.280	0.284	1.000							
CMC2	0.438	0.431	0.453	0.293	0.297	0.978	1.000						
CMC3	0.453	0.446	0.469	0.273	0.276	0.957	0.938	1.000					
CMC4	0.444	0.438	0.460	0.280	0.284	1.000	0.978	0.957	1.000				
CMK1	0.403	0.418	0.439	0.269	0.273	0.860	0.843	0.863	0.860	1.000			
CMK2	0.391	0.406	0.426	0.278	0.282	0.827	0.846	0.830	0.827	0.960	1.000		
CMK3	0.424	0.439	0.462	0.269	0.273	0.860	0.843	0.863	0.860	0.959	0.922	1.000	
CMK4	0.397	0.412	0.433	0.282	0.269	0.843	0.827	0.846	0.843	0.980	0.941	0.940	1.000

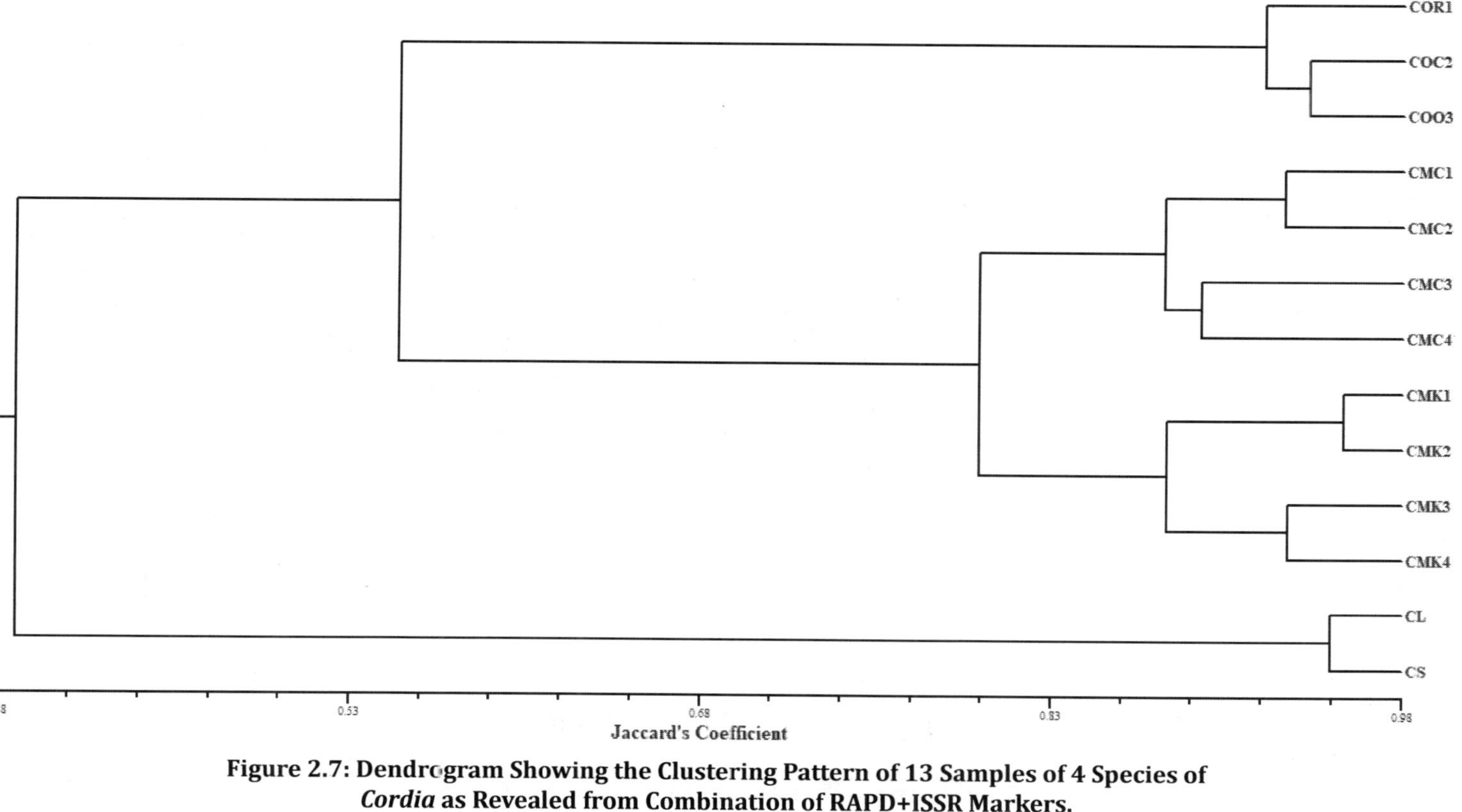

Figure 2.7: Dendrogram Showing the Clustering Pattern of 13 Samples of 4 Species of *Cordia* as Revealed from Combination of RAPD+ISSR Markers.

Table 2.6: Jaccard's Table Revealed from RAPD+ISSR Markers Showing the Similarity among and within the 13 Samples of 4 *Cordia* Species

	COR1	*COC2*	*COO3*	*CL*	*CS*	*CMC1*	*CMC2*	*CMC3*	*CMC4*	*CMK1*	*CMK2*	*CMK3*	*CMK4*
COR1	1.000												
COC2	0.926	1.000											
COO3	0.916	0.940	1.000										
CL	0.404	0.396	0.398	1.000									
CS	0.409	0.402	0.404	0.949	1.000								
CMC1	0.548	0.552	0.581	0.391	0.397	1.000							
CMC2	0.524	0.544	0.558	0.381	0.386	0.929	1.000						
CMC3	0.580	0.569	0.598	0.377	0.383	0.897	0.872	1.000					
CMC4	0.539	0.544	0.573	0.381	0.398	0.885	0.860	0.894	1.000				
CMK1	0.538	0.557	0.570	0.373	0.378	0.828	0.785	0.835	0.824	1.000			
CMK2	0.514	0.532	0.545	0.378	0.383	0.811	0.787	0.798	0.787	0.955	1.000		
CMK3	0.529	0.563	0.562	0.374	0.379	0.804	0.761	0.791	0.780	0.888	0.848	1.000	
CMK4	0.509	0.542	0.556	0.385	0.378	0.828	0.785	0.796	0.785	0.911	0.871	0.931	1.000

two accessions of *Cordia myxa* (*i.e.* CMK1 and CMK2) had the highest similarity coefficient of 0.955, whereas minimum similarity (0.373) was between *C. lutea* and *Cordia myxa* (CMK1). The jaccard's similarity among the species is represented in Table 2.6.

Discussion

The family Boraginaceae has traditionally been subdivided into four taxonomic units Boraginoideae, Heliotropioideae, Cordioideae and Ehretioideae (de Candolle, 1845; Gurke, 1893). Cordiaceae has been established as a monophyletic group based both on molecular (Gottschling *et al.*, 2001) and morphological data (multilayered endocarp; Diane *et al.*, 2002) and constitutes the Primarily Woody Boraginales (*sensu* Gottschling, 2003). The phylogenetic relationships of Cordiaceae were investigated using ITS1 sequences from 50 in-group taxa and three out-group taxa (Gottschling *et al.*, 2005). The study segregated Cordiaceae into four major monophyletic assemblages, more or less corresponding to established taxonomic units as previously inferred from morphology, ecology and biogeography such as- (i) Varronia (currently recognized as *Cordia* sect. *Varronia*) (ii) Collococcus clade comprises the New World species of *Cordia* sect. *Myxa*), all of *Cordia* sect. *Superbiflorae* and *Cordia varronifolia,* (iii) Myxa clade including all of the Old World species of *Cordia* sect. *Myxa* and the New World species *Cordia dentata* and *Cordia lutea* and (iv) Sebestena clade including *Cordia* sects. *Cordia* and *Gerascanthus* etc. A set of ancestral characters (including apomorphies and plesiomorphies) is reconstructed for Cordiaceae based on the phylogenetic hypotheses as inferred from the congruence between morphological and ITS1 data.

In the present study, based on RAPD and ISSR data *Cordia lutea* and *C. sebestena* formed a separate cluster sharing a node with C. *dichotoma- C. myxa* accessions. The closeness of *Cordia lutea* and *C. sebestena* was not in accordance with the infra-generic classification of Gottschling *et al.* (2005) based on molecular analysis and earlier inference from taxonomy and biogeography of the genus. While *C. lutea* belongs to Collococcus clade, *C. sebestena* belongs to Sebestena clade of *Cordia*. The segregation of *C. sebestena* from C. *dichotoma* - *C. myxa* cluster was supported by the trichome study of Wafaa (2006), who reported the presence of sparse trichomes on the upper leaf surface of *C. myxa* and *C. dichotoma* but moderately distributed trichomes in *C. sebestena.* Similarly, absence of pyrrolizidine alkaloids (PAs) in *C. lutea* separates the species from other taxa of the genus (Mayevych *et al.,* 2015).

It was observed that the accessions of two species *C. dictotoma* and *C. myxa* shared a common node and all accessions a particular species formed clear clusters justifying their taxonomic identity as species. Both the species belong to *Cordia* sect. *Myxa* as per the ITS1 sequence data of Gottschling *et al.* (2005) and also based on previous morphological and bio-geographical studies and therefore, their clustering in the dendrogram appears justified. Though *C. dictotoma* and *C. myxa* have been established as two distinct taxonomic entities, they have been considered synonymous in several floras and monographs. The problem arose when the name *C. myxa* was used in "Flora of British India" (Hooker, 1883) for a species from India, which is same as *C. obliqua* and now correctly as known *C. dichotoma.* This distinct plant now understood to be confined to Asia minor and Egypt. *Cordia dictotoma* and *C. myxa* can be distinguished from each other by the colour of bark; size and shape of leaves and their venation; size of fruits and seeds as shown in Figure 2.4. While the leaves of *C. myxa* are ovate or rotund with 5 basal nerves, the leaves of *C. dichotoma* are oblong to obovate with one basal mid-vein. The diameter of frits ranges from 2.5 to 4.0 cm in *C. myxa* and from 1.5 to 2.0 cm in *C. dichotoma;* the seed size also varies proportionately. All the accessions of *Cordia dictotoma* and *C. myxa* formed compact clusters justifying the fact that they represent two biological species. Under each species, the accessions grouped according to their geographical locations of occurrence or collection. Nandedkar and Mulani (2016) analyzed the genetic diversity of *Cordia dichotoma* using RAPD markers and Latifi *et al.* (2018) studied the inter and intra-population genetic variability in *Cordia myxa* with application of Conserved DNA-Derived Polymorphism (CDDP) molecular markers and made similar observations on genetic variability in these two species. However, assessment of genetic diversity taking large number of species and populations and application of more consistent molecular markers would throw light on the genomic relationship and phylogeny of the genus *Cordia*. As remarked by Gottschling *et al.* (2005), the taxonomy of *Cordia* is complex and controversial and this is partly due to "extraordinarily high intra-specific variability" in some groups of species, making identification difficult, and partly due to new taxa having been "airily described on the basis of poorly preserved herbarium specimens".

References

Bhattacharya, P. and Saha, A. (2013). Evaluation of reversible contraceptive of *Cordia dichotoma* leaves extract. *Rev. Bras. Farmacogn*. 23: 342-350.

Candolle, A. P. de. (1845). Borraginese. *In*: A. P. de Candolle (Ed). Prodromus Systematis Naturalis Regni Vegetabilis 9: 466-566. Treuttel and Wttrtz, Paris.

Costa, J. F. O., David, J. P., David, J. M., Giulietti, A. M., Queiroz, L. P., Santos, R. R. and Soares, M. B. P. (2008). Immunomodulatory activity of extracts from *Cordia superba* Cham and *Cordia rufescens* A. DC. (Boraginaceae), plant species native from Brazilian Semi-arid. *Brazilian Journ. Pharmacognos.* 18: 11-15.

Derero, A., Gailing, O., Finkeldey, R. (2011). Maintenance of genetic diversity in *Cordia africana* Lam., a declining forest tree species in Ethiopia. *Tree Genet. Genomes*, 7: 1-9.

Diane, N., Forther, H. and Hilger, H. H. (2002). A systematic analysis of Heliotropium, Tournefortia and allied taxa of the Heliotropaceae (Boraginales) based on ITS1 sequences and morphological data. *Am. J. Bot.* 89: 287-295.

Doyle, J. J. and Doyle, J. L. (1990). Isolation of plant DNA from fresh tissue. *Focus,* 12: 13-15.

Gherardi, M., Mangin, B., Goffinet, B., Bonnet, D. and Huguet, T. (1998). A method to measure genetic distance between allogamous populations of alfalfa (Medicago sativa) using RAPD molecular markers. *Theor. Appl. Genet.* 96: 406–412.

Gottschling, M. and Hilger, H.H. (2001). Phylogenetic analysis and character evolution of Ehretia and Bourreria (Ehretiaceae, Boraginales) and their allies based on ITS1 sequences. *Bot. Jahrb. Syst.* 123: 249–268.

Gottschling, M. (2003). Phylogenetic Analysis of Selected Boraginales. Ph.D. Thesis, Freie Universitat Berlin, Berlin <http: //www.diss.fu-berlin.de/2003/30/>.

Gottschling, M., Hilger, H.H., Wolf, M. and Diane, N. (2001). Secondary structure of the ITS1 transcript and its application in a reconstruction of the phylogeny of Boraginales. *Pl. Biol.* 3: 629–636.

Gottschling, M., Miller, J.S., Weigend M. and Hartmut H. Hilger. (2005). Congruence of a Phylogeny of Cordiaceae (Boraginales) Inferred from ITS1 sequence data with morphology ecology, and biogeography. *Ann. Missouri Bot. Gard.* 92: 425-437.

Gurke, M. (1893). Borraginaceae. *In*: A. Engler and K. Prantl (Ed.), Die Natiirlichen Pflanzenfamilien 4 (3): 71-131. Engelmann, Leipzig.

Hooker, J. D. (1883). Boraginaceae. *In*: Fl. Brit. India, J. D. Hooker (Ed.). L. Reeve and Co., London, 4(X): 1-78.

Jaccard, P. (1908). Nouvelles rescherches sur la distribution florale. *Bull. Soc. Vaud. Sci. Nat.*, 4: 223-270.

Latifi, E., Yousofi, M. and Haerinasab, M. (2018). Study of genetic diversity in some populations of *Cordia myxa* L. in Iran by using CDDP molecular marker. *Iranian Journ. Plant Biol.* 9 (4): 1-16.

Mabberley, David J. (2017). Mabberley's plant-book: a portable dictionary of plants, their classification and uses. Fourth Edition. Cambridge University Press. New York.

Matias, E. F. F., Alves, E. F., do Nascimento Silva, M. K., de Alencar Carvalho, V. R., Coutinho, H. D. M., and da Costa, J. G. M. (2015). The genus *Cordia*: botanists, ethno, chemical and pharmacological aspects. *Revista Brasileira de Farmacognosia* 25: 542–552.

Matias, E. F. F., Santos, K. K., Costa, J. G. M. and Coutinha, H. D. (2010a). Screening for *in vitro* phototoxic activity of methanol extracts of *Croton campestris* A., *Ocimum gratissimum* L. and *Cordia verbenaceae* DC. *Indian J. Med. Res.* 132(5): 520–522.

Matias, E. F. F., Santos, K.K., Ammeida, T.S., Costa, J.G.M. and Coutinho, H.D.M. (2010b). Atiridadeantibacteriana*in vitro* de *Croton campestria* A., *Ocimum gratissimum* L. e *Cordia verbenacea* DC. *Rev. Bras. Biosci.* 8: 294-298.

Mayevych I, López-Romero JM, Cabanillas J (2015). Chemical Composition of Cordia lutea L.: Absence of Pyrrolizidine Alkaloids. *Nat. Prod. Chem. Res.* 3: 194. doi: 10.4172/2329-6836.1000194.

Medeiros, R., Passos, G., Vitor, C., Koepp, J., Mazzuco, T., Pianowski, L., Campos, M. and Calixto, J. (2007). Effect of two active compounds obtained from the essential oil of *Cordia verbenacea* on the acute inflammatory responses elicited by LPS in the rat paw. *Br. J. Pharmacol.* 151: 618-627.

Nandedkar, P. H. and Mulani, R. M. (2016). Morphological and molecular diversity of Cordia dichotoma G. Frost populations from Nanded district in Maharashtra, India. *Int. J. Curr. Microbiol. Appl. Sci.* 5(6): 135-143.

Paulino, R. D. C., Henriques, G. P. D. S. A., Neto, H. D.S. L., Coelho, M. D. F. B., Dombroski, J. L. D. and de Araujo Lopes, C. P. (2011). Diferentes substratos na propagação por estaquia de *Cordia globosa* e *Cordia leucocephala. Revista Verde* (Mossoró – RN – Brasil) 6 (4): 274 – 278.

Prevost, A. and Wilkinson, M. J. (1999). A new system of comparing PCR primers applied to ISSR fingerprinting of potato 2393©FUNPEC-RP. *Genet. Mol. Res.* 10 (4): 2382-2393.

Rohlf, F.J. (1997). NTSYS-pc Numerical taxonomy and multivariate analysis system. Version 2.02e. Exeter Software, Setauket, New York.

Scheeren, L.W., Schnider, P.S.P., Schnider, P.R., Finger, C.A.G. (2002). Growth of Louro, *Cordia trichotoma* (Vell.) Arrab. ex Steud., in the center depression of Rio Grande do Sul. *Ciênc. Florest.* 12 (2): 169-176.

Sertie, J.A., Basile, A.C., Panizza, S., Matida, A.K., Zelnik, R. (1990). Anti-inflammatory activity and sub-acute toxicity of artemetin. *Planta Med.*56: 36-40.

Sneath, P. H. A. and Sokal, R. R. (1973). Numerical taxonomy - the principles and practice of numerical classification, W. H. Freeman, San Francisco.

The Plant List (2013). Version 1.1. Published on the Internet; http: //www. theplantlist.org/(accessed 10th April, 2020).

Thirupathi, K., Sathesh Kumar, S., Raju, V.S., Ravikumar, B., Krishna, D. R. and Krishna Mohan, G. (2008). A review of medicinal plants of the genus *Cordia*: Their chemistry and pharmacological uses. *J. Nat. Remed.* 8 (1): 1 – 10.

Williams, J. G. K., Kubelik, A. R., Livak, K. J., Rafalski, J. A. and Tingey, S. V. (1990). DNA polymorphisms amplified by arbitrary primers are useful as genetic markers. *Nucleic Acids Research,* 18: 6631-6635.

Zietkiewicz, E., Rafalski, A. and Labuda, D. (1994). Genome fingerprinting by simple sequence repeat (SSR)-anchored polymerase chain reaction amplification. *Genomics,* 20: 176-183.

Chapter 3

Polypetalous Plant Diversity of Mayurbhanj District, Odisha with Special Reference to RET Species

K.K. Mandal[1], T. Kar[2] and A.K. Biswal[1]*

[1]Department of Botany North Orissa University, Baripada – 757 003, Odisha
[2]Department of Botany, Fakir Mohan University, Balasore – 756 089, Odisha
**e-Mail: kumar.kishore718@gmail.com*

ABSTRACT

The sub-class Polypetalae under dicotyledons are characterised by distinct sepals and petals in their floral structure. The present study is aimed to document the diversity of polypetalous plants occurring in Mayurbhanj district, Odisha, India. During the study, emphasis was given to determine the distribution status of polypetalous plants with special reference to Rare, Endemic and Threatened (RET) taxa. The diversity of polypetalous plants exhibit the structure and composition of forests in Mayurbhanj. Their proper knowledge will be immensely helpful in planning for conservation and sustainable use of available plant resources of the district. As many as 450 taxa belonging to 242 genera under 72 families are reported. Out of which five species are found to be rare, while two are coming under endangered and four belongs to vulnerable category.

Keywords: *Conservation, Endangered, Flora, Mayurbhanj, Polypetalous plant.*

Introduction

Collection and documentation of the floristic wealth of a region is pre-requisite for planning and proper utilization of its potential plant resources in general and to conserve the depleting genetic resources in particular. The flora of India is both rich and diverse due to wide range of variations in climate, altitude, topography and ecological habitats. The Indian flora accounts for 11 per cent of the total plant species of the World. About 28 per cent of Indian plants are endemic to the country. Angiosperms comprises ca 17,643 species under 3000 genera belonging to 248 families, representing roughly 7 per cent of the World angiospermic flora. (Anon, 2010).

The knowledge of plant diversity is a pre-requisite for assessment and conservation of biodiversity. It helps us to understand the overall structure and function of an ecosystem (Sumeet *et al.,* 2010). Accurate and precise information of the plant species occurring in a region is highly essential for their utilisation and for conservation strategies. The information also allows us to prevent or to avoid the potential chances of biodiversity loss and to plan future policy for the protection of our environment. Rare, endemic and threatened taxa are evaluated to know the present status of biodiversity. Invasive alien species which are second greatest threat to biodiversity (Wilcove *et al.,* 1998) can be better managed only if, proper knowledge about them especially their distribution and reproductive biology are available. The sub-class Polypetalae under dicotyledons are characterised by distinct sepals and petals in their floral structure. Study on diversity of polypetalous plants in Mayurbhanj district of Odisha with special reference to Rare, Endemic and Threatened taxa was undertaken in the first phase.

Mayurbhanj is a land- locked district with a total geographical area of 10,418 Sq.km. and is situated in the Northern boundary of the state with district Head quarters at Baripada. The district lies between 21° 16' and 22° 34' North latitude and 85°40' and 87° 11' East longitudes. The Similipal hill ranges are located in the central part of the district. The geological formations of the district chiefly belong to Meso Proterozoic age, Mio-Pliocene age and recent geological periods. The soil of the district may be broadly classified into Red soil and Laterite soil. The red soil is further classified in to three subgroups namely, typical soil, red loamy soil and clay, loam soil.The reddish and sandy soil favours the luxuriant growth of Sal, the dominant tree of the district.

Mayurbhanj district enjoys tropical monsoon climate. The average rainfall of Odisha is 142.2cm, where as Mayurbhanj receives the maximum rainfall of 158 cm per annum. An oppressive hot and humid summer followed by the monsoon between June and November and a severe cold winter characterizes the general climate of Mayurbhanj. The higher reaches of the Similipal hills experience frosting during the peak of winter. Relative Humidity is quite high throughout the year. At places, it reaches upto 90 per cent during the months of June and July.

As per the classification of Champion and Seth, 1968 the forests of Mayurbhanj belong to two broad categories namely Northern Tropical Moist Deciduous Forests and Northern Tropical Semi-evergreen Forests.

The following forest types have been recognised in the district (Biswal, 2009). Such as Orissa Semi Evergreen Forests, Northern Tropical Moist Deciduous Forests, Tropical Dry Deciduous Forests, Central Indian Hill Forests, Dry Deciduous Sal Forests and Grass lands.

Materials and Methods

The present study was under taken during the year 2009-2011 covering different forest types of the district. Extensive field work, literature survey, Herbarium scrutiny and critical examination of plant specimen were done. The entire vegetation was thoroughly scanned by repeated visit in different season of the year covering all ecological habitats. Normally five specimens of each species with reproductive parts were collected to prepare complete herbarium for future reference. Relevant field notes were made on the spot including diagnostic features, distribution status of each species in terms of abundance *etc.* were noted. All the collected specimens were properly processed and housed in Herbarium, P.G Department of Botany, North Orissa University. Plants were identified in consultation with taxonomic treatments Haines (1961), Mooney (1950), Saxena and Brahmam (1994-96), Singh *et al.* (2001), Murty and Panigrahi (1989).

Results and Discussion

The flora of the district is considerably diverse in terms of taxa. There are 450 Polypetalous plant species distributed within 242 genera under 72 families. The family Fabaceae exhibits highest number of species *i.e.* 112 belongs to 46 genera followed by Malvaceae (24 species of 8 genera), Cucurbitaceae (22 species of 15 genera) and Caesalpiniaceae (20 species of 5 genera). Interestingly, Tiliaceae and Vitaceae contribute 4 genera and 15 species each. Magnoliaceae, Bixaceae, Cochlospermaceae, Pittosporaceae, Protulacaceae, Dipterocarpaceae, Balsaminaceae, Simaroubaceae, Caricaceae, Olacaceae, Siphonodontaceae, Staphylaceae, Moringaceae, Crassulaceae, Rhizophoraceae, Lecythidaceae, Punicaceae, Trpaceae, Turneraceae, Passifloraceae, Caricaceae, Begoniaceae, Aizoaceae and Alangiaceae *etc.* are having single species. Among the invasive alien species belonging to polypetalae, frequently observed taxa are *Argemone mexicana* L., *Cassia alata* L., *Cassia hirsuta* L., *Cassia occidentalis* L., *Cassia tora* L., *Cleome monophylla* L., *Cleome viscosa* L., *Corchorus aestuans* L., *Crotalaria pallida* Dryand,., *Indigofera astragalina* DC., *Indigofera linifolia* (L.f.) Retz., *Indigofera linnaei* Ali, *Ludwigia adscendens* (L.) Hara, *Ludwigia perennis* L., *Melochia corchorifolia* L., *Mimosa pudica* L., *Oxalis corniculata* L., *Passiflora foetida* L., *Portulaca oleracea* L., *Sida acuta* Burm.f., *Triumfetta rhomboidea* Jacq., *Turnera ulmifolia* L., *Urena lobata* L.,and *Walthera indica* L. *etc.* The taxon *Cassia alata* L. is spreading rapidly in wasteland and degraded forests. During the present study as many as 11 no.

of flowering plants have been enlisted as Rare, Endemic, Threatened (RET) taxa within the geographical boundaries of Mayurbhanj district, Odisha.

Table 3.1: List of RET Species of Mayurbhanj

Sl.No.	*Name of the Plants*	*Family*	*Status*
1	*Alphonsea ventricosa* (Roxb.) Hook.f and Thoms	Annonaceae	R
2	*Garcinia cowa* Roxb.ex Dc	Clusiaceae	R
3	*Garcinia xanthochymus* Hook. f. Ex T. Anders	Clusiaceae	R
4	*Hypericum gaitii* Haines	Hypericaceae	R
5	*Mesua ferrea* L	Clusiaceae	EN
6	*Polyalthia simiarum* (Buch.- Ham. ex Hook.f. and Thoms.)Hook. f. and Thoms	Annonaceae	VU
7	*Pueraria tuberose* (Willd.)Dc. Ann.	Fabaceae	VU
8	*Polyalthia simiarum* (Buch.- Ham. Ex Hook.f. and Thoms.) Hook.	Annonaceae	VU
9	*Pterocarpus marsupium* Roxb.	Fabaceae	VU
10	*Saraca asoca* Roxb.	Caesalpiniaceae	CR
11	*Thalictrum foliolosum* Kunth.	Ranunculaceae	VU

Abbreviations: Vu = Vulnerable, CR = Critically Endangered, EN = Endangered, R = Rare

Conclusion

The present study will serve as primary input towards monitoring and sustaining the plant diversity of largest district of Odisha. The success of any conservation programme, irrespective of any region, depends fundamentally on the participatory role of the local people. It is therefore, important to educate the country people about the endangered species, their importance in the ecosystem and how these can be protected and multiplied.

Acknowledgement

The authors are grateful to the Head, P.G. Department of Botany, North Orissa University, Takatpur, Baripada for providing necessary facilities to carry out this piece of work. Special thanks to local community and forest dept. staff who cooperated the authors to carry out the field work.

REFERENCES

Anonymous, (2010). Global Forest Resources Assessment, FAO, Rome.

Bal, S.N.(1942). Useful plants of Mayurbhanj state in Orissa. *Rec. Bot. Surv. India* 6: 1-119.

Biswal, A.K. (2009). A Glimpse into vegetation of Mayurbhanj in Souvenir. *33rd Annual Conf.* OBS. NOU

Bentham, G. and J.D. Hooker, (1862-1883). Genera plantarum, Vol. I-IV. London.

Haines, H.H. (1921-25).The Botany of Bihar and Orissa, Vol. I-III. London.

Mooney, H.F. (1950).Supplements to the Botany of Bihar and Orissa.,Catholic Press, Ranchi

Saxena, H.O. and Brahmam, M. (1994-96). The Flora of Orissa Vol I-IV. RRL and OFDC, Bhubaneswar.

Senapati, N. and N.K. Sahu (1969). Gazetteer of India Orissa, Mayurbhanj.

Singh *et al.* (2001). Flora of Bihar Analysis. BSI, Calcutta.

Sumeet, G., C. M. Sharma, C. S. Rana, S. K. Ghildiyal, and S. Suyal, (2010), Phytodiversity (Angiosperms and Gymnosperms) in Mandal – Chopta Forest of Garhwal Himalaya, Uttarakhand, India, *Nature and Science*, Vol. 8(1): 1 – 17.

Wilcove, D. S., D. Rothstein, J. Dubow, A. Phillips, and E. Losos, (1998), Quantifying Threats to imperilled Species in the United States, *Bioscience*, Vol. 48(8): 607-615.

Chapter 4

Plant Species Selection for Restoration of Degraded Habitats of Jajang Iron Ore Mines, Keonjhar, Odisha, India

R.K. Mishra[1] and B.K. Behera[2]*

[1]Department of Wildlife and Biodiversity Conservation, North Orissa University, Takatpur, Baripada – 757 003, Odisha
[2]Jajang Iron and Mn. Mines, Rungta Mines Limited, Keonjhar – 758 304, Odisha
**e-mail:rabikumishra@gmail.com*

ABSTRACT

The study was carried out with an aim to select indigenous plant species for restoration of degraded habitats at Jajanga Khesra and District Level Committee (DLC) forest area formed due to mining by Rungta Mines Limited in Keonjhar district of Odisha. Species for restoration were selected from mine forest area through systematic phytosociological analysis that involved measurement of Importance Value Index (IVI) and regeneration values of tree species. However, species selection from plantation was made based on their growth, productivity, and adaptation to terrain and soil types. Shrubs and grasses were selected based on relative index and relative abundance, respectively. The top 15 tree and 16 grass species, and the entire six shrub species were selected from mine forest area and plantation were considered for restoration of dump areas. Since soils of dump were of similar nature as that of the forest area, the findings will be helpful in faster restoration of the degraded habitats with initial human facilitation. After care and monitoring has also been suggested to enhance the recovery process.

Keywords: *Growth, Importance value index (IVI), Productivity, Restoration, Vegetation.*

Introduction

Ecological restoration aims to recreate, initiate, or accelerate the recovery of habitats that has been disturbed. Disturbances are environmental changes that alter ecosystem structure and function. Common disturbances include mining, logging, damming rivers, intense grazing and fire. Out of these disturbances degradation of natural habitats due to opencast mining leads towards clearing of vegetation is a common phenomenon. Normal recovery of distributed habitats takes times through colonization of plant and animal species (Pandey, 2000; Cole *et al.,* 2010; Vibhash *et al.,* 2015) and can be accelerated through human intervention. Till yet, not much attention was paid to restore these habitats. Recent awareness followed by stringent enforcement of rules and regulations has generated consciousness about the environment. Restoration has been attempted on an experimental basis in various parts of India (Jha *et al.,* 2000; Dongmei and Changqun, 2008; Sharma *et al.,* 1998; Sing *et al.,* 2000) through implementation of plantation programme. In such practices plantation of exotic and horticultural species has attempted for restoration and as a result the degraded land has been made productive (Sharma and Sunderraj, 2005).

Restoration is the recreation of entire communities of organisms, closely modelled on those occurring naturally, while reclamation refers to any deliberate attempt to return a damaged ecosystem to some kind of productive use (Parrota and Knowles, 1999; Kumar *et al.,* 1998; Jim, 2001). It would appear that in many cases, it is reclamation rather than restoration which is attempted. Restoration needs knowledge of indigenous biological resources for understanding the species response to micro-environmental features and edaphic conditions (Rao and Tak, 2002; Singh *et al.,* 2004; Wong, 2003). Information on these aspects not only helps in selecting species adapted to local conditions (Sing *et al.,* 2002) but also in the fast recovery of the ecosystem when such species are planted there (Boyd, 2002; Verma and Verma, 2017). This article represents the outcome of a study that included these aspects to help in ecological restoration of disturbed sits at Jajang DLC and Khesra forest area in Keonjhar district, Odisha.

The Jajang DLC and Khesra forest in Keonjhar district, Odisha has been disturbed due to mining since 1957. The area with dry deciduous forests contains good quality of minable iron ore deposits. To undertake the mining activities some part of the forest covers of such forests were cleared. Presently dumping is being done in well vegetated valleys adversely affected the vegetation. Thus the present study was an attempt to select plant species based on a set of criteria in restoring the degraded habitat of Jajang Iron Ore Mine, Odisha.

Study Area

The Jajang Iron Ore Mine (JIOM) 21° 55/00//to 21° 56/35//N latitude and 85° 25/00//to 85° 27/10//E longitude having a lease area of about 666.15 ha includes 247 ha of forest area and 41.97 ha of dump area (Figure 4.1).

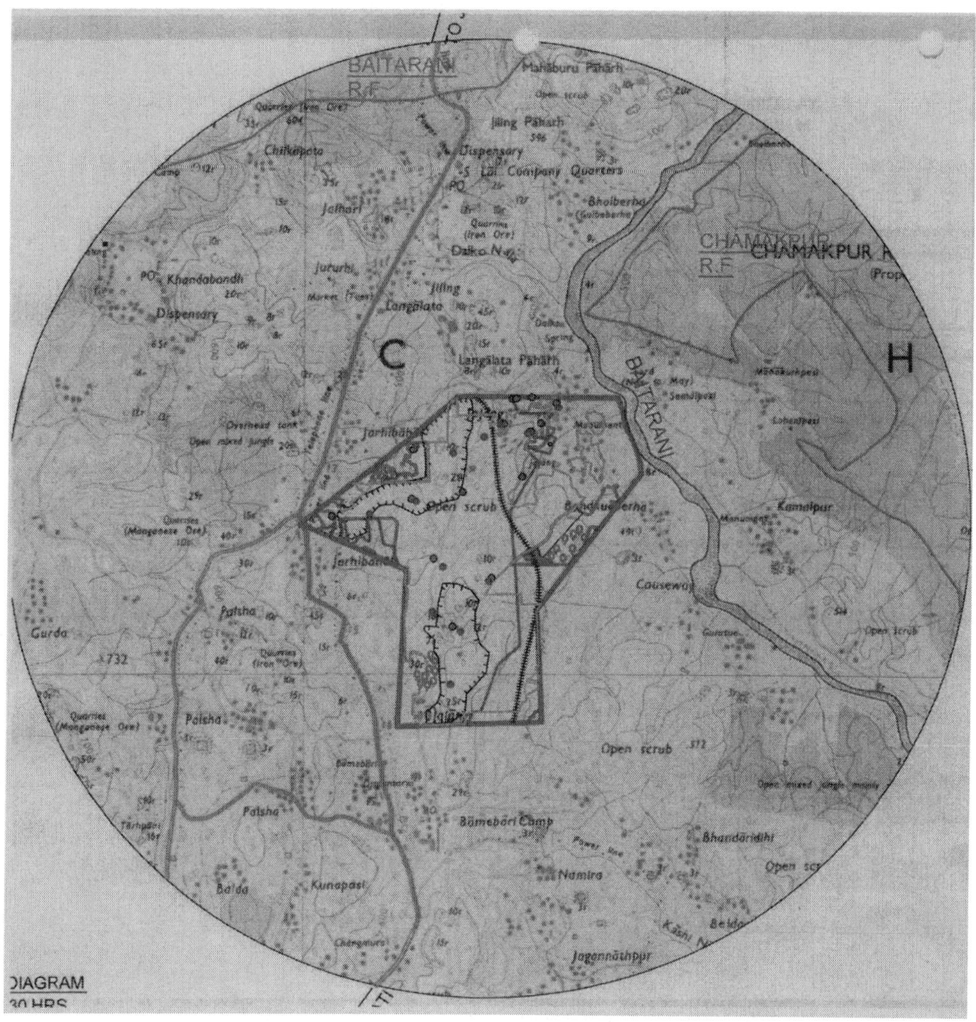

Figure 4.1: Outline Map of Jajanga Iron Ore Mine.

In this area mining has been done in 377.18 ha and 41.97 ha or nearly about 42 ha have been used as dump. The rate of extraction of ore is around 16.5 million ton per annum (MTPA) and the ore to waste ratio is 1:0.043. The mines lease area includes the following:

1. Forest area (FA): 247 ha, where mining activities have not yet been taken up and have natural vegetation in some parts (Figure 4.2).
2. Dump Area (DA): 41.967 ha *i.e.* equal to 42 ha, where the waste and top soil was dumped in the past (Figure 4.3).
3. Mining Area (MA): 377.18 ha, where mining has been done (Figure 4.4).

Figure 4.2: Forest Area (FA) of JIOM.

Figure 4.3: Dump Area (DA) of JIOM.

Figure 4.4: Mining Activity at JIOM.

The soil of the area is slightly acidic (pH: 6.42 ± 0.46), loamy with some proportions of silt and clay. The depth varies from shallow (0-30cm) and medium (31-50cm) to moderately deep (51-80cm). Most areas have medium depth soil. More than 50 per cent of the area has moderate to good water holding capacity. Average value of some important soil physico-chemical characters of the study area is presented in Table 4.1.

Table 4.1: Soil Physico-chemical Characters (Mean ± SD) of the Study Site

Parameters	*Values (Mean ± SD)*
Sand (per cent)	68.5 ± 3.18
Silt (per cent)	15.3 ± 0.76
Clay (per cent)	16.2 ± 1.32
Bulk density (g/cm^3)	1.18 ± 0.065
Soil Moisture (per cent)	16.73 ± 1.17
WHC (per cent)	48.3 ± 3.6
pH	6.42 ± 0.46
Nitrogen (kg/ha)	163.07 ± 18.62
Phosphorus (kg/ha)	4.25 ± 0.66
Potassium (kg/ha)	279.92 ± 24.3
Organic Carbon (per cent)	3.8 ± 0.12

Source: Soil analysis report of Rungta Iron Ore Mine (RIOM); WHC: Water Holding Capacity.

Methods

A reconnaissance survey was carried out from June, 2014 to May, 2016 in FA, DA and plantation (P). FA was studied in 150 random quadrats, each one having 20m × 20m size. The stratification was based on the habitats and major topographic variation. In each quadrat, data on trees (≥ 30cm gbh) such as girth at breast height (137cm above ground), height, canopy height, canopy width and regeneration (plants <10cm gbh) were recorded. Shrub species and their frequency were recorded on these quadrats, while in case of herbaceous species, abundance was recorded. DA having plant life forms was estimated quantitatively at regular intervals and categorised as rare, common and abundant. Further abundance rating (AR) was given to each plant life form. The rare species was rated as 1, while the common and abundant species were given a score of 2 and 3, respectively.

Species Selection from Forest

Tree species for restoration were selected in two phases. In the first phase, the top 15 species were selected based on Importance Value Index (IVI). Remarking was given to each species based on the IVI. IVI was calculated as a sum of relative density (RD), relative frequency (RF), and relative dominance (RDo) (Curtis and

Mc Intosh, 1950). During the second phase of selection, regeneration potential was included. Tree species represented by more than 15 individuals in the regeneration stage was ranked 1, and species having no regeneration were ranked 3. Adding IVI and regeneration ranks, the top ten tree species were selected for restoration. Shrub species were selected using relative index (RI) computed as a sum of RD and RF (Sharma and Sunderraj, 2005). For herbaceous species only abundance was calculated as per Kershaw (1973).

Species Selection from Plantation

Tree species in "P" were selected based on their growth, productivity and adaptation. The growth was estimated using Size Index (Sharma and Sunderraj, 2005) as:

SI=Tree height (cm) × gbh (cm).

Productivity potential of tree species was estimated using Canopy Index (Sharma and Sunderraj, 2005) as:

CI= (Average canopy height + Average canopy width) × no. of sample trees.

Adaptation potential of species was evaluated as; species grown on the slopes were ranked 1, species grown on flat terrain were ranked 2, while those species grown in valleys, stream bed or river bed were ranked 3 and 4, respectively. The species growing on shallow soils were ranked 1, while those on medium and deep soil were ranked 2 and 3, respectively. Adding the ranks for growth, productivity, terrain and soil, the top ten species were selected. Some species were included based on their properties to enrich the soil.

Results and Discussion

Species Richness and Abundance

A total of 60 species were recorded in FA and includes 32 trees, 6 shrubs and 22 herbaceous species (Table 4.2). However, in plantation only 12 tree species were recorded. Tree species enumerated in FA were nearly equal to 2.5 times more than in "P" (Table 4.4). Seven tree species were common to both FA and "P". Of these, *Samanea saman*, *Flindersia australis* and *Albizia lebbeck* performed better in "P" only. On the other hand *Xylia xylocarpa* performed well in both. In shrubs their diversity was low with respect to number of species as compared to trees and herbaceous species. Only six shrub species were recorded in the study area. However, 22 numbers of herbaceous species were recorded and their diversity in terms of number of species was intermediate between tree and shrub species enumerated from the study area (Table 4.2). Plant species diversity at tree level was more in FA then "P". This is natural because plantations are generally monocultures or at best include few species of interest and are single layered. Forests, on the other hand, grow with natural succession have stratification resulting in higher species diversity or not only plants but also animals [20].

Table 4.2: Importance Value Index (IVI) of Trees Recorded in Forest Area (FA), Relative Index (R.I.) of Shrub Species and Abundance Rating (AR) of Herbaceous Species of Jajang Iron Ore Mine

Tree Species	*IVI*	*Rank*	*Shrub Species*	*RI*	*Rank*	*Herbaceous Species*	*Abun-dance*	*Rank*
Mitragyna parviflora	15.40	1*	*Woodfordia fruticosa*	23.71	1	*Apluda mutica*	4	6*
Anogeissus latifolia	12.12	2*	*Helicteres isora*	20.81	2	*Aristida setacea*	6	4*
Adina cordifolia	10.96	3*	*Nyctanthes arbortristis*	17.31	3	*Cynodon dactylon*	3	7*
Lannea coromandelica	10.44	4*	*Wendlandia exerta*	15.29	4	*Chloris barbata*	6	4*
Miliusa velutina	9.91	5*	*Flacourtia indica*	13.17	5	*Chloris virgata*	8	2*
Bridelia retusa	7.19	6*	*Cassia auriculata*	11.87	6	*Cryperus compressus*	9	1*
Schleichera oleosa	5.78	7*				*Cyperus difformis*	4	6*
Syzygium cumini	4.89	8*				*Cyperus eleusiniodes*	3	7*
Garuga pinnata	4.65	9*				*Cyperus iria*	3	7*
Wendlandia tinctoria	4.34	10*				*Cyperus sp.*	2	8
Shorea robusta	4.23	11*				*Dactyloc-tenium aegyptium*	7	3*
Terminalia bellirica	2.73	12*				*Digitaria marginata*	7	2*
Terminalia alata	2.50	13*				*Oplismenus burmannii*	2	8
Dalbergia latifolia	2.41	14*				*Isellema laxum*	5	5*
Elaeodendron glauca	2.21	15*				*Dichantus sp.*	2	8
Morinda tinctoria	2.13	16				*Dichanthium annulatum*	4	6*
Cassia fistula	1.86	17				*Paspalidium punctatum*	3	7*
Holarrhena antidysenterica	1.73	18				*Pennisetum sp.*	1	9
Madhuca latifolia	1.55	19				*Setaria verticillata*	5	5*

Tree Species	IVI	Rank	Shrub Species	RI	Rank	Herbaceous Species	Abun-dance	Rank
Ficus benghalensis	1.43	20				*Sporobolus pallidum*	2	8
Lagerstroemia parviflora	1.37	21				*Sporobolus sp.*	2	8
Butea monosperma	1.22	22				*Themeda arundinaceae*	3	7*
Terminalia arjuna	1.2	23						
Trema orientalis	1.18	24						
Albizia lebbeck	1.15	25						
Alangium salvifolium	1.11	26						
Pongamia pinnata	1.08	27						
Phyllanthus emblica	1.05	28						
Xylia xylocarpa	0.58	29						
Bauhinia racemosa	0.46	30						
Terminalia chebula	0.41	31						
Gmelina arborea	0.31	32						

* Species selected for restoration.

Tree Species Selection from FA and "P"

Selection of tree species from FA for plantation was made based on IVI and regeneration rank. *Mitragyna parviflora* had maximum IVI followed by *Anogeissus latifolia, Adina cordifolia, Lannea coromandelica, Miliusa velutina* and *Bridelia retusa.* These tree species had IVI greater than 7. Rest of the species had IVI ranging from 0.31 to 5.78. The top 15 species with higher IVI values were selected for regeneration studies. Taking both the parameters (IVI and regeneration rank) into account tree species selected from FA for plantation is presented in Table 4.3.

In plantation instead of IVI, growth rate and productivity of tree species was considered along with terrain and soil depth for species selection. The age of tree species (years after plantation) ranged from 4 to 6 years. *G. arborea* was the fast-growing species with maximum growth rate of 3842 cm^2/year while *Xylia xylocarpa* gained maximum productivity of 46 cm^2/year (Table 4.4). When adaptation to terrain and soil depth was included in the selection parameter, *Acacia farnesiana* and *Flindersia australis* shared the top rank.

Table 4.3: Importance Value Index and Regeneration Ranks of Trees Species in Mine Forest at Jajang

Name of the Species	*IVI*		*Regeneration Rank*	*Total Rank*
	Value	*Rank*		
Mitragyna parviflora	15.40	1*	1	2
Anogeissus latifolia	12.12	2*	1	3
Adina cordifolia	10.96	3*	2	5
Lannea coromandelica	10.44	4*	2	6
Miliusa velutina	9.91	5*	1	6
Bridelia retusa	7.19	6*	1	7
Schleichera oleosa	5.75	7*	1	8
Syzygium cumini	4.89	8*	3	11
Garuga pinnata	4.65	9*	2	11
Wendlandia tinctoria	4.34	10*	1	11
Shorea robusta	4.23	11*	3	14
Terminalia bellirica	2.73	12*	2	14
Terminalia alata	2.50	13*	2	15
Dalbergia latifolia	2.41	14*	3	17
Elaeodendron glauca	2.21	15*	3	18

Table 4.4: Growth Productivity, Terrain and Soil Ranks of Tree Species in RIOM Plantations

Tree Species		*Growth Rate*		*Productivity Rate*				*Total Rank*
	Age	*GR/Y*	*Rank*	*P/Y*	*Rank*	*T*	*S*	
						Rank	*Rank*	
Acacia auriculiformis	4	1856	11	26	8	2	1	22
Acacia nilotica	6	1963	10	24	9	2	1	22
Samanea saman	5	3556	2	33	4	2	1	9
Albizia lebbeck	3	2948	5	29	6	2	1	14
Caassia siamea	6	1421	12	29	6	1	2	21
Dalbergia sissoo	3	2580	7	42	2	1	2	12
Delonix regia	6	2130	9	33	4	2	3	18
Flindersia australis	3	3269	3	40	3	2	1	9
Madhuca latifolia	5	2873	6	27	7	1	1	15

Tree Species		*Growth Rate*		*Productivity Rate*				*Total Rank*
	Age	*GR/Y*	*Rank*	*P/Y*	*Rank*	*T Rank*	*S Rank*	
Gmelina arborea	5	3842	1	24	9	2	3	15
Peltophorum pterocarpum	5	3173	4	31	5	2	3	14
Xylia xylocarpa	3	2553	8	46	1	2	3	14

GR/Y: Growth rate/year; P/Y: Productivity/year; T: Terrain; S: Soil.

Parameters used for selection of species were decided in order to measure their adaptation and robustness to grow in difficult conditions. Inclusion of more than one parameter ensure that even if some species scored equal points on one parameter, the difference in scores in other parameters helped in selecting the better species. For example, tree species in plantation showed similarity in productivity but there was a variation in growth rate. In addition to selection of parameters, selection of species for plantation from different sampling areas (FA, DA and P) helped in developing a comprehensive list of species based on their performance in diverse habitats. Only 12 tree species were grown in plantation, out of which 7 species did occur in FA. On the other hand, a relatively large number species in the FA helped in providing more options for the species selection. The cumulative list of 20 tree species, 10 each from FA and plantation provided best performing local species in the vicinity, to boost natural colonization (Table 4.5).

Further, the study envisages that to increase the fertility level of dump, growing leguminous plant would be more useful, as they supply nitrogen to the root zone in mine spoil (Singh *et al.,* 2004). *Acacia* sp. and *A. lebbeck* are some of the species that have high nitrogenous activity in root nodules (Cole *et al.,* 2010; Sharma *et al.,* 2004). Therefore, these tree species are suggested to grow on dumps as well as in the degraded areas. In addition to these *Pongamia pinnata* is considered as one of the promising species for overburden of mining areas (Sekhar, 2000) and should be planted on the dumps to enrich soil. The findings also suggest that tree species selected from 'P –may perform well when planted on dumps as indicated by better growth of *Samanea saman, Flindersia australis, Gmelina arborea, Peltophorum pterocarpum*. The list of tree species suggested for plantation is given in Table 4.5.

Shrub and Herbaceous Species Selection

The RI of shrub species in the FA ranged from 11.87 to 23.71. *Woodfordia fruticosa* had maximum RI followed by *Helicteres isora* and *Nyctanthes arbor-tristis,* while C*assia auriculata* had the minimum RI (Table 4.2). All the six species found were recommended for use in restoration activity. Abundance of herbaceous species in the FA ranged from 1 to 9. Maximum abundance was experienced by *Cyperus compressus* followed by *Chloris virgata, Dactyloctenium aegypticum = Digitaria*

Table 4.5: Tree Species Suggested for Plantation on Dumps

Species	*From Forest Area*	*From Plantations*
Mitragyna parviflora	?	
Anogeissus latifolia	?	
Adina cordifolia	?	
Lannea coromandelica	?	
Miliusa velutina	?	
Bridelia retusa	?	
Schleichera oleosa	?	
Syzygium cumini	?	
Garuga pinnata	?	
Wendlandia tinctoria	?	
Samanea saman		?
Flindersia australis		?
Albizia lebbeck		?
Cassia fistula		?
Dalbergia sissoo		?
Morinda tinctoria		?
Madhuca latifolia		?
Gmelina arborea		?
Peltophorum pterocarpum		?
Xylia xylocarpa	?	?

marginata, Chloris barbata = Aristida setacea, Setaria verticillata= Isellema taxum, etc. However, the minimum was experienced by *Pennisetum* sp. (Table 4.2). On the basis of abundance rating the top sixteen herbaceous species were selected from the mining forest area for restoration activity (Table 4.2). Generally natural succession commences with herbaceous species among the higher plant groups. Thus introducing the herbaceous species from the FA can enhance succession rate on DA. Promoting selected native shrub and herb species on the periphery of dumps and on stream-bed can act as vegetative filter to reduce soil and water pollution in the surrounding areas as observed by Sharma *et al.* (1998) and Chaubey *et al.* (2014). Shrubs are more effective in halting runoff and erosion (Sharma *et al.,* 2004) than the agro forestry crops. Therefore, in this study, native shrub species are suggested for the dumps as the first phase of restoration. One of the parameters considered to select plant species was the slope gradient. Two shrubs named as *N. arbor-tristis and W. exerta* and three herbaceous species named as *S. verticillata, Themeda arundinaceae* and *D. aegyptium* were found abundant on the slopes. Therefore, these species should be grown on dump slopes. The soils of DA were of similar nature as that of FA and P. Moreover 70cm of the topsoil is stocked separately for spreading it on the dumps during resolution. Therefore, the selected plant species will successively grow on dumps. Some part of DA has already been successfully planted with *A. Farnesiana* and other parts with other dominant species like *A. nilotica, A. lebbeck, Cassia siamea, Madhuca latifolia, etc.*

Conclusion

To restore the biological quality of the degraded habitats of Jajang Iron Ore Mine (JIOM), selection of native plant life forms of different strata from FA and P based on the study of their IVI, regeneration potential, growth rate, productivity, relative index and abundance rank for plantation may be an ideal method. In achieving better growth of planted species on dumps certain steps should be followed which includes water management to promote growth of perennial species, protection from anthropogenic disturbances *i.e.* grazing and cutting. In addition, ecological study at regular intervals is needed to monitor the survival and growth of plant species and to find out the effects of various plant species on the habitat conditions. At regular intervals study on soil physico- chemical parameters indicate about the change in soil nutrient status of the restored area. Further, monitoring of vertebrate faunal communities and their association with plant life forms in restored areas may also be important in understanding the overall effects of restoration work. Such restoration practice is also important to reduce high rates of forest destruction, energy shortages and steps towards conservation of biological resources.

Acknowledgments

We thank Director, Jajang Iron Ore Mines (JIOM), Keonjhar for providing permission and necessary support during the study.

REFERENCES

Boyd, E. (2002). The paradox of the CDM and sustainable development. International Forestry Review, 4: 77-80.

Chaubey, O.P., Jamaluddin, O. P., and Ram Prakash, R. (2014). Bio-reclamation of degraded ecosystem. International Journal of Bio-Science and Bio-Technology, 6(4): 145-154

Cole, R. J., Holl, K. D. And Zahawi, R. A. (2010). Seed rain under tree islands planted to restore degraded lands in a tropical agricultural landscape. Ecological Applications, 20: 1255-1269.

Curtis, J.T. and McIntosh, R.P. (1950). The interrelations of certain analytic and synthetic phytosociological characters. Ecology, 31: 434-455.

Dongmei, L.E.I. and Changqun, D. (2008). Restoration potential of pioneer plants growing on land-zinc mine tailings in Lanping, south-west China. Journal of Environmental science, 20(10): 1202-1209.

Jha, A.K., Singh, A., Singh, A.N. and Singh, J.S. (2000). Evolution of direct seeding of tree species as a means of re-vegetation of coal mine spoils. *Indian Forester,* 126 (11): 1217-1221.

Jim, C.Y. (2001). Ecological and landscape rehabilitation of a quarry site in Hong Kong. Restoration Ecology, 9: 85-94.

Kershaw K.A. (1973). *Quantitative and Dyamic Plant Ecology.* Edward Arnold Ltd. (Second Edition), London, U.K

Kumar, S. S., Sharma, B.K. and Gough, L. P. (1998). Richness, diversity and evenness of vegetation upon rehabilitation of gypsum mine spoiled lands in the Indian arid zone. Annals of Arid Zone, 37**:** 139-145.

Mensah, A. (2015). Role of re-vegetation in restoring fertility of degraded mine soils in Ghana: a review. International Journal of Biodiversity and Conservation, 7(2): 57-80.

Pandey, D.N (2002) Sustainability science for mine- spoil restoration. *Current Science,* **83**: 792-793.

Parrotta, J.A. and Knowles, O.H. (1999). Restoration of tropical moist forests on bauxite mined lands in the Brazilian Amazon. Restoration Ecology, 7: 103-116.

Rao, A.V. and Tak, R. (2002). Growth of different tree species and their nutrient uptake in limestone mine spoil as influenced by arbuscular mycorrhizal (AM) fungi in India arid zone. Journal of Arid Environments, 51: 113-119.

Sayer, J. Chokkalingam, U. and Poulsen, J. (2004) The restoration of forest biodiversity and ecological value. *Forest Ecological and Management,* 201: 3-14.

Sekhar, N.U. (2000) Decentralized natural resource management: From state to co-management in India. *Journal of Environmental Planning and Management,* 43: 123-138.

Sharma, D. and Sunderraj, S.F.W. (2005) Species selection for improving disturbed habitats in western India. *Current Science,* 88(3): 462-467.

Sharma, K.D., Kumar, P., Gough, L.P. and Sanfilipo, J. R. (2004). Rehabilitation of a lignite mine disturbed area in the Indian desert. Land Degradation and development, 15 (2): 163- 176.

Sharma, K.D., Kumar, S. and Gough, L. (2000). Rehabilitation of lands mined for limestone in the Indian desert. Land Degradation and Development, 11: 563-574.

Sharma, S. D., Soni, P. and Chandra, V. (1998). Amelioration of the overburden of iron ore mine for free establishment. Annals of forestry, 6 (2): 239-244.

Singh, A. N., Raghubanshi, A. S. and Singh, J. S. (2002). Plantation as a tool for mine spoil restoration. *Current Science,* 82: 1436-1441.

Singh, A.N., Raghubanshi, A. S. and Singh, J. S. (2004). Impact of native tree plantations on mine spoil in a dry tropical environment. Forest Ecology and Management, 187: 49-60.

Verma, P. and Verma, R.K. (2017). Species diversity of arbuscular mycorrhizal (AM) fungi in Dalli-Rajhara Iron Mine overburden dump of Chhattisgarh (Central India). International Journal of Current Microbiology and Applied Sciences, 6 (4): 2766-2781.

Vibhash, R., Sen, P., Kumar, D. and Sarsawat, A. (2015). A review on dump slope stabilization by revegetation with reference to indigenous plant. Ecological Processes, 4: 1-14.

Wong, M. H. (2003). Ecological restoration of mine degraded soils, with emphasis on metal contaminated soils. Chemosphere, 50: 775-780.

Chapter 5

Wild Edible Plants of Similipal Biosphere Reserve, Odisha

Manas R. Mohanta,
*Sushree S. Dash and Sudam C. Sahu**

Department of Botany,
Maharaja Sriram Chandra Bhanja Deo University,
*Baripada – 757003, Odisha *e-mail: sudamsahu.bdk@gmail.com*

ABSTRACT

The present study documented the wild edible plants used by the tribal people of Similipal Biosphere Reserve (SBR) through semi-structural interviews, field observations, group discussions, and local weekly surveys. A total of 166 species of higher plants belong to 126 genera and 63 families were reported. The edible parts of plants were tuber/root, fruit, seed, leaf, flower and whole plant which were taken directly as raw or cooked food or medicines. Most of the plants were fruit edible plants (87 species) followed by leaf edible plants (36 species), seed edible plants (18 species), tuber/root edible plants (15 species), flower edible plants (7 species) and whole plant edible (3 species). Maximum number of edible plant species were trees (78 species) followed by herbs (35 species), climbers (29 species) and shrubs (24 species). The information in the present study will be useful in further value addition and commercialization of these wild edible plant species for the livelihood security of the tribal people and also in development strategies for better management of wild edible plants of SBR.

Introduction

The reliance of mankind on plant resources is ineluctable. Since the early eras living close to nature, man has learnt to use plants for food, fodder, gum, medicines, and other purposes. Over the years man has exploited these resources for own advantages and survival. Millions of people in many developing countries do not get enough food to meet their daily requirements, and many more suffer from deficiency of micronutrients. The rural and tribal communities even today depend on wild edible plants to meet their daily food requirements.

There are umpteen wild plant species available in our forest having excellent food values and medicinal properties. Therefore, these plants are life support plant species for the tribal ones. Wild plants of forests have been poorly addressed, confining it into mostly traditional and local use only. A considerable proportion of our people are still un-nourished and people living in remote areas cannot produce enough food grains to meet their requirements. So a large portion of population meets its food requirement through non-conventional means by consuming various wild plants resources especially in periods of food scarcity (FAO, 2004; Balemie and kebebew, 2006). Wild edible plants are valuable nutritional supplements for a balanced diet. It is also one of the alternative sources of income for many poor communities and the source of domestication (Shresta and Dhillon, 2006). For the ever increasing population, wild foods act as potential sources of consumption (Rashid *et al.,* 2008). Though the use of wild foods is not much commercially accepted but people still use them during the periods of scarcity. Sometimes they are sold in urban markets and are then in competition with exotic fruits (Manyafu,1971).Wild plants are the major copying mechanism at the times of food shortage and famine is widely documented(Abbink,1993; Guinand and Dechassa,2000; Redzic, 2007; Fentahun and Hager,2009).

Much work has also been done on ethnobotany of wild plants over the world (Ali shtayeh *et al.,* 2008; Abubakar *et al.,* 2012; Ju *et al.,* 2014). Many attempts has been made worldwide on compositional evaluation and functional properties of various types of wild edible plants (Bello *et al.,* 2008; Mohan and kalidas, 2010; Abdualrahaman *et al.,* 2011; Olujobi, 2012; Deshmukh and Rathod, 2013). In India,the nutritional value of various types of wild edible fruit plants has been reported by various authors (Dahot 1993; Nazarudeen 2010; Deshmukh and Waghmode, 2011; Gupta *et al.,* 2011; Vishwakarma and Dubey, 2011; Jain and Tiwari, 2012; Mahapatra *et al.,* 2012; Seal, 2012; Rathi *et al.,* 2012; Anjum *et al.,* 2013).The value of wild edible plants in food security has not been given sufficient attention in India and particularly in Odisha. During last few years some attempts has been made to document the rich indigenous knowledge on the use of wild plants but socio–economic,traditional,nutritional aspect of wild food plants still lack adequate attention. There are about 1532 edible wild food species available in India mostly in Eastern Ghats, WesternGhats and Himalayan ranges (Reddy *et al.,* 2007).

Similipal Biosphere Reserve (SBR), in the Mayurbhanj district of Odisha (India) is one of the tropical forest ecosystem rich in diversity of flora. Some of the ethnic tribe *viz.* Kolha, Santal, Khadia, Bhuyan, Bathudi, Kudumi, and Mankdias are residing in the Similipal forest region. These ethnic groups are the traditional collectors of different edible plants. The different kinds of edible food plants are consumed by them vary with locality and with tribes. The study of Anthropology, Ethnobotany, and Sociology of tribal groups of SBR shows many events that they are consuming the wild foods. Hence, the present study was carried out to document the hidden knowledge from tribes of SBR about wild edible plants.

Materials and Methods

Study Area

Similipal Biosphere Reserve (SBR) is a part of biotic province of Chhotanagpur pleateau. It has a representative ecosystem under Mahanadian biogeographic region. SBR is located in the central part of the Mayurbhanj district of Odisha, close to the inter-state boundary with West Bengal in the North eastern and Jharkhand in the Northwest (Figure 5.1). The SBR is having a compact mass of natural forests spread over an area of 5,569 sq km lying between 21° 10' to 22° 12' N and 85° 58' to 87° 45' E, ranging between 300 m to 1,185 m above sea level with numerous rolling hills with tropical semi evergreen forests, moist deciduous forests, dry deciduous forests, Sal forests, grassland and savannah. The average temperature ranges between minimum 2° C in winter to a maximum of 48° C in summer with an average rainfall of 2200 mm. Three distinct seasons are felt during the year. They include the rainy season (mid June to mid October), winter (mid October to February) and summer (march to mid June).

The core area occupies 845 sq km, which is intensively protected and absolutely undisturbed, secured legally and managed scientifically where as there is 2,212 sq km of buffer zone surrounds the core zone. The core and buffer zones of the biosphere reserve are under the administrative control of the three forest divisions: Karanjia, Baripada, and Rairangpur (Mishra *et al.,* 2011). There are four villages inside the core area; Jenabil is one of them, having a great beauty of Sal (*Shorea robusta*) forest and a rich diversity of flora and fauna. There are 61 villages in the buffer zone and 1200 villages in the transitional zone of SBR. Here, tribes occupy a big chunk of the population constituting 73 per cent of the total population; 53 communities both aboriginal and migrated are found in the SBR (Naik, 1998). The dominated tribes of the SBR are Santal, Kolha, Munda, Bhumija, Kudumi, Ho, Mankdias, Bathudi, Bhuyan, *etc.* (Panda *et al.,* 2011).They are mainly dependent on agriculture for food andalso collect roots,tubers,leaves,flowers, fruits, and mushrooms as food supplements. These tribal communities reside in the forest areas and depend on forest resources for their livelihood.

Collection of Data

In the view of the phytogeographic interest in Similipal Biosphere Reserve, the study was conducted using semi-structural interviews, field observations, group discussions, and local weekly survey. In this study, efforts had been made to document the diversity of species and mode of consumption of wild edible plants of the Biosphere Reserve. The standard participatory rural appraisal method

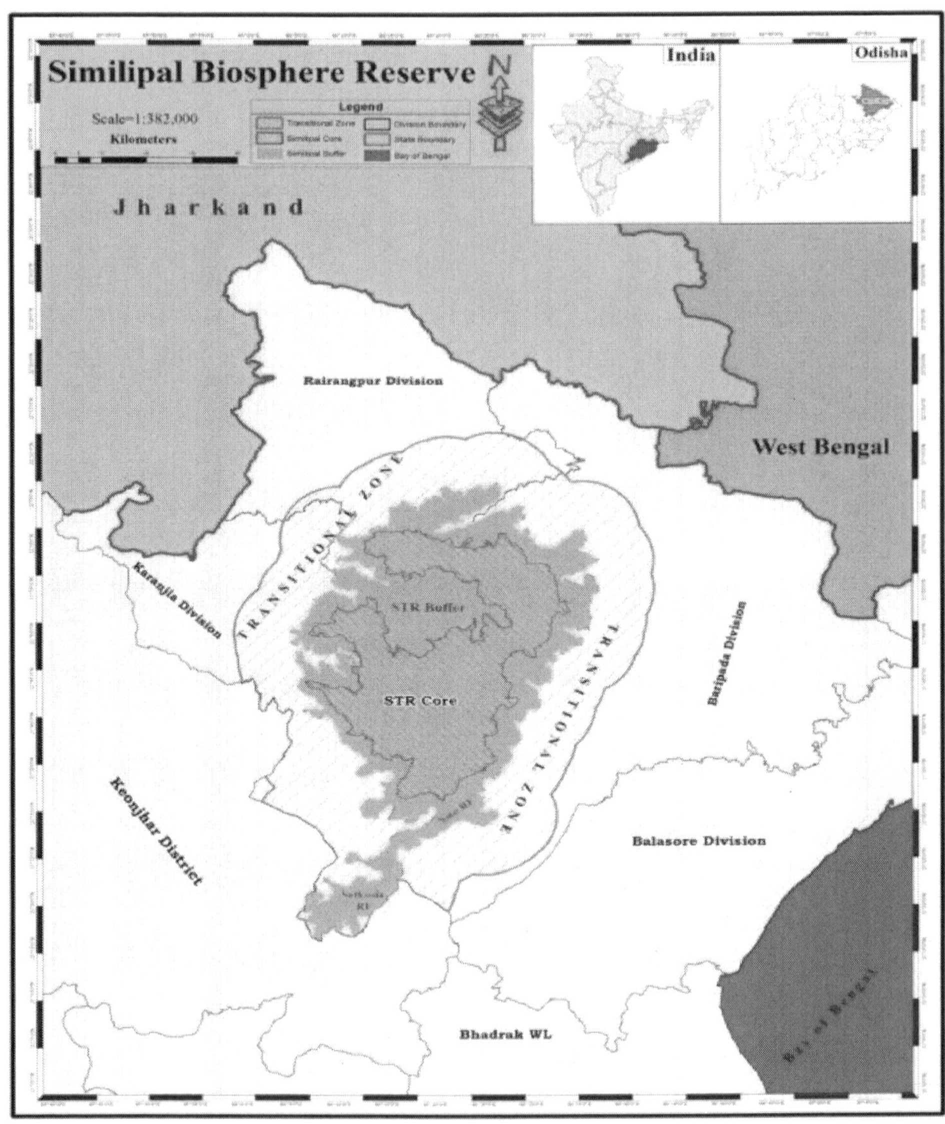

Figure 5.1: Map of Study Area.

(Cunningham 2001, Gerique 2006) was adopted for sampling and data collection to document the indigenous knowledge on wild edible plants. Field visits were conducted and raw data was collected from the area during September 2017 to march 2018. The information was accrued after discussion with several tribal communities and local inhabitants. Repeated interviews through questionnaires were made in different villages of SBR, such as Pithabata, Kadhiabasa, Samakhunta, Kuchilaghati, Betna *etc.* The collected plant materials were identified by the authors following the Botany of Bihar and Orissa and Flora of Orissa (Haines, 1921-25; Saxena and Brahmam, 1994-96).

Results

The present study reported a total of 166 higher plant species belong to 126 genera and 63 families which were consumed by the tribal people of SBR either as raw or after cooking. Majority of plant species were trees followed by herbs, climbers and shrubs, respectively (Figure 5.2). There were 63 families, among which Moraceae (10 tree species + 1 shrub species), Cucurbitaceae (10 climber species), Fabaceae (1 tree species + 2 herb species + 6 climber species), Rutaceae (6 tree species + 1 climber species) and Rubiaceae (5 tree species + 1 herb species) were most specious families. Edibility of plant parts vary from species to species starting from tubers to fruits. However, fruit edible plants were reported highest followed by leaf, seed, root/tuber, flower in decreasing order (Figure 5.3). There were 4 tuber edible species namely *Colocasia esculenta* (L.) Schott, *Dioscorea glabra* Roxb., *Dioscorea bulbifera* L. and *Pueraria tuberose* (Willd.) DC. eaten after cooking whereas others are eaten either as raw or as medicine by adding with several other plant products. Maximum numbers of plants reported were fruit edible plants (87 species). Out of these 87 plant species, 57 were trees, 10 were shrubs (including 2 aquatic floras), 7 were herbs and 13 were climbers. Followed by fruit edible plants, leafy edible plants (36 species) were the second most in number, among which 11 are trees, 6 are shrubs, 15 are herbs and 4 were climbers. There were also 18 seed edible plants, 15 tuber edible plants and 7 flower edible plants found to be eaten by the tribal people of SBR. There were also some plants (3 species) whose whole plant body was found to be edible in some way or other. The complete list of plants and their made of use is given in Table 5.1.

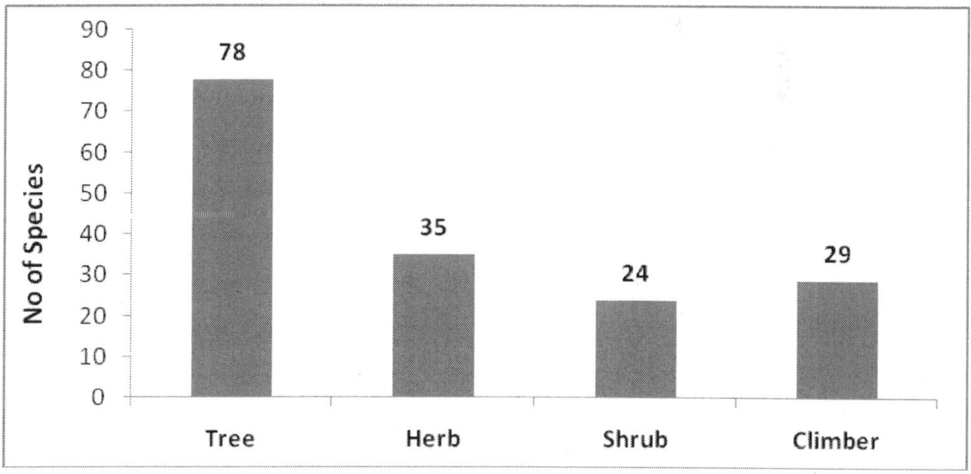

Figure 5.2: Number of Edible Plant Species in different Life Forms of the Study Area.

Table 5.1: List of Edible Plant Species of Study Area with their Local Name, Family and Mode of Uses

Sl.No.	Plant Species	Local Name	Family	Mode of Use
1	*Abelmosschus moschatus* Medix.	Jangali bhindi	Malvaceae	Tender fruits are eaten
2	*Abrus precatorius* L.	Kaincha	Fabaceae	Leaves are edible
3	*Acacia nilotica* (L.) Delile	Babul	Mimosaceae	Eaten for stomach problems
4	*Achyranthes aspera* L.	Apamaranga	Amaranthaceae	Seeds are used as chutney
5	*Adenanthera pavonina* L.	Manda kaincha	Mimosaceae	Seed is eaten raw
6	*Aegle marmelos* (L.) Correa.	Belo	Rutaceae	Fruit pulp is eaten
7	*Aerva lanata* L. Juss.	Paunsia	Amaranthaceae	Leaves are eaten as vegetables
8	*Agave angustifolia* var. Nivea (Trel.) Gentry.	Baramasi	Agavaceae	Young leaves are eaten
9	*Allium cepa* var. Bifolium Alef.	Jangali piyaja	Liliaceae	Rhizomes are eaten
10	*Allophylus serratus* (Hiern) Kurz	Khandakoli	Sapindaceae	Ripe fruits are eaten
11	*Alternanthera sessils* (L.) R.Br. ex DC.	Madaranga sago	Amaranthaceae	Young leaves are eaten
12	*Amaranthus caudatus* L.	Khaadasago	Amaranthaceae	Leaves are eaten
13	*Amaranthus spinosus* L.	Leutia	Amaranthaceae	Leaf and shoots are eaten
14	*Amaranthus tricolor* L.	Nautia	Amaranthaceae	Young tender leaves are cooked
15	*Amaranthus viridis* L.	Marsha saga	Amaranthaceae	Eaten as curry
16	*Amophophallus aculatum* Hook.f.	Olua	Araceae	Tubers are eaten
17	*Antidesma ghaesembilla* Gaertn.	Katha murmuri	Euphorbiaceae	Raw fruits are edible
18	*Aporosa octandra* var. malesiana Schot	Masania	Euphorbiaceae	Ripe
19	*Aporosa octandra* L.	Masania	Euphorbiaceae	Ripe fruits are eaten
20	*Ardisia solanaceae* (Poir). Roxb.	Tinikoli	Mysinaceae	Young leaves are eaten as vegetables
21	*Argemone maxicana* var. albiflora A. Grey	Kanta kusuma	Papaveraceae	Seed oil is edible
22	*Arisaema tortuosum* Wall. Schott.	Kushopheni	Araceae	Young leaves are eaten

Sl.No.	Plant Species	Local Name	Family	Mode of Use
23	*Artocarpus bifaria* Wall. ex Miq.	Dahu	Moraceae	Ripen fruits are eaten
24	*Artocarpus heterophyllus* Lam.	Panasoo	Moracea	Fruits are eaten
25	*Asparagus racemosus* Willd.	Satabari	Liliaceae	Root tubers are edible
26	*Bambusa arundinacea* Willd.	Kantabanso	Poaceae	Grains are eaten as food
27	*Basella alba* L.	Pui	Basellaceae	Leaves are cooked as curry
28	*Basella rubra* var. Virscens Moq.	Bana poi	Basellaceae	Eaten as curry
29	*Bauhinia purpurea* L.	Debakanchana	Caesalpiniaceae	Flowers are eaten
30	*Bauhinia vahlii* Wight and Arn.	Siali	Caesalpiniaceae	Cotyledons are eaten
31	*Boerhaavia elongata* Brandegee.	Purnisago	Nyctaginaceae	Leaves are eaten
32	*Boerhavia diffusa* var. diffusa	Ghodapuruni	Nyctaginaceae	Leaves are eaten
33	*Bombax ceiba* Burm.f.	Simuli	Bombacaceae	Uripe green fruits are cooked
34	*Borassus flabellifer* L.	Tala	Arecaceae	Fruits are edible
35	*Bridelia retusa* (L.) A. Juss.	Khashiphal	Phyllanthaceae	fruits are eaten
36	*Cajanus scarabaeoides* L. Thouars.	Kulthia	Fabaceae	Seeds are eaten raw
37	*Calamus guruba* Buch. Ham. Ex Mart.	Kanta beeta	Arecaceae	Pulp of ripe fruits are eaten
38	*Calamus viminalis* Willd.	Panibeta	Ripe fruits are eaten	arecaceae
39	*Capparis zeylanica* L.	Asadua	Capparaceae	Ripen fruits are eaten
40	*Careya arborea* Roxb.	Kumbhi	Lecythidaceae	Ripen fruits are cooked
41	*Carissa carandas* Lour.	Ankukolia	Apocynaceae	Berries are eaten
42	*Carissa spinarum* L.	Khirikoli	Apocynaceae	Ripen fruits are eaten
43	*Cassia tora* L.	Chakor	Piniaceae	Leaves are eaten
44	*Celastrus paniculatus* Willd.	Kujri	Celastaceae	Leaves are edible
45	*Chenopodium album* var. album	Bathua	Chenopodiaceae	Leaves and stems are eaten

Sl.No.	Plant Species	Local Name	Family	Mode of Use
46	*Cipadessa baccifera* (Roth) Miq.	Pittamari	Meliaceae	Fruits are eaten raw
47	*Cleome monophylla* L.	Rangosorisa	Capparaceae	Leaf and shoots are eaten
48	*Cleome umini* L.	Banosarisa	Capparaceae	Tender leaves are eaten
49	*Cochlospermum religiosum* (L.) Alston	Hupu	Bixaceae	Flowers are edible
50	*Colocasia esculenta* L. Schoot.	Banasaru	Araceae	Tubers are cooked
51	*Combretum nanum* Buch.-Ham.ex.D. Don	Juljuli	Combretaceae	Seeds are edible
52	*Corchorus aestuans* var. aestuans.	Banakadali	Tiliaceae	Unripe fruits arecooked
53	*Corchorus capsularis* Salisb.	Nalita	Tiliaceae	Leaves are eaten as vegetables
54	*Cordia dichotoma* G. Forst.	Ambota	Boraginaceae	Raw fruits are eaten
55	*Costus speciosus* (J. Koenig) Sm.	Kauka	Liliaceae	Tubers are eaten
56	*Crateva religiosa* G. Frost.	Baruno	Capparaceae	Uripe fruits are eaten
57	*Cucumis melo* L.	Kharabhuja	Cucurbitaceae	Ripen fruits are eaten
58	*Cucurbita pepo* var. Texana (Scheele) D.S Decker	Kakharu	Cucurbitaceae	Fruit,young leaves are eaten as food
59	*Curcuma umini* Salisb.	Parbati palua	Zingiberaceae	White portion of the rhizome is eaten
60	*Cycas circinalis* f. Circinalis	Rengua	Cycadaceae	Dried fruits are eaten
61	*Dillenia pentagyna* Roxb. (Roxb.) Baill.	Rai	Dilleniaceae	Edible as spices
62	*Dioscorea glabra* Roxb.	Soranda kanda	Dioscoraceae	Eaten as curry
63	*Dioscorea hipsida* Dennst.	Kulia kanda	Dioscoreaceae	Tubers are eaten
64	*Dioscorea alata* L.	Khamba alu	Dioscoreaceae	Root tubers are eaten
65	*Dioscorea bulbifera* L.	Pita allu	Dioscoreaceae	Tubers eaten as vegetables
66	*Dioscorea daemona* Roxb.	Bana allu	Dioscoreaceae	Tubers are eaten
67	*Dioscorea pubera* Blume.	Danga allu	Dioscoreaceae	Root tubers are eaten

Sl.No.	Plant Species	Local Name	Family	Mode of Use
68	*Diospyros chloroxylon* Roxb.	Gaurakasa	Ebenaceae	Ripe fruits are eaten
69	*Diospyros melanoxylon* Roxb.	Kendu	Ebenaceae	Ripen fruits are edible
70	*Diosyros malabarica* (Desr.) Kostel.	Kostell	Ebenaceae	Ripe fruits are eaten
71	*Ehretia laevis* Roxb.	Masaniaa	Boraginaceae	Ripe fruits are eaten
72	*Erycibe paniculata* Roxb.	Joraikoli	Convolvulaceae	Ripen fruits are eaten
73	*Euphorbia hirta* var. glaberrima Koidz.	Chitkuti	Euphorbiaceae	Leaves are eaten
74	*Ficus acidula* King.	Amuta	Moraceae	Ripen fruits are eaten
75	*Ficus benghalensis L.*	Baroo	Moraceae	Ripe receptacles are eaten
76	*Ficus carica* L.	Anjeer	Moraceae	Seeds are edible
77	*Ficus glomerata* Roxb.	Lawa	Moraceae	Ripe fruits are eaten
78	*Ficus heterophylla* L.f.	Bhui dumur	Moraceae	Receptacles are eaten
79	*Ficus hispida* L.f.	Thendu	Moracea	Eaten as fruit
80	*Ficus racemosa* L.	Dimiri	Moraceae	Ripen fruits are edible
81	*Ficus semicordata* Buch. -Ham.ex Sm.	Kujidimiri	Moraceae	Receptacles are eaten
82	*Gardenia gummifera* Roxb.	Bhurudukoli	Rubiaceae	Ripen fruits are eaten
83	*Garuga pinnata* Roxb.	Sarupatri	Burseraceae	Fruits and seeds are eaten
84	*Glycosmis pentaphylla* (Retz.) DC.	Chauladhua	Rutaceae	Ripe fruits are eaten
85	*Grewia serrata* Blanco	Burso	Tiliaceae	Ripe fruits are eaten raw
86	*Hibiscus sabdariffa* L.	Khatta palang	Malvaceae	Flowers are eaten
87	*Holoptelea integrifolia* Roxb.	Charla	Ulmaceae	Ripe fruits are eaten
88	*Indigofera pulchella* Roxb.	Jerhul	Fabaceae	Fruits are eaten raw
89	*Ipomoea umini* Forssk.	Kalama saga	Convolvulacea	Leaves and young shoots are eaten
90	*Ipomoea batatus* (L.) Lam.	Kanda mula	Convolvulaceae	Root tubers are eaten

Sl.No.	Plant Species	Local Name	Family	Mode of Use
91	*Ipomoea reptans* Poir.	Tutiisago	Convulvulacea	Tender leaves are eaten as fruits
92	*Ixora undulate* Roxb.	Karuna	Rubiaceae	Ripen fruits are eaten
93	*Kavalama urens* (Roxb.) Raf.	Genduli	Sterculiaceae	Roasted seeds are eaten
94	*Lablab purpureus* (L.) Sweet	Simbo	Fabacea	Young fruits are cooked
95	*Lagenaria siceraria* (Molina) Standl.	Lau	Cucurbitacea	Fruits and young leaves are cooked
96	*Lannea coromandelica* (Houtt.) Merr.	Mahi	Anacardiaceae	Fruits are eaten
97	*Lantana camara* L.	Jaikoli	Verbenaceae	Ripe fruits are eaten
98	*Lepisanthes maculosum* (L.) Raf.	Nanga	Sapindaceae	Fruits are eaten
99	*Limonia acidissima* Groff.	Kayitha	Rutaceae	Pulp of unripe fruit is eaten
100	*Luffa acutangula* L. Roxb.	Janhi	Cucurbitaceae	Young fruitcooked
101	*Luffa cylindrical* L. M. Roem.	Tadhari	Cucurbitaceae	Eaten as vegetables
102	*Macrotyloma unifloram* Lam. Verdc.	Kolatha	Fabaceae	Seeds are edible
103	*Madhuca longifera* var. latifolia (Roxb.)	Mahula	Sapotaceae	Used for beverages
144	*Mangifrea indica* L.	Amba	Anacardiaceae	Raw and ripe fruits are eaten
105	*Marsilea minuta* var. Incurve (A.Braun) Launert	Sunsunia	Marsiliaceae	Leaves are cooked
106	*Maytenus emarginata* Willd.	Bainchokoli	Celastraceae	Fruits are eaten
107	*Melia azadirachta* A. Juss.	Limba	Meliaceae	Young leaves and flowers are eaten
108	*Mentha spicata* var. Longifolia L.	Podina	Lamiacea	Leaves are used for chutney, medicine
109	*Mesua ferrea* L.	Nageswar	Cluslaceae	Ripen fruits are eaten
110	*Meyna spinosa* Roxb.ex	Solora	Rubiaceae	Roasted seeds are eaten
111	*Momoedica ovate* Cogn.	Kankada	Cucurbitaceae	Fruit is cooked
112	*Momordica charantia* var. abbreviatan Ser.	Tusikalara	Cucurbitaceae	Fruit is used as vegetable

Sl.No.	Plant Species	Local Name	Family	Mode of Use
113	*Moringa olifera* Lam.	Sajana	Moringaceae	Used for curry
114	*Mucuna pruriens* L. DC.	Baidanka	Fabaceae	Young fruits are eaten
115	*Mukia maderaspatana* (L.) M. Roem.	Pahari kakharu	Cucurbitaceae	Ripe berries are eaten
116	*Murraya koenigii* L.	Bhusunga	Rutaceae	Leaves are used in food
117	*Narangi crenulata* L.	Benta	Rutaceae	Ripe fruits are eaten
118	*Nelumbo nucifera* Gaertn.	Kayinn	Nymphaeaceae	Immature seeds are eaten
119	*Neolamarkia cadamba* (Roxb.)	Kadamba	Rubiaceae	Fruits are eaten
120	*Ocimum americanum* L.	Banatulasi	Lamiaceae	Seeds are eaten raw
121	*Olax scandens* Roxb.	Bhadadia	Olacaceae	Leaves are eaten
122	*Opilia amentacea* Roxb.	Drrrakuli	Opiliaceae	Tender leaves are eaten
123	*Pachyrhizus erosus* (l.) Urb.	Chini alu	Fabaceae	Root tubers are edible
124	*Phoenix acaulis* Roxb.	Bana khajuri	Arecaceae	Flesh fruits are eaten
125	*Phyllochlamys taxoides* (Roth).	Phutkuli	Lamiaceae	Eaten as fruit
126	*Physalis angulata* var. angulata	Tipai	Solanaceae	Sweet mature seeds are eaten
127	*Pithecellobium dulce* (Roxb.) Benth.	Simakoina	Mimosaceae	Sweet seeds are eaten
128	*Polyalthia suberosa* Roxb.	Lohania mossu	Annonaceae	Ripe fruits are eaten
129	*Protium serratum* Wall.ex Colebr. Engl.	Rajamahi	Bursereceae	Raw fruits are eaten
130	*Pueraria tuberose* (Willd.) D.C	Bhui kakharu	Fabaceae	Tubers are used as vegetables
131	*Punica granatum* L.	Dalimbo	Punicacea	Ripen fruits are eaten
132	*Randia uliginosa* (Retz.) Poir.	Pinder	Rubiaceae	Fruits are eaten
133	*Rubus ellipticus* Sm.	Machdim	Rosaceae	Ripe fruits are eaten
134	*Russula emetic* L.	Bana chatu	Russulaceae	Eaten after cooked
135	*Saccharum spontaneum* L.	Kasatandi	Poaceae	Stem is eaten

Sl.No.	Plant Species	Local Name	Family	Mode of Use
136	*Schleichera oleosa* (Lour.) Merr.	Kusuma	Sapindaceae	Fruit pulp are eaten
137	*Semicarpus anacardium* L.	Bhalia	Anacardiaceae	Ripen receptacles are eaten
138	*Senna hirsute* (L.) H.S Irwin and Barneby	Bado chakunda	Caesalpiniaceae	Seeds are roasted and eaten
139	*Smilax zeylanica* L.	Mutri	Smilaceae	Ripe fruits are eaten
140	*Solanaceae nolana* L.	Lunikoli	Solanaceae	Ripen,raw fruits are eaten
141	*Solanum nigrum* L.	Lunikoli	Solanacea	Raw/ripen fruits are eaten
142	*Solanum nigrum* var. Virginicum L.	Nununia	Solanaceae	Ripe fruits are eaten
143	*Solanum rudepannum* Dunal	Kantua	Solanaceae	Ripe fruits are eaten
144	*Solena amplexicaulis* (Lam.) Gandhi	Mataka	Cucurbitaceae	Unripe fruits are eaten
145	*Spinacia oleracea* L.	Palanga	Chenopodiaceae	Tender leaves are eaten
146	*Spondias pinnata* (L.F.) Kurz	Ambada	Anacardiaceae	Ripe fruitys are eaten
147	*Steculia urens* Roxb.	Genduli	Sterculiaceae	Seed are eaten as roasted
148	*Streblus asper* Lour.	Sahada	Moraceae	Flowers and fruits are eaten
149	*Strychnos potatorum* L.f.	Nirmali	Strychnaceae	Roasted seateneeds are
150	*Swietienia chloroxylon* Roxb.	Bheru	Rutaceae	Leaves are eaten
151	*Syzygium umini* L. Skeels.	Badajamukoli	Myrtaceae	Ripen fruits are eaten
152	*Tamarindus indica* L.	Tentuli	Caesalpiniaceae	Leaves,fruits are eaten
153	*Tamilnadia uliginosa* Retz.	Telkur	Rubiaceae	Ripe fruits are eaten
154	*Terminalia alata* Wall.	Asana	Combretaceae	Used as medicine
155	*Terminalia bellirica* (Gaertn.) Roxb.	Bahada	Combretaceae	Fleshy seeds are edible
156	*Terminalia chebula* Retz.	Harida	Combretacea	Used as medicine
157	*Toddalia asiatica* (L.) Lam	Tundapora	Rutaceae	Ripen fruits are consumed

Sl.No.	Plant Species	Local Name	Family	Mode of Use
158	*Trapa natans* L.	Kantaphala	Trapaceae	Fruits are edible
159	*Trapa natans* var. bispinosa (Roxb.) Makino	Panisingara	Trapaceae	Kernel of fruit are eaten
160	*Trichosanthes cucumerina* L.	Bana potala	Cucurbitaceae	Fruits are eaten as vegetables
161	*Vigna unguiculata* L. Walp.	Jhuranga	Fabaceae	Seeds are edible
162	*Woodfordia fruiticosa* (L.) Kurz.	Dhatuki	Lythraceae	Fruits are eaten
163	*Ziziphus xylocarpa* Retz.	Ghat	Rhamnaceae	Fruits are edible
164	*Ziziphus mauritiana* Lam.	Barakoli	Rhamnaceae	Fruits are eaten
165	*Ziziphus oenoplia* L. Mill.	Kanteikoli	Rhamnaceae	Eaten as fruit
166	*Zizipus nummularia* (Burm.f.) Wight and Arn.	Jangalikoli	Rhamnaceae	Ripe fruits are eaten

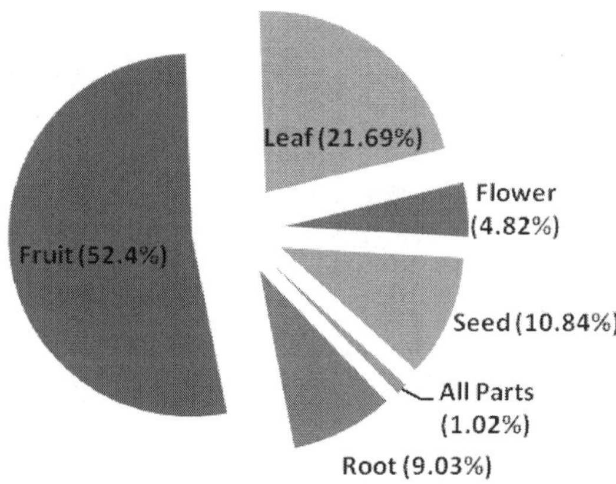

Figure 5.3: Contribution (in per cent) of Reported Plants Based on their Edible Part.

The mostly raw consumed fruits are *Aporusa octandus, Ardisia solanacea, Cajanus scarabaeoides, Ficus auriculata, Punica granatum, Ficus racemosa, Grewia hirsuta, G. Rothii, Glycosmis pentaphylla, Lantana camara,Mangifera indica, Tamarindus indicus,Nelumbo nucifera etc.* The ripen fruits consumed are mainly *Cucurbita maxima, Ficus hipsida, Carissa carandas, C. Spinarum, Garuga pinnata, Lepisanthes rubiginosa, Phoenix acaulis, Physalis angulata, Bridelia retusa, Manilkara hexandra etc.*

The leafy vegetables such as *Amaranthus spinosus* and *A. Viridis* are collected from the village periphery. Other leafy parts consumed were *Moringa oleifera, Bauhinia purpurea, Lucas aspera, Opilia amentacea, Ipomoea aquatica, Hibiscus sabdariffa, Marsilea minuta, Mentha spicata, Polygonum plebeium, Murraya koenigii, Spinacia oleracea*,etc.

Dioscorea species are the major wild tuberous food resources being consumed by the tribals of SBR. Some of the most common *Dioscorea* species were *D. hipsida, D. tomentosa, D. glabra, D. wallichii, D. pentaphylla etc.* Seven species of *Dioscorea* had right twining stem, while four species had left twining stems. All the species except the *D. glabra* had woody rhizomes. *D. oppositifolia* was very common among kharia tribe due to its sweet taste. Tubers like *D. glabra, D. pentaphylla, D. oppositifolia,* and *D. wallichii* were consumed during the months of September and October (Sinha and Lakra, 2005).

The species such as *Vigna mung, V. radiata, V. ricciardiana, V. unguiculata, Sesamum spontaneum, Shorea robusta*, whose seeds were the popular ingredients for consumption by the tribal people.

The tubers of *Pueraria tuberosa* locally known as Bhui Kokharu was popular due to its medicinal values. The fleshy flowers of *Madhuca longifolia* (Mahula) was used to make beverages and many food items. Leaves of *Opilia amentacea*were consumed to enhance memory power. The leaf of *Lucas aspera*was dried and used with other vegetables to make curry. Species like *Capparis zeylanica, Carissa carandas, Garuga pinnata, Limonia acidissima, Mangifera indica, Protium serratum, Semecarpus anocardium, Toddalia asiatica, Ziziphus mauritiana etc.* were commonly used to make pickles. The species such as *Bauhinia vahlii, Bombax ceiba, Combretum album, Gardenia gummifera, Meyna spinosa, Smilax zeylanica, Terminalia bellirica, Xylia xylocarpa etc.*were roasted and fried before consuming. Further, species like *Calamus guruba, Capparis zeylanica*,were used to make pulp. *Cycas circinalis*and *Indigofera astragalina*were used to make cakes. Various types of chutney were prepared from species like *Mangifera indica* and *Phyllanthus emblica*. The tribes of SBR usespecies like *Mucuna pruriens, Ficus hipsida, F. racemosa, C. aromatica* and *E. hirta* var. glaberrima *etc.* were used for medicinal purposes. The stem bark of *Hibbiscus sabdoriffa* was used to make ropes. The species those had edibility of every body part in some way or other were *Russula emetic, Curcuma aromatica*, and *Euphorbia hirta* var. glaberrima. *Russula emetic* was a wild edible mushroom whereas other two were potential medicinal herbs.

Discussion

Towards the Eastern Ghats, particularly in Similipal Biosphere Reserve Forest, several rural and tribal communities still today depend on wild food plants (fruits, flowers, leafs, bulbils, tubers *etc.*) as their daily food requirements and also during the periods of food shortage and scarcity. These plants play a major role in supplementing staples foods with micronutrients and can constitute a "Safety Net" during adverse conditions (Kumar 2012). They represent cheap but quality nutrition for large segments of the population in both tribal and rural areas. Knowledge on wild edible food is gradually declined and even disappearing with increased modernization and migration to urban area due to poverty. Revalorizing and documentation indigenous knowledge on the wild edible plants is urgently needed to promote nutritional health of local the inhabitants and beyond to conserve the genetic and cultural diversity.

In Similipal Biosphere Reserve, tribal have their own socio-cultural practices, traditions and typical food practices. Although they grow food grains but it is not sufficient throughout year because of marginal land holdings. In this context, wild edible plants play a significant role not only in providing nutrient food supplement during scarcity but also generating side way income for the tribal people. Fruits collected by these local/tribal people from the natural forest are often seen on sale in rural markets. Their diet comprises variety of unconventional foods, *viz.* edible forms of flowers, fruits, leaves tubers, and seeds. It is estimated that about 80 per cent of the forest dwellers in the Similipal Biosphere Reserve (SBR), depends on forest for their annual food supplements. At present, the practice of collection by the tribal people in SBR is vanishing as a result of rampant deforestation and the displacement of the tribal forms from their traditional habitats. The negligence of these tribal foods may seriously deteriorate the health and wellbeing of the tribal people. Furthermore, nature based traditional foods are generally viewed as interchangeable, diet,being highly regarded as the primary basis for sustaining and restoring health and wellbeing.

Conclusion

The neglected/underutilized species of various wild plants of SBR region have a great potentiality for sustainable livelihood of the local inhabitants. The systematic survey, collection and documentation of information on the wild edible plants reveal the socio- economic importance of the tribal people. This ethno-botanical information significantly contributes to our knowledge about the various types of edible forms that are being unnoticed by common man, but are extensively being used as food supplements by the tribal communities residing over here. An organised collection mission from diversity rich areas and characterisation/ evaluation of germplasm of the underutilized and agro-horticulture importance fruit species are urgently required in view of the loss of genetic resources and erosion of traditional knowledge associated with these species. Many wild varieties of species too have a high market demand such as *Cycas circinalis, Capparis zeylanica, Mangifera indica, Spondias pinnata, Dillenia aurea, Limonia acidissima etc.* These species are less available in common markets but are extensively available in tribal markets. Thus, the domestication of these wild edible plants will not only improve the economic conditions of the local communities but also helps in the conservation of biological diversity.

Acknowledgements

We are grateful to the tribal people of Similipal Biosphere Reserve, Odisha for sharing their valuable traditional knowledge. Thanks to the Head, Department of Botany, North OrissaUniversity, Baripada for providing infrastructure facility and

Forest Officers of Similipal Biosphere Reserve for extending their cooperation during the field survey.

REFERENCES

Abbink, J. (1993). Famine, gold, and guns: the Suri of southwestern Ethiopia, 1985-91. Disasters: the international journal of disaster studies and practice. 17: 218-225.

Abdualrahman, M.A.Y., Ali, A.O. and Suliman, A.M.A. (2011). Nutritional evaluation of Guddaim fruits (*Grewia tenax*) and its utilization in ice cream production. J. Sci. Technol. 12: 38-43.

Abubakar, M., Arshed, M.J., Ali, Q. and Hussain, M.(2012). Spatial trend of Foot and Mouth Disease virus (FMDV) serotypes in cattle and buffaloes, Pakistan. Virologica Sinica, 27: 320-323.

Ali-Shtayeh, M.S., Jamous, R.M., Al-Shafie, J.H., *et al.* (2008). Traditional knowledge of wild edible plants used in Palestine (Northern West Bank): A comparative study. J Ethnobiol Ethnomed. 4: 1–13.

Anjum, Nishat, Vikash and Tripathy, Y.C. (2013). Physico-chemical and phytochemical investigation on Callicarpa macrophylla Vahl. Leaves from Dehradun Uttarakhand. University Journal of Phytochemistry and ayurvedic heights. 2(15): 13-16.

Balemie, K. and Kebebew, F. (2006).Ethnobotanical Study of Wild Edible Plants in Derashe and Kucha Districts, South Ethiopia. Journal of Ethnobiology and Ethnomedicine, 2, 53.http: //dx.doi.org/10.1186/1746-4269-2-53

Bello, M. O., Falade, O. S., Adewusi, S. R. and Olawore, N.O. (2008). Studies on the Chemical Compositions and Anti-nutrients of Some Lesser Known Nigeria Fruits. African Journal of Biotechnology. 7(21): 3972-3979.

Cunningham, A.B. (2001). Applied Ethnobotany. People, Wild Plant Use and Conservation.WWF, UNESCO, Royal Botanical Gardens, Kew.Earthscan Publications, London and Sterling.

Dahot, M.U. (1993). Cultivation of *Penicillium expansum* on rice husk powder for protease production. J. Islamic Acad. Sci. 6: 193-196.

Deshmukh, B.S. and Waghmode, A. (2011). Role of wild edible fruits as a food resource: Traditional knowledge. Int. J. of Pharm. and Life Sci. 2(7): 919-924.

Deshmukh, S.R. and Rathod, V. (2013). Nutritional composition of wild edible ceropegia tubers. Adv App Sci Res. 4(1): 178-181.

FAO, (2004). FAO Statistics. Food and Agriculture Organization, Rome, Italy.

Fentahun, M. and Hager, H. (2009). Exploiting locally available resources for food and nutritional security enhancement: wild fruits diversity, potential and state of exploitation in the Amhara region of Ethiopia, Food Security: The Science,

Sociology and Economics of Food Production and Access to Food, Springer; The International Society for Plant Pathology, 1(2): 207-219.

Gerique, A. and Pohle, P. (2006). Use of biodiversity along the altitudinal gradient: Plant knowledge and plant use among indigenous Shuar and Saraguro communities in southern Ecuador. In: Ecological Studies. Springer.

Guinand, Y. and Dechassa, (2000). Indigenous Food Plants in southern Ethiopia: Reflections on the Role of 'Famine Foods' at the time of drought. United Nations Emergencies Unit for Ethiopia (UNEUE), Addis Ababa.

Gupta, D., Bhardwaj, R. and Gupta, R.K. (2011). In Vitro Antioxidant activity of extracts from the leaves of Abies pindrow Royle. Afr J Tradit Complement Altern Med. 8(4): 391 397.

Haines, H. H.(1921 – 25). The Botany of Bihar and Orissa, 6 parts. London.

Jain, A.K., Tiwari, P. (2012). Nutritional value of some traditional edible plants used by tribal communities during emergency with reference to Central India. Indian journal of traditional medicines. 11(1): 51-57.

Ju, Y.S., Lucey, B.P. and Holtzman, D.M. (2014). Sleep and Alzheimer disease pathology- A bidirectional relationship. Nat Rev Neurol.10: 115–9.

Mahapatra, A.K., Mishra, S., Basak, U.C. and Panda, P.C. (2012). Nutrient Analysis of Some Selected Wild Edible Fruits of Deciduous Forests of India: an Explorative Study towards Non Conventional Bio-Nutrition. Advance Journal of Food Science and Technology. 4(1): 15-21.

Manyafu, J.L. (1971). The role of wild fruits and vegetables in Tanzania. Acta Hort. 21: 93-96.

Mohan, V.R. and Kalidas, C. (2010). Nutritional and antinutritional evaluation of some unconventional wild edible plants.Tropical Subtropical Agroecosystem, 12: 495-506.

Naik, D.(1998). Tribal culture in the context of Similipal. Workshop Journal. District Environmental Society, Mayurbhanj, Orissa, India. pp. 63-64.

Nazarudeen, A. (2010). Nutritional composition of some lesser-known fruits used by the ethnic communities and local folks of Kerala. Indian journal of traditional knowledge. 9(2): 398-402.

Olujobi, O.B. (2012). Comparative evaluation of nutritional composition of African Locust bean fruits fromtwo locations, Nigerian Journal of Basic and Applied Science. 20(3): 195-198.

Panda, S.K., Rout, S.D., Mishra, N. and Panda, T. (2011). Phytotherapy and traditional know-ledge of tribal communities of Mayyurbhanj District, Orissa, India. Journal of Phrmacognosy and Phytotherapy. 3(7): 101-113.

Rashid, U., Anwar, F., Moser, B.R.and Knothe, G. (2008). *Moringa oleifera* oil: a possible source of biodiesel. Bioresour Technol. 99(17): 8175-9.

Reddy, C.S., Pattanaik, C., Mohapatra, A. and Biswal, A.K. (2007). Phytosociological Observations on Tree Diversity of Tropical Forest of Similipal Biosphere Reserve, Orissa, India. Taiwania, 52(4): 352-359.

Redzic, S.S. (2007). The ecological aspect of ethnobotany and ethnopharmacology of population in Bosnia and Herzegovina. Coll. Antropol. 31, 869–890.

Saxena, H.O. and Brahmam, M. (1994-1996). The Flora of Orissa. Vol. I-IV. Orissa Forest Development Corporation Ltd, Bhubaneswar, India.

Seal, T. (2012).Evaluation of Nutritional Potential of Wild Edible Plants, Traditionally Used by the Tribal People of Meghalaya State in India. American Journal of Plant Nutrition and Fertilization Technology.2(1): 19-26.

Shrestha, P.M. and Dhillion, S.S. (2016). Diversity and traditional knowledge concerning wild food species in a locally managed forest in Nepal. Agroforestry Systems. 66: 55–63.

Sinha, R. and Lakra, V. (2005). Wild tribal food plants of Orissa. *Indian Journal of Traditional Knowledge.* 4(3): 246–252.

Vishwakarma, K.L. and Dubey, V. (2011). Nutritional analysis of indigenous wild edible herbs used in eastern Chhattisgarh, India. Emir. J. Food Agric. 23 (6): 554-560.

Chapter 6

Bryofloral Diversity in Jajpur District of Odisha, India

*Gouri Sankar Juga Prakash Jena[1], Ramakanta Mishra[2], Ashirbad Mohapatra[3] and Kunja Bihari Satapathy[4]**

[1]*Department of Botany, Sadhu Goureswar College, Kanikapada, Jajpur – 755 011, Odisha*
[2]*Post Graduate Department of Botany, Utkal University, Vani Vihar, Bhubaneswar – 7510 04, Odisha*
[3]*Department of Botany, Sri Jayadev College of Education and Technology, Naharkanta, Bhubaneswar, Odisha*
[4]*Department of Botany, School of Applied Sciences, Centurion University of Technology and Management, Bhubaneswar – 752 050, Odisha*
**e-mail: kbs_bot@rediffmailmail.com*

ABSTRACT

The present study deals with the investigation of bryophytic flora of Jajpur district, Odisha. During the course of extensive survey and collection of plant materials a total of 37 taxa of Bryophytes distributed under 29 genera and 19 families were recorded. Of these, the mosses were represented by 23 species belonging to 20 genera and 13 families, while Liverworts were comprised of 11 species of 7 genera and 5 families, besides the Hornworts were represented by 3 species of 2 genus belonging to 1 family. The present study revealed that Bryaceae and Dicranaceae were dominant families in the study area having 4 species each followed by Aytoniaceae, Marchantiaceae, Funariaceae and Anthocerotaceae with 3 species each.

Keywords: *Diversity, Jajpur, Odisha, Bryophytes.*

Introduction

The bryophytes which comprise liverworts, hornworts and mosses are widely distributed predominantly between the altitudes of 1000-8000 metres and considered important components of the vegetation in many regions of the world including the forest ecosystems. They constitute a major part of the biodiversity in moist environments, wetland, and mountain ecosystems (Hallingback and Hotsgetts, 2000). Being pioneers of the terrestrial vegetation, they usually grow as varieties of life forms contributing to the main component of mountain forest in view of their high degree of soil binding capacity besides the water retention characteristics (Alam, 20011). Bryophytes play a significant role as indicators of environmental pollution (Bates, 2000). They are ecologically significant in playing a key role in ecosystem dynamics (Vanderpoorten and Goffinet, 2009). Bryophytes are also potential sources of new antibiotics and anticancerous substances (Chopra and Vasishtha, 1994). They constitute one of the richest groups of plants in India and occupy a wide range of substrata. Currently, about 2489 taxa of bryophytes are reported from India comprising about 1786 species under 355 genera of mosses, 675 species in 121 genera of liverworts and 25 species in six genera of hornworts (Dandotiya *et al.,* 2011). The present knowledge on the bryophytes of India began with studies by European bryologists such as Montagne (1842a), Müller (1853), Mitten (1859, 1861) and Dixon (1914). The works of Benedix (1953), Chopra (1938), Asthana and Srivastava (1991), Asthana *et al.* (1995), Nath and Asthana (1998), Parihar *et al.* (1994), Singh (1994, 2002) and Lal, (2005) and many others, contain records of collections from India. Recently, many workers have made important contribution to the Bryophyte flora of the eastern region of India and adjacent plains (Gangulee, 1980), Meghalaya (Singh and Nath, 2007), Nagaland (Bansal *et al.,* 2011). Most recently, Singh and Barbhuiya (2012) reported several liverworts and hornworts from Assam.

Material and Methods

Study Area

Jajpur an interior district of Odisha (India), situated between 86° 332 663$^{\emptyset}$ East longitude and 20° 85$^{\times}$ 145$^{\emptyset}$ North latitudes, covers an area of 2,888 sq. kms. with normal climate of Indian standard.

The average height of the district from the sea level is 330 m with an average rain fall of 1015.5 mm, besides a maximum and minimum temperatures of 40 °C and 10 °C respectively. It is surrounded by other four districts such as Bhadrak in east, Dhenkanal in west, Keonjhar in north and Cuttack in the south. The types of soil found in the district are alluvial, saline and red laterite. The district is wildly festooned with many diversified flora, which are completely unique in their habitat and shows idiosyncrasy in ecological condition. Even though this district is bestowed with a quite rich bryodiversity, no attempt had yet been made to study and document the diversity of Bryophytes of this area. To cover the knowledge gap and in view of the need for generation of floral diversity lists of this region an important tool for species and ecosystem conservation as well as for the detailed taxonomic studies the present investigation was undertaken.

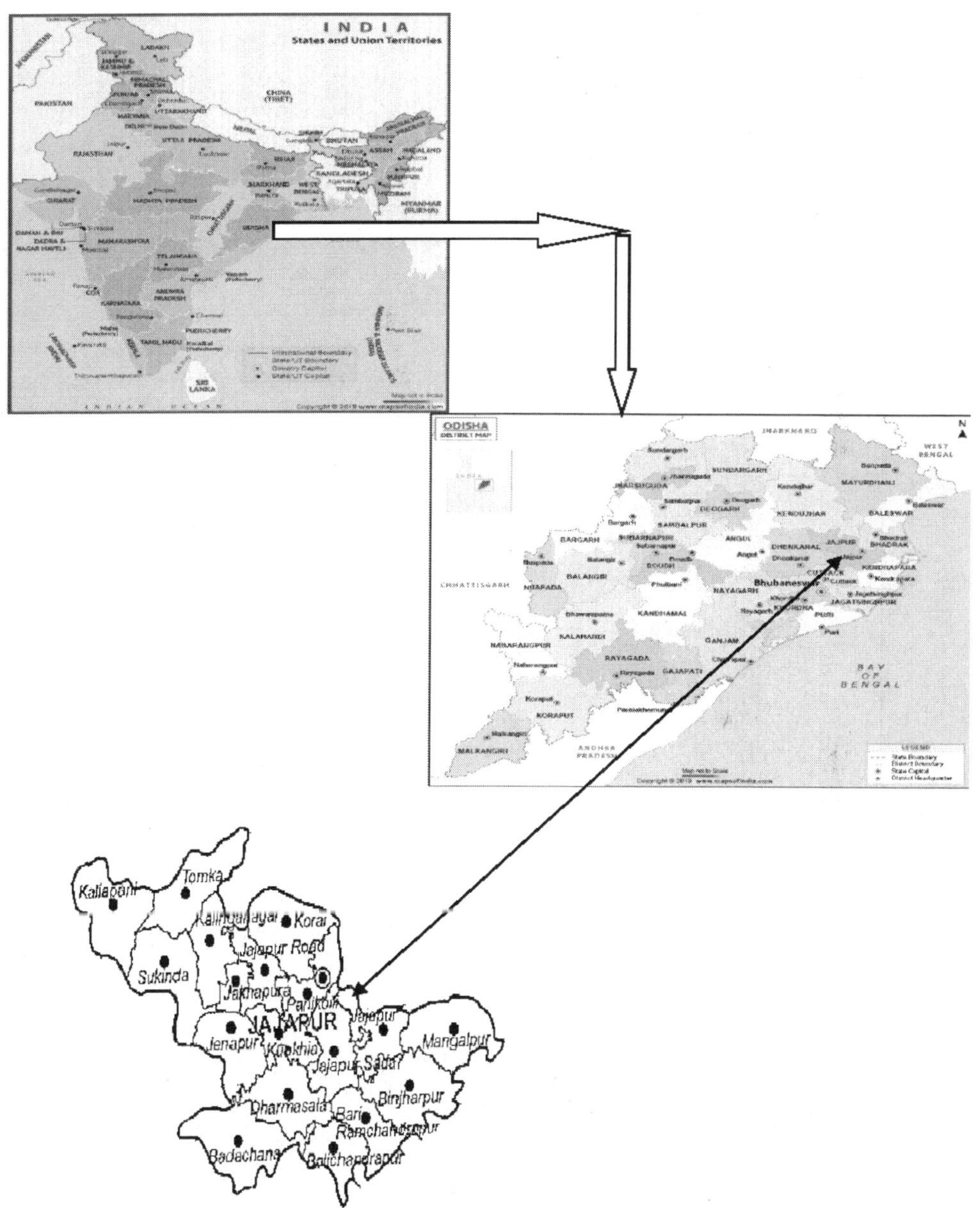

Figure 6.1: Map Showing Location of Study Area, Jajpur District, Odisha, India.

Methodology

An extensive Bryological Inventory was carried out in different parts of Jajpur district of Odisha during 2017-2018 which included areas of Mahavinayak forest, Gunthuni, Gokarneswar, Langudi, Kaima, Ashokjhar and Lembo. Samples were also collected from different habitats including natural forests, agro-forests, urban ecosystems and road sides from different localities in the district, during March, 2018 to November, 2018. Simple methodology was adapted to collect specimens in the field using a knife to peel off specimens from the substratum. The collected materials were dried at room temperature on blotting paper and placed carefully in simple thick paper envelopes. Date of collection, locality and habitat along with the substratum type were recorded in the field note-book and also on the packets containing the materials. The collected bryophytes were identified carefully and deposited to Bryological Laboratory of NBRI, Lucknow. The identification of taxa has been done in the laboratory by studying the specimens and consulting various relevant literatures, flora, monographs and research publications (Bansal and Nath, 2012a, 2012b; Schuster, 1979; Schuster, 1984a; Schuster1984b; Singh and Singh 2003).

Table 6.1: List of the Species of Bryophytes with Family, Habitat and Altitude of Occurrence.

Sl.No.	*Name of the Species*	*Family*	*Habitat*	*Altitude*
1.	*Anthoceros erectus* Kashyap.	Anthocerotaceae	Soil	930m
2.	*Anthoceros formosae* Stephani.	Anthocerotaceae	Soil	802m
3.	*Asterella khasiana* (Griff.) Grolle.	Aytoniaceae	Soil	756m
4.	*Asterella leptophylla* (Mont.) Grolle.	Aytoniaceae	Soil	730m
5.	*Atrichum undulatum* (Hedw.) P. Beauv.	Polytrichaceae	Soil	669m
6.	*Bryum alpinum* Huds. ex With.	Bryaceae	Rock	450m
7.	*Bryum argenteum* Hedw.	Bryaceae	Rock	779m
8.	*Bryum coronatum* Schwägr.	Bryaceae	Rock	900m
9.	*Campylopus introflexus* (Hedw.) Brid.	Dicranaceae	Soil	676m
10.	*Cyathodium cavernarum* Kunze.	Targioniaceae	Rock	822m
11.	*Dumortiera hirsuta* (Sw.) Nees	Marchantiaceae	Rock	390m
12.	*Entosthodon wichurae* M. Fleisch.	Funariaceae	Soil	885m
13.	*Funaria hygrometrica* Hedw.	Funariaceae	Soil	669m
14.	*Hyophila involuta* (Hook.) A. Jaeger	Pottiaceae	Rock	450m
15.	*Leucobryum crispum* Müll. Hal.	Dicranaceae	Soil	633m
16.	*Leucobryum glaucum* (Hedw.) Ångstr.	Dicranaceae	Soil	689m

Sl.No.	Name of the Species	Family	Habitat	Altitude
17.	*Macromitrium involutifolium* (Hook. and Grev.) Schwägr.	Orthotrichaceae	Tree bark	989m
18.	*Marchantia emarginata* Blume and Nees.	Marchantiaceae	Soil	856m
19.	*Marchantia paleacea* Bertol.	Marchantiaceae	Soil	789m
20.	*Meteoriopsis reclinata* (Müll. Hal.) M. Fleisch.	Meteorioceae	Tree bark	667m
21.	*Metzgeria furcata* (L.) Corda.	Metzgeriaceae	Tree bark	956m
22.	*Neckera pennata* Hedw.	Neckeraceae	Tree bark	756m
23.	*Neckeropsis semperiana* (Hampe ex C. Müll.) Touw.	Neckeraceae	Tree bark	789m
24.	*Octoblepharum albidum* Hedw.	Dicranaceae	Tree bark	635m
25.	*Phaeoceros laevis* (L.) Prosk.	Anthocerotaceae	Soil	684m
26.	*Physcomitrium pyriforme* (Hedwig) Hampe	Funariaceae	Soil	650m
27.	*Plagiochasma appendiculatum* Lehm. and Lindenb.	Aytoniaceae	Rock	880m
28.	*Plagiomnium ellipticum* (Brid.) T. Kop.	Mniaceae	Soil	420m
29.	*Ptychanthus striatus* (Lehm. et Lindenb.) Nees	Lejeuneaceae	Tree Bark	575m
30.	*Racopilum cuspidigerum* (Schwägr.) Ångström.	Racopilaceae	Tree Bark	986m
31.	*Rhodobryum giganteum* (Schwägr.) Paris.	Bryaceae	Soil	450m
32.	*Riccardia chamedryfolia* (With.) Grolle.	Aneuraceae	Rotten log	450m
33.	*Riccia fluitans* L.	Ricciaceae	Rock	480m
34.	*Riccia glauca* L.	Ricciaceae	Soil	400m
35.	*Targionia hypophylla* L.	Targioniaceae	Rock	880m
36.	*Taxiphyllum barbieri* (Cardot and Copp.) Z. Iwats.	Hypnaceae	Rock	900m
37.	*Trematodon longescens* Müll. Hal.	Bruchiaceae	Soil	800m

Result and Discussion

As a result of the extensive floristic exploration followed by detailed taxonomic study a total of 37 taxa of Bryophytes distributed under 29 genera and 19 families have been recorded. Out of them, the mosses comprised of 23 species belonging to 20 genera and 13 families, while Liverworts were represented by 11 species of 7genera and 5 families. 3 species of Hornworts were recorded belonging to

2 genera and 1 family. The data presented in the Table-1 revealed that Bryaceae and Dicranaceae were the two dominant families in the study area representing 4 species each followed by Aytoniaceae, Marchantiaceae, Funariaceae and Anthocerotaceae with 3 species. Habitat wise categorization indicated that 18 of the reported species were found thriving on the surface of soil, 9 species were found growing on the surface of rocks, and 9 species namely *Cyathophorella tonkinensis, Macromitrium involutifolium, Meteoriopsis reclinata, Metzgeria furcata, Neckera pennata, Neckeropsis semperiana, Octoblepharum albidum, Ptychanthus striatus* and *Racopilum cuspidigerum* were observed to be epiphytic and 1 species such as *Riccardia chamedryfolia* was found growing on rotten bamboo surface. All the bryophytes specimens collected during the present investigation were located within the altitudinal range of 390m to 1000m above the sea level (Table 6.1). Some of the recorded species namely *Rhodobryum giganteum, Dumortiera hirsuta, Atrichum undulatum, Philonotis fontana* and *Marchantia* sp. were reported to possess medicinal value being used by the local inhabitants. It is pertinent to mention that the district of Jajpur harbours a rich bryodiversity. It is has been observed from the study that Jajpur district, Odisha harbours rich bryodiversity and no systematic data on them in this region were earlier available. It is expected that the present study may serve as valuable contribution to the knowledge of the bryoflora of the state and in Indian context as well to use the present data-base for the future conservation aspects.

Conclusion

Field observations at regular intervals could reveal that the disturbances and threat to the bryodiversity of this region is associated with many human activities like expansion of agricultural land, the construction of road, destruction of forests, execution of unplanned developmental activities; and habitat destruction through over exploitation of resources, shifting cultivation and other anthropogenic interference have resulted in depletion of bryodiversity of the area. Therefore, serious attentions and efforts are required to save these valuable bryoflora of this region.

REFERENCES

Alam, A. (2011). Diversity and distribution of terrestrial liverworts (Hepaticeae) in Nilgiri, Tamil Nadu, India. Proc. Nat. Acad. Sci.India. Sec. B, Vol. 81 Pt. II.

Asthana, A. K. and Srivastava, S.C. (1991). Indian hornworts. Bryophytorum Bibliotheca, 42: 1-158.

Asthana, G., Srivastava, S.C. and Asthana, A.K. (1995). The genus *Cheilolejeunea* in India. Lindbergia, 20: 125-145.

Bansal, P., Nath, V. and Chaturvedi, S.K. (2011). Epiphytic Bryophytes on *Thuja orientalis* in Nagaland North- Eastern India. Bangladesh Journal of Plant Taxonomy,18: 163-167.

Bansal, P. and Nath, V. (2012a). A new record of *Bryum coronatum* Schwaegr. (Bryophyte) in Meghalaya, India. Taiwania, 57: 294-299.

Bansal, P. and Nath, V. (2012b). Bryum bessonii Ren. and Card. new to eastern Himalaya in new national and regional bryophyte records 32. J. Bryol. 34: 231-246.

Bates, J.W. (2012b). Mineral nutrition, substratum ecology and pollution. In: Shaw, A. J. and Goffinet, B (eds), Bryophyte Biology, pp. 248-311. Cambridge University Press. Cambridge, U.K.

Benedix, E.H. (1953). Indomalayische Cololejeuneen. Feddes Repertorium, Beihefte, 134-188.

Chopra, R.N. and Vashistha, B.D. (1994). Bryophyte Morphogenetic Study, In: Botany of India (History and Progress), I edited by Johri BM (Oxford IBH Publishing Co. Pvt. Ltd., New Delhi), pp. 437-453.

Chopra, R.S. (1938). Notes on Indian Hepaticae. I.South India. Proceedings of the Indian Academy of Science 7 B: 239-251.

Dandotiya, D., Govindapyari, H., Suman, S. and Uniyal, P.L. (2011). Checklist of the bryophytes of India. Archive for bryology, 88.

Dixon, H.N. (1914). Report on the mosses collected by C.E.C. Fischer and others from south India and Ceylon.Records of the Botanical Survey of India, 6(3): 75-91.

Gangulee, H.C. (1969-1980). Mosses of Eastern India and adajacent regions. Fascicles, pp. 1-8. Books and Allied Limited, Calcutta.

Hallingbäck, T. and Hodgetts, N. (2000). Status survey and conservation action plan for bryophytes: mosses, liverworts and hornworts. IUCN/SSC Bryophyte Specialist Group. IUCN, Gland, Switzerland and Cambridge, U.K., 106.

Lal, J. (2005). A checklist of Indian Mosses, Bishen Singh Mahendra Pal Singh. Dehra Dun, India, 1-164.

Mitten, W. (1859). Musci Indiae Orientalis, an enumeration of the mosses of the East Indies. Journal of the Proceedings of the Linnean Society, supplement to Botany, 1: 1-171.

Mitten, W. (1861). Hepaticae Indiae Orientalis: an enumeration of the Hepaticae of the EastIndies. Journal of the proceedings of the Linnean Society. Botany, 5: 385-392.

Montagne, C. (1842a). Cryptogamae Nilgherienses seu plantarum cellularium in montibus peninsulae indicae Neel-Gherries dictis a Cl. Perrottet collectarum enumeratio. Musci. Annales des Sciences Naturelles; Botanique, 2: 17(2): 243-256.

Muller, C. (1853). Musci Neilgherienses I. Bot. Ztg. Regenberg, 11: 17-21; 33 - 40; 57- 92.

Nath, V. and Asthana, A. (1998). Diversity and distribution of genus FrullaniaRaddi in South India. Journal Hattori Botanical Laboratory, 85: 63-82.

Parihar, N.S., Lal, B. and Katiyar, N. (1994). Hepatics and Anthocerotes of India. A new annotated checklist. Central book depot. Allahabad.

Schuster, R.M. (1979). The Phylogeny of the Hepaticae In: Bryophyte Systematics. Systematic Association Special 14 Edited by Clarker IGC and Duchett JG, 1979; 41-82.

Schuster, R.M. (1984). Evolution, phylogeny and classification of the Hepaticae. New Manual of BryologyMichinan, Japan, 892-1070.

Singh, A.P. and Nath, V. (2007). Hepaticae of Khasi and Jaintia Hills: Esatern Himalayas. Bishen Singh Mahendra Pal Singh, Dehradun, India.

Singh, D.K. (1994). Distribution of family Notothylaceae in India and its phytogeographical significance. Advances in Plant Sciences Research, 2: 28-43.

Singh, D.K. (2002). Notothylaceae of India and Nepal (A morpho-taxonomic revision). Bishen Singh Mahendra Pal Singh, Dehradun.

Singh, S.K. and Singh, D.K. (2003). Contribution to the bryoflora of Great Himalayan National Park, Kullu, Himachal Pradesh, India-II. Hepaticae. Phytotaxonomy, 3: 35-52.

Smith, A.J.E. (1982). Epiphytes and epiliths. In: Bryophyte ecology. London, Chapman and Hall, 191-227.

Chapter 7

Floristic and Ethnobotanical Studies in Loktak Lake, Manipur: A Source of Potential Food, Nutraceuticals and Drugs

Rajkumari Supriya Devi[1],
Kunja Bihari Satapathy[1]* *and Sanjeet Kumar*[2]

[1]*School of Applied Sciences,*
Centurion University of Technology and Management, Odisha
[2]*Biodiversity and Conservation Lab.,*
Ambika Prasad Research Foundation, Odisha
**e-mail: kbs_bot@rediffmailmail.com*

ABSTRACT

Loktak lake is the largest natural freshwater lake in the North-Eastern region of India playing an important role in the ecological and economic security of the region. It is also known for 'Phumdi', a heterogeneous mass of soil, vegetation and organic matter at various stages of decomposition. For fishing purposes, local communities make a ring with 'Phumdi' mass called 'Athaphum' measuring 500 ft in diameter, a traditional practice popular since last 200 years. It is further interesting to report that these also harbour a rich flora of medicinal herbs. Keeping the importance of its medicinal wealth, an attempt was made to survey 20 different 'Athaphum' of the lake to explore and document the diversity of medicinal plants of the fresh water lake. Extensive floristic survey in the study area revealed the occurrence of over 89 plant species in and around 'Athaphum' and 'Phumdis'. Among them, five plants (Blumea mollis, Ludwigia repens, Rumex nepalensis, Oenanthe javanica and Carex riparia) were selected for phytochemical screening to validate their therapeutic claims. All the test plants were found to be enriched with phenolic compounds. On the basis of their ethnobotanical claims and results of phytochemical analysis, Rumex nepalensis and Oenanthe javanica could be recommended as potential nutraceutical for being a popular food plant among the inhabitants in the lake region.

Keywords: *Nutraceutical, Secondary metabolites, Traditional therapeutic values, Wetlands.*

Introduction

Loktak lake (Figure 7.1) is the largest natural freshwater lake in the north-eastern region of India and plays an important role in the ecological and economic security of the region. People of Manipur are socially, economically, culturally and ecologically connected with the lake (Clarke, 1889; Deb, 1957; Devi, 2009; Devi *et al.,* 2014). The lake has been the source of water for domestic generation of hydro-electric power, irrigation, habitat for several plants used as food, fishing ground for local people, fodder, fuel, medicines, biodiversity, recreation *etc.* Hence, it has been referred to as the Lifeline of Manipur. It is also known for its floating island 'Phumdi' which is mainly made up of plant biomass. The main plants grown in 'phumdi' are fragmitis and paragrass, besides sparsely developed other plants like *Mikania, Colocasia, etc.* (Devi *et al.,* 2014). Indigenous people make use of the hard matured 'phumdi' to make an artificial ring locally known as 'Athaphum' (Figure 7.2) which is mainly constructed for the purpose of fishing and one athaphum is owned by a particular family at a time or shared among two or three families. The matured phumdi of thickness 1.5 ft is cut with 5ft wide and joined to form a ring structure of 500 ft in diameter (Devi and Singh, 2017). The fish feed like rice husk or grasses are spread to attract the fish for entering the ring area. Pieces of matured phumdis are also placed inside the ring to make an artificial habitat for the fish. Presence of fishes is indicated by faeces of fishes that float on the surface of the water which normally takes around 10 days from the freshly constructed athaphum. With such indications, fish net are set around the ring such that heavy stone are tied at the edges so as to touch the floor of the lake. The net is entangled with a rope and is pressed against the floor of the lake. Patches of phumdis are later removed from the middle of the ring to have easy access of fishing (Plate 7.1.5). The cost of construction of athaphum is around 1 lakh rupees which can be reused from time to time for more than 4-5 years. According to the villagers, this practice of fishing with athaphum is common since 200 years. Fishing is one of the main sources of livelihood of the villagers around the Loktak lake. The ringed masses of Phumdi are the home of many floral wealth (Deb, 1961; Sinha, 1987; Devi *et al.,* 2012; Devi *et al.,* 2013). The local communities use them for different purposes for which the present attempt has been made to enumerate the floral wealth available in and around Athaphum and phytochemical analysis of selected plants. The present study highlights the medicinal plant wealth available in and around Athaphum for their further application in drug discovery.

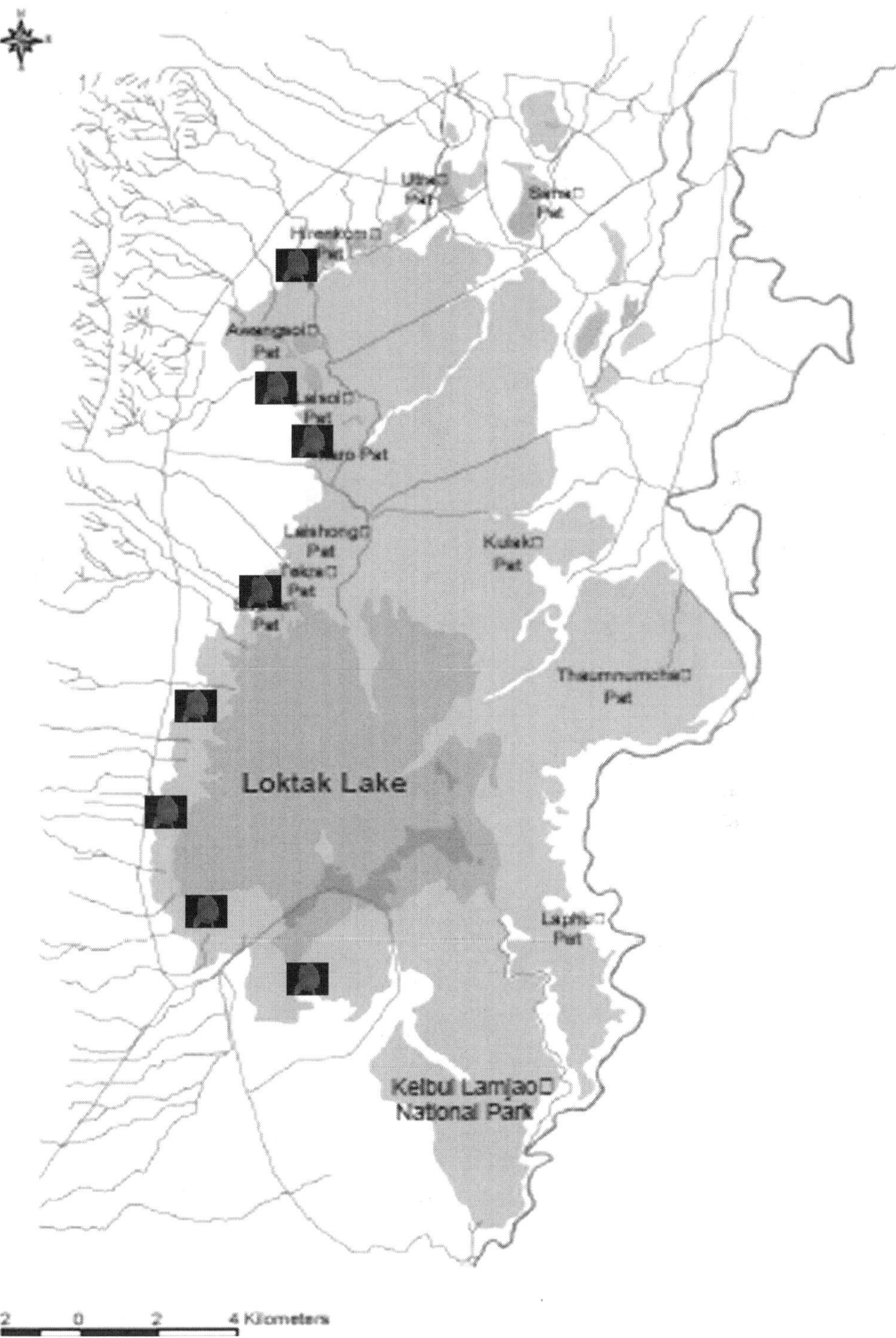

Figure 7.1: Geographical Map of Study Area and Collection Sites.

1

2

3

4

Figure 7.2: Formation of Athaphum,
1: Collection of Biomass; 2-3: Making ring;
4: Ringed biomass in the form of Athaphum.

Materials and Methods

Floristic survey was carried out at regular intervals during 2017 - 2019 in Loktak lake, Manipur. The collected plant specimens were identified by following available literature and recent publications (Deb, 1961; Sinha, 1987; Devi *et al.*,

2012; Devi *et al.*, 2013). After taxonomic studies the plant samples were identified and arranged alphabetically as per their botanical name. For ethnobotanical information, Passport Data Form (PDF) was used, developed by Biodiversity and Conservation Lab., Ambika Prasad Research Foundation, Odisha and methodological framework was followed as per standard technique of ethno-biological approaches of Christian and Brigitte (2004).

Plate 7.1: Survey for Floral Diversity and Ethnobotanical Data Collection in Loktak Lake:
1: *Spiranthes spiralis;* 2: *Oenanthe javanica;* 3: Local community selling vegetative parts of *Trapa natans;* 4: *a. Salvinia cucullata; b. Trapa natans; c. Hydrilla verticillata;* 5: Athaphum in Loktak lake, Manipur.

Collection of Selected Plants for Experimental Work

The samples were collected from different Athaphum of Loktak lake (24° 48' 22.67 " N; 93° 56' 53.5" E), Manipur and kept in poly bags tagged with the botanical name and sorted out as per standard sampling procedure and passport description (Koppar, 1998).

Preparation of Extracts

Soxhlet method and percolation were adopted to obtain the different extracts (Tiwari *et al.*, 2011; Kumar *et al.*, 2017a; Kumar *et al.*, 2017b). The collected experimental plant materials were dried at room temperature under shade and were powdered after drying using mechanical devices. The powdered material of the experimental plant was kept in thimble and extraction was carried out using the Soxhlet apparatus. The residues were collected and left for air drying and dried crude extracts were stored in refrigerator for further phytochemical analysis and antibacterial activities.

Phytochemical Analysis

Phyto-chemical analyses were carried using standard procedure to identify the bioactive compounds (Kumar *et al.,* 2013) as per polarity index.

Results and Discussion

Loktak lake being the life line of Manipur provides all life stuffs to the populace of the state and maintain the ecological balance of Indo-Burma Biodiversity hotspot of the country. The survey and experimental works unwind the diverse things.

Floral Diversity

The data collected during the floristic survey revealed that about 89 plant species naturally occur that belong to 67 genera and 28 families. Among them, 47 species were herbs, 14 grass species, 4 floating herb, 7 shrubs and one parasitic plant. Further, it is reported that a rich population of one aquatic ground orchid (*Spiranthes spiralis*: Plate 7.1.1) and two carnivorous plants (*Utricularia aurea* and *U. exoleta: Table* 7.1) are also found. Tuboi *et al.* (2012) reported 185 plant species from Keibul Lamjao National Park, Manipur

Table 7.1: Diversity of Plant Species in and around 'Phumdi' and 'Athaphum'

Botanical Name	*Family*	*Habitat*
Acmella paniculata (Wall. ex DC.) R.K. Jansen	Asteraceae	Herb
Acorus calamus L.	Araceae	Herb
Ageratum conyzoides (L.) L.	Asteraceae	Herb
Alocasia macrorrhiza Schott	Araceae	Shrub
Alpinia galanga (L.) Willd.	Lythraceae	Herb
Alternanthera philoxeroides (Mart.) Griseb.	Amaranthaceae	Herb
Alternanthera sessilis (L.) R.Br. ex DC.	Amaranthaceae	Herb
Amomum aromaticum Roxb.	Zingibereaceae	Herb
Garnotia stricta Brongn	Poaceae	Grass
Argyreia nervosa (Burm.f.) Bojer	Convolvulaceae	Grass
Arundo donax L.	Poaceae	Grass
Azolla pinnata R.Br.	Salviniaceae	Floating herb
Blumea molis (D. Don). Merr.	Asteraceae	Herb
Brachiaria mutica (Forssk.) Stapf.	Poaceae	Grass
Carex riparia Curtis.	Cyperaceaae	Herb

Botanical Name	**Family**	**Habitat**
Centella asiatica (L.) Urb.	Apiaceae	Herb
Ceratophyllum demersum L.	Ceratophyllaceae	Herb
Cissampelos pareira L.	Menispermaceae	Climber
Alocasia cucullata (Lour.) G.Don.	Araceae	Herb
Commelina benghalensis L.	Commelinaceae	Herb
Cuscuta reflexa Roxb.	Convolvulaceae	Climber
Cynodon dactylon (L.) Pers.	Poaceae	Grass
Cyperus difformis L.	Cyperaceae	Herb
Cyperus haspan L.	Cyperaceae	Herb
Cyperus iria L.	Cyperaceae	Herb
Cyperus rotundus L.	Cyperaceae	Herb
Dactyloctenium aegypticum (L.) Willd.	Poaceae	Grass
Euryale ferox Salisb.	Nymphaeaceae	Aquatic herb
Eclipta prostrata (L.) L.	Asteraceae	Herb
Eichhornia crassipes (Mart.) Solms	Pontederiaceae	Herb
Emilia sonchifolia (L.) DC. ex DC.	Asteraceae	Herb
Enhydra fluctuans Lour.	Asteraceae	Herb
Fagopyrum esculentum Moench.	Polygonaceae	Climber
Fuirena umbellate Rottb.	Cyperaceae	Herb
Gynura cusimbua (D.Don) S.Moore.	Asteraceae	Herb
Hedychium coronarium J. Koenig	Zingiberaceae	Shrub
Hydrilla verticillata (L.f.) Royle. (Plate 7.1.4c)	Hydrocharitaceae	Submerged floating herb
Imperata cylindrica (L.) Raeusch.	Poaceae	Grass
Ipomoea alba L.	Convolvulaceae	Climber
Ipomoea cairica (L.) Sweet.	Convolvulaceae	Climber
Ipomoea aquatic Forssk.	Convolvulaceae	Climber
Jussiaea repens L.	Onagraceae	Herb
Kyllinga brevifolia Rottb.	Cyperaceae	Herb
Lemna minor L.	Lemnaceae	Herb
Leersia hexandra Sw.	Poaceae	Grass
Limnophila aquatica (Roxb.) Alston	Plantaginaceae	Herb
Limnophila aromatica (Lam.) Merr.	Plantaginaceae	Herb
Limnophyton obtusifolium (L.) Miq.	Alismataceae	Shrub
Lindernia procumbens (Krock.) Philcox.	Linderniaceae	Herb
Ludwigia adscendens (L.) H.Hara.	Onagraceae	Herb
Ludwigia octovalvis (Jacq.) P.H. Raven.	Onagraceae	Shrub
Ludwigia repens J.R. Forst.	Onagraceae	Herb

Botanical Name	Family	Habitat
Marsilea minuta L.	Marsileaceae	Herb
Melastoma malabathricum L.	Melastomataceae	Shrub
Merremia aegyptia (L.) Urb.	Convolvulaceae	Climber
Mikania cordata (Burm.f.) B.L. Roxb.	Asteraceae	Climber
Mikania micrantha Kunth.	Asteraceae	Climber
Mikania scandens (L.) Willd.	Asteraceae	Climber
Mimosa pudica L.	Mimosaceae	Herb
Oenanthe javanica (Blume) DC.	Apiaceae	Herb
Parthenium hysterophorus L.	Asteraceae	Shrub
Persicaria sagittata (L.) H. Gross.	polygonaceae	Climber
Phragmites karka (Retz.) Trin ex. Steud.	Poaceae	Grass
Pistia stratiotes L.	Araceae	Herb
Polygonum barbatum L.	Polygonaceae	Herb
Polygonum capitatum Buch.-Ham. ex. D.Don.	Polygonaceae	Herb
Polygonum hydropiper L.	Polygonaceae	Herb
Polygonum odoratum Lour.	Polygonaceae	Herb
Polygonum perfoliatum L.	Polygonaceae	Herb
Polygonum persicaria L.	Polygonaceae	Herb
Polypogon monspeliensis (L.) Desf.	Poaceae	Grass
Polystichum sp.	Dryopteridaceae	Herb
Ranunculus sceleratus L.	Ranunculaceae	Shrub
Rotala rotundifolia (Buch.-Ham. ex Roxb.) Koehne	Lythraceae	Herb
Rumex maritimus L.	Polygonaceae	Herb
Rumex nepalensis Spreng.	Polygonaceae	Herb
Saccharum spontaneum L.	Poaceae	Grass
Sacciolepis indica (L.) Chase.	Poaceae	Grass
Sagittaria sagittifolia L.	Alismataceae	Herb
Salvinia cucullata Roxb. (Plate 7.1.4a)	Salvinaceae	Floating herb
Salvinia natans (L.) All.	Salvinaceae	Floating herb
Setaria sp.	Poaceae	Grass
Spiranthes spiralis (L.) Chevall.	Orchidaceae	Aquatic ground orchid
Trapa natans L.	Trapaceae	Floating herb
Utricularia exoleta R.Br.	Lentibulariaceae	Carnivorous herb
Utricularia aurea Lour.	Lentibulariaceae	Carnivorous herb
Vallisneria spiralis L.	Hydrocharitaceae	Submerged floating herb
Zizania latifolia (Griseb.) Turcz. ex Stapf.	Poaceae	Grass

Ethnomedicinal Practices

The questionnaire with local people using PDF revealed that different plant species found in 'Phumdis' and 'Athaphum' were used as food as well as medicine. The data on ethnobotanical survey revealed (Plate 7.1.3) that *Centella asiatica, Oenanthe javanica* (Plate 7.1.2), *Euryale ferox, Alpinia galanga, Alternanthera philoxeroides, A. sessilis, Commelina benghalensis, Hedychium coronarium, Ipomoea aquatica, Rumex nepalensis* and *Trapa natans* (Plate 7.1.4b) were used as food whereas *Hedychium coronarium, Acorus calamus, Ageratum conyzoides, Argyreia nervosa, Blumea mollis, Carex riparia, Centella asiatica, Eclipta alba, Ludwigia adscendens, Mimosa pudica* and *Trapa natans* were used for medicinal purposes. *Arundo donax* and *Phragmites karka* are used as fuel while *Zizania latifolia, Polygonum barbatum, Limnophila aromatica* and *Alternanthera philoxeroides* are used as fodder. Tuboi *et al.* (2012) documented the plant species available in Keibul Lamjao National Park, Manipur.

Table 7.2: Qualitative Analysis of Secondary Metabolites Present in Selected Ethnomedicinal Plants of Loktak Lake, Manipur

Plant Name	*Solvent*	*Bioactive Compounds Detected*
Blumea mollis	n-hexane	Not detected
	Methanol	Tannin, Flavonoids and Phenolic compounds
	Aqueous	Tannin, Flavonoids, Saponin and Phenolic compounds
Carex riparia	n-hexane	Not detected
	Methanol	Phenolic compounds
	Aqueous	Tannin and Phenolic compounds
Ludwigia repens	n-hexane	Not detected
	Methanol	Flavonoids, Phenolic compounds
	Aqueous	Tannin, Saponin, Phenolic compounds
Oenanthe javanica	n-hexane	Not detected
	Methanol	Phenolic compounds
	Aqueous	Flavonoids and Phenolic compounds
Rumex nepalensis	n-hexane	Not detected
	Methanol	Tannin, Flavonoids, Saponin and Phenolic compounds
	Aqueous	Tannin, Flavonoids, Saponin and Phenolic compounds

Phytochemical Screening

Keeping in view the species richness and its ethnic claims, *Blumea mollis, Ludwigia repens, Rumes nepalensis, Oenanthe javanica* and *Carex riparia* were selected for qualitative analysis of bioactive compounds. The results of phytochemical analysis confirmed the presence of phenolic compounds in all the plants extracts (Table 7.2) indicating that the extracts might be having chemopreventive properties, antioxidant (Balasundaram *et al.,* 2006), anticancer, anti-inflammatory, organoleptic properties (Connell and Fox, 2001), molecular anti-cancer mechanisms and anti-bacterial activity (Huang *et al.,* 2010).

Conclusion

World's only floating lake, Loktak is considered as the lifeline of Manipur state of India. Phumdi, a heterogenous mass of biotic and abiotic components and Athaphum, a ringed mass of such bio-wealth are the heart of the lake and a good repository of many plant species About 89 species of plant are available in and around of Athaphum which is used in different purposes by the local communities. The plants have medicinal values and possess diverse bioactive compounds like saponin, tannin, phenolic compounds *etc.* Due to presence of diverse secondary metabolites, they could be used to formulate new drugs to fight against lethal diseases as well as anti-microbial agents (Figure 7.3).

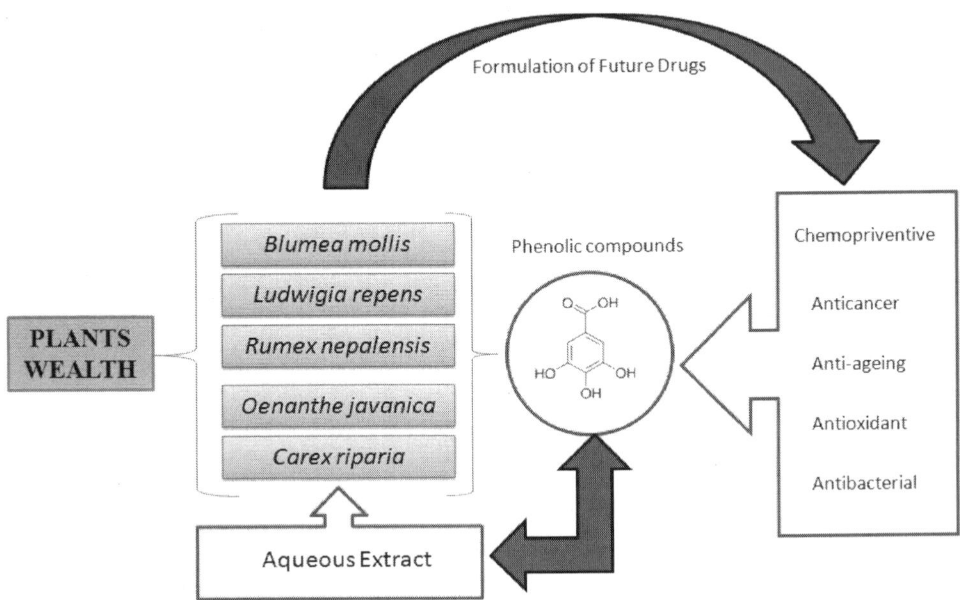

Figure 7.3: Plant Wealth of Loktak Lake: Source of Bioactive Compounds.

Acknowledgements

Authors are thankful to the authorities of Centurion University of Technology and Management, Odisha and office-in-charge, Ambika Prasad Research Foundation, Regional Centre, Imphal, Manipur for providing facilities to conduct the investigation.

REFERENCES

Balasundram, N., Sundram, K. and Samman, S. (2006). Phenolic compounds in plants and agri-industrial by-products: antioxidant activity. Occurrence and potential uses. Food. Chem. 99(1): 191-203.

Clarke, C.B. (1889). On the plants of Kohima and Manipur. J. Zinn. Soc. 25: 1-107.

Connell, J.E.O. and Fox, P.F. (2001). Significance and applications of phenolic compounds in the production and quality of milk and dairy products: a review. Int. Dai. J. 11(3): 103-120.

Deb, D.B. (1957). Studies on the flora of Manipur. Bull. Bot. Soc. Beng. 11(1): 15-24.

Deb, D.B. (1961). Monocotyledonous plants of Manipur Territory. Bull. Bot. Surv. Ind. 3(2): 115-138.

Devi, M.H. (2009). Consumable parts of wild edible plants of Loktak lake, Manipur. J. North. Nort. Eas. Food. 8(1 and 2): 15-18.

Devi, M.H. and Singh, P.K. (2017). Flowering calendar of the macrophytes of Keibul Lamjao National Park, Loktak Lake, Manipur. India. Res. J. Bot. 12: 14-22.

Devi, M.H., Salam, J.S., Joylani, S.D., Singh, P.K. and Choudhury, M.D. (2013). Biochemical study of ten selected fodder plants of critically endangered Sangai (*Rucervus eldii eldii* Mc Clelland). Env. Ecol. 31(2): 573-581.

Devi, M.H., Singh, P.K. and Choudhury, M.D. (2012). Traditional medicines from wetland plants of Keibul Lamjao National Park, Manipur. In: Choudhury, M.D., Sharma, G. D., Talukdar, A.D. and Choudhury, S. (eds.) Researches in Medicinal and aromatic plants. Swastic Publications, Delhi, India, pp. 1-10.

Devi, M.F., Singh, P.K. and Choudhury. M.D. (2014). Income generating plants of Keibul Lamjao National Park, Loktak Lake, Manipur and man-animal conflicts. Pleion. 8(1): 30-36.

Huang, W.Y., Cai, Y.Z. and Zhang, Y. (2010). Natural phenolic compounds from medicinal herbs and dietary plants: potential use for cancer prevention. Nutr. Canc. 62(1): 1-20.

Kumar, S., Behera, S.P. and Jena, P.K. (2013). Validation of tribal claims on *Dioscorea pentaphylla* L. through phytochemical screening and evaluation of antibacterial activity. Plant. Sci. Res. 35(1 and 2): 55-61.

Kumar, S., Das, G., Shin, H. S. and Patra, J.K. (2017a). *Dioscorea* spp. (A wild edible tuber): a study on its ethnopharmacological potential and traditional use by the local people of Similipal Biosphere Reserve, India. Front. Pharmacol. DOI: 10.3389/fphar.2017.00052.

Kumar, S., Mahanti, P., Singh, N.R., Rath, S.K., Jena, P.K. and Patra, J.K. (2017b). Antioxidant, antibacterial potential and characterization of active fraction of *Dioscorea pentaphylla* L. tuber extract collected from Similipal Biosphere

Reserve, Odisha, India. Braz. J. Pharma. Sci. 53(4), DOI: 10.1590/s2175-97902017000417006.

Sinha, S.C. (1987). Ethno botanical of Manipur-Medicinal plants. Front. Bot. 1: 123-152.

Tiwari, P., Kumar, B., Kaur, M., Kaur, G. and Kaur, H. (2011). Phytochemical screening and extraction: a review. Int. Pharma. Sci. 1(1): 98-106.

Tuboi, C., Babu, M.M. and Hussain, A.A. (2012). Plant species composition of the floating meadows of Keibul Lamjao National Park, Manipur. NeBI. 3(4): 1-11.

Chapter 8

Quantitative Assessment of Angiospermic Taxa in Similipal Biosphere Reserve, Odisha

T. Kar[1] and A.K. Biswal[2]*

[1] P.G. Department of Botany, Fakir Mohan University, Nuapadhi Balasore – 756 089, Odisha
[2]P.G. Department of Botany, North Orissa University, Takatpur, Baipada – 757 003, Odisha
**e-mail: truptirekha86@gmail.com*

ABSTRACT

Assessment of biodiversity is essential for estimating the conservation value of ecosystem in order to develop appropriate management measures. Similipal Biosphere Reserve is located in the Mayurbhanj district at the north-eastern corner of Odisha, which lies in the tri-junction of three states– Odisha, West Bengal and Jharkhand. The present study discusses the floristic composition and diversity of three main forest types (Semi-evergreen, Moist deciduous and Dry deciduous) of Similipal Biosphere Reserve. A total of 615 species belonging to 108 families were included in the analysis. The dominant families were Leguminosae, Orchidaceae, Rubiaceae, Poaceae, Euphorbiaceae etc. The Shannon-Wiener index (H') ranges from 0.96 to 4.03 and Simpson Index of dominance ranges from 0.03 to 0.43 showing species diverse system.

Keywords: *Assessment, Biosphere reserve, Floristic, Mayurbhanj, Similipal.*

Introduction

Phytodiversity refers to the diversity in plant species, their richness, abundance, distribution *etc.* Vegetation is the most precious gift, nature has provided to us, as it is meeting all kind of essential requirements of the humans in the form of food, fodder, fuel,medicine,timber,resins and oil *etc.* (Gaur, 1999). Plant community plays a pivotal role in sustainable management by maintaining biodiversity and conserving the environment. (Farooquee and Saxena, 1996). The knowledge of floristic composition of a plant community is a prerequisite to understand the overall structure and function of any ecosystem.

The Indian subcontinent, with its rich biodiversity, is one of the twelve mega –diversity centres of the world. Primary forests of Asia, particularly those of the Western Ghats and the Eastern Ghats of peninsular India are disappearing at an alarming rate due to anthropogenic activities and are replaced by forests comprising inferior species or their land use pattern changed (Bahuguna, 1999). In peninsular India quantitative phytodiversity inventories are available from the forest of the western Ghats whereas study of Eastern Ghats is meagre. Eastern Ghats are along chain of broken hills and elevated plateaus, running along the Indian east coast and passes through the state of Odisha, Andhra Pradesh, Tamil Nadu and two district of Karnataka state. The wide range of topography, varied climatic favour luxurious growth of diverse plant community and forest with number of invaluable medicinal plant species (Rawat, 1997). Despite a grand repository of phytodiversity, the studies on quantitative assessment of Eastern Ghats are not adequate. However few attempts have been made for such studies in Eastern Ghats (Kadavul and Parthasarathy, 1999a and b; Chittibabu and Parthasarathy 2000; Jaykumar *et al.,* 2002; Natarajan *et al.,* 2004; Reddy *et al.,* 2007; Sahu *et al.,* 2007; Mishra *et al.,* 2008; Reddy *et al.,* 2008; Sahu *et al.*, 2010). These kind of studies are poorly explored in the State of Odisha which covers 46 per cent of the forest in Eastern Ghats (Anonymous 2006). Hence the present study was under taken to determine the structure and floristic composition of different forest types in Similipal Biosphere Reserve, Eastern Ghats of Odisha India with special reference to rare,endemic and threatened plants.

Similipal Biosphere Reserve (SBR) is located in the Mayurbhanj district at the north-eastern corner of Odisha, which lies in the tri-junction of three states – Odisha, West Bengal and Jharkhand. It lies between 20° 17' to 22° 34'N latitude and 85° 40' to 87° 10' E longitude covers major portion in the central part of Mayurbhanj district (Figure 8.1). The Similipal Biosphere Reserve extends over an area of 5569.00 sq. Km. with a core area of *ca.* 1194.75 sq. Km., buffer area *ca.* 1335.88 sq. Km and transition area *ca.* 3038.39 sq.km. Similipal was declared as a biosphere reserve by Govt. of India on 22nd June 1994 due to its rich biodiversity and natural heritage. The hills, with their innumerable crests and valleys, interspersed with countless streams and rivers exhibits a great degree of topographic variation, ranging from 200 to 1166m above sea level. Geological formation of the region consists of sub-metamorphic sandstones and quartzite haematites. They produce reddish and sandy soils which favour the luxuriant growth of *Shorea robusta* most dominant tree of Similipal The climate of Similipal is warm and humid. The average rainfall of the area is 173 cm. Relative humidity is generally high throughout the year. The mean temperature varies from 20° to 28° C. (Saxena and Brahmam 1989).

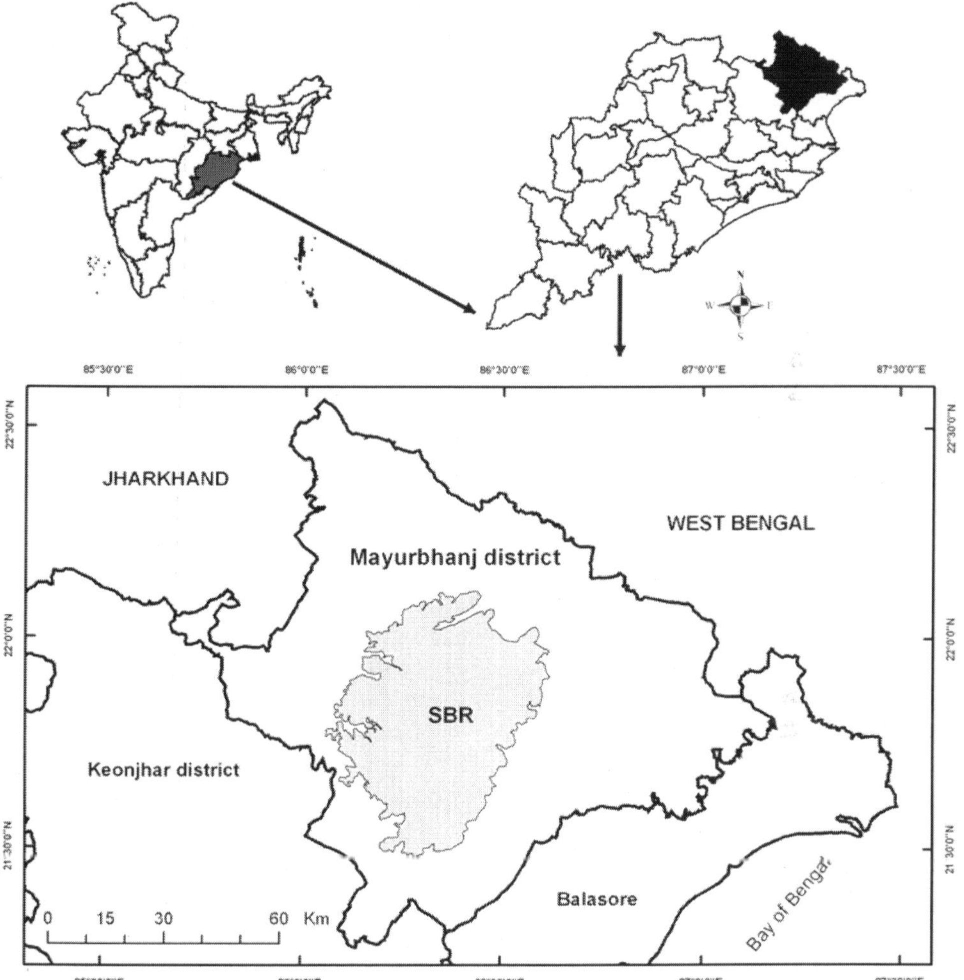

Figure 8.1: Map of Similipal Biosphere Reserve.

Vegetation of Similipal Biosphere Reserve is basically a tropical moist deciduous forest interspersed by some patches of semi-evergreen and mixed deciduous forest. Regular annual fires, forest clearing and perhaps damage of forest cover by storms and cyclones in the past has resulted grassland formation in several places and abundances of grasses in Similipal. As per the broad forest types of India by Champion and Seth (1968), the forest of Similipal is broadly classified as follows:

i. Northern Tropical Moist Deciduous Forest
ii. Northern Tropical Semi-Evergreen Forest
iii. Tropical Dry Deciduous Hill Forest
iv. Moist Peninsular High Level Sal Forest
v. Grassland and Savannah

Materials and Methods

Field Sampling

Vegetation analysis was carried out during the months of September 2013 to August 2015 in Similipal Biosphere Reserve. The number and size of quadrats needed were determined using the species area curve (Misra,1968). Sample plots (quadrat) size of 20x20m were laid down and systematically surveyed for all trees ≥30 cm girth at breast height (gbh above 130cm from the ground). Individual with less than 15cm were considered as saplings. The shrubs were sampled through 5x5m and herbs, climbers and saplings were sampled by 1x1m quadrats, which are nested within the quadrat of trees. Thus data were obtained from a total of 259 sample plots. Voucher specimens were collected for preparation of Herbarium and identified with the help of floras (Haines 1961; Mooney, 1950; Murty and Panigrahi, 1989; Saxena and Brahmam, 1994-96; Singh *et al.,* 2001) and deposited in the P.G. Department of Botany, North Orissa University, Takatpur, Baripada. The spatial data (Latitude, longitude and altitude) of each quadrat was collected using a Global Positioning System (GPS). Emphasis has been given to cover different elevation, slope, aspects, rainfall and temperature gradients to study overall spectrum of vegetation.

Data Analysis

The field data were quantitatively analysed for basal area, relative density, relative frequency and relative dominance (Phillip, 1959). The important value index (IVI) was determined as the sum of the relative frequency, relative density, and relative dominance (Cottam and Curtis, 1956). Some of the formulas are given below

Relative density = (Density of the species/Total density of all species) x 100.

Relative frequency = (Frequency of the species/Total frequency of all species) x 100.

Relative dominance = (Basal area of the species/Total basal area for all species) x 100.

Importance Value Index (IVI) = Sum of relative density + relative frequency + relative dominance.

Species diversity of this each forest type was determined (Shannon and Weiner, 1963)

$$H2 = -\Sigma\ [(ni/N)\ \log2\ (ni/N)]$$

Where, ni = Total no. of individuals of species i.

N = Total no. of individuals of all species in that vegetation type.

Concentration of dominance was also measured (Simpon, 1949)

$$C = -\ S\ (ni/N)$$

Where ni and N are the same as those for the Shannon– Weiner information function.

Result and Discussion

Floristic Composition

The floristic composition of Similipal Biosphere Reserve is unique and remarkable diversity in terms of taxa and habit. As many as 259 numbers of sample plots (*ca* 10.36 ha.) were selected in three main forest types *i.e.* Moist deciduous (160 plots), Semi-evergreen (65 plots) and Dry deciduous (34 plots). During the study, a total no. of 615 species belonging to 108 families was recorded in different forest types. Within the three forest types, the family with highest number of species was Leguminosae (43 species) followed by Orchidaceae (36 species), Rubiaceae (36 species), Poaceae (33 species), Euphorbiaceae (31 species), Acanthaceae (28 species) and Asteraceae (26 species). The number of species in different forest types is shown in Figure 8.2.

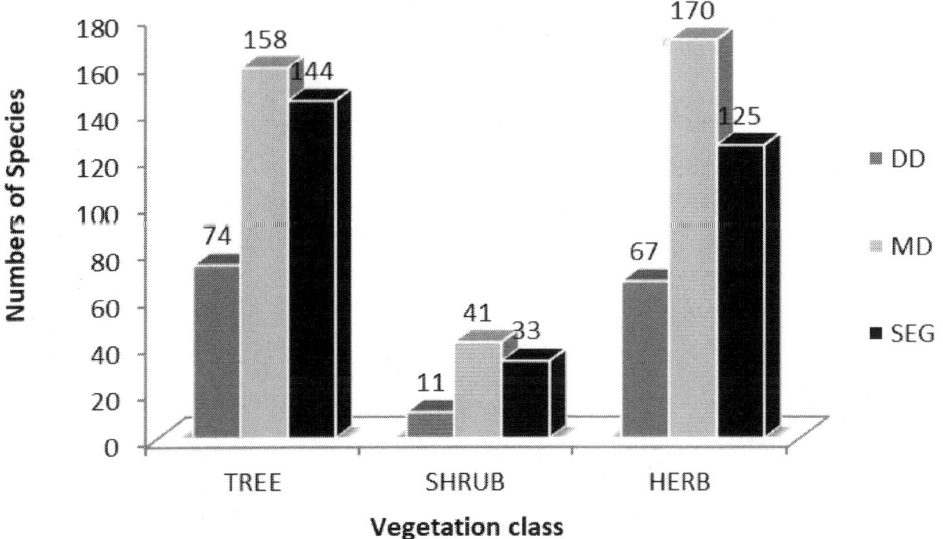

Figure 8.2: Number of Species in different Forest Types.

The Shannon-Weinner index (H') for various vegetation types in three different forest ranges from 0.96 to 4.03. Species diversity of tree layer is found to be 3.05, 3.38 and 3.9 for Dry deciduous, Moist deciduous and Semi-evergreen forest types respectively. The value of Simpson dominance index (Cd) for various vegetation types in three different forest ranges from 0.03 to 0.43. Dominance index of tree layer is found to be 0.13, 0.11 and 0.05 for Dry deciduous, Moist deciduous and Semi-evergreen forest types respectively.

Vegetation Analysis

The mean basal area of tree vegetation per hector was found to be 63.45 cm^2/ha for Semi-evergreen, 48.66 cm^2/ha for Moist deciduous and 36.01 cm^2/ha in Dry deciduous forest types. Mean density and mean basal area of tree species are found to be very high in Semi-evergreen forest type which are mostly located in the core region of Similipal Biosphere Reserve which has got adequate protection for their luxuriant growth. The mean basal area was found lower in Dry deciduous forest types which are located largely in buffer and transition zone of Similipal Biosphere Reserve.

Importance Value Index of *Shorea robusta* was found to be highest in all three forest types of Similipal Biosphere Reserve due to the higher contribution related factors such as relative frequency, relative density, and relative dominance (Table 8. 1). The IVI values of shrubs and herbs in three different forest types are shown in Tables 8.2 and 8.3 respectively.

Table 8.1:- IVI of the Ten Most Important Trees in Three different Forest Types

Plant Species	*Relative Density*	*Relative Frequency*	*Relative Dominance*	*IVI*
	Semi-evergreen			
Shorea robusta Gaertn.f.	17.06	5.29	13.82	36.16
Syzygium cumini (L.) Skeels	9.42	5.56	8.78	23.77
Michelia champaca L.	3.68	2.92	9.09	15.69
Terminalia alata Heyne ex Roth	5.05	4.45	4.12	13.62
Dillenia pentagyna Roxb.	3.73	4.17	4.54	12.45
Mangifera indica L.	2.58	2.50	6.83	11.92
Protium serratum (Wall.ex Colebr.) Engl.	3.96	3.20	3.67	10.83
Schleichera oleosa (Lour.) Oken	2.13	2.23	3.01	7.36
Diospyros malabarica (Desr.) Kostel.	2.35	1.81	2.33	6.49
Haldinia cordifolia (Roxb.) Ridsd.	1.61	1.95	1.66	5.21
	Moist Deciduous			
Shorea robusta Gaertn.f.	29.49	8.28	34.95	72.72
Terminalia alata Heyne ex Roth	8.64	7.16	8.11	23.90
Anogeissus latifolia (Roxb.ex DC) Wall.ex Guill. and Perr.	6.52	4.64	4.76	15.91

Plant Species	***Relative Density***	***Relative Frequency***	***Relative Dominance***	***IVI***
Syzygium cumini (L.) Skeels	4.02	4.17	2.84	11.04
Protium serratum (Wall.ex Colebr.) Engl.	2.76	3.18	3.35	9.29
Xylia xylocarpa (Roxb.) Taub.	2.27	2.25	3.64	8.16
Dillenia pentagyna Roxb.	2.12	3.05	2.86	8.04
Schleichera oleosa (Lour.) Oken	1.53	2.45	3.34	7.32
Haldinia cordifolia (Roxb.) Ridsd.	1.97	2.85	2.50	7.32
Madhuca indica Gmel.	1.88	2.39	2.99	7.25
	Dry Deciduous			
Shorea robusta Gaertn.f.	33.00	9.43	37.51	79.95
Terminalia alata Heyne ex Roth	5.71	6.92	5.47	18.09
Madhuca indica Gmel.	4.22	5.03	7.80	17.05
Buchanania lanzan Spreng.	5.83	5.66	4.08	15.57
Syzygium cumini (*L.*) Skeels	4.59	4.09	4.75	13.43
Anogeissus latifolia (Roxb.ex DC) Wall.ex Guill. and Perr.	3.97	3.46	5.82	13.25
Diospyros melanoxylon Roxb.	2.61	4.09	3.09	9.78
Schleichera oleosa (Lour.) Oken	1.61	2.52	4.20	8.33
Haldinia cordifolia (Roxb.) Ridsd.	2.23	3.46	2.54	8.23
Pterocarpus marsupium Roxb.	2.23	2.83	2.46	7.53

Table 8.2: IVI of the Ten most Important Shrubs in Three different Forest Types

Plant Species	***Relative Density***	***Relative Frequency***	***Relative Dominance***	***IVI***
	Semi-evergreen			
Ardisia solanacea Roxb.	23.77	24.62	3.01	52.20
Homonoia riparia Lour.	28.30	10.77	10.36	49.43
Flemingia semialata Roxb. Ex Aiton f.	10.19	7.69	5.22	23.10
Clerodendrum viscosum Vent	5.66	1.54	14.51	21.71
Gomphostemma parviflorum Wall. Ex Benth	4.91	1.54	12.57	19.02
Clerodendrum indicum (*L.*) Kuntze	3.40	1.54	8.71	13.64
Leea asiatica (*L.*) Ridsdale	3.77	3.08	4.84	11.69
Chassalia curviflora (Wall.) Thw.	1.89	1.54	4.84	8.26
Indigofera cassioides Rottl.ex DC.	1.89	4.62	1.61	8.11
Phoenix acaulis Buch.-Ham.ex Roxb.	1.89	4.62	1.61	8.11

Plant Species	*Relative Density*	*Relative Frequency*	*Relative Dominance*	*IVI*
	Moist Deciduous			
Ardisia solanacea Roxb.	17.73	17.12	3.01	37.85
Phoenix acaulis Buch.-Ham.ex Roxb.	14.64	10.81	3.94	29.39
Flemingia chappar Buch.-Ham.ex.Benth	12.91	8.11	4.63	25.64
Homonoia riparia Lour.	7.13	1.80	11.50	20.43
Clerodendrum viscosum Vent	5.78	7.21	2.33	15.32
Indigofera cassioides Rottl.ex DC.	4.82	7.21	1.94	13.97
Crotolaria spectabilis Roth	3.08	0.90	9.94	13.93
Hypericum gaitii Haines	4.43	1.80	7.15	13.38
Petalidium barlerioides (Roth) Nees	2.70	0.90	8.70	12.30
Flemingia strobilifera (L.) Aiton and Aiton f.	4.62	3.60	3.73	11.96
	Dry Deciduous			
Flemingia chappar Buch.-Ham.ex.Benth	17.24	6.25	23.95	47.44
Indigofera cassioides Rottl.ex DC.	15.52	25.00	5.39	45.91
Phoenix acaulis Buch.-Ham.ex Roxb.	17.24	12.50	11.98	41.72
Chromolaena odorata (L.) R.King and H.Robins	13.79	6.25	19.16	39.20
Mimosa himalayana Gamble	15.52	12.50	10.78	38.80
Clausena excavate Burm.f.	6.90	6.25	9.58	22.73
Breynia vitis-idaea (Burm.f.) Fischer	3.45	6.25	4.79	14.49
Ardisia solanacea Roxb.	3.45	6.25	4.79	14.49
Flemingia semialata Roxb. Ex Aiton f.	3.45	6.25	4.79	14.49
Grewia hirsuta Vahl.	1.72	6.25	2.40	10.37

Table 8.3: IVI of the Ten most Important Herbs in Three different Forest Types

Plant Species	*Relative Density*	*Relative Frequency*	*Relative Dominance*	*IVI*
	Semi-evergreen			
Rungia pectinata (L.) Nees	10.21	7.30	1.12	18.63
Elephantopus scaber L.	8.22	7.66	0.86	16.74
Alternanthera sessilis (L.) R.Br. ex DC.	4.93	0.36	10.86	16.15
Justicia betonica L.	4.33	2.55	1.36	12.00
Canscora diffusa (Vahl) R.Br. ex Roem. and Schult.	4.15	2.55	1.31	8.78
Vernonia cinerea (L.) Less.	3.72	4.38	0.68	8.67
Imperata cylindrica (L.) Raeuschel.	3.63	0.36	8.00	8.24

Plant Species	*Relative Density*	*Relative Frequency*	*Relative Dominance*	*IVI*
Andrographis paniculata (Burm.f.) Wallich ex Nees	3.37	3.65	0.74	8.01
Curculigo orchioides Gaertn.	2.85	1.46	1.57	7.77
Triumfetta rhomboidei Jacq.	2.60	0.36	5.71	7.13
	Moist Deciduous			
Elephantopus scaber L.	16.34	11.15	0.89	28.38
Rungia pectinata (L.) Nees	7.55	9.03	0.50	17.08
Cyperus niveus Retz.	2.80	0.18	9.54	12.52
Vernonia cinerea (L.) Less.	3.07	4.96	0.37	8.40
Curculigo orchioides Gaertn.	3.54	3.54	0.60	7.68
Cyrtococcum oxyphyllum (Steud.) Stapf	1.68	0.18	5.75	7.61
Andrographis paniculata (Burm.f.) Wallich ex Nees	3.30	3.72	0.54	7.56
Curcuma amada Roxb.	2.36	3.19	0.45	5.99
Ageratum conyzoides L.	2.76	1.95	0.86	5.57
Brachiaria distachiya (L.) Stapf	2.26	0.71	1.93	4.89
	Dry Deciduous			
Elephantopus scaber L.	11.84	9.52	1.90	23.26
Rungia pectinata (L.) Nees	9.60	9.52	1.54	20.66
Hybanthus enneaspermus (L.) F.v. Muell.	4.80	0.79	9.22	14.82
Vernonia cinerea (L.) Less.	4.80	7.14	1.02	12.97
Evolvulus nummularius (L.) L.	5.76	3.17	2.77	11.70
Evolvulus alsinoides (L.) L.	3.20	0.79	6.15	10.14
Canscora diffusa (Vahl) R.Br. ex Roem. and Schult.	4.16	1.59	4.00	9.74
Andrographis paniculata (Burm.f.) Wallich ex Nees	3.52	3.97	1.35	8.84
Imperata cylindrica (L.) Raeuschel.	3.68	1.59	3.54	8.80
Ajuga macrosperma Wall. ex Benth.	3.20	1.59	3.07	7.86

Conclusion

Vegetation is an integral component ecosystem, which indicates the effect of changing environmental conditions in an obvious and easily measurable manner. Thus, careful analysis of vegetation types is used as a mean of evaluating useful information about other component of ecosystem. The presence of high species richness and diversity indicates the uniqueness and potentiality of Similipal Biosphere Reserve for conservation of ecosystem. The problem of forest fire, grazing, fuel wood extraction, medicinal plants collection is observed in study area, which must be checked practically and highest priority should be given for habitat conservation. The present study can serve as baseline information for biodiversity conservation in Similipal Biosphere Reserve in particular and Eastern Ghats of Odisha in general. Further detailed study on biogeographic and ecological sketch of the Biosphere Reserve is suggested for better management of this world heritage.

Acknowledgements

The authors are thankful to the Regional Chief Conservator- cum- Field Director Similipal Tiger Reserve, Baripada for granting permission for the field work and Head P.G. department of Botany, North Orissa University, Takatpur, Baripada for providing infrastructure facilities to carry out this work successfully. We would also like to thank forest officials and local people for assisting during the field study.

REFERENCES

Anonymous. 2006: Biodiversity Characterization at Landscape Level in Eastern Ghats using Satellite Remote Sensing and Geographic Information System. Technical Report. Department of Space and Department of Biotechnology, Govt. of India. pp. 8-9.

Bahuguna, V.K. 1999: Forest fire prevention and control strategies in India. Int.For. Fire News, 20: 5-9

Champion, H.G. and Seth S.K. 1968: A Revised Survey of the Forest Types of India. Manager publications. Delhi, India. pp 100-150

Chittibabu, C. V. and Parthasarathy, N. 2000: Attenuated tree species diversity in human-impacted tropical evergreen forest sites at Kolli hills, Eastern Ghats, India. Biod. Cons. 9: 1493-1519.

Cottam, G. and Curtis, J.T. 1956: The use of distance measurement in phytosociologycal sampling. Ecol.37: 451-460.

Farooquee, N.A., and Saxena, K.G., 1996: Conservation and utilisation of medicinal plants in high hills of the central Himalayas. Environ. Conserve.23, 75-80

Gaur R.D., 1999: Flora of the district Garhwal northwest Himalaya (With ethnobotanical notes). Transmedia publication, Srinagar (Garhwal) India.

Haines, H.H. 1921-25: The Botany of Bihar and Orissa. Vol. I-III. London. Botanical Survey of India, Calcutta (Repn.Edn. 1961).

Jain, S.K. and Sastry, A.R.V. 1980: Threatened Plants of India A State-of-the-Art Report, BSI, and MAB committee, New Delhi.

Jayakumar, S., Arockiasamy, D. I. and Britto, S. J., 2002: Conserving forests in the Eastern Ghats through remote sensing and GIS – A case study in Kolli hills. Curr. Sci. 82: 1259-1267.

Kadavul, K. and Parthasarathy N. (1999 b): Structure and composition of wood species in tropical semi-evergreen forest of kalrayan hills, Eastern Ghats, India. Trop. Ecol. 40: 247-260.

Kadavul,K and N. Parthasarathy (1999 a): Plant biodiversity and conservation of tropical semi-evergreen forests in the Shervarayan hills of Eastern Ghats, India. Biod. Vons. 8: 421-439

Mace, G.M. and Stuart, S.N. 1994: Draft IUCN Red list Categories, Version 2.2 Species 21-22: 13-24.

Maruti, S.K. and Panigrahi, G. 1989 and 1999: Flora of Bilaspur District,M.P. Vol. I and II BSI, Kolkata.

Mishra, R. 1968: Ecology Workbook. Oxford and IBH Publishing Co., New Delhi, India. pp. 244-245.

Mishra, R.K., Upadhyay, V.P., and Mohanty, R.C. 2008: Vegetation ecology of Similipal Biosphere Reserve,Orisa,India. Appl. Ecol. and Env.Res. (2): 89-99.

Mooney, H.F. 1950: Supplement to the Botany of Bihar and Orissa, Catholic Press, Ranchi.

Natarajan, D., Britto, S. J., Nagamurugan, N., Soosairaj, S. and Arockiasamy, D. I. 2004: Identification of conservation priority sites using remote sensing and GIS – A casa study from Chitteri hills, Eastern Ghats, Tamil Nadu. Curr. Sci. 86: 1316-1323.

Nayar, M. P. and Sastry, A. R. K. 2000: Red data book of Indian Plants. Vol. I, II and III BSI, Kolkata.

Phillips, E. A.1959: Methods of Vegetation Study,Henri Holt Co Inc.

Rawat, G.S., 1997: Conservation status of forests and wildlife in Eastern Ghats, India. Environmental Conservation 24: 307-315

Reddy, C.S., Pattanaik, C.,Mahapatra, A., Biswal, A.K. 2007: Phytosociological observation on tree diversity of tropical forest of Similipal Biosphere Reserve, Orissa, India, Taiwania 52: 352-359

Reddy, C.S., Ugle, Prachi., Murthy, M.S.R., Sudhakar, S. 2008: Quantitative structure and composition of tropical forests of Mudumalai Wildlife Sanctuary, Western Ghats, India, Taiwania, 53(2): 150-156

Sahu, S.C., Dhal, N.K. and Bhadra, A.K. 2010: Arboreal Taxa Diversity of Tropical Forests of Gandhamardan Hill Range, Eastern Ghats, India: An approach to sustainable Biodiversity Conservation, Taiwania, 55(3): 208-215

Sahu, S.C., Dhal, N.K., Reddy, C.S., Pattanaik, C. and Brahmam,M. 2007: Phytosociological study of tropical dry deciduous forest of Boudh district,Orissa,India. Res. Jour.For.1: 66-72.

Saxena, H.O and Brahmam M.1994-96: Flora of Orissa Vol. I-IV. Orissa Forest Development Corporation Limited, Bhubaneswar. India.

Saxena, H.O. and Brahmam, M. 1989: Flora of Similipal Orissa. Regional Research Laboratory, Bhubaneswar.

Shannon, C.E. and Wiener, W. 1963: The Mathematical theory of communication. Univ. of Illinois Press, Urbana.

Simpson, E. M. 1949: Measurement of diversity. Nature 163: 688.

Singh, N.P., Mudgal, V., Khanna, K.K. 2001: Flora of Bihar, An analysis. BSI, Kolkata.

Chapter 9

High Frequency Plant Regeneration in *Curcuma aromatica* Ensuring Genetic Fidelity of Micropropagated Plants

*Reena Parida[1], Sakuntala Behura[1], Sujata Mohanty[2] and Sanghamitra Nayak[1]**

[1]*Centre for Biotechnology,*
Siksha 'O' Anusandhan (Deemed to be University), Bhubaneswar, Odisha
[2]*Ramadevi Women University, Bhubaneswar, Odisha*
**e-mail: sanghamitranayak@soa.ac.in*

ABSTRACT

An efficient high frequency clonal propagation protocol was developed for Curcuma aromatica Salisb., a highly valued traditional medicinal plant. The axillary bud explants of C. aromatica were explanted on Murashige and Skoog (MS) medium with 1-7 mg/l of 6-benzyladenine (BA), 0.5-1 mg/l of indole-3-acetic acid (IAA), 0.5 mg/l of indole-3-butyric acid (IBA), 0.5 mg/l of kinetin (KIN), 0.5 mg/l of naphthalene acetic acid (NAA) and 100 mg/l of adenine sulphate (Ads) in different combinations. The maximum number of shoots per explant was 7.4±0.3 achieved on media containing BA 1 mg/l and KIN 0.5 mg/l and roots per shoot was 5.1±0.5 on BA 1 mg/l and IBA 0.5 mg/l. The stability in phytomedicinal yield potential of micropropagated plants was assessed through its rhizome essential oil and curcumin content compared to conventional plants which had similar profile. Genetic integrity of in vitro and conventional plants was evaluated using random amplified polymorphic DNA (RAPD) and inter simple sequence repeats (ISSR) primers. A total number of 4020 bands in 19 RAPD primers and 3960 bands in 9 ISSR primers were produced showing monomorphic banding patterns across all plants analyzed. There was no somaclonal variations found up to two generations in tissue culture plantlets. The present micropropagation protocol could be applied for in vitro conservation and large scale production of true-to-type Curcuma aromatica for commercial use.

***Keywords**: Micropropagation, Curcuma aromatica, Genetic stability, Essential oil, Curcumin.*

Introduction

*Curcuma aromatic*a Salisb., commonly known as wild turmeric, is an under exploited medicinal plant belonging to family Zingiberaceae. This species is mainly found in the South Asian regions, predominantly in Eastern Himalayas and South India. Its rhizome is orange yellow in colour and bears yellow colour flower. The rhizome powder is used in flavouring food and as colouring agent throughout the world. The rhizome is a rich source of volatile essential oil responsible for its pleasant aroma and its colour is due to the presence of curcumin content. *C. aromatica* has been reported for various medicinal activities such as promoting blood circulation to remove blood stasis and is also useful in treatment of early stages of cervical cancer due to its active principle of curcumin (Hasmeda and Polya, 1996; Wu *et al.,* 2000). Its rhizome is also useful as anti-dote for snakebite, cure sprains, bronchitis, cough and skin infections. The rhizome essential oil is antihelmintic, antifungal, antimicrobial and bears antioxidant activity (Asolkar *et al.,* 1992; Srvidya *et al.,* 2009). The conventional method for vegetative propagation of *C. aromatica* is not satisfactory because of slow multiplication rate, rhizome rot diseases in field and storage conditions. Due to the growing demand of this important plant and indiscriminate collection there is unavailability of requisite quality planting material. Thus it was felt necessary to develop a high frequency plant regeneration system for commercial production of *C. aromatica* through tissue culture method. The regenerants arising after long term *in vitro* conservation are susceptible to genetic changes due to culture stress and mode of regeneration. For this reason, monitoring the degree of genetic stability of *in vitro* grown plants is of utmost importance for commercial utilization of tissue cultured plants of the desired genotype. Polymerase chain reaction (PCR) based techniques like randomly amplified polymorphic DNA (RAPD), inter simple sequence repeats (ISSR), simple sequence repeats (SSR), amplified fragment length polymorphism (AFLP) are being widely used to confirm the genetic fidelity of micropropagated plants (Bairu *et al.,* 2011; Aversano *et al.,* 2011; Nayak *et al.,* 2011). Among these DNA markers, RAPD and ISSR markers are simpler, quicker and powerful methods for the evaluation of genetic integrity of tissue culture derived plantlets. Efficient micropropagation protocols have been reported in several *Curcuma* species including *C. longa, C. amada, C. aromatica* and *C. zedoaria* (De Souza Ferrari *et al.,* 2016; Barthakur and Bordoloi 1992; Sharmin *et al.,* 2013; Anisuzzaman *et al.,* 2008). Ideally no single technique being sufficient to evaluate the stability of micropropagated plants, a combination of several techniques should be used (Peredo *et al.,* 2009). The present work was aimed to develop high frequency plant regeneration protocol through axillary bud formation in *C. aromatica*. Further evaluation of genetic stability of tissue culture derived plants was carried out by RAPD and ISSR analysis. The gas chromatography and mass spectrometry (GC-MS) of essential oil and curcumin content of both micropropagated and conventionally grown plants were analyzed. Thus this technique would be of enough significance for large scale production of genetically uniform plantlets of *Curcuma aromatica* with stable phytomedicinal yield potential for commercial use.

Materials and Methods

Plant Material and Culture Conditions

Healthy rhizomes of *C. aromatica* were collected from Pottangi, Koraput district, Odisha and were maintained in the greenhouse of Centre for Biotechnology, Siksha 'O' Anusandhan University. The immature dormant rhizome axillary buds were excised with sharp blade and were properly washed with distilled water to remove dirt and then soaked in liquid detergent (Extran, Merck, India) for 3 min and subsequently rinsed thrice with distilled water, which were used as the explants. The surface sterilization was done by 0.1 per cent mercuric chloride for 3-5 min in a laminar flow cabinet and rinsed thrice in sterile distilled water to remove traces of sterilants prior to inoculation. These explants were inoculated on basal Murashige and Skoog (MS) (1962) medium supplemented with various combination of benzyl adenine (BA), kinetin (KIN), indole-3-acetic acid (IAA), indole-3-butyric acid (IBA), naphthalene acetic acid (NAA) and adenine sulphate (Ads) with 30 g/l of sucrose and 0.8 per cent of agar. The pH of the medium was adjusted to 5.7 before adding agar and the media were autoclaved at 121 °C and 105 kg/cm^2 of pressure for 20 min. A total of 15 replications per treatment were used. Micropropagated plantlets were maintained with regular sub culturing at 90 days interval. *In vitro* regenerated plants with well-developed shoots and roots were then transferred to pots containing soil, cow dung and sand mixture in 1:1:1 ratio and then to greenhouse for acclimatization and maturity. The percentage of survival was noted. Two years of field grown tissue culture derived and conventionally propagated plantlets of *C. aromatica* were compared for different biochemical and molecular analysis.

Extraction of Essential Oil and GC-MS Analysis

Matured rhizomes on harvest were collected from both conventional and micropropagated plants of *C. aromatica* and washed under running tap water to remove soil and sliced for immediate extraction of rhizome essential oil. Sliced rhizomes of about 100 g were mixed with distilled water and its essential oil was extracted by hydro-distillation using a Clevenger's apparatus following the method (Guenther, 1972). A flask containing the sliced rhizomes was heated up to 6 h and the condensed vapour was separated in oil-water surface. The oil present at the upper most layers was collected in a micro centrifuge tube through the collecting chambers and was stored at 4 °C prior to analysis. The component identification was achieved by the GC-MS analysis using HP 6890 series GC (Hewlett-Packard, USA) coupled with a mass selective detector (MSD), HP 5973 series (Hewlett-Packard). Helium was used as a carrier gas and the sample was injected in split less mode in a column HP5 phenyl methyl siloxane (30 m length of the column × 25 μm film thickness × 320 μm internal diameter). Mass spectra were acquired over a 40-400 atomic mass unit range. Compounds were identified by comparing the mass spectral data with those in the NIST library provided with software and with commercially available data. Temperature programming was: initial temperature 60 °C, ramping rate 3 °C, final temperature 243 °C, run time 61 min. For GC-MS evaluation, 15 plants from micropropagated and conventionally grown were randomly selected.

Curcumin Analysis

The rhizomes from both conventional and micropropagated plants were dried in shade condition and were powdered to uniform mesh and subsequently extracted with methanol for 6 h using a soxhlet apparatus. The prepared extract was filtered, concentrated using rotary evaporator and stored in airtight container at 4 °C until analysis. The dried extract (0.5 g) was mixed in 50 ml of HPLC grade methanol. The stock solution of standard (curcumin, 0.2 mg/ml) was prepared by dissolving 1 mg of curcumin in 5 ml of methanol. All the solvent used in this study were analytical grade. Standard and sample were filtered through 0.22 micron membrane filter prior analysis.

A modular HPLC system (Shimadzu, Kyoto, Japan) consisting of two LC-20 AD pump, a Rheodyne 8125 injector, an SPD-20A diode array detector, a CBM-20A controller and CT0-20AC column oven was used for estimating curcumin content. The separation was carried out on a reverse phase Shimpak C18 column (Shimadzu, Kyoto, Japan, 250×4.6 mm ID, 5 μm). The mobile phase was composed of acetonitrile (A) and water containing 0.4 per cent (v/v) acetic acid (B). The gradient program was as follows: 45 per cent A at 0–13 min, 45–56 per cent A at 13–16 min, 56 per cent A at 16–50 min, and 56–100 per cent A at 50–55 min. Elution was carried out at a column temperature of 35 °C, a flow rate of 1.0 ml/min and a run time of 55 min. The injection volume was 20 μl. The detection wavelength was set at 430 nm.

Isolation of Genomic DNA, RAPD and ISSR Analysis

Genomic DNA from fresh 2 g of young leaves was isolated from conventional and micropropagated plants using CTAB DNA extraction method (Doyle an Doyle, 1996). The quantification of DNA was done based on spectrophotometric analysis and the samples were diluted to a concentration of 25 ng lL-1, kept at –20 °C for further analysis. Amplifications were carried out in a reaction mixture of 25 μl containing 25 ng of genomic DNA as a template, 10X PCR buffer mixed with 15 mM $MgCl_2$ (Bangalore Genei), 200 μM dNTPs (Bangalore Genei), 3 U/μl of Taq DNA polymerase (Bangalore Genei) and 5 pM of each primer. The amplification of the reaction was carried out in a thermal cycler (Applied Biosystems Veriti 96 Well Thermal Cycler) under the following program: initial denaturation for 5 min at 94 °C, followed by 40 cycles at 94 °C for 1 min (denaturation) then annealing temperature for 1 min and then 72 °C for 2 min (extension), followed by final extension at 72 °C for 7 min. A total of 25 primers for RAPD and 12 primers for ISSR were used for analysis out of which 19 RAPD primers and 9 ISSR primers were selected based on their reproducibility in banding pattern following the standard protocol (Williams *et al.*, 1990; Zeitkiewicz *et al.*, 1994). The PCR products were electrophoresed on 1.5 per cent and 2 per cent agarose gel for RAPD and ISSR with 3000 bp DNA marker (MBI Fermentas, Lithuania), documented using Gel doc system (Germany). The analysis was repeated twice using each primer to establish reproducibility of the banding pattern from different DNA samples of *C. aromatica.*

Results

Establishment of *In vitro* Culture

In *Curcuma aromatica,* the dormant axillary buds from rhizome of field grown plants were taken as explants (Figure 9.1A) for the culture initiation on MS basal medium supplemented with BA alone and with different combinations of BA, KIN, IAA, IBA, NAA and Ads. Three to five lateral shoots were developed from the basal region of the explant within 20-25 days of culture (Figure 9.1B). The MS media

Table 9.1: *In vitro* Shoot Multiplication in *Curcuma aromatica* using different Plant Growth Regulators

Sl.No.	MS Media + Growth Regulators (mg/l)	Per cent of Shoot Initiation (Mean ± SE)	No.of Shoots/ Explant (Mean ± SE)	No. of Roots/ Plant (Mean ± SE)
1	BA(1)	61.8 ± 0.4	4.1 ± 0.3	1.2 ± 0.4
2	BA(3)	71.8 ± 0.5	6.3 ± 0.2	2.4 ± 0.2
3	BA(5)	87.2 ± 0.8	3.2 ± 0.2	4.4 ± 0.4
4	BA(7)	59.8 ± 0.4	2.4 ± 0.3	1.8 ± 0.5
5	BA(1)+IAA(0.5)	41.6 ± 0.6	2.2 ± 0.2	1.8 ± 0.2
6	BA(3) +IAA(0.5)	66.6 ± 0.5	4.6 ± 0.6	1.8 ± 0.5
7	BA(3)+IAA(1)	73.2 ± 0.5	3.8 ± 0.5	2.2 ± 0.2
8	BA(1)+KIN(0.5)	62.2 ± 0.6	7.4 ± 0.3	1.4 ± 0.4
9	BA(3)+ KIN(0.5)	68.2 ± 0.5	6.8 ± 0.2	2.6 ± 0.2
10	BA(1)+IBA(0.5)	28.8 ± 0.6	2.4 ± 0.3	5.1 ± 0.5
11	BA(3)+IBA(0.5)	32.6 ± 0.4	3.4 ± 0.2	4.8 ± 0.3
12	BA(1)+NAA(0.5)	51.8 ± 0.8	3.5 ± 0.6	1.0 ± 0.3
13	BA(3) +NAA(0.5)	56.8 ± 0.5	5.4 ± 0.6	2.8 ± 0.2
14	BA(3)+IAA(0.5)+ADS(100)	52.6 ± 0.8	3.1 ± 0.3	0.6 ± 0.2
15	BA(3)+IAA(1)+ADS(100)	58.6 ± 0.5	4.0 ± 0.3	0.8 ± 0.2

Data represent the mean of 15 replicates for each treatment.

Figure 9.1: A: *Curcuma aromatica* Explant; B: Shoot Induction; C: Shoot Multiplication with Roots; D: Potted Plant; E: Tissue Cultured Plants in Greenhouse Condition.

with BA alone was not enough to produce maximum number of shoots in *Curcuma aromatica*. The medium supplemented with BA 1 mg/l and KIN 0.5 mg/l was the most effective for multiple shoot formation producing an average of 7.4±0.3 number of shoots (Table 9.1). When the concentration of BA was increased the initiation of shoots decreased. The number of roots was maximum on BA 1 mg/l and IBA 0.5 mg/l was 5.1±0.5 (Figure 9.1C). The multiplication rate did not change during the subsequent subculture up to 2 years and about 95 per cent of plants survived, had normal growth after transferring them to the field during the planting season (Figures 9.1D-E).

Assessment of Phytomedicinal Yield Potential of Regenerants

Stable phytomedicinal yield potential of micropropagated *Curcuma aromatica* was found through GC-MS and curcumin content analysis. The yield of rhizome oil in both conventional and micropropagated *C. aromatica* plantlets were 0.51 per cent and 0.53 per cent respectively. The GC-MS analysis identified 26 compounds in conventionally propagated and micropropagated plants details (Table 9.2). Compounds were identified by comparing the mass spectral data with those in the NIST Wiley library provided with software and with commercially available data. The compounds present in both cases were quite similar to each other revealing camphor to be the major compound in both conventional and micropropagated *C. aromatica*, followed by curdione, 1, 8- cineole, borneol, camphene and others. Similarly the percentage of mean curcumin content was found to be 0.013 per cent in conventionally propagated and 0.016 per cent in micropropagated plants of *C. aromatica*, respectively.

Table 9.2: Chemical Composition of Rhizome Oil in Conventional and Micropropagated *Curcuma aromatica*

Sl.No.	*Compound Name*	*RT*	*Peak Area per cent*	
			Conventionally Propagated (Mean ± SD)	*Micropropagated (Mean ± SD)*
1	α-pinene	5.63	1.42 ± 0.01	1.46 ±001
2	Camphene	6.11	6.03 ±0.01	6.81 ± 0.01
3	Sabinene	6.86	0.45 ± 0.03	0.51 ± 0.02
4	β-pinene	7.14	0.78 ± 0.01	0.71 ± 0.01
5	Limonene	8.53	1.82 ± 0.06	2.15 ± 0.03
6	1,8-cineole	8.69	6.51 ±0.03	6.11 ± 0.05
7	Linalool	11.16	2.33 ± 0.01	3.14 ± 0.01
8	Camphor	13.20	16.02 ± 0.13	16.51 ± 0.95
9	Isoborneol	13.40	0.65 ± 0.01	0.62 ± 0.02
10	Borneol	13.81	6.08 ± 0.09	6.03 ± 0.05
11	Terpinen-4-ol	14.44	0.38 ± 0.01	0.31 ±0.03

Sl.No.	Compound Name	RT	Peak Area per cent	
			Conventionally Propagated (Mean ± SD)	Micropropagated (Mean ± SD)
12	α-terpineol	15.06	0.95 ± 0.03	0.91 ±0.01
13	β-elemene	23.28	1.50 ± 0.31	1.20 ±0.27
14	α-humelene	25.92	0.44 ± 0.01	0.52 ±0.02
15	Cis-muurolene	26.75	0.41 ±0.03	0.53 ±0.02
16	γ-muurolene	26.98	0.86 ± 0.01	0.91 ± 0.01
17	Cis-β-guaiene	27.27	0.55 ± 0.02	0.51 ±0.07
18	Curzerene	27.53	2.06 ± 0.15	2.02 ± 0.01
19	β-Himachalene	28.03	0.59 ± 0.01	0.52 ±0.02
20	α-muurolene	28.49	0.68 ±0.06	0.71 ± 0.03
21	Germacrene-B	30.02	0.71 ±0.02	0.76 ± 0.01
22	Cis-β-elemene	32.28	0.94 ± 0.01	0.86 ± 0.02
23	Cubinol	33.81	0.98 ± 0.01	0.91 ± 0.01
24	Germacrone	35.19	5.38 ± 0.15	4.95 ±0.09
25	Curdione	36.06	12.13 ± 0.07	12.36 ± 0.05
26	Neo-curdione	37.19	2.77 ± 0.12	2.82 ±0.06

RAPD and ISSR Analysis

Genetic stability of micropropagated plants of *C. aromatica* was assessed through RAPD and ISSR analysis. Out of 25 RAPD primers used, 19 primers gave a total of 67 scorable bands ranging from 200-1900 bp (Table 3A). Number of bands were varied from 1-10, with an average of 3.5 per primer. A total of 4020 bands [(number of plantlets analyzed) x (number of bands with all primers)] generated by the RAPD techniques showed monomorphic patterns across all 60 plantlets analyzed (Figures 9.2A,B). Highest number of monomorphic band was 10 in primer OPC11 (ranging from 600-1900bp) whereas lowest number of band was 1 in primer OPA7 (1475bp), OPD12 (900bp) and OPAF15 (650bp). In ISSR analysis 9 primer gave rise to a total of 66 bands ranging from 250-2000bp which were all monomorphic without any variations. Number of bands in all primers were varied from 5-11, with an average of 7.3 bands per primer (Table 9.3B). A total of 3960 bands [(number of plantlets analyzed) x (number of bands with all primers)] were generated by the ISSR techniques, giving rise to monomorphic patterns across all 60 plantlets analyzed (Figure 9.2C), highest 11 in primer (GGA)4 (ranging from 300-1500bp) and lowest 5 in (GTGC)4 (ranging from 300-1050bp) and T(GA)9 (ranging from 600-1100bp).

Table 9.3A: List of Primers, their Sequences, Total Bands and Size of Amplified Fragments Generated by RAPD Primers in both Micropropagated and Conventionally grown Plants of *Curcuma aromatica*

Primer Name	*Primer Sequence*	*Total Bands Amplified*	*Range of Amplicons [bp]*
OPA4	AATCGGGCTG	4	600-1500
OPA7	GAAACGGGTG	1	1475
OPA9	GGGTAACGCC	2	1200-1400
OPA18	AGGTGACCGT	5	200-1075
OPC2	GTGAGGCGTC	5	650-1050
OPC5	GATGACCGCC	2	1700-1900
OPC11	AAAGCTGCGG	10	600-1900
OPD3	GTCGCCGTCA	2	1031-1350
OPD7	TTGGCACGGG	5	450-1250
OPD8	GTGTGCCCCA	3	1031-1800
OPD12	CACCGTATCC	1	900
OPD18	GAGAGCCAAC	2	750-1031
OPD20	ACCCGGTCAC	7	400-1900
OPN4	GACCGACCCA	3	500-1900
OPN16	AAGCGACCTG	5	300-1200
OPN18	GGTGAGGTCA	4	450-1050
OPAF5	CCCGATCAGA	2	200-1050
OPAF14	GGTGCGCACT	3	900-1200
OPAF15	CACGAACCTC	1	650
	Total	**67**	

Discussion

Several Zingiberaceae species have been reported for successful micropropagation (Tyagi *et al.*, 2006; Anish *et al.*, 2008; Lincy *et al.*, 2009; Mohanty *et al.*, 2011). Maximum shoot regeneration potential of *C. aromatica* was on MS media in BA and KIN combinations. The use of BA in development of healthy root system in *C. aromatica* has been reported by Nayak (2000, 2002). In the present study, BA (1 mg/l) and KIN (0.5 mg/l) induced maximum shoot multiplication. In the study of Roy and Raychaudhuri, (2004) 4 mg/l of NAA and 5 mg/l of KIN was responsible for best multiplication in *C. aromatica.* Mohanty *et al.* (2008) reported plant regeneration in *C. aromatica* from callus culture and *in vitro* detection of somaclonal variation through cytophotometric analysis. In their study 2,4D and KIN was responsible for callus formation whereas BA and NAA produced maximum number of shoots. The efficiency for BA for shoot regeneration has been reported earlier in several Zingiberaceae species (Balachandra *et al.*, 1990;

Table 9.3B: List of Primers, their Sequences, Total Bands and Size of Amplified Fragments Generated by ISSR Primers in both Micropropagated and Conventionally Grown Plants of *Curcuma aromatica*

Primer Name	*Primer Sequence*	*Total Bands Amplified*	*Range of Amplicons [bp]*
SPS1	(GAC)5	9	350 - 1031
SPS 2	(GTGC)4	5	300 - 1050
SPS 3	(GACA)4	8	475 - 900
SPS 4	(AGG)6	6	400 - 1031
SPS 5	(GA)9T	6	450 - 975
SPS 6	T(GA)9	5	600 - 1100
SPS 7	(GTG)5	8	250 - 850
SPS 8	(GGA)4	11	300 - 1500
SPS 9	(CAA)5	8	600 - 2000
	Total	**66**	

Tyagi *et al.,* 2004; Shirin *et al.,* 2000; Nayak 2002; Sushma *et al.,* 2005). Similarly KIN is also efficient for shoot multiplication in Zingiberaceae as reported by many workers (Borthakur *et al.,* 1999; Selvakkumar *et al.,* 2007; Badoni *et al.,* 2010). MS media with different combinations of KIN and NAA when tried for *in vitro* shoot multiplication in presence or absence of BA in medium showed no significant increase in multiplication rate as reported in *Z. officinale* (Sharma and Singh, 1997). In our study rooting was observed in 4 week old culture on the multiplication media tested for this species which is in close agreement with reports available in other species of Zingiberaceae (Borthakur *et al.,* 1999; Nayak 2000; Tyagi *et al.,* 2006; Panda *et al.,* 2007). Further our results showed similarity in yield of essential oil and curcumin content in both micropropagated and conventionally grown plants. Similarly, the higher uniformity of chemical components was found in rhizome essential oil of *Curcuma longa* micropropagated and conventional plants (Nayak *et al.,* 2011; Singh *et al.,* 2011). Essential oil from rhizome contained camphor as the major compound in both micropropagated (18.91 per cent) and conventional plants (16.02 per cent). Our report is similar with others who also reported camphor as the major compound (28 per cent) followed by 1, 8-cineole (20 per cent) in *C. aromatica* essential oil. This variation might be due to climatic, geographic or genotypic differences that affects the chemical composition of plant essential oil. There were no remarkable changes in essential oil yield percentage and curcumin content between micropropagated and conventionally propagated plants of *C. aromatica*, thus indicating the stability in phytomedicinal yield potential of the regenerants. Similarly, biochemical characterization of tissue cultured plants has been documented by other researchers revealing stability in phytomedicinal yield potential (Nayak *et al.,* 2011; Singh *et al.,* 2011). Curcumin an active biological

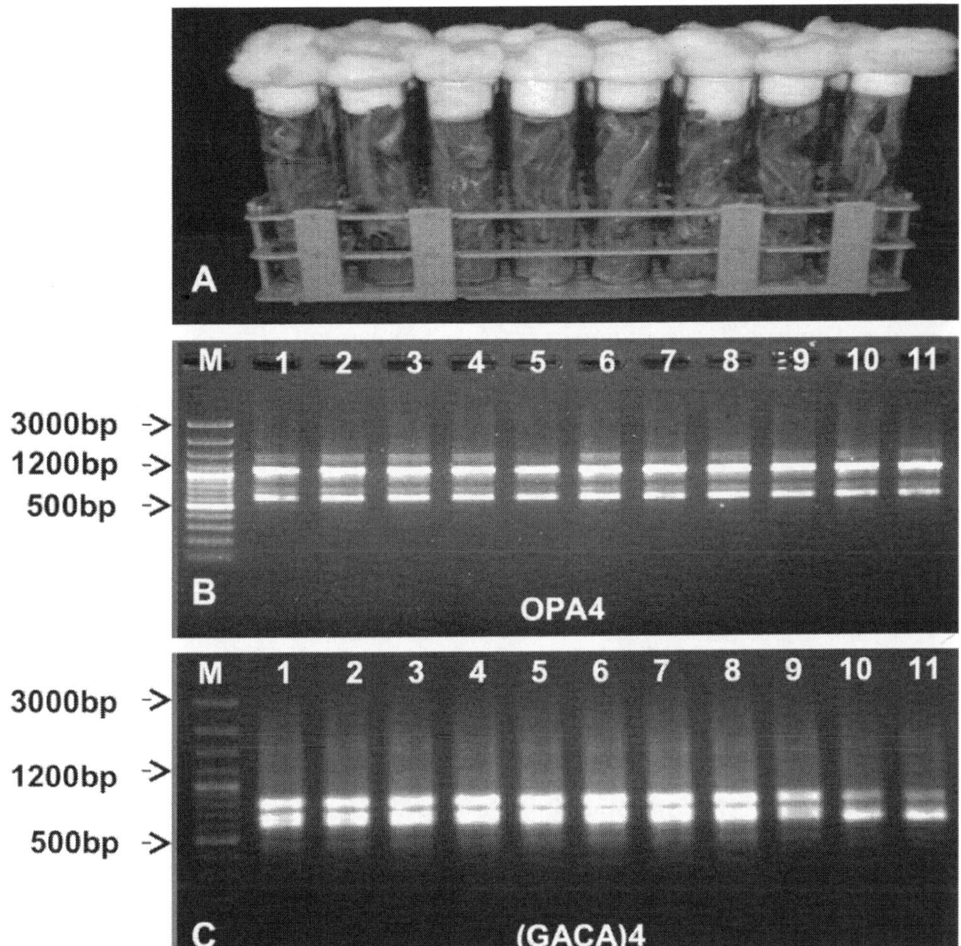

Figure 9.2: A: *In vitro* Conserved Plantlets; B, C: RAPD and ISSR Banding Pattern in both Micropropagated and Field grown Mother Plants of *Curcuma aromatica* (Lane 1, mother plant; Lane 2-11 micropropagated plants).

component extracted from rhizomes of several *Curcuma* species bears potent biological activities (Lin and Lin-Shiau 2001; Woo *et al.*, 2003; Anand *et al.*, 2008).

Genetic stability of microprapagated *C. aromatica* was confirmed through molecular analysis using RAPD and ISSR primers which revealed monomorphic banding pattern in each case and were similar to that of source plants. RAPD and ISSR marker based genetic integrity of *in vitro* regenerants have been documented in several other important medicinal plants such as *Kaempferia galanga* (Mohanty *et al.*, 2011), *Gerbera jamesonii* (Bhatia *et al.*, 2009) and *Swertia chirayita* (Joshi and Dhawan, 2007).

In conclusion, the present protocol of rapid plant regeneration with genetically uniform plantlets and stable phytomedicinal yield potential can be recommended for effective mass propagation of true-to-type *C. aromatica* for its commercial utilization.

Acknowledgements

The authors are grateful to Prof (Dr.) M.R. Nayak, President and Prof (Dr.) S.C. Si, Dean, Centre for Biotechnology, Siksha 'O' Anusandhan University for providing facilities and encouraging throughout.

REFERENCES

Anand, P., Thomas, S. G., Kunnumakkara, A. B., Sundaram, C., Hari- kumar, K. B., Sung, B., Tharakan, S. T., Misra, K., Priyadarsini, I. K., Rajasekharan, K. N. and Aggarwal, B. B. (2008). Biological activities of curcumin and its analogues (Congeners) made by man and mother nature. Biochem Pharmacol 76: 1590–1611.

Anish, N. P., Dan, M. and Bejoy, M. (2008). Conservation using *in vitro* progenies of the threatened ginger-*Boesenbergia pulcherrima* (Wall.) Kuntze. Int J of Botany, 4(1): 93-98.

Anisuzzaman, M., Sharmin S. A., Mondal, S. C., Sultana, R., Khalekuzzaman, M., Alam. I., Alam, M. F. (2008). *In vitro* microrhizome induction in *Curcuma zedoaria* (Christm.) Roscoe-a conservation prioritized medicinal plant. J Biol Sci, 8: 1216–1220.

Asolkar, L. V., Kakkar, K. K. and Chakre, O. J. (1992). In: Second supplement to Glossory of Indian Medicinal plants with active principles, Part-I. New Delhi: Publication and Information Directorate, CSIR: 246-248.

Aversano, R., Di Dato, F., Di Matteo, A., Frusciante, L., and Carputo, D. (2011). AFLP analysis to assess genomic stability in Solanum regenerants derived from wild and cultivated species. Plant Biotechnol Rep, 5: 265-271.

Badoni, A., Bisht, C. and Chauhan, J. S. (2010). Micropropagation of *Hedychium spicatum* Smith using *in vitro* shoot tip. Stem cell, 1(1): 11-13.

Bairu, M. W., Aremu, A. O. and Van Staden, J. (2011). Somaclonal variation in plants: causes and detection methods. Plant Growth Regul, 63: 147–173.

Balachandran, S. M., Bhat, S. R and Chandel, K. P. S. (1990). *In vitro* clonal multiplication of turmeric (*Curcuma longa*) and ginger (*Zingiber officinale* Rosc.). Plant Cell Rep, 3: 521–524.

Barthakur, M. P. and Bordoloi, D. N. (1992). Micropropagation of *Curcuma amada* (Roxb.). J Spices Aromat. Crop, 1: 154–156.

Bhatia, R., Singh, K. P., Jhang, T. and Sharma, T. R. (2009). Assessment of clonal fidelity of micropropagated gerbera plants by ISSR markers. Sci Hort 119: 208–211.

Borthakur, M., Hazarika, J. and Singh, R.S. (1999). A protocol for micropropagation of *A.galanga*. Plant Cell, Tiss and Org Cult, 55: 231-233.

De Souza Ferrari, M. P., Antoniazzi, D., Nascimento, A. B., Franz, L. F., Bezerra, C. S. and Magalhães, H. E. M. (2016). Evaluation of new protocols to *Curcuma longa* micropropagation: a medicinal and ornamental specie. J Med Plant Res, 10: 367–376

Doyle, J. J. and Doyle, J. L. (1987). A rapid DNA isolation procedure for small quantities of fresh leaf tissue. Phytochem Bull 19: 11–15.

Guenther, E. (1972). The production of essential oils. In: Robert E (ed) The essential oils, vol I. Krieger, New York, pp 361–391.

Hasmeda, M. and Polya, G.M. (1996). Inhibition of cyclic AMP-dependent protein kinase by curcumin. Phytochem, 42(3): 599-605.

Joshi, P. and Dhawan, V. (2007). Assessment of genetic fidelity of micropro- pagated *Swertia chirayita* plantlets by ISSR marker assay. Biol Plant 51: 22–26.

Lin, J. K. and Lin-Shiau, S. Y. (2001). Mechanisms of cancer chemoprevention by Curcumin. Proc Natl Sci Counc Repub China B 25: 59–66.

Lincy, A.K, Remashree, A.B. and Sasikumar, B. (2009). Indirect and direct somatic embryogenesis from aerial stem explants of ginger (*Zingiber officinale* Rosc.), 68(1): 93-103.

Mohanty, S., Panda, M. K., Sahoo, S. and Nayak, S. (2011a). Micropropagation of *Zingiber rubens* and assessment of genetic stability through RAPD and ISSR markers. Biol Plant, 55: 16–20.

Mohanty, S., Panda, M.K., Subudhi, E., Acharya,L. and Nayak, S.(2008). Genetic stability of micropropagated ginger derived from axillary bud through cytophotometric and RAPD analysis. Z Naturforsch., 63: 747-754.

Mohanty, S., Parida, R., Singh, S., Joshi, R. K., Subudhi, E. and Nayak, S. (2011b). Biochemical and molecular profiling of micropropagated and con- ventionally grown *Kaempferia galanga*. Plant Cell Tissue Organ Cult, 106: 39–46.

Murashige, T. and Skoog, F. (1962). A revised medium for rapid growth and bioassays with tobacco tissue culture. Physiol Plant, 15: 473-79.

Nayak, S. (2000). *In vitro* multiplication and microrhizome induction in *Curcuma aromatica* Salisb. Plant Growth Reg. 32: 41-47.

Nayak, S. (2002). *In vitro* clonal multiplication and short term conservation of four promising cultivars of turmeric. Plant Sci. Res., 24: 7-10.

Nayak, S., Kaur, T., Mohanty, S., Ghosh, G., Choudhury, R., Acharya, L. and Subudhi, E. (2011). *In vitro* and *ex vitro* evaluation of long-term micropropagated turmeric as analyzed through cytophotometry, phytoconstituents, biochemical and molecular markers. Plant Growth Regul, 64: 91–98

Nayak, S., Kaur, T., Mohanty, S., Ghosh, G., Choudhury, R., Acharya, L. and Subudhi, E. (2011). *In vitro* and *ex vitro* evaluation of long-term micropropagated turmeric as analyzed through cytophotometry, phytoconstituents, biochemical and molecular markers. Plant Growth Regul 64: 91–98.

Panda, M. K., Mohanty, S., Subudhi, E., Acharya, L. and Nayak, S. (2007). Assessment of genetic stability of micropropagated plants of *Curcuma longa* L. by cytophotometry and RAPD analysis. Inter J Int Biol., 1(3): 189-195.

Peredo, E. L., Arroyo-Garcia, R. and Revilla, M. A. (2009) Epigenetic changes detected in micropropagated hop plants. J Plant Physiol, 166: 1101–1111.

Roy, S. and Raychaudhuri, S. (2004). *In vitro* regeneration and estimation of curcumin content in four species of *Curcuma*, 21(4): 299-302.

Selvakkumar, C., Balakrishnan, A. and Lakshmi, B.S. (2007). Rapid *in vitro* micropropagation of *Alpinia officinarum* Hance,an important medicinal plant through rhizome bud explants. Asian Journal of Plant sciences, 6(8): 1251-1255.

Sharma, T. R. and Singh, B.M. (1997). High frequency *in vitro* multiplication of disease free *Zingiber officinale* Rosc. Plant Cell Rep., 17: 68-72.

Sharmin, S. A., Alam, M. J., Sheikh, M. M. I., Zaman, R., Khalekuzzaman, M., Mondal, S. C., Haque, M. A., Alam, M. F., Alam, I. (2013). Micropropaga- tion and antimicrobial activity of *Curcuma aromatica* Salisb., a threatened aromatic medicinal plant. Turkish J Biol, 37: 698–708.

Shirin, F., Kumar, S. and Mishra, Y. (2000). *In vitro* plantlet production system for *Kaempferia galanga*,a rare Indian medicinal herb. Plant Cell Tiss and Org Cult., 63: 193-197.

Singh, S., Kuanar, A., Mohanty, S., Subudhi, E. and Nayak, S. (2011). Evaluation of phytomedicinal yield potential and molecular profiling of micropropagated and conventionally grown turmeric (*Curcuma longa* L.). Plant Cell Tiss Organ Cult 104: 263–269.

Srvidya, A. R., Yadev, A. K. and Dhanbal, S. P. (2009). Antioxidant and antimicrobial activity of rhizome of *Curcuma aromatica* and *Curcuma zeodaria*, leaves of Abutilon indicum. Arch Pharm Sci and Res, 1(1): 14-19.

Sushma, K., Vishal, R. and Krishen, S.S. (2005). Conservation and propagation of high altitude medicinal and aromatic plant: *Hedychium spicatum*. J Plant Biochem and Biotechnol, 14 (1).

Tyagi, R. K., Agrawal, A. and Yusuf, A. (2006). Conservation of *Zingiber* germplasm through *in vitro* rhizome formation. Scientia Horticult., 108: 210-219.

Tyagi, R. K., Yusuf, A., Dua, P. Agrawal, A. (2004). *In vitro* plant regeneration and genotype conservation of eight wild species of *Curcuma*. Biol Plant., 48: 129–132.

Tyagi, R. K., Agrawal, A. and Yusuf, A. (2006). Conservation of *Zingiber* germplasm through *in vitro* rhizome formation. Scientia Horticult., 108: 210-219.

Williams, J. G. K., Kubelik, A. R., Liva, K. J., Rafalski, J. A., Tingey, S. V. (1990). DNA polymorphisms amplified by arbitrary primers are useful as genetic markers. Nucl Acid Res, 18: 6531–6535.

Woo, J. H., Kim, Y. H., Choi, Y. J., Kim, D. G., Lee, K. S., Bae, J. H., Min, D. S., Chang, J. S., Jeong, Y. J., Lee, Y. H. and Park, J. W. (2003). Molecular mechanisms of curcumin-induced cytotoxicity: induction of apoptosis through generation of reactive oxygen species, down-regulation of Bcl-X L and IAP, the release of cytochrome c and inhibition of Akt. Carcinogenesis 24: 1199–1208.

Wu, W. Y., Xu, Q., Shi, L. C. and Zhang, W. B. (2000). Inhibitory effects of *Curcuma aromatica* oil on proliferation of hepatoma in mice. World J of Gastroenterology, 6 (2): 216–219.

Zeitkiewicz, E., Rafalski, A. and Labuda, D. (1994). Genome finger printing by simple sequence repeat (SSR)-anchored PCR amplification. Genomics 20: 176–183.

Chapter 10

Conservation of Orchids in Odisha

Sulagna S. Jena[1], *Prasad Kumar Dash*[2] *and*
Nihar Ranjan Nayak[1]

[1]*Regional Plant Resource Centre, Nayapalli,*
Bhubaneswar – 751 015, Odisha
[2]*Odisha Biodiversity Board, Regional Plant Resource Centre Campus,*
Nayapalli, Bhubaneswar – 751 015, Odisha

ABSTRACT

Information from published literature confirms the presence of 151 species of Orchids from Odisha. Even though they are sparsely distributed in all parts of the state, maximum diversity is reported from Similipal Biosphere Reserve followed by Rebena Reserve Forest. From various studies, it is found that the orchid populations in Odisha are facing serious threats and, in many cases, populations are severely depleting. There exist many forest areas in the state which are unexplored, those may yield more orchid species. Deforestation, habitat fragmentation, unsustainable collections of wild orchids from its natural habitat, land-use changes, climatic variations and problem in the seed formation and germination are some of the major causes of the depletions of the orchid populations. As orchid biology is highly complex, it is required to understand the behavior of each species under natural conditions including maintenance of appropriate insect and mycorrhizal populations associated with propagation. Conservation approaches need to be implemented immediately through preservations of their natural habitats in-situ., Propagation and re-introduction of the plants into the natural habitat proved to be the ideal method for the restoration of the orchid populations in Odisha. As all the orchids are listed under Appendix I and II of CITES (Convention on International Trade in Endangered Species of Wild Fauna and Flora), trade of wild orchids should be banned.

Introduction

Orchids are known because of their extraordinary flower features. The flowers are available in a wide range of shapes, sizes and, colors. One of the major attractions of the flowers is the excellent vase life; some flowers can stay fresh more than 60 days in the blooming stage under controlled conditions. In many cases, the added feature is the production of fragrance in flowers. *Phalaenopsis, Dendrobium, Cattleya, Vanda, Aerides, Oncidium, Cypripedium, Paphiopedilum* are known worldwide and being used as potted flowers (Chugh *et al.,* 2009). For cut flower production, some of the *Dendrobium* hybrids are being used extensively. The orchid *Vanilla planifolia* is known for the production of the widely used 'Vanilla' flavor. Some of the tuber producing orchids are used for the food in Africa, the tubers are a rich source of carbohydrates, used for the production of *chinaka, chikande* and *kikande* (Kasulo *et al.,* 2009). For this purpose, orchids like *Disa robusta*and *Satyrium amblyosaccos*are being used. Orchids are also used for their medicinal properties; the number of *Dendrobium* species are extensively used in traditional Chinese medicine (Hu *et al.,* 1970). In India, some of the terrestrial orchids (*Malaxis muscifera, Habenaria intermedia* and, *Habenaria edgeworthii*) are used in the preparation of *Ashtavarga*. They are thetreasure trove for the students, gardeners, floriculturists, ethnobotanists and, naturalists. They are the best entities for plant-animal mutualism and plant-fungi studies and last but not the least on their role in phylogeny and evolutionary biology.

The members of the plant group are belonging to the family Orchidaceae comprising of 28,000 species, thus considered to be one of the largest families of flowering plants (Fay, 2018). Using these wild species, more than 100,000 hybrids have been developed, many of them are playing a major role in the flower production system. The family is divided into five subfamilies, among these, Apostasioideae is composed of only 16 numbers of genus, whereas Epidendroideae composed of 19,785 numbers of the genus of the orchid reported so far. The origin of the group is predicted to be 125 million years ago, however, the members are still evolving (Chase, 2001). Orchids are growing naturally in major parts of the world except in the cold areas covered with snow. However, the majority of the orchids are growing in the tropicalpart of the world. In South America, particularly in the Andes of Columbia and Ecuador are known to be the richest orchid biodiversity center of the world, estimated that one-third of the orchids of the world are reported from the area.

The plants can grow either on the soil as that of the other monocotyledon plants (terrestrial orchids) or need the support of the tree trunks (epiphytic orchids). Some of the orchids can grow on the organic matter accumulated on the rock fissures are known as the lithophytic orchids. Among the terrestrial orchids, those are not able to produce food through photosynthesis, rather obtaining foods from dead organic matters is known as the saprophytic orchids. The members of the family have some unique morphological features, like the presence of labellum in the flower, pollinia, gynostegium, extremely small size seeds without any endosperms.

Orchids propagate through vegetative means and seeds. Separation of shoots or the production of offshoots are the common methods used for vegetative propagation. The process is extremely slow and only a few plants could be produced in a year. Production of seedlings in orchids is equally acomplex process. For the fertilization process, majority of the group requires the association of insect under natural conditions and again the seeds produced are devoid of endosperm to support the germination process, hence it requires the association of a mycorrhizal association that provides nutrients as well as the growth regulators required for germination (Rasmussen *et al.*, 2015). However, the efficiency of the process is very less, only a few percentages of the seeds could complete the germination. In recent years, using tissue culture techniques, seedlings could be produced in mass scale in a short period (Yam and Arditti, 2017).In the nutrient medium, adequate nutrients and growth regulators can be provided for achieving a high rate of germination and seedling production. Several vegetative parts could also be used as the initial tissue for the production of plants in mass scale with the applications of plant growth regulators (Chugh *et al.*, 2009).

The members of Orchidaceae are facing severe threats in the natural habitats in every corner of the world. The survey conducted by the Global IUCN on 948 orchid species, 56.5 per cent those included in the Global Red List (IUCN, 2012). Still,a large number of orchids need to be evaluated. It has been observed that the populations are depleting, in many cases at a faster rate. Recently, all orchids are included in appendix I and II of the Convention of the International Trade in Endangered Species of Wild Fauna and Flora (CITES, 2017). The major causes of the rarity might be the complex biology of orchids; dependence of insects for pollination, mycorrhiza for seed germination and seedling development, and an appropriate host tree (Fay, 2018). Deforestation, over-collection from the natural habitats, and environmental degradation are found to be the major factors depleting the orchid population (Gale *et al.*, 2018).

Conserving orchid species is a complicated process. It is extremely important to understand the biology of each orchid. Knowledge on their interaction with the host in cases of epiphytes, pollinators as well as with the mycorrhizal fungus need to be available before implementing any conservation approaches(Swartsand Dixon, 2009). It is also important to know the genetic structure of the individuals to conserve the biodiversity effectively. As the climatic condition such as rainfall, temperature, *etc.* are also playing a major role in the growth and development of orchids, it is required to have the information in hand before the implementation of different approaches. Conservation of habitats through orchid reserves, propagation, and reintroduction, translocation, and conservation through botanic gardens are effective methods being used worldwide (Swarts and Dixon, 2009).

Success for the conservation of orchids is limited. *Ipsea malabarica* is an endangered orchid of Western Ghats of Kerala which requires immediate attention for conservation. Efficient protocols have been developed for the production of planting materials under *in vitro* conditions using rhizomes as the explant. These plants were successfully introduced into their natural habitat (Martin, 2003). *Efforts were made for the conservation of Renanthera imschootiana* an endangered orchid through the reintroduction of seedling. This orchid is threatened due to the over-collection and loss of the natural habitat. Seedlings were produced through *in vitro* techniques. The survival rate of the seedlings was found to be of 79.67 per cent (Wu *et al.,* 2014).Different approaches are being implemented for the conservation of orchids of Australia. The western part of Australia is highly diverse in orchid flora representing over 400 taxa of which 76 are declared to be rare. The major threats have been analyzed and found that habitat loss, salinity, activities of feral animals, and in many cases severe drought conditions are the major factors. Since orchids depend on mycorrhiza fungus for seed germination, the detail orchid-fungus diversity and specificity in western Australia have been thoroughly studied. Orchid flower pollinators are also important components for the conservation plans of orchids; in this direction, significant improvement has been made and number pollinating agents have been identified (Brundrett, 2007).

In this article,we discussed current issues related to the conservation of orchids of Odisha and suggested possible conservation measures. Information on the distribution of the orchids in different parts of the state is provided. Detail information available on the orchids, those are extremely rare in the state are provided for future conservation plans. Different threats because of which orchids are depleting are provided. Finally, progress made on the conservation of orchids are discussed.

Orchids of Odisha

From different parts of India, so far 1331 orchid species have been reported, of these 400 are endemic to the country and many species have ornamental and commercial importance. Though distributed across the country, the majority are growing in the forest areas of the North Eastern States as well in the parts of the Western Ghats. *Dendrobium* orchids are the dominant members comprising of 116 number species. From the state of Odisha, 151 number of orchids have been reported. Maximum 17 species are from *Habenaria* genus followed by 12 species from *Dendrobium* genus. Twenty-three endemic orchids of India are growing in the state and again of these seven species (*Habenaria panigrahiana* var. *parviloba, Cirrhopetalum panigrahianum, Eria meghasaniensis, Liparis espeevijii, Liparis udaii, Zeuxine mooneyi* and *Odisha cleistantha*) are exclusive to the state. Apart from above 7 species, 10 more endemic Indian Orchids such as *Habenaria barbata, Aerides maculosa, Eulophia ochreata, Habenaria crassifolia, Habenaria foliosa, Habenaria grandifloriformis, Habenaria panigrahiana, Obeneria proudlockii, Zeuxine lindleyana and Cheirostylis parviflora*also occurs in Odisha (Mishra, 2004). Even though orchids have sporadic occurrences in the state, maximum populations and species diversity occurred in high-rainfall forested landscapes of Similipal, Mahendrgiri –Singharaju-Debgiri hill complex in Gajapati, Deomali hills and Govindpalli Ghats of Koraput, Rebena Reserve Forests and Khandadhar waterfall of Keonjhar and

Kandhamal district. The climatic condition of Similipal Biosphere Reserve is most ideal for the growth and development of orchid, thus 96 species are reported. The reserve is the house of extremely rare orchids. Rebana forest of Keonjhar is another best habitat of orchids, representing 52 species. In the Gajapati district,37 species reported from Mahendragiri and Singharaj hills. Orchids like *Dendrobium regium,Phaius tankervilleae,Cymbidium aloifolium,Aerides odorata*and, many more have ornamental value are growing at different habitats of the state.

Threatened Orchids of Odisha

Though all orchids are considered to be rare, in Odisha 10 orchids are considered to be extremely rare and conservation approaches need to be taken immediately. Brief information of these orchids are provided (Figure 10.1).

Eria meghasaniensis (S.Mishra) S. Mishra

This orchid first time reported from the Odisha and considered endemic to Similipal. Growing only at the Meghasani hillsof Similipal Biosphere Reserve of Mayurbhanj district. It is an epiphytic Orchid, small in size, (6 cm) with 2-4 leaves on the top of the stem. Flower appears during September – October and capsules are formed from October to March. Capsules are small in size. In each inflorescence 10-16 flowers are produced. Total no. of plants maybe around 200 no. The propagation rate is extremely slow. As this orchid grows in matured and old Sal trees, protection of its natural habitat is of utmost significance for its future conservation.

Dendrobium regium Prain

This orchid is endemic to India and reported to occur in only three states, Chhattisgarh, Jharkhand, and Odisha. Epiphytic orchid, growing in tropical semi-evergreen to moist deciduous forest under shade condition. In Odisha, though major populations are found at Similipal, it is also reported from Satkosia, Karlapat, and Kotgarh. Stem stout, somewhat compressed, narrowed to the base and slightly swollen at the base. Leaves are present at the upper portion of the stem. Inflorescences are sub-terminal and each used to have 2-4 flowers. Flowers are pink in color with a yellowish-white lip. Capsules are drooping, robust, ringed at the time of maturity. A maximum of 300 plants are growing in different habitats. Population growth is extremely low.

Pomatocalpa decipiens (Lindl.) J.J.Sm.

Distribution of the species is reported from Sri Lanka, India, and Bangladesh. In India,the populations are reported to be confined to Barbara Reserve Forest, Hadgarh Wildlife Sanctuary, and Kapilash Wildlife Sanctuary of Odisha. The number of individuals in all the population are low. It is short epiphytic orchid; alternate, distichous, larate leaves; flowers yellow with reddish blotches, an extra axillary spike; lios with jug shaped spur, capsules sessile, oblong – ellipsed. Flowers appearing during January to March and fruits are produced from March to May. Population growth under natural condition is extremely low.

Pecteilis gigantea — ***Pomatocalpa decipiens***

Cymbidium bicolor — ***Phaius tankervilleae***

Dendrobium regium — ***Bulbophyllum guttulatum***

Figure 10.1: Rare Orchids of Odisha.

Phaius tankervilleae (Banks ex L. Her Blame)

This orchid is a terrestrial evergreen herb and reported from Deoghati, and Bhaunighati of Sundergarh district, growing mostly in deciduous forest particularly near the moist valleys under deep shade. Stems with pseudobulbs, leaves strongly ribbed and prominent veins below. Inflorescences are produced from the lower node of the pseudobulbs. Flowers are large, downwardly faced and mildly scented. Flowers appeared during February to March and fruits are produced during April to June. Currently, few plants are available. Though seeds are produced every year, seedling production under natural conditions not reported.

Tainia penangiana Hooks

It is a terrestrial orchid and in Odisha it is reported from Mayurbhanj and Dhenkanal. Major population are growing near the perennial streams of Jenabil village of Similipal Biosphere Reserve in semi-evergreen forests. Pseudobulbs are stout and covered with sheath. Leaves oblong, lanceolate or broadly elliptic. Flowers pale yellow to greenish yellow with brown hairs. Flowering occurs during the month of January to March and capsules are available from March to April. Total number of plants not more than 150 in Odisha. Considered to be extremely rare.

Cymbidium bicolor Lindl.

This is an epiphytic Orchid, growing in tropical semi-evergreen to evergreen forests in Odisha, growing at Similipal Biosphere Reserve, Satkosia Tiger Reserve and Mahendragiri Hills. Stems are pseudobulbs. Flowers pale yellow at base and maroon towards tip. Inflorescence are raceme arising from the base of the pseudobulbs. Flowering time under natural condition is from February to April and fruits are available from May to December.

Cirrhopetalum panigrahinum (S. misra) S.Misra

Epiphytic or Lithophytic orchid, growing in tropical semi-evergreen or evergreen forests. In Odisha,it is growing in Similipal Biosphere Reserve and Rebana forest. The population of Rebena is almost destroyed due to habitat loss and a new population of around 20 individuals are recently observed by the authors in Mandasaru gorge of Kandhamal district. Small pseudobulbs, covered with dark brown foliar sheath; small waxy white flowers with maroon flushing in a sub-umbellate spike. Flowering time in natural condition is June and vase life of this flower is about two weeks. Fruits generally seen in July and January. Considered to be extremely rare.

Bulbophyllum guttulatum (Hook. F.) Balkr.

Epiphytic orchid growing in tropical semi-evergreen forest on hill slope. In Odisha growing at Kutia hills and Deomali hills, considered to be extremely rare. Small erect pseudobulbous, creeping rhizomes, one leaf, oblong, inflorescence is longer than leaf, large flowers, densely maroon spotted, lateral sepals are unequal and longer. Flowering time under natural condition is March end.

Plecteilis gigantea (J.E.Sm.) Rafin.

It is a tall terrestrial herb growing up to 4 ft otherwise called butterfly orchids due to its winged sepals, fragrance flowers and fringed lip. It usually occur in swampy grasslands above 3000 ft. Five to six large terminal flowers appears during September-November and fruiting comes during December. The lip is trilobed and it has a long, tubular greenish cylindrical spur near base and the coulm has a single anther with two pollinia. The major threat to this orchid is from wild boar as this animal love to eat its tuber which was observed by the authors in Karlapat. Earlier it was reported from Koraput district, but the wild population is now restricted to Karlapat only.

Didymoplexis pallens Griff.

This orchid is terrestrial habitat growing in with loamy soil covered with humus decaying leaves and twigs. In Odisha, growing at Siddheswari Hill and Dhuanali. This species is leafless, slender tiny herbs. The plant is only seen when flowering, rest of the time it perennates underground through its modified stem. Flowering in natural condition is May-June and fruiting in June-Jul. Considered to be rare. Recently two new populations which contain 30 individuals each have been discovered by the authors from Debrigarh wildlife sanctuary, Sambalpur which is protected and conserved by Debrigarh wildlife Division.

Causes of Rarity

Habitat Destruction

As the biology of orchid is highly complex and requires specific ecological niche for their survival; any change in the habitat, severely affect the growth, development and reproduction of orchids (Crib *et al.,* 2003). During last 50 years, with the rapid growth of human population; exploitation of natural resources has occurred in area of high biological diversity(Kennedy, 2006). Land clearing for agriculture, mining and urban development are the major factors of habitat destruction. Fragmentation of habitats, removal of the key species at the ecosystem and introduction of feral animals are documented to be the key factors of decline in orchid population and diversity (Coates and Dixon, 2007) The 200 years of European settlement in Australia significantly affected the orchid population particularly in the South West Australian Floristic Region (SWAFR) (Hopper and Gioia 2004). During this period, Australian has recorded with extinction of 83 plants and 43 animals(Beeton *et al.,* 2006). Wild fire and drought reported to be affecting the natural habitat of orchids. In a study carried out in *Pterostylis revoluta* shown that prescribed burning changes the orchid population and endophytic fungi (Jasinge, 2018).

After independence, in Odisha, mining activities are high particularly in the district of Kendujhar, Sundargarh, Angul, Jharsuguda and Koraput. Depletion of the orchid population in these districts might be due to the land clearing activity. *Dendrobium regium* reported earlier from the Keonjhar district from number of locations; has now reduced to few restricted pockets. Similarly,it is now very tough to find single population of *Phaius tankervilleae* reported earlier from Sundargarh district. Forest fire also happening in different forest naturally, might be affecting the growth and reproduction of the terrestrial orchids. The wild population of Kingidium *deliciosum*, earlier reported to occur in several populations in Kotagarh wildlife sanctuary has now reduced to only two populations (in Bandaka forest area and Subarnagiri Reserve Forest) due to rampant forest fire. The wild population of *Cymbidium bicolor* in Mahendragiri has drastically reduced to only few individuals due to habitat change and road expansion. *Habenaria stenopetala*, another extremely rare terrestrial orchid seen frequently in Saptasajya has now become locally extinct from that locality due to habitat loss.

Over Collection

For horticulture purpose as well as for food, reports exist on the indiscriminate collection of orchids from the natural habitats. *Cattleya, Laelia, Renanthera, Cypripedium, Paphiopedilum* were over collected, at some places stripped near to the point of extinction (Fay, 2018), These species are highly important for horticultural purpose. However, in many cases for horticulture purpose collection are made in small quantity may not affect the conservation status of orchids (Cribb *et al.*, 2003). Collection of orchids for food and medicine severely affecting population of orchids. The species of genera *Anacamptis, Dactylorhiza* and *Orchis* are collected from the natural habitat for production of salep in several Mediterranean countries threatening natural populations (Kreziou *et al.*, 2016). In Odisha, none of the orchids are being used for preparation of food, thus may not be a threat to orchids. However, Wild boar consuming the tubers of *Pecteilis gigantea* and *Habenaria barbata*-two rare orchids of Odisha observed by the authors in 2015 at Krishnamali hills (above 3000 ft amsl) of Karlapat wildlife sanctuary. This has resulted in destroying several individuals of both species. Wild orchids locally called, *Rasna jadi or Madanga*in general have been collected from Gupteswar, Koraput, Mahendragiri hills of Gajapati and other non-protected areas in large quantities and sold in neighboring states of south India is also a major threat to wild orchid populations in the state.

Climate Change

Climate change particularly increase in temperature affects orchid population growth (Willmer, 2014). In a study carried out on *Ophrys sphegodes* has shown that climate warming has changed the timing of flower production, ultimately mismatching at the availability of male bee for pollination. Warming affect might load to complete reproductive failure (Hutchings *et al.*, 2018). The depletion of orchid population in Odisha might not be related to climate change. Two studies carried out, one of the Similipal Biosphere Reserve and another at Barbara Reserve Forest, temperature and rainfall data collected since 1903, and found that average temperature in these areas has not changed significantly. Recently, the super cyclone Titli has devastated the Mahendragiri ecosystem and thousands of evergreen trees were uprooted due to land slide and wind action. This has resulted in loss of almost all populations of *Dendrobium bicameratum, Gastrochilus acaulis, Oberonia proudlockii* and *Flickingeria macraei.*

Pollination

In orchids, pollination requires the involvement of a biological agent and in majority of the cases an insect, in some species autogamy is also reported. Because of the complex flower structures, some of orchids might be depending on a particular pollinating agent. Orchid fails to produce seeds when disruption to pollination system occurred. In absence of propagation through seeds, the population may lose the degree of genetic diversity, affects their long-term survival and evolutionary potential (Swarts and Dixon, 2009). In western Australia, nine species of *Drakaea* orchids are pollinated with different species of thynnine wasps because of the specialized pollination system. These wasps depend on the larvae of scarab beetles for their survival. Finally, these beetles need the substrate from the roots of a native plant. Any disturbance in these linked components may lead to severe depletion of the local population (Stewart and Dixon, 2009). Though extensive studies have not been carried out with effect of limitation of pollination on orchids of Odisha, it seems like that there are severely problems in production of seeds in some orchids. In a study carried out on the *Tainia penangiana* growing at Similipal, shown that, the plants are producing flower every year, however seeds are not produced might be due to the non-availability of the required pollinating agent in the area. Flowers are highly fertile and seed-bearing capsules were formed by hand pollination. This might be the fact that the species is confined to only 73 induvial and located only at Jenabila area-the core region of the reserve.

Mycorrhizal Association

Seed germination and seedling establishment are the major factors under *in-situ* conditions for population growth and sustainability of orchids. As they require the symbiotic association of mycorrhizal fungus for seed germination and seedling establishment, the presence of the fungus is essential under natural conditions (Rasmussen *et al,,* 2015). The interaction of orchid with the symbiotic partner is quite complex; some orchid requires the association of a particular fungus whereas others can manage with a range of fungus (Stewart and Dixon, 2009). In a study carried out on three terrestrial orchids such as *Goodyera pubescens, Liparis liliifolia,* and *Tipularia discolor*has shown that seed germination under *in-situ* condition is directly related to the abundant mycorrhizal fungi (McCirmick *et al.,* 2012). No studies have so far been carried out on the symbiotic association of orchids of Odisha. Studies carried out to understand the causes of the rarity of *Dendrobium regium* shown that the plants were producing flowers and fruits every year, however, seeds are not able to germinate under natural conditions. Seeds are highly viable, not associated with any kind of biochemical deficiency, under *in vitro* condition, 90 per cent of the seeds able to germinate. The failure of seed germination under natural conditions might be due to the non-availability of a symbiotic partner.

Conservation Approaches

From different studies (Saxena and Brahman, 1994; Haines, 1924, Misra 2004,

Misra 2007 and Mishra, 2012), the total number of orchid species growing in the state of Odisha and their distribution to some extent is known. Information on the flowering and fruiting period of the reported 151 species is available. Details morphological features of 139 species have been thoroughly characterized (Misra, 2004). Though all orchids considered to be rare, conservation approaches need to be implemented on priority basis on those species which are facing greater risk. Priority should be given to the threatened and endemic orchids of Odisha as the populations are only growing in the state. The best example is the *Eria meghasaniensis,* populations are only available in the Meghasani hills of Similipal Biosphere Reserve. The second priority should be given to the endemic orchids of India and growing in the state of Odisha. *Dendrobium regium* is coming under this category. The major population of the orchid is growing at Similipal Biosphere Reserve, though reported from other parts of the country like Jharkhand and Andhra Pradesh. The flower has high ornamental value, can be used for future breeding purposes. The third category would be orchids growing in different parts of the world; the population of those species in the state of Odisha confined to few individuals, orchids like *Phaius tankervilleae* and *Tainia penangiana* are falling under this category. In the state, individuals of *Phaius tankervilleae* though reported from Sundargarh forest areas, it is becoming difficult to find the flowering plants. The populations of *Tainia penangiana* are reported from different places in the state so far only two populations reported and in each only a few plants are growing. Conservation approaches can also be implemented in groups; base on their similar habit, habitat, and propagation methods. For protection and conservation of orchids *in-situ*, the frontline staff of forest departments like forest guards and foresters of all the 53 forest divisions have been trained by the scientists of Odisha Biodiversity Board especially on the identification of wild orchids of economic, ecological and phytogeographic significance. The board is also in a process of establishing small orchidarium near forest rest houses of respective wildlife sanctuaries harboring native orchids collected from their native habitats for creating and disseminating public awareness.

Systematics

For the conservation of the orchids of Odisha,it is highly important to know the details of each species growing in the state. The identification of each species needs to be thoroughly carried out. It is also required to have the information from different studies on the number of plants for each species growing at different regions of the state. Their morphological diversity and their distribution within the state will help to implement different conservation approaches. For example, though we know that *Eria meghasaniensis* is endemic to Similipal Biosphere Reserve, we do not have the information on the number of individuals currently available in the area and the status of the individuals. Similarly, though we know that the orchid *Dendrobium regium* is extremely rare and the viable populations are only growing in the state of Odisha, detailed information on the distribution and their population structure is not available. *Phaius tankervilleae*, another threatened orchid seems to be no more available in Odisha; however, it is extremely important to carryout surveys in other nearby areas regarding the existence of the species.

Genetics

Genetic studies are highly essential for the conservation of the orchids of Odisha, no studies have been carried out so far. In the state, from *Dendrobium* genus, 11 species have been reported and all are considered to be rare, however, it is highly important to understand the phylogenetic relationship of the group. The species which falls out of the group need to be prioritized for immediate conservation. In a study carried out on the subtribe Spiranthinae in Brazil has shown that the genus *Cotylolabium*falls in a separate position as compared to that of the other members, thus considered as the most important group for conservation (Borba *et al.,* 2014).Selecting the population of an orchid to be conserved, genetics studies also plays a major role. In any conservation approach, it is highly essential to conserve the population where more genetic diversity exists between the individuals as it is speculated that more the genetic diversity, more the survival of the species. In recent years, molecular markers have been developed, could be used for the determination of genetic diversity. While implementing the conservation approaches, in *Dendrobium regium,* studies were carried out to understand the genetic diversity between the individual of two populations in Similipal Biosphere Reserve using molecular markers and found that both the populations are not genetically different. For any reintroduction program, it is required to introduce genetically different seedlings, thus it is required to study more orchid populations in the state.

Habitat Conservation

The climatic conditions, host trees in cases of the epiphytic orchids, insect populations and the microflora of a habitat are the major factors affecting the growth and propagation of orchids. Conservation of the natural habitats is the most important method for the conservation of orchids. In many places, thus orchid reserves have been developed in which all the components are preserved. In Odisha, it thus required to develop such reserves, Similipal Biosphere Reserve is the most ideal place for this purpose as so far 96 orchids have been reported comprising of terrestrials, epiphytes, lithophytes as well as the saprophytic ones. The temperature, rainfall, host trees, diversity of insect populations and the microflora of the reserve seems to be the most ideal for the growth and development of orchids. Similarly special habitats like Mahendragiri, Deomali and Kapilash are suitable for high altitude loving species and as Mandasaru Gorge in Phulabani is recently protected after declaration of first Biodiversity Heritage Site (BHS) of Odisha under Biological Diversity Act, 2002, some threatened and endemic Orchids of the state can be conserved there.

Propagation and Reintroduction

This method is being used for the restoration of populations of threatened plants in recent years. For any reintroduction programs, seedlings are ideal because of the existence of genetic diversity among them. Using tissue culture techniques, seedlings could be produced in mass scale. At the Regional Plant Resource Centre, the orchid research program has standardized the protocols for the efficient production of seedling of 25 wild orchids of Odisha (Figure 10.2). The seedlings were also successfully introduced at different natural habitats of the state. Seedlings of *Tainia penangiana, Dendrobium regium,* and *Cymbidium bicolor* were introduced at Similipal Biosphere Reserve near their mother plants. Similarly, the seedlings of *Cymbidium aloifolium, Aerides odorata* and *Vanda tessellate* have been introduced at the Chandaka Reserve Forest, Bhubaneswar, and seedlings of *Pomatocalpa*

Figure 10.2: *In vitro* Production of Seedlings of Wild Orchids of Odisha
(a and b) Seedlings of 25 wild orchids being produced under *in vitro* conditions;
(c) *In vitro* production of seedlings of *Phaius tankervilleae;*
(d) *In vitro* production of seedlings of *Tania penangiana;*
(e) *In vitro* production of seedlings of *Cymbidium bicolor;*
(f) Acclimated orchid seedlings to be used for reintroduction purpose.

decipiens have been introduced in Mandasaru Gorge of Kandhamal and Berbera and Dhuanali Reserve Forest of Khordha district.

Development of Seed Bank

Seed bank development offers tremendous scope for the conservation of orchids. As the seeds are small in size could be stored easily in the seed banks with minimum expenditure. Other parts like protocorms, pollens, cell suspension lines, shoot meristems also could be used for long-term preservation of different species and hybrids. From different efforts, more than 100 wild species and hybrids have been preserved. The study carried out on *Hadrolaelia grandis* an endangered orchid has shown that the seeds could be stored at – 80 °C for three years without any detrimental effect on seed germination and seedling development (Vudala and Ribas, 2017). No such efforts have yet to be made in this direction for Orchids in Odisha. The national facility developed at the National Bureau of Plant Genetic Resources could be effectively used for long term conservation of Orchids.

Conclusion

Conserving the orchids of Odisha is a highly complex process; requires the involvement of experts of different filed of science like plant taxonomists, ecologists, genetic, and microbiology. In the state, training need to be provided to the scientist focusing on different aspects of orchid biology; this will help to understand the requirement of the orchids. One of the important aspects is the financial support for implementing different activities; in this regard, the Forest and Environment Department of Govt. of Odisha needs to come forward. It is also required to involve different private agencies, Non-Government Organizations as well as individuals for the conservation of orchids.

REFERENCES

Beeton, R.J.S., Buckley, K.I., Jones, G.J., Morgan, D., Reichelt, R.E., Trewin, D., 2006. State of the environment report; Australia. Canberra: Department of the Environment and Heritage. http: //www.environment.gov.au/soe/2006/ index.html.

Borba, E.L., Salazar, G.A., Mazzoni-Viveiros, S., Batista, J.A.N., 2014. Phylogenetic position and floral morphology of the Brazilian endemic, monospecific genus *Cotylolabium*: a sister group for the remaining Spiranthinae (Orchidaceae). Bot. J. Linn. Soc. 175, 29–46.

Brundrett, M.C., 2007. Scientific approaches to Australian temperate terrestrial orchid conservation. Aust. J. Bot. 55(3), 293–307.

Chase, M.W., 2001. The origin and biogeography of Orchidaceae. In Pridgeon, A.M., Cribb, P.J., Chase M.W. and Rasmussen, F. eds. *Genera orchidacearum, Vol. II: Orchidoideae (part I)*. Oxford University Press, Oxford. 1-5.

Chugh, S., GuhaI, S., Rao, U., 2009. Micropropagation of orchids: A review on the potential of different explantsScientia Horticul, 122, 4, 507-520.

CITES. 2017.*The CITES Appendices*. Published on the internet at https: //cites.org/ eng/app/appendices.php. Retrieved on 31 October 2017.

Coates, D.J., Dixon, K.W., 2007. Current perspectives in plant conservation biology. Aust. J. Bot., 55, 187–193.

Cribb, P.J., Kell, S.P., Dixon, K.W., Barrett, R.L., 2003. Orchid conservation: aglobal perspective. In: Dixon KW, Kell SP, Barrett RL, Cribb PJ, eds. Orchid conservation. Natur. History Public., 1–24.

Fay, M.F., 2018. Orchid conservation: how can we meet the challenges in the twenty first centuryÿþ. Bot. Stud., 59,16.

Gale, S.W., Fischer, G.A., Cribband, P.J., Fay M.F., 2018. Orchid conservation: bridging the gap between science and practice. *Bot. J. Linn. Soc.*, 186, 425–434.

Haines, H.H., 1924. Botany of Bihar and Orissa. Part II. *Botany of Bihar and Orissa. Part II.*

Hopper, S.D., Gioia, P., 2004. The Southwest Australian Floristic Region: evolution and conservation of a global hot spot of biodiversity. Ann. Review Ecol. Evolu. Systema., 35, 623–650.

Hu, S.Y., 1970. *Dendrobium* in Chinese medicine. Econo Botany, 24, 165-74.

Hutchings, M.J., Robbirt, K.M., Roberts, D.L., Davy, A.J., 2018. Vulnerability of a specialized pollination mechanism to climate change revealed by a 356-year analysis. *Bot. J. Linn Soc*, 186, 4, 498–509.

Jasinge, N.U., Huynh, T., Lawrie, A.C., 2018. Changes in orchid populations and endophytic fungi with rainfall and prescribed burning in *Pterostylis revoluta* in Victoria, Australia. *Ann. of Bot.*, *121*(2), 321-334.

Kasulo, V., Mwabumba, L., Cry, M., 2009. A review of edible orchids in Malawi. J. of Horticul and Forest, 1(7), 133-139.

Kennedy, D., 2006. Life on a human-dominated planet. State of the planet, Washington: The American Association for the Advancement of Science.

Kreziou, A., de Boer, H., Gravendeel, B., 2016. Harvesting of salep orchids in north-western Greece continues to threaten natural populations. Oryx. 50, 393–396.

Martin, K.P., 2003. Clonal Propagation, Encapsulation, and Reintroduction of *Ipsea malabarica* (Reichb. f.) J. D. Hook., an Endangered Orchid. *In Vitro Cellular and Developmental Biology. Plant*Vol. 39, No. 3 (May - Jun., 2003), pp. 322-326.

McCormick, M.K., Lee Taylor, D., Juhaszova, K., Burnett Jr, R.K., Whigham, D.F. and O'NEILL, J.P., 2012. Limitations on orchid recruitment: not a simple picture. *Molecular Ecology*, *21*(6),1511-1523.

Misra, S., 2004. Orchids of Orissa. Bishen Singh and Mahendra Pal Singh, Dehra Dun, India.

Misra, S., 2007. Orchids of India. Bishen Singh and Mahendra Pal Singh, Dehra Dun, India

Popova, E., Kim, H.H., Saxena, P.K., Engelmann, F., Pritchard, H.W., 2016. Frozen beauty: The cryobiotechnology of orchid diversity. Biotechnol. Adv., 34(4), 380-403.

Rasmussen, H.N., Dixon, K.W., Jersakova, J., Tesitelova, T., 2015. Germination and seedling establishment in orchids: a complex of requirements. *Ann. Bot.*, 116(3), 391-402.

Reiter, N., Whitfield, J., Pollard, G., Bedggood, W., Argall, M., Dixon, K., Davis, B., Swarts, N., 2016 Orchid re-introductions: an evaluation of success and ecological considerations using key comparative studies from Australia. Plant Ecol., 217, 81-95.

Saxena, H.O. and Brahman, M., 1994. Flora of Orissa, Orissa Forest Dev. *Corp., Bhubaneswar, 2,* 993-995.

Seaton, P.T., Hu, H., Perner, H., Pritchard H.W., 2010. Ex Situ Conservation of Orchids in a Warming World. Bot. Rev., 76,193-203.

Swarts N.D., Dixon K.W., 2009. Perspectives on orchid conservation in botanic gardens. Trends Plant Scien., 1360-1385.

Vudala, S.M., Ribas, L.L.F., 2017. Seed storage and asymbiotic germination of *Hadrolaelia grandis* (Orchidaceae). Sou. Afri. J. Bot., 108, 1-7.

Willmer, P., 2014. Climate Change: Bees and Orchids Lose Touch. Curr. Biol., 24, 1133-1135.

Wing, Y.T., Joseph A., 2017. *Micropropagation of orchids* (Third edition). John Wiley and Sons, Inc, Chichester, West Sussex.

Wu, K., Zeng, S., Lin, D., Teixeira da Silva, J.A., Bu, Z., Zhang, J., Duan, J., 2014. *In Vitro* Propagation and Reintroduction of the Endangered *Renanthera imschootiana* Rolfe. PLoS One.9(10).

Chapter 11

Micropropagation of *Mucuna pruriens* var. *utilis*

*Abhipsa Panda[1], Sanjay Kumar Madkami[1], Arpita Moharana[1] and Durga Prasad Barik[1]**

[1]Department of Botany and Biotechnology, Ravenshaw University, Cuttack-753 003, Odisha
**e-mail: barikdp@gmail.com*

ABSTRACT

The purpose of this study was to develop an efficient and reliable micropropagation protocol for Mucuna pruriens var. utilis an important edible medicinal legume plant in India using cotyledonary node explants derived from axenic seedlings. Conventional propagation via seed poses problems due to low germination ability and poor viability. Highest percentage (100 per cent) of seed germination was found using fresh seeds on MS medium without any PGRs. MS medium supplemented with BAP 1.0 mg/L was found most effective for multiple shoot induction and proliferation. The response of shoot multiplication percentage was found to be 100 per cent with highest 9.1 numbers of shoots per explant with an average 3.12 cm shoot length. Shoot cultures with the explant were sub-cultured on fresh medium of same composition for better shoot multiplication and elongation. Rooting was best induced in vitro derived shoots on ⅛ MS medium without any PGRs and developed 7 no of roots with 5.32cm length. Plantlets with well developed roots were acclimatized in sand and soil (1:1) and established in pot with garden soil. The result of this study tends to give credence to the use of in vitro derived plants by the pharmaceutical industries without disturbing the natural stocks and will be a supplement as edible legume.

Keywords: *Anti-Parkinson effect, Cotyledonary node, Cultivated edible legume, In vitro multiple shoots, Velvet bean.*

Introduction

The genus *Mucuna* which includes about 150 species of perennial and annual legumes belongs to the family Fabaceae; old name Leguminoceae (Sathyanarayana *et al.*, 2008). Many species of the genus offer an excellent source as cover crop and green manure, in addition to their traditional use as feed and food (Janardhanan and Lakshmanan, 1985; Mohan and Janardhanan, 1993; Capo-chichi *et al.*, 2003; Sathyanarayana *et al.*, 2008) and as dye (Standley and Steyermark, 1946). In addition, *Mucuna* is also traditionally used in treatment of pain and numbness of joints, and irregular menstruation (Ding *et al.*, 1991). Almost all the species are reported to contain an amino acid 3-(3, 4-dihydroxyphenyl)-L-alanine (L-dopa) acid that acts as precursor for the neurotransmitter dopamine, used in the treatment of Parkinson's disease (Manyam, 1995).

Genus *Mucuna* comprises 9 species and 4 varieties through-out India. Unfortunately most of the species remained neglected for their investigation regarding their valuable properties (Sathyanarayana *et al.*, 2008). Among the different species of *Mucuna*, one most important tropical legume is *M. pruriens* (L.) DC., which is commonly called as 'the cowhage' or 'velvet' bean found in bushes and hedges at damp places, ravines and scrap jungles throughout the plains of India (Faisal *et al.*, 2006a). All the parts of this plant possess valuable medicinal properties. The seeds of *M. pruriens* are used as a tonic and aphrodisiac for male virility in Ayurvedic medicine (Anon, 1962; Faisal *et al.*, 2006b). Roots are useful to relieve constipation, dropsy, ulcers and fever. Leaves are useful in inflammation, cephalagia and general debility (Faisal *et al.*, 2006a). The plant and its extracts are used in tribal communities as a toxin antagonist for various snakebites (Singh *et al.*, 1995). The presence of L-dopa (neurotransmitter precursor) attracted attention of researchers for utilization of the plant as the source for it (Bell and Janzen, 1971; Daxenbichler *et al.*, 1971; Faisal *et al.*, 2006a).

Besides, the seeds are good source of protein as they contain higher amounts of crude protein and lipids when compared with other most of the commonly consumed legumes. According to the report of Balogun and Olatidoye, 2012 the nutritive compositions of *Mucuna utilis* is as follows, crude protein (25.65), crude carbohydrate (42.98 per cent), crude fibre (7.23 per cent) and ash (3.6 per cent), fatty acid 12.4 per cent. The fatty acid profile of the seeds revealed that the seed lipids contained higher concentration of palmitic acid and Oleic acid. The seed protein contained higher levels of essential amino acid such as valin, tyrosine, threonine, histidine *etc.* So Balogun and Olatidoye, 2012 advised utilis seeds for domestic consumption in Nigeria. According to their investigation the seeds of this under-utilised crop can be explored as an alternative protein source to alleviate protein-energy-mal nutrition among economically weaker section of people.

(Balogun and Olatidoye, 2012). Another report said boiled seeds with reduced L-DOPA are known to be consumed by Oceanic group of tribal's like Onges, Great Andamans, Sompens *etc.* The presence of L-dopa coupled with rich protein and amino acid marked *Mucuna pruriens* var. utilis as an important under-utilised legume.

Due to large-scale, unrestricted exploitation of the whole plant to meet its ever-increasing demand by the pharmaceutical industry, coupled with limited cultivation and insufficient attempts for its replenishment, the wild stock of this valuable medicinal plant has markedly depleted. In nature, this species propagates only through seed. Conventional propagation via seed poses problems due to the high allergic properties of pods which causes uncontrolled itching, low germinating ability and poor viability (Verma *et al.,* 1993; Krishnamurthy *et al.,* 1996). Thus, conventional propagation through seed is not an adequate solution to meet the demand for this plant. Therefore, it is useful to develop a method for large-scale multiplication for commercial production. Tissue culture technology is successfully utilized in propagation of plants with poor and uncertain response to conventional propagation. However, to the best of our knowledge, there is only one in vitro propagation protocol for *Mucuna pruriens* var. utilis, where Sathyanarayana *et al.,* 2008 used axenic axillary bud to produce multiple shoots which were very low in number. So attempt has been made here to standardize high fidelity, rapid and reliable micropropagation protocol using a special type of explant *i.e.* cotyledonary node.

Materials and Methods

In this present work, three types of seeds were tasted *i.e.* one year old seeds, six months old seeds and fresh seeds. The seeds were collected from the Botanical garden, Department of Botany, Ravenshaw University, Cuttack-3. The one year old and six months old seeds were kept under running tap water for about 30 minutes followed by 15 minutes treatment with 5 per cent (v/v) aqueous solution of Teepol (Reckilt Benckiser Ltd., India) and rinsed 5 times with distilled water and then surface sterilized with 0.1 per cent (w/v) aqueous solution of mercuric chloride ($HgCl_2$, Hi-media, Mumbai, India) for 20 minutes followed by five rinses with sterile double distilled water. But in case of fresh seeds, the seeds were kept under running tap water for 30 minutes and then treated with 5 per cent (v/v) aqueous solution of labolin for 15 minutes and rinsed with tap water. Then the seeds were treated for 15 minutes with 5 per cent (w/v) Megastin (K.P.R. fertilizers limited, Tata Nagar, India) and rinsed thoroughly with distilled water followed by surface sterilization with 0.1 per cent (w/v) aqueous solution of mercuric chloride ($HgCl_2$, Hi-media, Mumbai, India) for 20 minutes. After that the seeds were rinsed five times with sterile double distilled water and inoculated to seed germination media.

For seed germination, different strengths of Murashige and Skoog's (1962) basal medium (MS) was used *i.e.* MS, ½ MS, ¼ MS and ⅛ MS without any growth regulator. For establishment of multiple shoot culture and rooting of *in vitro* regenerated shoots, MS basal media of different strength were augmented with or without a range of concentration of plant growth regulators in single or in combinations. The seed germination percentage and days require for seed germination was evaluated for different strengths of basal media. At the same time seed germination percentage was compared among the three different types of seeds *i.e.* one year old seeds, six months old seeds and fresh seeds.

Cotyledonary nodes excised from 10-11 days old *in vitro* germinated seedlings were inoculated into the MS basal media supplemented with different concentrations of BAP or Kin (0.5- 5.0 mg/L) for multiple shoot induction. The cotyledonary node explants were sub-cultured on same fresh media at 15 days intervals for shoot multiplication and elongation. Well developed *in vitro* shoots were excised and transferred to culture tubes containing MS, ½ MS, ¼ MS, ⅛ MS, $^{1}/_{16}$ MS media without any growth regulators and gelled with 0.6 per cent (w/v) agar (Hi-media, Mumbai). For rooting, culture tubes were placed in vertical position on wooden racks. The cultures were maintained in culture room at 25±1°C temperature less than 16h photoperiod of 35-50 μM m^{-2} s^{-1} photon flux density provided by cool white fluorescent tube and 55-60 per cent relative humidity.

The *in vitro* derived plantlets were washed cleanly and carefully with tap water to get rid of agar gelled medium to avoid contamination. Then the plastic glasses were wiped with alcohol and filled with autoclaved sand and soil (1:1) mixture, where the plantlets were planted with care and sufficient water was poured to the glass till the sand and soil mixture became fully wet. Then the plastic glasses were covered with polythene bags and kept in culture room. After 2 days, small holes were done in polythene bags to give water to the plants and plantlets became adaptable to environmental humidity. The polythene bags were removed from the plastic glasses on 12th day, watered sufficiently. After 3-4 days the plantlets were transferred to the pots containing garden soil avoiding direct expose to sunlight. After 10-12 days the plants were transferred from pot to field.

Mean data pooled from a total 03 number of replications each comprising of 05 culture flasks containing 02 explants per flask for seed germination and shoot regeneration experiment. For *in vitro* rooting mean data was pooled from 03 separate experiments each with 05 replicates, each replications consisting of one shoot/culture tube.

Results and Discussion

Seed viability of the three types of seeds were studied, the fresh seeds collected from the plant grown in the Botanical garden of the Department of Botany showed 100 per cent germination (7-8 days) whereas, the stored seeds showed germination 93.3 per cent (22-24 days) and 80.0 per cent (28-30 days) respectively at six months and one year. Out of four different strengths of MS basal media evaluated for seed germination full strength MS was found to be most suitable (Table 11.1; Figure

11.1A). In case of six month and one year old seeds, the seeds were germinated after 20-25 and 25-30 days respectively and the percentage of germination of these seeds were lower (93.3 per cent and 80.0 per cent respectively) than that of fresh seeds on MS and other media also (Table 11.1). So cotyledonary nodes from the fresh seeds were taken for the shoot multiplication experiments. Likewise, Vishwakarma *et al.* (2009), Sanatombi and Sharma (2007) have taken fresh mature seeds of *Desmodium gangeticum* and *Capsicum annuum* respectively for better germination. Kambizia *et al.* (2006) also reported the use of current season seeds to obtain maximum germination.

Table 11.1: Effect of Storage on Seed Germination

Media Strength	*Fresh Seeds*		*Six Months Old Seeds*		*One Year Old Seeds*	
	Mean per cent of Germination	*Days for Germination*	*Mean per cent of Germination*	*Days for Germination*	*Mean per cent of Germination*	*Days for Germination*
MS	100	7-8	93.3	22-24	80	28-30
½ MS	60	8-9	56.6	22-24	53.3	30-32
¼ MS	43.3	9-10	40	25-27	33.3	36-38
[! MS	40	11-12	36.6	26-30	30	40-42

In our experiment we have evaluated four concentrations of medium *i.e.* MS, ½ MS, ¼ MS and ⅛ MS without any plant growth regulators (PGRs) for germination of seeds of *Mucuna gigantea*, out of which the MS showed 100 % of germination in 7-8 days (Figure 11.1 A). In case of another species of *Mucuna* genus *i.e. Mucuna pruriens,* Faisal *et al.* (2006a) has germinated the seeds using MS without any PGRs. Similarly, *in Arachis hypogaea, Clitoria ternatea, Desmodium gangeticum, Lathyrus ochrus, Psoralea corylifolia, Thermopsis turcica* MS medium without any PGRs were used by Venkatachalam and Kavipriya (2012), Mukhtar *et al.* (2012), Vishwakarma *et al.* (2009), Saglam (2012), Shinde *et al.* (2009), Cenkci *et al.* (2008) respectively for germination of seed. Some workers like Indravathi and Pullaiah, 2013, Parveen and Shahzad, 2010 and Barik *et al.,* 2007 have also used ½ MS without PGRS for seed germination in *Albizia amara, Cassia sophera, Clitoria ternatea* respectively.

PGRs are numerous chemical substances that profoundly promote and influence plant growth. Addition of various PGRs to the media promotes the growth of different plant parts. The cotyledonary nodes derived from 10-11 day's old axenic seedling was experimented to various concentrations of PGRs to get the morphogenic response and the results were documented in the (Table 11.2). On MS medium without any PGRs the cotyledonary node showed only 2 shoots on either side. So we have essentially added cytokinin for the induction of multiple shoots from the cotyledonary node explants. Though two different types of cytokinins

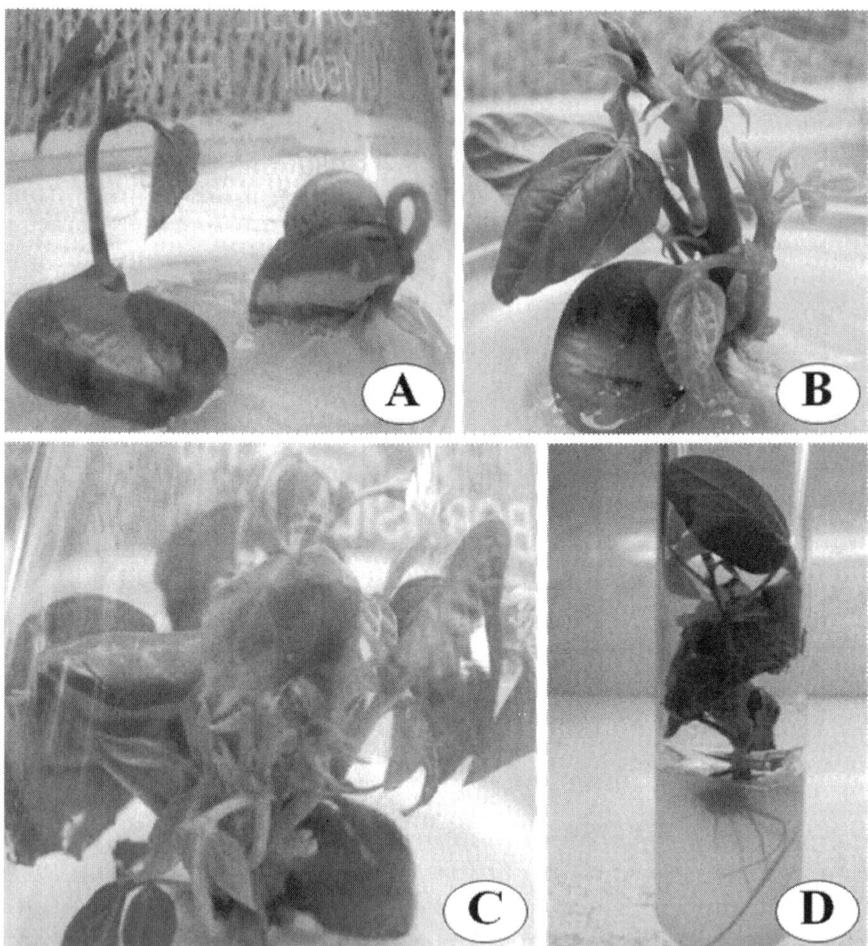

Figure 11.1: A: Seed germination on MS basal media; B: Multiple shoots induction and proliferation on MS + Kin 1.0 mg/L at day 35; C: Multiple shoot proliferation and elongation on MS + BAP 1.0 mg/l at day 35; D: Rooting of *in vitro* shoots on 1/8 MS at day 15.

(BAP and Kin) were tested, the best result was recorded with BAP. To induce multiple shoot proliferation in cotyledonary node, MS supplemented with BAP at a concentration of 1.0 mg/L was found to be optimum medium. The average highest (*c.a.* 9.1) number of shoots was found per explants (with maximum average shoot length 3.12 cm) with a frequency of 100 per cent (averaging) on 30-35 days on the optimum medium (Figure 11.1C) whereas, MS with addition of Kin showed poor response in comparison to BAP. Among different concentrations of Kin, Kin 1.0mg/L showed highest (*c.a.* 4.8) number of shoots with 3.74 cm length (Figure 11.1B).

Table 11.2: Multiple Shoot Induction from Cotyedonary Node with two Whole Cotyledons

Cytokinins mg/L	*Mean Per cent of Shoot Response*	*Average No. of Shoots/Explants*	*Average Shoot Length in (cm)*
Control (MS)	100	2	4.78
	MS + BAP		
0.5	93.3	2.8	2.00
1.0	100	9.1	3.12
2.0	100	8.5	3.85
3.0	100	6.7	3.00
4.0	100	6.4	4.71
5.0	100	4.5	2.50
	MS + Kin		
0.5	93.3	3.9	2.31
1.0	100	4.8	3.74
2.0	100	4.6	3.64
3.0	100	3.5	3.61
4.0	100	3.3	3.78
5.0	100	3.0	3.86

In the same plant Sathyanarayana *et al.* (2008) obtained 6.7 no of shoots on MS + BAP 3.5 µM using axenic axillary bud which was lower than we obtained in the first harvest. Similarly in other plants like *Mucuna pruriens* (Faisal *et al.*, 2006b) have founded ½ MS supplemented with BAP (5.0 µM) and NAA (0.5 µM) as the most suitable medium for maximum shoot induction and proliferation with 17.8 shoots with 5.9 cm length. Further my results are consistent with the findings of Indravathi and Pullaiah (2013) and Barik *et al.* (2007) who have reported MS + BAP (1.0 mg/L) as best medium for shoot culture and got fairly good result in *Albizia amara* (shoot number 6.84, length 3.20cm), *Clitoria ternatea* (shoot number 5.2, length 6.4 cm) respectively.

Result found with cotyledonary node with two whole cotyledons was similarly observed in *Lathyrus sativus* (Barik *et al.,* 2004), blackgram (Das *et al.,* 1998) and mungbean (Sen and Mukherjee, 1998). Complete removal of both cotyledons caused a delayed shoot regeneration response and produced fewer shoots. Similar observations were also reported in mungbean (Gulati and Jaiwal, 1994; Polisetty *et al.,* 1997). Also in *Lathyrus ochrus* (L.) DC., Saglam (2012) used longitudinally sliced half CN (maximum 5.92 shoots) and in *Lotononis bainesii* Baker, Vidoz *et al.* (2012) used longitudinally cut cotyledons along the midrib and got the best responses.

Rooting of *in vitro* regenerated shoots was experimented on five different concentrations of medium and out of those ⅛ MS medium devoid of any PGRs responded faster in terms of the day and facilitated better rhizogenesis. Root induction was started within 5-6 days and in 14-15 days the average root number and length were 7 and 6.67cm respectively. Though ¼ MS and [! MS gave response at a time but later one found to be the optimum medium in terms of root number and length. MS, ½ MS responded slower than the above mediums followed by $^1/_{16}$ MS and resulted in lesser number of roots (Table 11.3; Figure 11.1D).

Table 11.3: Rooting of *in vitro* Shoots

Media	*Per cent of Rooting*	*No. of Roots/ Shoots*	*Mean Root Length in cm*
MS	100	2.7	5.10
½ MS	100	3.0	5.23
¼ MS	100	3.5	5.27
⅛ MS	100	7.0	5.32
$^1/_{16}$ MS	100	4.0	5.08

In our experiment it seems as if PGRs are not essential for rooting. Likewise another species of Fabaceae *i.e. Psoralea corylifolia, developed* maximum rooting (highest number 5.4 roots per explant) and highest length (10.1 cm) in MS without any PGR (Shinde *et al.*, 2009). But *Mucuna pruriens,* Faisal *et al.* (2006) have found satisfactory improvement in rooting on ½ MS medium containing IBA (2.0 mM) with good length (5.5 cm) and number (7.8) of roots per shoots. Luna *et al.* (2003) also developed maximum number of roots in *Ilex dumosa in* ¼ MS medium but supplemented with IBA (7.3 mM).

Plantlets with leaflets and well developed roots were successfully acclimatized by transferring it to plastic glasses containing soil and sand in 1:1 and eventually established in pots containing garden soil. The survival rate of the plantlets after transfer to soil: sand (1:1) was 100 per cent, and 58.33 of the plants transferred to garden soil survived.

To acclimatize the *in vitro* rooted plantlets of *Mucuna pruriens,* Faisal *et al.* (2006a) transferred them to the plastic pots filled with sterile soilrite and 90 per cent of plantlets survived successfully. In *Sinningia speciosa* Kashyap and Dhiman (2011) have acclimatized the *in vitro* raised plantlets by taking four different hardening media, out of which sand and soil in 1:1 showed 45.16 survival rates. Venkatachalam and Kavipriya (2012) used Soil and sand in 2:1 for successful acclimatization in *Arachis hypogae.* However, Stanly and Keng (2007) found that *in vitro* rooted plantlets of *Curcuma zedoaria* and *Zingiber zerwnbet* cultured in pots containing mixture of organic soil and sand (1:1) without direct sunlight were successfully acclimatized with 100 per cent of the plantlet survived. In *Dracaena sanderiana,* Aslam *et al.* (2013) had transferred the *in vitro* rooted plantlets to plastic pots containing sand and soil rite (1:1) and transferred plantlets showed normal morphological appearance.

Conclusion

In this study the protocol developed for *Mucuna pruriens* var. utilis using cotyledonary node as explants found to be most efficient reproducible and reliable one which can be use for the large scale production and conservation of the elite genotype.

REFERENCES

Anon, A. (1962). The wealth of India. New Delhi: Publication and Information Directorate, p, 34.

Aslam, J., Mujib, A. and Sharma, M.P. (2013). *In vitro* micropropagation of *Dracaena sanderiana* Sander ex Mast: An important indoor ornamental plant. Saudi Journal of Biological Sciences, 20: 63-68.

Balogun, I. O. and Olatidoye, O. P. (2012). Chemical composition and nutritional evaluation of velvet bean seeds (Mucuna utilis) for domestic consumption and industrial utilization in Nigeria. Pakistan Journal of Nutrition, 11(2), 116.

Barik, D.P., Naik, S.K., Mohapatra, U. and Chand, P.K. (2004). High-frequency plant regeneration by *in vitro* shoot proliferation in cotyledonary node explants of grass pea (*Lathyrus sativus* L.). *In vitro* Cellular and Developmental Biology-Plant, 40: 467-470.

Barik, D.P., Naik, S.K., Mudgal, A. and Chand, P.K. (2007). Rapid plant regeneration through *in vitro* axillary shoot proliferation of butterfly pea (*Clitoria ternatea* L.)- A twinning legume. *In vitro* Cellular and Developmental Biology-Plant, 43: 144-148.

Bell, E. A. and Janzen, D. H. (1971). Medical and ecological considerations of L-dopa and 5-HTP in seeds. Nature, 229(5280), 136-137.

Capo-chichi, L.J.A., Weaver, D.B. and Morton, C.M. (2003). The use of molecular markers in the study of genetic diversity in *Mucuna.* Tropical and Subtropical Agroecosystems, 1: 309-318.

Cenkci, S., Kargioglu, M., Dayan, S. and Konuk, M. (2008). *In vitro* propagation of an endangered plant species, *Thermopsis turcica* (Fabaceae). Biologia, 63: 652-657.

Das, D.K., Prakash, N.S. and Sarin, N.B. (1998). An efficient regeneration system of black gram (*Vigna mungo* L.) through organogenesis. Plant Science, 134: 199-206.

Daxenbichler, M. E., VanEtten, C. H., Hallinan, E. A., Earle, F. R. and Barclay, A. S. (1971). Seeds as sources of L-DOPA. Journal of medicinal chemistry, 14(5), 463-465.

Ding, Y., Kinjo, J., Yang, C. R. and Nohara, T. (1991). Triterpenes from Mucuna birdwoodiana. Phytochemistry, 30(11), 3703-3707.

Faisal, M., Siddique, I. and Anis, M. (2006a). An efficient plant regeneration system for *Mucuna pruriens* L. (dc.) using cotyledonary node explants. *In vitro* Cellular and Developmental Biology-Plant, 42: 59-64.

Faisal, M., Siddique, I. and Anis, M. (2006b). *In vitro* rapid regeneration of plantlets from nodal explants of *Mucuna pruriens* - A valuable medicinal plant. Annals of Applied Biology, 148: 1-6.

Gulati, A. and Jaiwal, P.K. (1994). Plant regeneration from cotyledonary node explants of mung bean [*Vigna radiata* (L.) Wilczek]. Plant Cell Reports, 13: 523-527.

Indravathi, G. and Pullaiah, T. (2013). *In vitro* propagation studies of *Albizia amara* (Roxb.) Boiv. African Journal of Plant Science, 7: 1-8.

Janardhanan, K. and Lakshmanan, K.K. (1985). Studies on the pulse, *Mucuna utilis.* Chemical composition and anti-nutritional factors. Journal of Food Science and Technology, 22: 369-37.

Kambizia, L., Adebolab, P.O. and Afolayana, A.J. (2006). Effects of temperature, pre-chilling and light on seed germination of *Withania somnifera*, a high value medicinal plant. South African Journal of Botany, 72: 11-14.

Kashyap, B. and Dhiman, S.R. (2011). Effect of media on hardening of in vitro multiplied plantlets of gloxinia and saintpaulia under low cost polytunnels. International Journal of Farm Sciences, 1: 63-67.

Krishnamurthy, R., Suvagia, P., Pathak, J. M., Parikh, K. M. and Bhikne, N. S. (1996, June). Selection of Mucuna lines for higher L-Dopa and improved yield through conventional breeding. In Proceedings of the Conference on Breeding Research on Medicinal Plants (Vol. 2, pp. 231-234).

Luna, C., Sansberro, P., Mroginski, L. and Tarrago, J. (2003). Micropropagation of *Ilex dumosa* (Aquifoliaceae) from nodal segments in a tissue culture system. Biocell, 27: 205-212.

Manyam, B.V. (1995). An alternative medicine treatment for Parkinson's disease: results of a multicenter clinical trial – HP-200 in Parkinson's Disease Study Group. Journal of Alternative and Complimentary Medicine, 1: 249-255.

Mohan, V.R. and Janardhanan, K. (1993). Studies on the Indian tribal pulses. In proceedings of the national seminar on biodiversity: strategies for conservation and future challenges. Department of Botany. Bharathiar University. Ed. by Udaiyan, K., Janardhanan, K., Manian, S., Reddy, V.R.K.P. 116-125.

Mukhtar, S., Ahmad, N., Khan, M.I., Ania, M. and Aref, I.M. (2012). Influencing micropropagation in *Clitoria ternatea* L. through the manipulation of TDZ levels and use of different explant types. Physiology and Molecular Biology of Plants, 18: 381-386.

Murashige, T. and Skoog, F. (1962). A revised medium for rapid growth and bio assays with tobacco tissue cultures. Physiologia Plantarum, 15(3), 473-497.

Parveen, S. and Shahzad, A. (2010). TDZ–induced high frequency shoot regeneration in *Cassia sophera* Linn. via cotyledonary node explants. Physiology and Molecular Biology of Plants, 16: 201-206.

Polisetty, R., Paul, V., Deveshwar, J.J., Khetarpal, S., Suresh, K. and Chandra, R. (1997). Multiple shoot induction by benzyladenine and complete plant regeneration from seed explants of chick pea (*Cicer arietinum* L.). Plant Cell Reports, 16: 565-571.

Saglam, S. (2012). Plant regeneration from pulse-treated longitudinally sliced half cotyledon node explants of turkish ochrus chuckling [*Lathyrus ochrus* (L.) DC.]. Archives of Biological Science, Belgrade, 64: 525-529.

Sanatombi, K. and Sharma, G.J. (2007). Micropropagation of *Capsicum annuum* L. Notulae Botanicae Horti Agrobotanici Cluj, 35: 57-64.

Sathyanarayana, N., kumar, T.N.B., Vikas, P.B. and Rajesha R. (2008). *In vitro* clonal propagation of *Mucuna pruriens* var. utilis and its evaluation of genetic stability through RAPD markers. African Journal of Biotechnology, 7: 973-980.

Shinde, A.N., Malpathak, N. and Fulzele, D.P. (2009). Induced high frequency shoot regeneration and enhanced isoflavones production in *Psoralea corylifolia.* Records of Natural Products, 3: 38-45.

Singh, B. M., Srivastava, V. K., Kidwai, M. A. and Gupta, S. (1995). Aloe (Aloe barbadensis Mill.), Psoralea and Mucuna. Advances in Horticulture, 11, 513-25.

Standley, P. C. and Steyermark, J. A. (1946). Leguminosae. Flora of Guatemala. Fieldiana, Botany, 24(5), 1-368.

Stanly, C. and Keng, C.L. (2007). Micropropagation of *Curcuma zedoaria* Roscoe and *Zingiber zerwnbet Smith. Biotechnology, 6: 550-560.*

Venkatachalam, P. and Kavipriya, V. (2012). Efficient Method for *in vitro* Plant Regeneration from cotyledonary node explants of Peanut (*Arachis hypogaea* L.). International Conference on Nuclear Energy, Environmental and Biological Sciences, Bangkok (Thailand).

Verma, I. M., Polak, M., Scharfmann, R., Seilheimer, B., Eisenbarth, G., Dressler, D. and Potter, H. (1993). Nerve growth factor induces neuron-like differentiation of an insulin-secreting pancreatic beta cell line. Proceedings of the National Academy of Sciences, 90(12), 5781-5785.

Vidoz, M.L., Quesenberry, K.H., Real, D. and Gallo, M. (2012). Plant regeneration of *Lotononis bainesii* Baker (Fabaceae) through cotyledon and leaf culture. African Journal of Biotechnology, 11: 9724- 9731.

Vishwakarma, U.R., Gurav, A.M. and Sharma, P.C. (2009). *In vitro* Propagation of *Desmodium gangeticum* (L.) DC. from cotyledonary nodal explants. Pharmacognosy Magazine, 5: 145-150.

Chapter 12

Refinement of Callus-Mediated *In vitro* Plant Regeneration Protocol of *Paederia foetida* L.

*Biswaranjan Behera[2], Shashikanta Behera[3], Durga Prasad Barik[1] and Soumendra Kumar Naik[1]**

[1]Department of Botany, Ravenshaw University, Cuttack-753003, Odisha
[2]ICAR-National Rice Research Institute (NRRI), Cuttack-753006, Odisha
[3]ICAR-Central Tuber Crops Research Institute (CTCRI), Bhubaneswar-751019, Odisha
**Corresponding author; Email: sknuu@yahoo.com*

ABSTRACT

Paederia foetida L. (Family: Rubiaceae) is well known for its various folkloric medicinal usages. The leaf of the plant is widely used by the tribal people of India and Bangladesh for the treatment of gastrointestinal disorder including dysentery, diarrhoea, gastritis, and abdominal pain. Various studies on biological activities of P. foetida also validate its medicinal potential which includes anti-diarrhoeal, gastro-protective, and anti-inflammatory activities. The leaves are reported to have antioxidant properties also. This medicinal plant is under stress in its natural habitat due to various anthropogenic activities. In fact, it has been placed among 41 conservation concern plant species of Odisha. A refined plant regeneration protocol for complete plantlets by callus-mediated shoot organogenesis has been reported here from axenic internode explant for P. foetida. Callus induction was best observed from axenic internodes on Murashige and Skoog's (MS) medium supplemented with 1.5 mg/l N^6-benzylaminopurine (BAP) and 0.3 mg/l naphthalene acetic acid. Callus induced from the internode explants were multiplied on the same medium. Highest shoot organogenesis (46.7 per cent; 6 shoot per callus) *was observed by culturing the callus on MS medium supplemented with 1.0 mg/l BAP and 0.5 mg/l indole-3-acetic acid. About 86.7 per cent of the in vitro formed shoots were rooted on ½ MS medium and an average 10.6 roots per shoot were recorded on the above medium. Plantlets were successfully acclimatized (86.7 per cent) using soil and sand (1:1) ratio and subsequently established in pot containing soil and field condition with zero mortality.*

Keywords: *Paederia foetida, Medicinal plant, Callus culture, Organogenesis.*

Introduction

Paederia foetida L. is one of 50 species in the genus *Paederia* belonging to family Rubiaceae with immense medicinal value (Wagner *et al.,* 1999). It is commonly known as Skunk vine or Stink vine or Chinese fever vine (Wagner *et al.,* 1999; Aquilar, 2001). This plant is used in traditional system of medicine for the treatment of various types of diseases. Traditionally this plant is famous for its use for intestinal or digestive disorders (Afroz *et al.,* 2006; Behera *et al.,* 2018). It acts as immunity enhancer and provides strength to the body against various infections (Chunekar, 1998). All most all the parts of the plants including leaf, root, bark and fruit are known to have medicinal uses (Chopra *et al.,* 1969; Anonymous, 2001; Barua *et al.,* 2007; Chanda *et al.,* 2013). Due to over-exploitation, the plant is under stress in its natural habitat. In fact, it has been placed among 41 conservation concern plant species of Odisha. To overcome the limitation of conventional propagation, a number of *in vitro* plant regeneration protocols have already been reported for this valuable medicinal plant species. Here, development of a refined plant regeneration protocol for *P. foetida* by callus-mediated shoot organogenesis using axenic internode explants has been reported.

Materials and Methods

Source and Preparation of Explants

Young twigs were collected from the plant maintained in the garden of the Department of Botany, Ravenshaw University, Cuttack, Odisha. Leaves and internode were excised from the twigs, cut into small pieces and used as explants for callus culture. The explants (leaves and internode segments) were kept under running tap water for 30 min followed by 8 min treatment with 5 per cent (v/v) aqueous solution of Teepol (Reckitt Benckiser Ltd, India). They were then rinsed with double distilled water for 5 min. Finally, surface sterilization of the leaf and internode segment was carried out inside the laminar air flow chamber with 0.1 per cent (w/v) of $HgCl_2$ for 3 min and 5 min respectively. The explants were then washed thoroughly, about 4-5 times, with sterilized double distilled water prior to inoculation on culture media.

To compare the efficacy of the axenic internode and leaf explants with that of the mature ones, another experiment was carried out. The *in vitro* shoots developed by axillary shoot proliferation of nodal segments on MS + 3.0 mg/l BAP medium

(Behera *et al.*, 2018) were the source of axenic leaf and internode explants. The leaf and internode portions were cut into small segments and used for callus induction studies.

Culture of Explants for Induction and Proliferation of Callus

For callus induction and proliferation, leaf and internodal explants collected from mature plant were inoculated on Murashige and Skoog's (1962) (MS) basal medium alone or supplemented with (0.25-4.0 mg/l) 2, 4 - dichlorophenoxyacetic acid (2, 4-D). The explants were also cultured on MS + 0.5 mg/l 2, 4-D supplemented with different concentrations of thidiazuron (TDZ) (0.01-0.1 mg/l). In a previous report from our laboratory (Naik *et al.*, 2013) it was observed that 1.5 mg/l N^6-benzylaminopurine (BAP) in combination with 0.1 mg/l 1-naphthalene acetic acid (NAA) was found to be suitable for callus induction of *P. foetida* from leaf and internode explants. Thus in this experiment different types and concentrations of auxin such as indole-3-butyric acid (IBA; 0.1-1.0 mg/l) or indole-3-acetic acid (IAA; 0.1-1.0 mg/l) or NAA (0.1-0.5 mg/l) were also tested in combination with 1.5 mg/l BAP to check their influence on callus induction. The axenic leaf and internode explants collected from the *in vitro* regenerated shoots obtained from the axillary shoot proliferation were also inoculated in the same callus induction media that were used for explants collected from young growth flush of mature plant. All the culture media were augmented with 3 per cent sucrose and gelled with 0.8 per cent agar. After the establishment of callus, they were sub-cultured on the same medium at an interval of 3 weeks.

Shoot Organogenesis from Callus

Among the explants tested, the axenic internode showed better results and the percentage of callus formation from them was found to be comparatively higher on MS augmented with BAP (1.5 mg/l) and NAA (0.1-0.5 mg/l) media. Thus to find out the best organogenic potential of callus induced from axenic internode on different combination of BAP (1.5 mg/l) and NAA (0.1-0.5 mg/l) were transferred to MS + 1.0 mg/l BAP + 0.5 mg/l IBA, which was reported to be the best regeneration medium in our previous study (Naik *et al.*, 2013). The best responded callus (induced on MS + 1.5 mg/l BAP + 0.3 mg/l NAA) in terms of its regeneration capacity was subsequently tested for enhanced regeneration of shoots from the callus, if any, by culturing them on MS supplemented with a single cytokinin like zeatin (Z; 0.5-4.0 mg/l) or meta-Topolin (mT; 0.5-4.0 mg/l) or TDZ (0.1-1.0 mg/l) or different combinations of BAP (1.0-3.0 mg/l) and IAA (0.1 -1.0 mg/l).

Rooting of *In vitro* Formed Shoots

The *in vitro* regenerated shoots were excised and inoculated in different rooting medium such as MS, ½ MS and ¼ MS. The rooting media were supplemented with 3 per cent sucrose and gelled with 0.6 per cent agar.

Culture Conditions

All the cultures for callus induction, shoot regeneration as well as subsequent rooting of shoots were maintained in a culture room at 25 ± 1 °C under 16 h of photoperiod with 50 μmol m^{-2} s^{-1} photon flux density provided by cool white fluorescent tubes.

Acclimatization of Plants

The regenerated plantlets were taken out from the culture tubes followed by washing under tap water to remove agar and traces of medium without hampering the roots. The shoots were then implanted in thermocol cups (7.5 cm dia) containing sterile sand and soil in 1:1 ratio and were covered with polyethylene bags. These plantlets were kept in the culture room. After a week's time, holes were punched in the polyethylene bags to reduce the humidity. Plantlets were watered once in 3 days with sterile distilled water. After three weeks, plants were transferred outdoors and kept under the shade for about one week. They were then transferred to pots having garden soil and kept for 2 weeks prior to transfer under full sun. Finally the potted plants were transplanted in field.

Data Analysis

For callus induction experiment eight culture vessels with two explants were used. Whereas five culture vessels each with two calli were used for shoot organogenesis from callus. Ten tubes containing one explant each were taken for rooting experiment. All these experiments were repeated thrice. Observations were recorded regarding the frequency of callus induction, shoot proliferation, and rooting as well as the number of shoots per callus clump, shoot length, roots per shoot and root length. Duncan's New Multiple Range Tests (Gomez and Gomez, 1984) was used to designate means with significant effect at $P \leq 0.05$.

Results

Evaluation of Explants for Callus Initiation and Proliferation

The efficacy of callusing varied with the explant type. The response of different explants for callus initiation and proliferation was different in different hormonal combinations (Table 12.1). Among the two different types of explants used from mature source (internode and leaf), internodal explants were found to be superior to leaf for callus induction. Further to determine the efficacy for callus induction and proliferation, these mature explants were compared with the explants derived from the axenic source. The axenic internode explants showed better results than axenic leaf explants. Overall the axenic internode was found to be the best explants for callus induction among all the four types of explants, where cent percent callus induction was recorded on the optimal medium (Table 12.1).

Table 12.1: Effect of plant growth regulators on callus induction from internode and leaf explants from different sources (mature and axenic) of *Paederia foetida*

MS Basal Medium + Plant Growth Regulators (mg/l)						*Callus Formation (per cent)*			
BAP	***2,4-D***	***TDZ***	***NAA***	***IAA***	***IBA***	***MI***	***AI***	***ML***	***AL***
-	-	-	-	-	-	0.0^{q}	0.0^{o}	0.0^{s}	0.0^{q}
	0.25					66.7^{i}	68.8^{h}	62.5^{i}	64.6^{ij}
	0.5					75.0^{e}	79.2^{d}	72.9^{e}	75.0^{de}
	1.5					72.9^{ef}	75.0^{e}	70.8^{ef}	77.0^{d}
	1.0					72.9^{ef}	72.9^{f}	72.9^{e}	77.0^{d}
	2.0					66.7^{i}	70.8^{g}	62.5^{i}	72.9^{ef}
	3.0					62.5^{j}	68.8^{h}	58.3^{ijk}	66.7^{hi}
	3.5					60.4^{jk}	64.6^{i}	54.2^{l}	64.6^{ij}
	4.0					50.0^{m}	50.0^{k}	45.8^{m}	50.0^{l}
	0.5	0.01				39.6^{p}	43.8^{n}	31.3^{qr}	41.7^{no}
	0.5	0.03				45.8^{n}	50.0^{k}	35.4^{p}	43.8^{mn}
	0.5	0.05				54.2^{l}	60.4^{j}	43.8^{n}	45.8^{m}
	0.5	0.08				43.8^{no}	47.9^{l}	39.6^{o}	41.8^{no}
	0.5	0.10				39.6^{p}	45.8^{m}	33.3^{po}	37.5^{p}
1.5					0.1	72.9^{ef}	75.0^{e}	68.8^{fg}	70.8^{fg}
1.5					0.5	79.2^{d}	81.3^{c}	70.8^{ef}	72.9^{ef}
1.5					1.0	70.8^{fg}	72.9^{f}	66.7^{gh}	68.8^{gh}
1.5				0.1		68.8^{gh}	72.9^{f}	62.5^{i}	64.6^{ij}
1.5				0.5		72.9^{ef}	79.2^{d}	68.8^{fg}	70.8^{fg}
1.5				1.0		66.7^{i}	75.0^{e}	60.4^{ij}	62.5^{jk}
1.5			0.1			91.7^{b}	100^{a}	83.3^{b}	89.6^{b}
1.5			0.2			91.7^{b}	100^{a}	83.3^{b}	93.8^{a}
1.5			0.3			95.8^{a}	100^{a}	87.5^{a}	93.8^{a}
1.5			0.4			89.6^{bc}	95.8^{b}	81.3^{bc}	89.6^{b}
1.5			0.5			89.6^{bc}	95.8^{b}	77.0^{d}	87.5^{bc}

MI: Mature Internode, AI: Axenic Internode; ML: Mature Leaf, AL: Axenic Leaf; Values within a column followed by same letters are not significantly different ($P \leq 0.05$; DMRT).

Evaluation of Plant Growth Regulators for Callus Formation

Plant growth regulator free MS medium failed to induce callusing irrespective of the explants (leaf and internode isolated from two different sources *i.e.* from mature plant and axenic shoot culture) used. All types of explant started swelling within 3 weeks of culture. Callus induction from the explants was seen from the cut ends and subsequently spreading towards the centre. Addition of plant growth regulators either alone or in combinations was necessary for the initiation and proliferation of callus (Table 12.1).

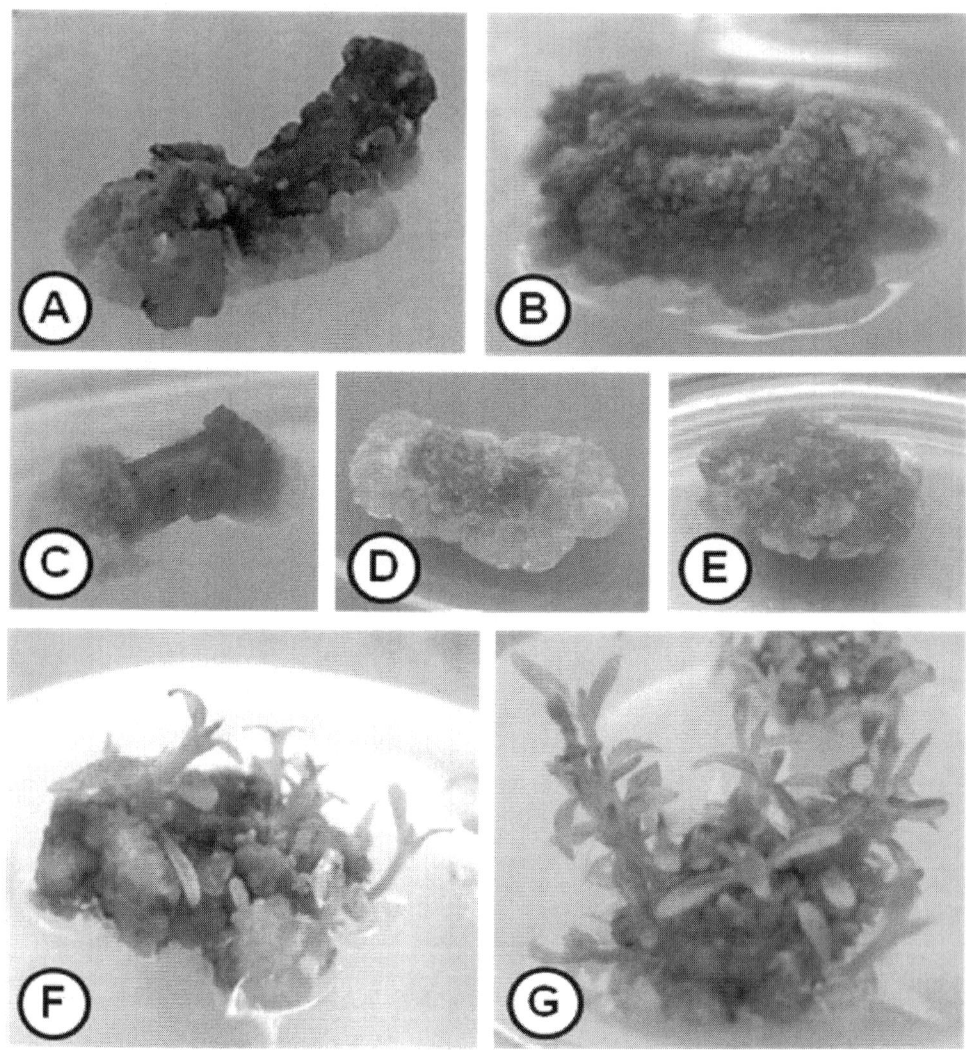

Figure 15.1: A: Callus initiation from mature internode explant on MS + 1.5 mg/l BAP + 0.3 mg/l NAA; B: Proliferation of callus from mature internode explant on MS + 1.5 mg/l BAP + 0.3 mg/l NAA; C: Axenic internodal segment cultured on MS medium augmented with 1.5 mg/l BAP and 0.3 mg/l NAA showing callus induction from cut ends; D: Proliferation of axenic internode derived callus on MS medium augmented with 1.5 mg/l BAP and 0.3 mg/l NAA; E: Callus derived from axenic internode explant on MS + 1.5 mg/l BAP + 0.3 mg/l NAAl; F: Initiation of shoot regeneration from the axenic internode derived callus cultured on MS + 1.0 mg/l BAP + 0.5 mg/l IAA; G: Proliferation of shoots from the same callus on MS medium containing 1.0 mg/l BAP + 0.5 mg/l IAA.

Among the different types of plant growth regulators used, either singularly or in combination, best callus induction was recorded on MS supplemented with BAP (1.5 mg/l) + NAA (0.3 mg/l). On this medium, highest (100 per cent) callus induction and proliferation was observed from axenic internode explant while the mature internode yields 95.8 per cent of callus on a medium of same composition (Table 12.1; Figures 12.1 A-E). Cent percent callus formation was also observed from axenic internode explants on MS + 1.5 mg/l BAP + 0.1 or 0.2 mg/l NAA medium. At the same time callus induction frequency from leaf explants derived from mature plant and axenic shoot cultures were found to be 87.5 per cent and 93.8 per cent respectively on MS + 1.5 mg/l BAP + 0.3 mg/l NAA. The calli obtained on all these media were greenish yellow in colour. Other two auxins (IAA and IBA) when used along with 1.5 mg/l BAP showed poor response of callus formation than NAA, which is true for both the sources of explants (Table 12.1).

Augmentation of 2, 4-D to MS medium induced soft white callus from axenic internode explant. Among all the concentrations of 2, 4-D tested, 0.5 mg/l was found to be optimum with an average callus induction frequency of 75.0 per cent (Table 12.1). Shoot bud induction and elongation of shoots was observed from these calli on the same medium (data not shown).

Shoot Organogenesis from Callus

The percentage of callusing was comparatively low in leaf explants, so those calli were not used for the further shoot organogenesis experiment. As high frequency of callusing was observed from axenic internode explants on MS + 1.5 mg/l BAP + (0.1- 0.5 mg/l) NAA, the calli were individually used for further organogenesis experiments (Figure 12.2). To select the best callus for shoot regeneration, the callus obtained from all these combinations of BAP and NAA were cultured on MS medium supplemented with 1.0 mg/l BAP + 0.5 mg/l IBA, which was the optimum shoot regeneration medium as observed from our previous report (Naik *et al.,* 2013). A maximum shoot regeneration frequency of 43.3 per cent was recorded from the callus obtained from MS augmented with 1.5 mg/l BAP and 0.3 mg/l NAA (Figure 12.2). Thus, callus obtained from the above mentioned medium were chosen for further experimentation with an aim to upscale the shoot regeneration.

In order to determine the influence of other plant growth regulators, either alone or in combination on shoot regeneration, the axenic internode derived callus were cultured on MS medium supplemented with different concentrations of Z or mT or TDZ or with combinations of BAP + IAA. MS medium devoid of plant growth regulators resulted in no organogenesis (Table 12.2).

Among the three different cytokinins *i.e.* Z, TDZ and mT tested individually, TDZ was found to be superior for shoot organogenesis. On MS + 0.3 mg/l TDZ, 43.3 per cent of callus exhibited shoot regeneration with 3.1 shoots per callus. 'Z' was found to be least effective for callus mediated shoot organogenesis among the three cytokinins tested (Table 12.2). Highest number of shoots (6.0) per callus with a response percentage of 46.7 was observed on MS medium supplemented with

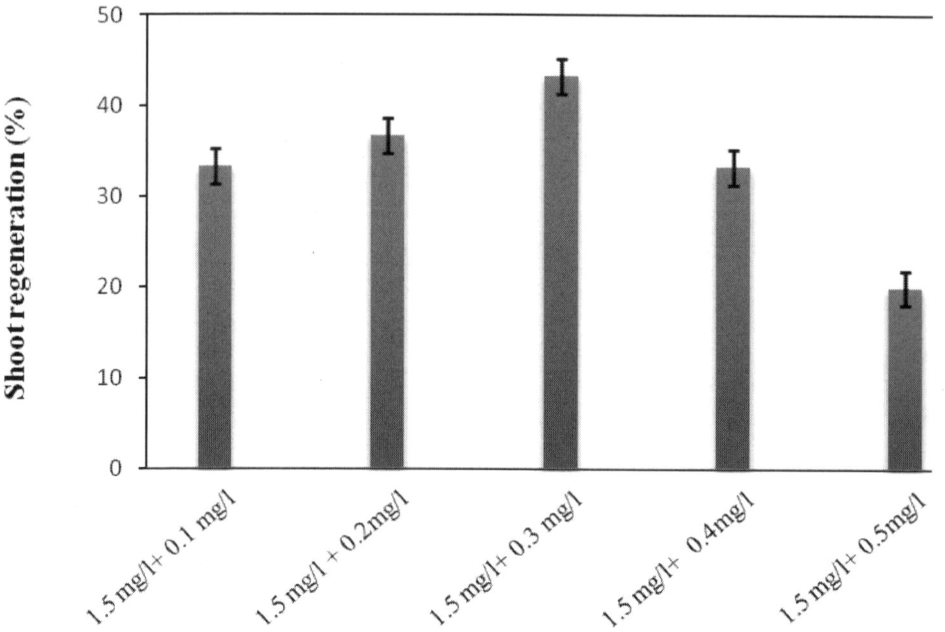

Figure 12.2: Frequency of Shoot Regeneration from Axenic Internode Derived Callus, Obtained from different Combinations of BAP and NAA, Cultured on MS + 1.0 mg/l BAP + 0.5 mg/l IBA.

1.0 mg/l BAP and 0.5 mg/l IAA in 7 weeks (Table 12.2; Figures 12.1F, G). Higher concentrations of BAP (2.0-3.0 mg/l) in combination with IAA were found to inhibit the regeneration percentage as well as the number of shoots from callus.

Rooting of *In vitro* Shoots

Among the different strengths of MS medium devoid of plant growth regulators tried for rooting, ½ MS showed highest percentage of rooting (86.7) with 10.6 roots per shoot (Table 12.3; Figures 12.3A, B).

Acclimatization of Regenerated Plantlets

Plantlets were successfully acclimatized in small thermocol cups containing a mixture of autoclaved sand and soil (1:1) with a survival percentage of 86.7 per cent (Figure 12.3C). Cent percent survival was recorded when these plantlets are transferred to larger plastic pots containing garden soil (Figure 12.3D). These acclimatized plants were successfully transferred to field condition.

Table 12.2: Shoot Regeneration from Axenic Internode Explant Derived Callus Obtained from MS + 1.5 mg/l BAP + 0.3 mg/l NAA of *Paederia foetida*

MS Basal Medium + Plant Growth Regulators (mg/l)					*Shoot Regeneration (per cent)*	*Shoots/ Callus*	*Shoot Length (cm)*
BAP	*Z*	*mT*	*TDZ*	*IAA*			
-	-	-	-	-	0.0^{l}	0.0^{p}	0.0^{j}
	0.5				20.0^{efghi}	2.3^{bcdefghi}	1.1^{fgh}
	1.0				23.3^{defgh}	2.0^{cdefghij}	1.3^{fg}
	2.0				26.7^{cdefg}	2.4^{bcdefgh}	1.8^{cde}
	3.0				20.0^{efghi}	1.8^{defghijk}	1.0^{ghi}
	4.0				16.7^{fghijk}	1.0^{hijklmno}	1.3^{fg}
		0.5			26.7^{cdefg}	2.8^{bcdef}	1.3^{fg}
		1.0			33.3^{abcde}	2.8^{bcdef}	1.8^{cde}
		2.0			30.0^{bcdef}	2.0^{cdefghij}	1.5^{ef}
		3.0			26.7^{cdefg}	1.5^{fghijklm}	1.3^{fg}
		4.0			20.0^{efghi}	1.0^{hijklmno}	1.0^{ghi}
			0.1		26.7^{cdefg}	1.8^{defghijk}	2.0^{cd}
			0.3		43.3^{ab}	3.1^{bcd}	2.2^{bc}
			0.5		36.7^{abcd}	1.7^{defghijkl}	2.0^{cd}
			0.8		33.3^{abcde}	1.2^{ghijklmn}	1.5^{ef}
			1.0		26.7^{cdefg}	1.0^{hijklmno}	1.0^{ghi}
1.0				0.1	30.0^{bcdef}	3.5^{b}	2.5^{ab}
1.0				0.5	46.7^{a}	6.0^{a}	2.9^{a}
1.0				1.0	43.3^{ab}	3.0^{bcde}	2.2^{bc}
2.0				0.1	40.0^{abc}	3.4^{bc}	2.0^{cd}
2.0				0.5	43.3^{ab}	3.4^{bc}	1.8^{cde}
2.0				1.0	36.7^{abcd}	2.8^{bcdef}	1.5^{ef}
3.0				0.1	33.3^{abcde}	2.8^{bcdef}	1.8^{cde}
3.0				0.5	36.7^{abcd}	2.5^{bcdefg}	1.3^{fg}
3.0				1.0	30.0^{bcdef}	2.0^{cdefghij}	1.3^{fg}

Values within a column followed by same letters are not significantly different ($P \leq 0.05$; DMRT).

Table 12.3: Rooting of *In vitro* Regenerated Shoots

Media	Rooting (per cent)	Roots/shoot	Root length (cm)
MS	66.7^{bc}	8.6^{b}	1.7^{c}
½ MS	86.7^{a}	10.6^{a}	3.0^{a}
¼ MS	73.3^{b}	6.8^{c}	2.0^{bc}

Values within a column followed by same letters are not significantly different ($P \leq 0.05$; DMRT).

Figure 12.3: A, B: Rooting of *in vitro* regenerated shoot(s) derived from callus culture on ½ MS; C: Acclimatized plantlet in sand and soil (1:1); D: Acclimatized plants growing in plastic pots containing garden soil.

Discussion

Regeneration of plants by callus mediated organogenesis has already been documented and found useful for large scale production of a number of plants species (Siddique *et al.,* 2010; Cheruvathur *et al.,* 2013; Sharma *et al.,* 2014; Abraham and Thomas, 2015; Jose and Thomas, 2015; Bolyard, 2018). Attempt has been made in this study to develop an efficient callus mediated plant regeneration protocol for *P. foetida.* Type or source of explants is one of the key factors influencing induction of callus. Hassan *et al.* (2012) reported callus induction (88.2 per cent) from axenic nodal explants of *P. foetida* and subsequent shoot organogenesis (14.4 shoots) from callus on MS + 0.5 mg/l BAP medium. The high shoot organogenesis response of nodal explant derived callus was probably due to the presence of some internal components from pre-existing meristem (Hassan *et al.,* 2012). In this study two different explants types *i.e.* internode and leaf from two different sources were tested for callus mediated shoot organogenesis. Results indicated that the frequency of callus induction was higher in internode explants than the leaf explants. Further the axenic internode was more responsive than the mature internode explants. Similar observation has also been observed in our earlier work on *P. foetida* (Naik *et al.,* 2013). Internode explant was also reported to be superior for callus induction than other explants including leaf in a number a medicinal plant species such as *Anoectochilces elatus* (Sherif *et al.,* 2016), *Caralluma fimbriata* (Rajaram *et al.,* 2012), *Caralluma stalagmifera* (Naik *et al.,* 2014), *Orthosiphon stamineus* (Elangomathavan *et al.,* 2017), *Salvia officinalis* (Jafari ***et al.,*** 2017) and *Thymus persicus* (Bakhtiar *et al.,* 2016). The different response of explants to the plant growth regulators could be due to presence of different amount of endogenous growth regulators in the source explant or due to different sensitivities of organs to various growth regulators (Lisowska and Wysonkinska, 2000; Jose and Thomas, 2015).

The choice of media for induction of callus in this study was based on previous report of our laboratory (Naik *et al.,* 2013) where we have already standardized the combination of BAP (1.5 mg/l) and NAA (0.1 mg/l) for induction of callus in *P. foetida*. In this thesis other types and combinations of plant growth regulators such as 2,4-D; 2,4-D + TDZ; BAP + IAA; BAP + IBA were tested to check their efficacy for better callus induction but no promotive effect was observed than the earlier report of Naik *et al.* (2013). Thus different other concentrations of NAA (0.1, 0.2, 0.3, 0.4 and 0.5 mg/l) was also tried in combination with BAP (1.5 mg/l, which was considered as optimum from earlier report of Naik *et al.,* 2013) to fine tune the concentrations of growth regulator (NAA) for enhancing the frequency of callus induction. In this study maximum callus induction frequency was observed when 0.3 mg/l NAA was used in combination with 1.5 mg/l BAP. A combination of BAP and NAA for callus induction from nodal explant of *P. foetida* was also reported by Hassan *et al.* (2012), which was in agreement with the present study but the optimum concentration used by them was different *i.e.* BAP (1.5 mg/l) + NAA (0.5 mg/l). The synergistic effect of BAP in conjunction with NAA for callus induction was also reported in various medicinal plants like *Aconitum heterophyllum* (Jabeen *et al.,* 2006), *Cinnamomum tamala* (Deb *et al.,* 2013), *Citrulla colocynthis* (Meena *et al.,* 2014), *Ophiorrhiza prostata* (Beegum *et al.,* 2007), *Rhaponiticum carthamoides* (Skala *et al.,* 2015), *Salvia officinalis* (Jafari *et al.,* 2017) and *Vitex leucoxylan* (Chordia *et al.,* 2010).

The differentiation of shoot from callus is known to be controled by different factors, of which plant growth regulator is one. For shoot regeneration from callus, inclusion of plant growth regulator(s) either alone or in combination to the basal medium is crucial and their requirement is tissue as well as species specific (Skoog and Miller, 1975; Dakah *et al.*, 2014). In the present study callus derived only from the *in vitro* internodal explants was used for shoot regeneration experiments. Among all the plant growth regulators tested either singularly or in combination, MS medium containing 1.0 mg/l BAP and 0.5 mg/l IAA was most effective for shoot regeneration from callus. On this medium 46.7 per cent regeneration frequency was observed in comparison to 33.3 per cent as reported in earlier report of *P. foetida* by Naik *et al.*, 2013. But the regeneration frequency (46.7 per cent) of this study was still considered to be low. In this study there was an increase in the number of shoot (ca. 6) regenerated per callus, which was significantly higher than earlier reports (2 shoots; Naik *et al.*, 2013). A balance between auxin and cytokinins is known to have a regulatory role in shoot organogenesis since long. A lower ratio of auxin to cytokinin usually promotes shoot formation during organogenesis from callus (Skoog and Miller 1957; Smigocki and Owens, 1989; Thorpe 1994; Tsai *et al.*, 2016). The result obtained in the present study clearly indicated that a combination of low auxin and high cytokinin is essential for getting best response for callus mediated shoot regeneration in *P. foetida*. Organogenic differentiation from callus using a combination of auxin and cytokinin has also been reported in medicinal plants like *Caralluma fimbriata* (Rajaram *et al.*, 2012), *Momordica charantia* (Thiruvengadam *et al.*, 2012) and *Orthosiphon stamineus* (Elangomathavan *et al.*, 2017).

Shoots regenerated from callus were best rooted on ½ MS medium which is similar to the previous reports of *P. foetida* (Behera *et al.*, 2017; Behera *et al.*, 2018). Poor initial growth and low transplantation rate were two major obstacles during acclimatization of *in vitro* generated plantlets (Abraham and Thomas, 2015). But in this study about 86.7 per cent of the plantlets produced were successfully acclimatized in field.

Conclusion

Callus mediated shoot regeneration is one of the common methods for production of plants. The plants regenerated in the present study were phenotypically similar to the mother plant and no morphological abnormalities have been noticed. Reports on production of clonal plants in a number of plant species using the technique of callus culture are not scanty. At the same time introduction of new genetic variations and obtaining useful somaclonal variant in plants regenerated through callus culture can't be ruled out (Lisowska and Wysokinska, 2000). However, in this plant regeneration system of *P. foetida* shoot production from callus occurs at low frequency (46.7 per cent) and further research is necessary to increase its efficiency before recommending this protocol for further use.

Acknowledgements

The first author gratefully acknowledges University Grants Commission (UGC), New Delhi for financial assistance through Rajiv Gandhi National Fellowship. The authors acknowledge the Head, Department of Botany for infrastructural support. DST-FIST support by Department of Science and Technology, Government of India to the Department of Botany, Ravenshaw University, Cuttack is also gratefully acknowledged.

REFERENCES

Abraham, J. and Thomas, T.D. (2015). Plant regeneration from organogenic callus and assessment of clonal fidelity in *Elephantopus scaber* Linn., an ethnomedicinal herb. Physiol. Mol. Biol. Plants. 21(2): 269-277.

Afroz, S., Alamgir, M., Khan, M.T., Jabbar, S., Nahar, N. and Choudhuri, M.S. (2006). Antidiarrhoeal activity of the ethanol extract of *Paederia foetida* Linn. (Rubiaceae). J. Ethnopharmacol. 105: 125-130.

Anonymous (2001). The Wealth of India, Raw Materials. New Delhi: National Institute of Science Communication, CSIR, India.

Aquilar, N. O. (2001). *Paederia foetida* L. In: van Valkenburg JLCH and Bunyapraphatsara N (Eds.) Plant resource of south-east Asia No. 12(2), Medicinal and poisonous plants. Ledien: Blackhuys Publisher, Netherlands, pp. 396-400.

Bakhtiar, Z., Hossein, M. M. and Ali, S. (2016). *In vitro* callus induction and micropropagation of *Thymus persicus* (Lamiaceae), an endangered medicinal plant. Crop Breed. Appl. Biotechnol. 16(1): 48-54.

Barua, U., Hore, D. K. and Sarma, R. (2007). Wild edible plants of Ajuli Island and Darrang districts of Assam. Indian J. Tradit. Know. 6: 191–194.

Beegum, A.S., Martin, K. P., Zhang, C., Nishitha, I. K., Ligimol, Slater, A. and Madhusoodanan, P. V. (2007). Organogenesis from leaf and inernode explants of *Ophiorrhiza prostrate*, an anticancer drug (camptothecin) producing plant. Electron. J. Biotechnol. 10(1): 115-123.

Behera, B., Behera, S., Jena, P. K., Barik, D. P. and Naik, S. K. (2017). Adventititious shoot organogenesis from internode explants of *Paederia foetida* L.: a valuable medicinal plant. Biosci. Biotechnol. Res. Asia. 14(3): 893-900.

Behera, B., Sinha, P., Gouda, S., Rath, S. K., Barik, D. P., Jena, P. K., Panda, P. C. and Naik, S. K. (2018). *In vitro* propagation by axillary shoot proliferation, assessment of antioxidant activity, and genetic fidelity of micropropagated *Paederia foetida* L. J. Appl. Biol. Biotechnol. 6(2): 41-49.

Bolyard, M. (2018). *In vitro* regeneration of *Artemisia abrotanum* L. by means of somatic organogenesis. In Vitro Cell. Dev. Biol. Plant. 54 (1): 127-130.

Chanda, S., Sarethy, I. P., De, B. and Singh, K. (2013). *Paederia foetida* - a promising ethnomedicinal tribal plant of north-eastern India. Int. J. For. Res. 24 (4): 801-808.

Cheruvathur, M. K., Abraham, J. and Thomas, T. D. (2013). Plant regeneration through callus organogenesis and true-to-type conformity of plants by RAPD analysis in *Desmodium gangeticum* (Linn.) DC. Appl. Biochem. Biotechnol. 169: 1799-1810.

Chopra, R. N., Chopra, I. C. and Verma, B. S. (1969). Supplement to glossary of Indian medicinal plants. New Delhi, CSIR.

Chordia, M. A., Kumari, R. S. and Kannan, R. R. (2010). *In vitro* regeneration of plantlets from internodal callus culture of *Vitex leucoxylon* L.-a rare medicinal plant. J. Med. Plants Res. 4(2): 2399-2403.

Chunekar, K. C. (1998). Medicinal Plants Used in Ayurveda. Rashtriya Ayurveda Vidyapeeth, New Delhi, India.

Dakah, A., Zaid, S., Suleiman, M., Abbas, S. and Wink, M. (2014). *In vitro* propagation of the medicinal plant *Ziziphora tenuior* L. and evaluation of its antioxidant activity. Saudi J. Biol. Sci. 21(4): 317-323.

Deb, C. R., Deb, M. S. and Saku Jamir, N. (2013). Callus mediated organogenesis and plant regeneration of *Cinnamomum tamala* Nees. (Lauraceae) from leaf and zygotic embryos. Int. J. Pharma. Bio Sci. 4(1): 614-622.

Elangomathavan, R., Kalaivanan, P., Hariharan, S. and Beaulah, S. N. (2017). Caulogenic response of *in vitro* raised nodal explants of *Orthosiphon stamineus* to selected auxins. Int. J. Adv. Multidiscip. Res. 4(2): 27-32.

Gomez, K. A. and Gomez, A. A. (1984). Statistical procedures for agricultural research. John Wiley and Sons Inc.

Hassan, A. K., Roy, C. K., Sultana, R. and Khatun, R. (2012). Callus induction and plant regeneration of *Paederia foetida* L., a widely used medicinal vine in Bangladesh. Bangladesh J. Sci. Ind. Res. 47(4): 373-378.

Jabeen, N., Shawl, A. S., Dar, G. H., Jan, A. and Sultan, P. (2006). Callus induction and organogenesis from explants of *Aconitum heterophyllum* medicinal plant. Biotechnology, 5(3): 287-291.

Jafari, S., Daneshvar, M. H., Salmi, M. S. and Jalal-Abadi, A. L. (2017). Indirect organogenesis and plant regeneration in common Sage (*Salvia officinalis* L.): an important medicinal plant of Iran. Mod. Appl. Sci. 11(5): 22-29.

Jose, S. and Thomas, T. D. (2015). High-frequency callus organogenesis, large-scale cultivation and assessment of clonal fidelity of regenerated plants of *Curcuma caesia* Roxb., an important source of camphor. Agrofor. Sys. 89(5): 779-788.

Lisowska, K. and Wysokinska, H. (2000). *In vitro* propagation of *Catalpa ovata* G. Don. Plant Cell Tiss. Organ Cult. 60(3): 171-176.

Meena, M. C., Meena, R. K. and Vidya, P. (2014). *In vitro* callus induction and shoot regeneration in *Citrullus colocynthis* (Linn.) Schrad. World J. Pharm. Pharm. Sci. 3(4): 1469- 480.

Murashige, T. and Skoog, F. (1962). A revised medium for rapid growth and bioassay with tobacco tissue cultures. Physiol. Plant. 15: 473-495.

Naik, N., Behera, B., Naik, S. K. and Barik, D. P. (2013). Callus mediated shoot proliferation from internode explant of *Paederia foetida* L. Plant Sci. Res. 35 (1 and 2): 44-48.

Naik, R., Karuppusamy, S. and Krishnamurty, L. (2014). Micropropagation of *Caralluma stalagmifera* var. Longipetala: a rare succulent medicinal plant from Karnataka, India. Afr. J. Biotechnol. 13(35): 3553-3559.

Rajaram, K., Priya, D., Sudarshana, D. V. and Suresh, K. P. (2012). *In vitro* regeneration of *Caralluma fimbriata* Wall. by organogenesis: a potent medicinal plant. Aust. J. Crop Sci. 6(1): 41-45.

Sharma, A., Kaur, R. and Sharma, N. (2014). *In vitro* morphogenic response of different explants of *Gentiana kurroo* Royle from Western Himalayas-an endangered medicinal plant. Physiol. Mol. Biol. Plants. 20(2): 249-256.

Sherif, N. A., Senthil Kumar, T. and Rao, M. V. (2016). *In vitro* regeneration by callus culture of *Anoectochilus elatus* Lindley, an endangered terrestrial jewel orchid. In Vitro Cell. Dev. Biol. Plant. 52: 72-80.

Siddique, I., Anis, M. and Aref, I. M. (2010). *In vitro* adventitious shoot regeneration via indirect organogenesis from petiole explants of *Cassia angustifolia* Vahl. - a potential medicinal plant. Appl. Biochem. Biotechnol. 162(7): 2067-2074.

Skala, E., Grabkowska, R., Sitarek, P., Kuzma, L., Blauz, A. and Wysokinska, H. (2015). *Rhaponticum carthamoides* regeneration through direct and indirect organogenesis, molecular profiles and secondary metabolite production. Plant Cell Tiss. Organ Cult. 123(1): 83-98.

Skoog, F. and Miller, C. O. (1957). Chemical regulation of growth and organ formation in plant tissue cultures in vitro. Symp. Soc. Exp. Biol. 11: 118-131.

Skoog, F. and Miller, C. O. (1975). Chemical regulation of growth and organ information of buds in plant tissues. *In*: Skoog, F. (Ed.), Plant growth Substances. Wisconsin Press, Wisconsin Univiversity, Madison, USA.

Smigocki, A. C. and Owens, L. D. (1989). Cytokinin-to-auxin ratios and morphology of shoots and tissues transformed by a chimeric isopentenyl transferase gene. Plant Physiol. 91(3): 808-811.

Thiruvengadam, M,. Nagella, P. and Chung, I. M. (2012). Plant regeneration from alginate encapsulated shoot tips of *Momordica dioica* for short term storage and germplasm exchange and distribution. Plant Omics J. 5(3): 266-270.

Thorpe, T. A. (1994). Morphogenesis and regeneration. *In*: Vasil, I. K. and Thorpe, T. A. (Eds.) Plant Cell and Tissue Culture. Kluwer Academic Publishers, Dordrecht.

Tsai, K.L., Chen, E. G. and Chen, J. T. (2016). Thidiazuron-induced efficient propagation of *Salvia mitiorrhiza* through *in vitro* organogenesis and medicinal constituents of regenerated plants. Acta Physiol. Plant. 38(29): 1-9.

Wagner, W. L., Herbst, D. R., and Sohmer, S. H. (1999). Manual of the flowering plants of Hawaii. Revised Edition, Vol. 2. University of Hawaii Press, Honolulu.

Chapter 13

Use of Isubgol and Semolina as Low Cost Alternative Gelling Agents in Tissue Culture of *Linum ussitatissimum* L. (Linseed)

Ashok Kumar Behera and U.B. Mohapatra*

Department of Botany, North Orissa University,
Baripada – 757 003, Odisha
**e-mail: akbbsg@gmail.com*

ABSTRACT

Tissue culture techniques have been developed for flax or Linseed, Linum usitatissimum L. by different workers. Hundreds of flax cultivars are being used in tissue culture throughout the world. Cotyledons, hypocotyls, meristems and stem segments have been used as explants for culture initiation in flax, but plant regeneration from hypocotyl segments has proved to be highly efficient. Tissue culture of four linseed cultivars such as "Sarada", "Padmini", "OL-98-15-2" and "OL-98-7-3" was initiated in a well equipped laboratory. For in vitro germination of seeds and for inoculation of the excised segments of hypocotyl from the germinated seeds, sodium hypochlorite (NaOCl) was used for surface sterilization. For callus proliferation and shoot regeneration on MS medium, agar as well as low cost alternative of agar such as Isubgol and Semolina were tested. Aauxins and cytokinins (IAA/BAP) were used as growth regulators, each at 4 different concentrations i.e. 0.5, 1.0, 1.5 and 2.0mgl^{-1}. After 28 days, the gram fresh weight of each callus was noted. All the calli were freshly inoculated in media with the same concentration of the growth regulators for the observation of number of shoots per explants. The overall findings of the investigations revealed fascinating results which would add enough *scope in the field of research for tissue culture of Linum usitatissimum L. with low-cost agar alternatives i.e. isubgol and semolina. On MS medium solidified by Agar, which we considered as standard, cultivar "Sarada" exhibited maximum callusing on 1.5mgl^{-1} IAA and 0.5mgl^{-1} BAP supplemented medium. Cultivar "Padmini" exhibited maximum shooting on 2.0 mgl^{-1} IAA and 1.0 mgl^{-1}BAP supplemented medium. On the MS medium solidified by Isubgol, cultivar 'Sarada" exhibited both maximum callusing on 0.5 mgl^{-1} IAA and 2.0 mgl^{-1} BAP supplemented medium and maximum shooting proliferations on 0.5 mgl^{-1} IAA and 2.0 mgl^{-1} BAP supplemented medium. On the MS medium solidified by semolina, cultivar "Sarada" attained maximum callusing on 0.5 mgl^{-1} IAA and 2.0 mgl^{-1} BAP supplemented medium and cultivar "Padmini" exhibited maximum shooting on 5.0 mgl^{-1}IAA and 0.8 mgl^{-1} BAP supplemented medium. Such response of isubgol and semolina corresponds to the results of cultures obtained by different workers using agar alternatives on different varieties of plants throughout the world.*

***Keywords**: Gelling agents, Isubgol, Semolina, Linseed, Tissue culture.*

Introduction

Flax or linseed, (*Linum usitatissimum* L., 2n=30, X=15) belongs to the order Malpibhiales, the family *Linaceae* and the tribe Linae. It is the most important *rabi* oilseed crop and stands next to rapeseed mustard in area of cultivation and seed production in India. Linseed is popularly known as *Atasi, Pesi, Phesi* or *Tisi* in Odia. Although linseed plants have several utilities, it is cultivated commercially for its seed, which is processed in to oil and after extraction of oil, a high protein stock feed is left (Sankari, 2000). Linseed oil has been used for centuries as drying oil whose oil content varies from 33-45 per cent (Gill, 1987). About 20 per cent of the total linseed oil produced in India is used by farmers and rest about 80 per cent goes to industries for manufacture of paints, varnishes, oilcloth, linoleum and printing ink *etc.* Keeping in view of such high economic importance, tissue culture of linseed has been done for large scale micropropagation of the same. Besides conventional uses of media used for tissue culture, some low cost alternative constituents of media *i.e.* agar alternatives, carbon source alternatives have been successfully done (Ganapathi *etal.* 1995; Prakash, 1993; Behera, 2016; Babbar, and Jain.1988).

Materials and Methods

Cultivars and Explants

Four cultivars of linseed *i.e.* cultivars "OL-98-7-3", "OL-98-15-2", "Padmini" and "Sarada" were taken for the study. Explant taken for the study was "hypocoty" because of it's higher performance in callogenesis, established by different workers. Flax can easily be regenerated from hypocotyl segments and less easily from cotyledon explants, callus and protoplasts (Mathews and Narayanaswamy, 1976). Response of different explants towards *in vitro* culture also varied. Among different explants like cotyledons, hypocotyls, stem segments and leaf cells, the hypocotyls exhibited a better response towards callusing and shoot induction (Friedt, 1990; Mathews and Narayanswamy, 1976; Yildiz and Cellar, 2002). They were surface sterilized with NaOCl (Sodium hypochlorite) with 0.2 per cent concentration as it corresponded to best result found by various workers in various crops *e.g.* Zhan *et al.,* 1989, Yildiz and Ozgen, 2006, *etc.*

Growth Regulator Combinations

Excised explants were inoculated in test tubes containing sterilized MS media as Growth regulator combinations "IAA-BAP" preferred to be better and hypocotyl explants proliferated better on MS medium with agar as gelling agent, culture were decided to be done with IAA-BAP combinations. Excised explants were inoculated in test tubes containing sterilized MS medium supplemented with both auxin (IAA) and cytokinin(BAP). Sixteen growth regulator combinations each containing one of the four concentrations (0.5, 1.0, 1.5 and 2.0 mgl^{-1}) of IAA and one of the four concentrations (0.5, 1.0, 1.5 and 2.0 mgl^{-1}) of BAP without repetition of any combination were evaluated. As growth regulator combinations "IAA-BAP" proved to be better and "hypocotyl" explants proliferated better on MS medium with agar as gelling agent, all somatic culture experiments were decided to be done with IAA-BAP combinations.

Gelling Agents

Instead of agar, two agar alternatives were chosen *i.e.* Isubgol and Semolina as low cost agar alternatives. Similar types of works were carried out by Pierik, 1989; Nagamori and Kobayashi, 2001; Stanley, 1995; Babbar and Jain, 1988, *etc.*

These agar alternatives were taken for testing their gelness by physical observations. Here, gelness observation was made by taking mass of agar alternatives in gram/volume of medium in milliliter (ml) at which they formed a semi-solid state. Here, gelness per cent means gram of agar alternative in 100ml of medium when a semi-solid state occurs.

Results

With MS medium, the agar alternatives exhibited the following gelness values. With Physical verification it was found that Isubgol of 5.0g per 100ml of MS medium attained semi-solid state or gelness. Similarly, Semolina of 14.0g per 100ml of MS medium attained gelness. Hence, 5.0g/100ml and 14.0g/100ml of Isubgol and Semolina were used in preparation of the low-cost tissue culture MS media.

The calli so formed on agar and agar alternatives taking hypocotyls as explants, varied in terms of their texture and appearances. After 28 days, the gram fresh weight of each callus was taken and noted. These calli were used for the next two-three weeks for fresh inoculation and to observe shooting responses. The results were found to be interesting.

Callusing and Shooting Response on 0.8 per cent (w/v) Agar

Callusing Response

CV "OL-98-7-3" exhibited maximum callusing of 2.7±1.0g in IAA/BAP combinations 0.5/1.0 mgl^{-1}. CV "OL-98-15-2" exhibited maximum callusing of 2.8±0.3g in IAA/BAP combinations 1.5/0.5 mgl^{-1}. CV "Padmini" exhibited maximum callusing of 2.6±0.1g in IAA/BAP combinations 2.0/1.0 mgl^{-1}. CV "Sarada" exhibited maximum callusing of 2.9±0.4g in IAA/BAP combinations 1.5/0.5 mgl^{-1} [Table 13.1, Figure 13.1(1)].

Table 13.1: Callusing and Shooting Response in Agar and Agar Alternatives with IAA-BAP

Agar and Agar Alternatives	*Cultivars*	*Maximum Callusing*		*Maximum Shooting*	
		Callus Weight in g.	*IAA-BAP Conc^n. in mgl^{-1}*	*No. of Shoots*	*IAA-BAP Conc^n. in mgl^{-1}*
Agar (0.8 per cent) w/v	OL-98-7-3	2.7±1.0	0.5/1.0	10.5±1.2	1.0/1.0
	OL-98-15-2	2.8±0.3	1.5/0.5	9.7±3.3	0.5/1.0
	PADMINI	2.6±0.1	2.0/1.0	12.2±2.8	2.0/1.0
	SARADA	2.9±0.4	1.5/0.5	11.2±2.2	0.5/0.5
Isubgol (5 per cent) w/v	OL-98-7-3	0.91±0.06	2.0/2.0	6.5±1.9	2.0/2.0
	OL-98-15-2	0.26±0.06	1.0/0.5	0.26±0.06	1.0/0.5
	PADMINI	0.33±0	2.0/0.5	0.33±0	2.0/0.5
	SARADA	1.60±0.01	0.5/1.5	1.60±0.01	0.5/1.5
Semolina (14 per cent) w/v	OL-98-7-3	0.10±0.012	0.5/1.5	2.0±0.8	0.5/1.5
	OL-98-15-2	0.04±0.020	1.0/1.5	2.5±0.5	1.0/1.5
	PADMINI	0.047±0.004	1.5/2.0	5.0±0.8	1.5/2.0
	SARADA	0.34±0.12	0.5/2.0	4.5±3.1	0.5/2.0

Shooting Response

CV "OL-98-7-3" exhibited maximum number of shooting 10.5±1.2 in IAA/BAP combinations 1.0/1.0 mgl^{-1}. CV "OL-98-15-2" exhibited maximum number of shooting 9.7±3.3 in IAA/BAP combinations 0.5/1.0 mgl^{-1}. CV "Padmini" exhibited maximum number of shooting 12.2±2.8 in IAA/BAP combinations 2.0/1.0 mgl^{-1}. CV "Sarada" exhibited maximum number of shooting 11.2±2.2 in IAA/BAP combinations 0.5/0.5 mgl^{-1} [Table 13.1, Figure 13.1(2)].

Callusing and Shooting Response on 5.0 per cent (w/v) Isubgol

Callusing Response

CV "OL-98-7-3" exhibited maximum callusing of 0.91±0.06g in IAA/BAP combinations 2.0/2.0mgl^{-1}. CV "OL-98-15-2" exhibited maximum callusing of 0.26±0.06g in IAA/BAP combinations 1.0/0.5 mgl^{-1}. CV "Padmini" exhibited maximum callusing of 0.33±0g in IAA/BAP combinations 2.0/0.5 mgl^{-1}. CV "Sarada" exhibited maximum callusing of 1.60±0.01g in IAA/BAP combinations 0.5/1.5 mgl^{-1} [Table 13.1, Figure 13.1(3)].

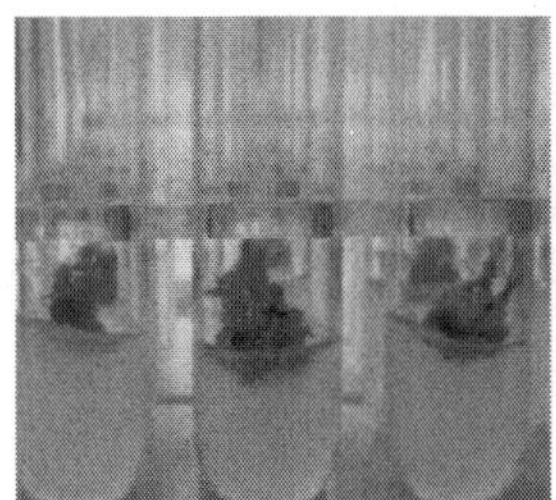

Calli on Agar

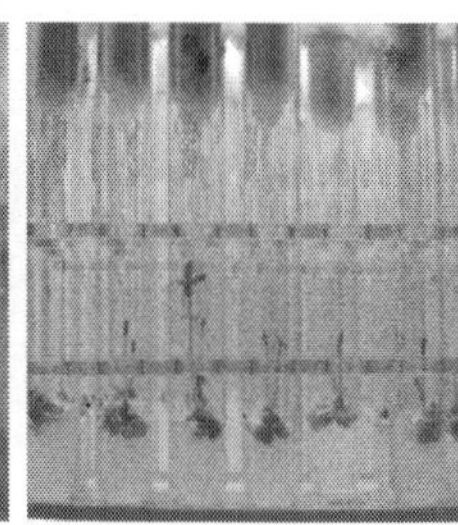

Calli with Shooting in Agar

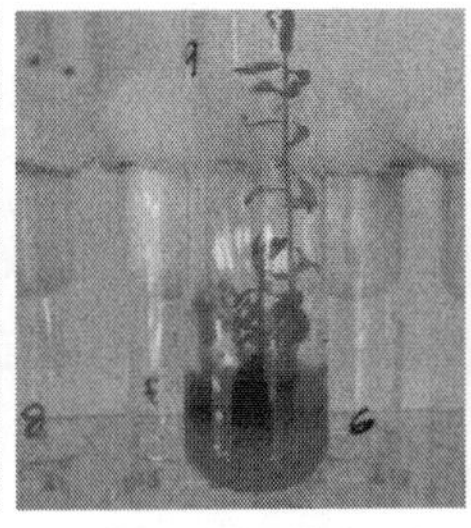

Calli with Shooting in Isubgol

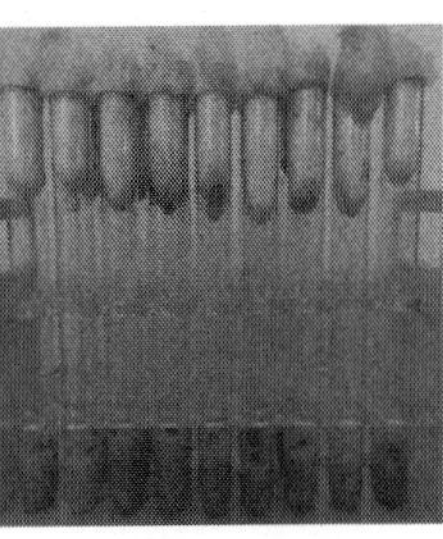

Calli in Semolina

Figure 13.1: Callus and Shooting in Agar and Agar Alternatives

Shooting Response

CV "OL-98-7-3" exhibited maximum number of shooting 6.5±1.9 in IAA/BAP combinations 2.0/2.0 mgl^{-1}. CV "OL-98-15-2" exhibited maximum number of shooting 0.26±0.06 in IAA/BAP combinations 1.0/0.5 mgl^{-1}. CV "Padmini" exhibited maximum number of shooting 0.33±0 in IAA/BAP combinations 2.0/0.5 mgl^{-1}. CV "Sarada" exhibited maximum number of shooting 1.60±0.01 in IAA/BAP combinations 0.5/1.5 mgl^{-1} [Table 13.1, Figure 13.1(3)].

Callusing and Shooting Response on 14.0 per cent (w/v) Semolina

Callusing Response

CV "OL-98-7-3" exhibited maximum callusing of 0.10±0.012g in IAA/BAP combinations 0.5/1.5mgl^{-1}. CV "OL-98-15-2" exhibited maximum callusing of 0.04±0.020g in IAA/BAP combinations 1.0/1.5 mgl^{-1}. CV "Padmini" exhibited maximum callusing of 0.047±0.004g in IAA/BAP combinations 1.5/2.0 mgl^{-1}. CV "Sarada" exhibited maximum callusing of 0.34±0.12g in IAA/BAP combinations 0.5/2.0 mgl^{-1} [Table 13.1, Figure 13.1(4)].

Shooting Response

CV "OL-98-7-3" exhibited maximum number of shooting 2.0±0.8 in IAA/BAP combinations 0.5/1.5 mgl^{-1}. CV "OL-98-15-2" exhibited maximum number of shooting 2.5±0.5 in IAA/BAP combinations 1.0/1.5 mgl^{-1}. CV "Padmini" exhibited maximum number of shooting 5.0±0.8 in IAA/BAP combinations 1.5/2.0 mgl^{-1}. CV "Sarada" exhibited maximum number of shooting 4.5±3.1 in IAA/BAP combinations 0.5/2.0 mgl^{-1} (Table 13.1).

Hence, the highest callusing and shooting responses of "hypocotyl" explants of all four linseed (*Linum usitatissimum* L.) cultivars *i.e.* "OL-98-7-3", "OL-98-15-2", "Padmini" and "Sarada", when cultured in "Agar" and "Agar alternatives" *i.e.* "Isubgol" and "Semolina" using growth regulators IAA-BAP combinations can be presented as a tabular form as follows.

Discussion

On the medium solidified by Agar, CV "Sarada" exhibited maximum callusing (concentration of IAA/BAP; 1.5/0.5) of 2.9 ± 0.4g wt. CV "Padmini" exhibited maximum shooting of 12.2± 2.8 at concentration 2.0/1.0 of IAA/BAP.

On the medium solidified by Isubgol, CV "Sarada" exhibited both maximum callusing (1.60±0.01g) and maximum shooting proliferations (1.60±0.01) in IAA-BAP concentrations of 0.5/1.5 mgl^{-1}. It corresponds to the findings of Babbar and Jain (1998) for propagation of Chrysanthemum.

On the medium solidified by Semolina, CV "Sarada" attained maximum callusing *i.e.* 0.34±0.12 g in IAA-BAP concentrations 0.5/2.0 mgl^{-1} and CV "Padmini" exhibited maximum shooting *i.e.* 5.0±0.8 shoots per calli. Such response of semolina corresponded to the average growth of cultures by Prakash (1993).

Comparing such a large number of data can be understood better by the help of the following graph.

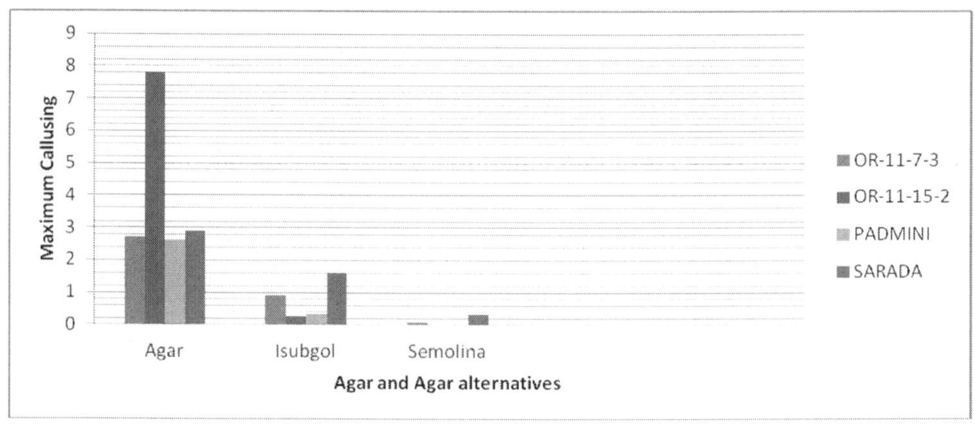

Figure 12.2: "Maximum Callusing" Response of Hypocotyl Explants of 4 Cultivars of *Linum usitatissimum* L. in 4 Agar Alternatives *i.e.* Isubgol and Semolina in Growth Regulator IAA-BAP.

Conclusion

Hence, it is concluded that hypocotyls as an explants is better than the cotyledon in tissue culture of *Linum usitatissimum* L. Even though agar is the best gelling agent so far for the micropropagation of tissue cultured plants, it is costly

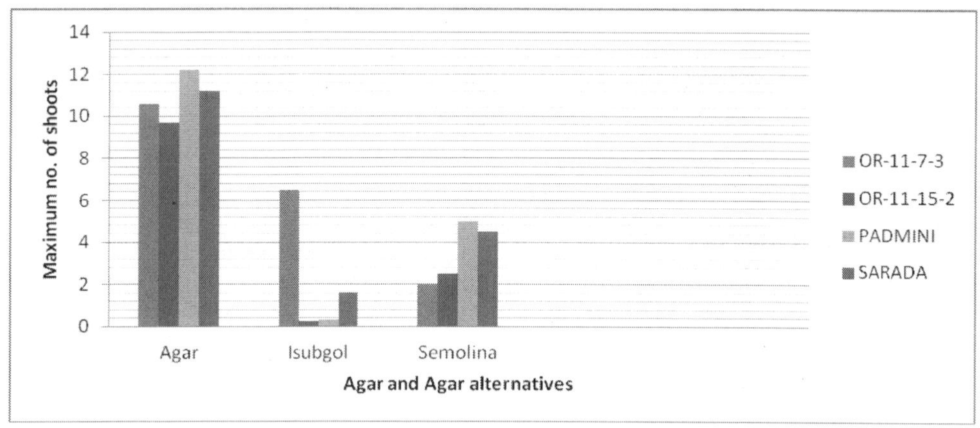

Figure 12.3: "Maximum Shooting" Response of Hypocotyl Explants of 4 Cultivars of *Linum usitatissimum* L. in 4 Agar Alternatives *i.e.* Isubgol and Semolina in Growth Regulator IAA-BAP.

and there is need of low cost alternative gelling agent. The market cost of tissue culture grade agar is approximately Rs 1800 to Rs 3500/Kg. Whereas, the cost of Isubgol and Semolina are Rs 150-180/Kg and Rs 40-50/Kg respectively. Hence, cost of Isubgol and Semolina are very much less than that of the Agar gel. But Isubgol and Semolina, both as alternative gelling agents are exhibiting promising results. Hence further efforts may be done to increase the efficiency of both of the alternative gelling agents in tissue culture.

REFERENCES

Babbar, S.B. and Jain. N. (1988). Isubgol as an alternative gelling agent in plant tissue culture media. Plant cell Rep.17: 318-322.

Behera, A.K. (2016). PhD Thesis, North Odissa University, *In vitro* culture of *Linum usitatissimum* L. with special emphasis on low cost materials and methods.

Friedt, W.(1990), Biotechnology in breeding of Industrial oil crops: the present status and future prospects. Fat Sci.Technol.90: 51-55.

Ganapathi, T.R., Mohan,T.S.S., Suprasanna, P, Bapat,V.A. and Rao. P.S. (1995). A low cost strategy for *in vitro* propagation of banana. *Curr. Sci.* 68: 646-665.

Gill, K.S. (1987). Linseed. Indian Council of Agricultural Research, New Delhi.

Koronfel, M. (1998). Development of Genotype independent and efficient regeneration procedure for flax (*Linum usitatissisum* L.). *Arab J. Biotechnol.* 1(1): 87-92.

Mathews, V.H. and Narayanswamy,S. (1976). Phytohormone control of regeneration in cultured tissue of flax. *Z. pflanzenphysiol.* 80, 436-442.

Nagamori,E.and Kobayashi,T.(2001).Viscous additive improves micropropagation in liquid medium. *J. Biosci. Bioeng.* 91: 283-287.

Prakash, S. (1993). Production of ginger and turmeric through tissue culture methods and investigations in to making tissue culture propagation less expensive. Ph.D. Thesis. Bangalore University, Bangalore.

Sankari, H.S.(2000). Linseed (*Linum usitatissimum* L.) cultivars and breeding lines as stem biomass producers. *J. Agron. Crop Sci.* 184; 225-231.

Stanley, D.(1995). Tissue culturing plants on corn starch. *Agr. Res.*, p-11,

Yildiz, M. and Cellar ER, (2002). Increasing injured area on hypocotyl explants of flax (*Linum usitatissisum* L.) leads to high frequency callus based shoot regeneration. *Turk J. Bio.*, 26, 95-98.

Yildiz, M., Ozgen, M. (2006). A comparison of growth regulators for adventitious shoot regeneration from hypocotyls of flax (*Linum usitatissisum* L.) *Journal of Food, Agriculture and Environment,* 4 (384), 171-174.

Zhan, X., Jones, D.A., Kerr, A. (1989). Regeneration of shoots on root explants of flax. *Ann. f Bot.* 297-299.

Chapter 14

Conservation of Rare and Endangered Plants through Micropropagation and other *In vitro* Approaches: An Indian Perspective

Rajashree Das and Ranindra Kumar Nayak*

P.G. Department of Environmental Science,
Fakir Mohan University, Balasore – 756 020, Odisha
**e-mail: rajashreedas23@gmail.com*

ABSTRACT

Biological diversity is the very basis of human survival and economic development as it provides food, housing, clothing, medicine and industrial raw materials and offers a potential for providing many more, yet unknown benefits to mankind. With ever increasing loss of biodiversity in recent years, that has posed a serious threat to the survival of mankind, biodiversity conservation has become a global concern. Advances in plant biotechnology, provide new options for collection, multiplication and short- to long-term conservation of plant biodiversity, using in vitro culture techniques. Significant progress has been made for conserving endangered, rare, crop, ornamental, medicinal and forest species, especially for non-orthodox seed- and vegetatively- propagated plants of temperate and tropical origin. Cell and tissue culture techniques ensure rapid multiplication and production of plant material under aseptic conditions.

Keywords: *Biodiversity, Micropropagation, Cryopreservation, In vitro seed germination, Somatic embryogenesis.*

Introduction

The variation of life forms within a given ecosystem, biome or an entire planet is termed as biodiversity. In other words, it is the totality of genes, species and ecosystem in a region. The world today is facing the loss of biodiversity which can be attributed to the influence of human beings on the world's ecosystem. Worldwide, many plant species are threatened because of the gradual disappearance and alteration of the natural ecosystems for various human activities. Often, this is due to the clearing of indigenous vegetation for anthropocentric land uses and the resulting erosion, salinization and invasion of alien species, but more recently climate change is looming as a significant new threat.

More than 50 per cent of the world's plant species are endemic to 36 global biodiversity hotspots (Conservation International, 2017), which once covered 15.7 per cent of the earth's land surface and which are now reduced to 2.3 per cent. It has been estimated that, one third of the global plant species are threatened in different levels according to International Union of Conservation of Nature (IUCN, 2013).Thus, biodiversity conservation and sustainable utilization of plant resources are a matter of great concern today. In this context, in order to bring about sustainable resource conservation and management, it is essential to adopt different approaches for managing biodiversity. The increasing awareness and great concern about global biodiversity conservation, the United Nations (UN) declared the current decade (2011-2020) as the "Decade of Biodiversity' and has set 2020 as the target for restoring at least 15 per cent of degraded ecosystems as well as conserving 17 and 10 per cent of terrestrial and inland water and marine and coastal areas respectively (Tscharntke *et al.,* 2012; SCBD, 2014).

In situ conservation alone is not sufficient to meet the challenges of saving endangered species. While seed banking can be utilized for *ex situ* conservation of the majority of endangered species, there are a significant number of species for which seed banking is not an option (Chauhan, 2016). Conserving the genetic diversity of species and populations is highly important in these situations, and micropropagation can provide large numbers of propagules from a cross-section of the genetic diversity of a region (Rogers, 2003).

India as a Mega-biodiversity Country

Forests are the main resource of species, habitat and genetic diversity (Noble and Dirzo, 1997). India occupies tenth position among the major plant diversity countries of the world. India has 2.4 per cent of world's area with 8 per cent of global biodiversity and it is one of the 12 mega diversity hotspot countries of the world with a rich diversity of biotic resources. Out of 36 hotspots recognized, India has four major hotspots (Conservation International, 2017).

i. Himalaya: Includes the entire Indian Himalayan region (and that is in Pakistan, Tibet, Nepal, Bhutan, China and Myanmar)

ii. Indo-Burma: Includes entire North-eastern India, except Assam and Andaman group of Islands (and Myanmar, Thailand, Vietnam, Laos, Cambodia and southern China)

iii. Sundalands: Includes Nicobar group of Islands (and Indonesia, Malaysia, Singapore, Brunei and Philippines)

iv. Western Ghats and Sri Lanka: Includes entire Western Ghats (and Sri Lanka)

The bio-geographic position of India is so unique that all known types of ecosystems range from coldest place like the dry cold deserts of Ladakh, temperate and Alpine and subtropical regions of the North-West and trans-Himalayas, rain forests with the world's highest rainfall in Mawsynram, in Meghalaya, wet evergreen humid tropics of Western Ghats, arid and semiarid conditions of Peninsular India, dry desert conditions of Rajasthan and Gujarat to the tidal mangroves of the Sunderbans and Bhitar Kanika (Sharma and Thokchom, 2014). Presently, biological diversity of India is facing various threats like deforestation, degradation, agriculture, commercial plantations and replacement of indigenous species with exotics (Rands *et al.,* 2010)

Conventional Approaches for Conservation of Rare and Endangered Plants and Associated Problems

There are various methods to conserve biodiversity. Both *in situ* and *ex situ* methods of biodiversity conservation are equally important. The choice of one or the other technique, or a combination of both, depends on the particular case (Khan *et al.,* 2012; Sharma and Sharma, 2013). Conservation of biodiversity through establishment of protected areas like National Park, Wild life sanctuary, Biosphere Reserves, Marine Reserves *etc.* is *in situ* conservation methods which are very effective in controlling the loss of biodiversity by protecting the habitat itself. Preserving diversity through botanical gardens, gene bank and seed bank are *ex situ* methods for biodiversity conservation.

However, the limited amount of plant materials available from rare and endangered species, poses major challenges for providing material for *ex situ* conservation methods (Sarasan *et al.,* 2006). Further, there are many instances in which seeds are sterile or not available, or so few plants remain that collecting whole plants would negatively impact the population (Chauhan, 2016). In addition, many tropical species have seeds that die when dried or frozen (recalcitrant seeds) and thus they cannot be stored using conventional seed-banking technologies (Pritchard *et al.,* 2014). Thus, alternative and advanced approaches are required to conserve such plants. In recent times, *in vitro* propagation techniques have been used as an alternative way to propagate and conserve a large number of rare (Bramwell, 1990; Holobiuc *et al.,* 2009) and threatened (AmoMarco and Lledo, 1996; Dhar *et al.,* 2000; Rajasekharan *et al.,* 2009) plant species with poor and uncertain responses to conventional methods of propagation (Sarasan *et al.,* 2006).

Recent Advances in Biotechnology for Conservation of Plant Genetic Resource

In vitro techniques have been increasingly applied for mass propagation and conservation of germplasm as it has some degree of superiority over conventional methods of propagation and offer some distinct advantages over alternative strategies. Thus, tissue culture has emerged as a promising technique for multiplying and conserving rare, endangered and endemic species within short period and limited space, which are difficult to regenerate by conventional methods and save them from extinction. Besides, *in vitro* propagation derived plants exhibit huge genetic variation that could be exploited for developing superior clones/varieties particularly in vegetatively propagated plant species (Shibli *et al.,* 2006).Many botanic gardens now have tissue culture laboratories for the micropropagation of plants that are difficult to propagate by conventional horticultural techniques. This review briefly presents the *in vitro* techniques, which can be efficiently used to improve the conservation of plant biodiversity. *In vitro* techniques can be considered as a means of complimentary method for the conservation of plant species together with conventional *ex situ* and *in situ* approaches in preserving plant biodiversity.

Utilization of biotechnological steps for conservation of endangered plants of India is still in its nascent stage. Several researchers have explored these techniques for conservation of agriculturally or medicinally important crops. But so far,a systematic approach for conservation of threatened species of India by biotechnological approach has not yet been made. Few of the *in vitro* studies for conservation of Indian rare and endangered plants are summarized in Table 14.1.

Table 14.1: Some *In vitro* Studies for Conservation of Threatened Species of India

Particulars of the Plants	*In vitro Method Employed*	*References*
Aeridesrosea sp. (an endemic orchid of Arunachal Pradesh)	*In vitro* Seed germination	Sinha *et al.,* 1998
Ipsea malabarica (Malabar daffodil orchid)	Micropropagation	Gangaprasad *et al.,* 1999
Pittosporum napaulensis (an endemic Himalayan medicinal tree)	Micropropagation	Dhar *et al.,* 2000
23 over-exploited of medicinal plant sp. of India	Micropropagation	Verma *et al.,* 2012
Two wild and endangered *Citrus* species	Cryopreservation	Malik and Chaudhury, 2006
*Tylophora indica*Threatened medicinal plant sp. of India	Micropropagation	Rajasekharan *etal.,* 2009
*Clerodendrum serratum*a rare, threatened medicinal plant of India	Micropropagation	Sharma *et al.,* 2009
*Aconitum violaceum*a threatened medicinal herb of India	Micropropagation	Rawat *et al.,* 2013

In vitro Seed Germination

In vitro seed germination has been extensively employed for multiplication of a large number of orchid species and could be a rapid means for multiplying rare and endangered orchids (Gangaprasad *et al.*, 1999; Stewart and Kane, 2006). Orchid seeds are characterized by lack of storage tissues required for seed germination and seedling development. In nature, association with a specific fungal partner is a pre-requisite for orchid seed germination (Mitra, 1986). However, the rate of vegetative propagation is very slow and the seed germination in nature is very poor, *i.e.*, 0.2 - 0.5 per cent (Vij, 2002). Thus, *in vitro* germination of seeds is an important aspect in the orchid multiplication and conservation program since the dust like tiny seeds have the capability of developing into complete seedlings without any fungal aid. Knudson (1922) demonstrated the possibility of by-passing the fungal requirement of orchid seeds during *in vitro* germination and since then non-symbiotic seed germination has been accepted as an important tool for propagating orchids (Arditti *et al.*, 1982).

Cryopreservation

Cryopreservation is the maintenance of living cells, tissues organs and microorganisms at ultralow temperature (usually that of liquid nitrogen, –196 °C). It is important for long-term preservation of plant species which produce recalcitrant seeds and cannot be readily conserved by conventional methods through seed preservation (Benson, 2004). Before cryopreservation of plant material, dehydration is necessary to protect the cells from lethal intracellular freezing and, dehydration can occur during rapid cooling in liquid nitrogen (Popova *et al.*, 2010; Chen *et al.*, 2011). The cooling rate is also an important factor for successful cryopreservation because ultra freezing rate helps to avoid intracellular freezing.

Sometimes genetic variations are generated in the cryopreserved material due to the free radical generation and it is presumed due to dehydration process (Sanchez *et al.*, 2008; Martin *et al.*, 2011). There are some limitations of this method for the preservation of such plant species. For instance, during the preservation of the plant tissues, methylation starts in the cells and it occurs in the plant mainly on cytosine (CpG and CpXpG) and forming 5-methyl-cytosine. Undoubtedly, genetic/ physiological factors determine outcomes of the cryopreservation for which a better understanding for their predisposing actions would benefit cryopreservation protocol development and germplasm responses following recovery (Johnston *et al.*, 2009). DNA methylation does not change the gene sequence and its function but its level of expression is changed. The genetic stability is maintained during cryopreservation and has been proved by molecular markers study (Liu *etal.*, 2008). However, the cryopreserved tissue is considered to be safer for exchange of germplasm between country and regions as it helps in eradication of virus and phytoplasm from plant tissues, and making them free from pathogen (Feng *et al.*, 2011). Cryopreservation of shoot tips is also being applied to eradicate systemic plant pathogens, a process termed cryotherapy (Cruz-Cruz *et al.*, 2013).

There are two types of cryopreservation protocols that basically differ in their physical mechanisms: classical cryopreservation procedures, in which cooling is performed in the presence of ice; and the procedures based on vitrification, in which cooling normally takes place without ice formation (Kami *et al.,* 2009).

A. Cryopreservation by Classical Methods

Classical freezing procedures involve cryoprotection by using different cryoprotective solutions and followed by slow cooling (0.5–2.0 °C/min) to a determined prefreezing temperature (usually around –40 °C) followed by rapid immersion of samples in liquid nitrogen. On the other hand, while recovering samples, rapid thawing is done. Cryopreservation following classical protocols induces a freeze-dehydration process using a slow freezing regime. During the slow temperature decrease, ice is initially formed in the extracellular solution and this external crystallization promotes the efflux of water from the cytoplasm and vacuoles to the outside of the cells where it finally freezes (Cruz-Cruz *et al.,* 2013). Classical cryopreservation techniques have been successfully applied to undifferentiated culture systems of different plant species (Gonzalez-Arnao*et al.* 2008)

B. Cryopreservation by Vitrification Methods

The vitrification-based procedures involve cell dehydration prior to cooling by exposure of samples to highly concentrated cryoprotective media (usually called plant vitrification solutions, PVS) and/or by air desiccation. Cooling rate may be rapid or ultra-rapid, depending on how samples are immersed into liquid nitrogen. Vitrification is a physical process, defined as the transition of the liquid phase to an amorphous glassy solid at the glass transition (Tg) temperature (Fahy *et al.,* 1984). This glass maycontribute to preventing tissue collapse, solute concentration and pH alterations during dehydration. Several vitrification-based procedures are well known at present: encapsulation-dehydration (Gonzalez-Arnao and Engelmann, 2006); encapsulation-vitrification (Sakai *et al.,* 2007); droplet-vitrification (Gonzalez-Arnao *et al.,* 2008) and more recently, a new procedure called Cryo-plate is emerging (Yamamoto *et al.,* 2011). Apart from preserving plant tissues, these methods are helpful for cryopreservation of artificial seeds.

Micropropagation

Plant micro-propagation is a biotechnological approach for the conservation of plant species and even for recalcitrant seeds of economically important crops. Micropropagation refers to *in vitro* mass production of plant propagules from any plant part or cell. Such propagules are used to raise whole plants (Sharma and Sharma, 2013). It may help in the production of virus free plants and is successful for those plant species which have difficulties in propagation using conventional methods. This is being utilized by industries in India for commercial production of mainly ornamental plants like orchids and fruit trees, *e.g.,* banana. Using this method, millions of genetically identical plants can be obtained from a single bud.

This method has, therefore, become an alternative to vegetative propagation. Shoot tip propagation is exploited intensively in horticulture and the nurseries for rapid clonal propagation of many dicots, monocots and gymnosperms. Verma *et al.* (2012) have employed the micropropagation techniques for *in vitro* conservation of twenty-three overexploited medicinal plants of the Indian sub continent.

During *in vitro* tissue culture technique, the cells receive sequential stimuli for proper growth and finally plantlets formation in a large number. However, in this process, the gene pool of the cells is altered may be due to insertion, deletion or substitution of genes which leads to the formation of somatic variations in the micro-propagules, also known as somaclonal variants. From conservation point of view, the regenerated plantlets should have minimum somatic variations. This can be achieved through micro-propagation method by reducing the number of sub-culturing and axillary bud or shoot tip culture (Khan *et al.,* 2012).

Somatic Embryogenesis and Production of Artificial Seed

An embryo derived from a somatic cell of a plant rather than the zygote is known as somatic embryo and the process of production of somatic embryo is known as somatic embryogenesis. The process of regeneration of embryos from somatic cells through micropropagation is regarded as somatic (or asexual) embryogenesis. Somatic embryos are also known as non-zygotic embryos, parthogenetic embryos (formed from unfertilized egg) or androgenic embryos (if formed from male gametophyte). Artificial seeds can be made by encapsulation of somatic embryos (Kriebel, 1995). The embryos, coated with sodium alginate and nutrient solution, are dipped in calcium chloride solution. The calcium ions induce rapid cross-linking of sodium alginate to produce small gel beads, each containing an encapsulated embryo. These artificial seeds (encapsulated embryos) can be maintained in a viable state till they are planted.

Inducible somatic embryogenesis was first observed in 1958 in *Daucus carota* (carrot) (Steward *et al.,* 1958), which resulted from exposure to the synthetic auxin 2,4 dichlorophenoxyacetic acid (2,4-D). Somatic embryogenesis is an important method for mass production of woody tree species for forestry and for the development of artificial seeds, making handling and direct planting easier (Carlson, 1995; Gupta and Timmis, 2005).

Conclusion

Biodiversity is the very basis of human survival and economic development. As the conservation of biodiversity is a global concern, several strategies have been adopted in understanding and conserving plant diversity throughout the world. *In vitro* plant propagation is a helpful technique in the conservation of genetic diversity of all types of plants (including rare, threatened and endangered plants) in a rapid and reliable way by maintaining the same clone or stock of plant material. It is just an efficient alternative method of plant propagation technique. It goes well to the natural process of plant conservation to restore the declined ecosystem with tremendous applicability. In this context, India, being a mega-biodiversity country has a tremendous potential and necessity to conserve its diverse biological resources for the posterity.

REFERENCES

AmoMarco, J.B. and Lledo, M.D. (1996)*In vitro* propagation of *Salix tarraconensis* Pau ex Font Quer, an endemic and threatened plant. *In Vitro Cell. and Dev. Biol. – Plant*, 32, 42–46.

Arditti, J., Clementsm, M. A., Fast, G., Hadley, G. and Nishimura, G. (1982) Orchid seed germination and seedling culture - A manual. In: J. Arditti (Ed.)Orchid Biology -Reviews and perspectives. Vol II. Cornell University Press, Ithaca, New York, 243-370.

Benson, E. E. (2004) Cryo-conserving algal and plant diversity: historical perspective and future challenges. In: B.Fuller, N. Lane and E. E. Benson (Eds.) Life in the Frozen State. CRC Press, London, UK, 299–328.

Bramwell, D. (1990) The role of *in vitro* cultivation in the conservation of endangered species. In: J. E. Hernandez Bermejo, M. Clement andV. Heywood (Eds.) Conservation techniques in botanic gardens. Koeltz Scientific Books, Koeningstein,3–15.

Carlson, W. C. and Hartle, J. E. (1995) Manufactured Seed of Woody Plants. In: S. Jain, P. Gupta and R.Newton (Eds.) Somatic Embryogenesis in Woody Plants; Kluwer Academic Press, London, UK, Volume 1, 253–263.

Chauhan, R. S. (2016) Biotechnological approaches for conservation of rare, endangered and threatened plants. International Journal of Scientific and Research Publication,6(12), 10-14.

Chen, X. L., Li, J. H., Xin, X., Zhang, Z.E., Xin, P. P. and Lu, X. X. (2011) Cryopreservation of *invitro*grown apical meristems of Lilium by droplet-vitrification. *S. Afr. J.Bot*, 77, 397–403.

ConservationInternational (2017), retrieved fromhttp: //www.cepf.net/resources/ hotspots/Pages/default.aspxon10/10/2017.

Cruz-Cruz, C. A., González-Arnao, M. T. and Engelmann, F. (2013) Bio. and Cons.Pl. Bio. Resources, 2, 73-95.

Dhar, U., Upreti, J., Bhatt, I. D. (2000) Micropropagation of *Pittosporum napaulensis* (DC.) Rehder and Wilson–a rare, endemic Himalayan medicinal tree. Plant Cell Tissue and Organ Culture, 63, 231–235.

Fahy, G. M., MacFarlane, D. R., Angell, C.A., Meryman, H.T. (1984) Vitrification as an approach to cryopreservation. Cryobiology, 21, 407–426.

Feng, C., Yin, Z., Ma, Y., Zhang, Z., Chen, L., Wang, B., Li, B., Huang, Y., Wang, Q. (2011) Cryopreservation of sweet potato (*Ipomoea batatas*) and its pathogen eradication by cryotherapy. Biotechnology Advances, 29, 84–93.

Gangaprasad, A., Decruse, S.W., Seeni, S., Menon, V.S. (1999) Micropropagation and restoration of endangered Malabar daffodil orchid. *Ipsea malabarica*. Lindleyana, 14: 47–56.

Gonzalez-Arnao, M. T., Engelmann, F. (2006) Cryopreservation of plant germplasm using the encapsulation-dehydration technique: Review and case study on sugarcane. Cryo Letters, 27, 155–168.

Gonzalez-Arnao, M. T., Panta, A., Roca, W. M., Escobar, R. H., Engelmann, F. (2008) Development and large scale application of cryopreservation techniques for shoot and somatic embryo cultures of tropical crops. Plant Cell Tissue and Organ Culture, 92, 1–13.

Gupta, P. K., Timmis, R. (2005) Mass propagation of conifer trees in liquid cultures-progress towards commercialization. Plant Cell Tissue and Organ Culture, 81, 339–46.

Holobiuc, M., Blindu, R., Mitoi, M., Heleciuc, F., Cristea, V. (2009) The establishment of an *in vitro* gene bank in *Dianthus spiculifolius* Schur and *D. glacialis* ssp. Gelidus (Schott Nym. et Kotschy) Tutin: I. The initiation of a tissue collection and the characterization of the cultures in minimal growth conditions. Annals of Forest Research, 52,117–128.

Johnston, J. W., Benson, E. E., Harding, K. (2009) Cryopreservation induces temporal DNA methylation epigenetic changes and differential transcriptional activity in Ribes germplasm. Plant Physiolgy and Biochemistry,47, 123–131.

Kami, D., Shi, L., Sato, T., Suzuki, T., Oosawa, K. (2009) Cryopreservation of shoot apices of hawthorn *in- vitro* cultures originating from East Asia.Scientia Horticulturae, 120, 84–88.

Khan, S., Al-Qurainy, F., Mohammad, N. (2012) Biotechnological approaches for conservation and improvement of rare and endangered plants of Saudi Arabia. Saudi Journal of Biological Sciences, 19, 1-11.

Knudson, L. (1922) Non-symbiotic germination of orchid seeds. Botanical Gazette, 73, 1-25.

Kriebel, H. B. (1995) Introduction. In: S. M. Jain, P. K. Gupta, R. J.Newton (Eds.) Somatic Embryogenesis in Woody Plants.Kluwer Academic, Dordrecht, The Netherlands,Volume 1, 1–8.

Liu, Y. G., Liu, L. X., Wang, L., Gao, A. Y. D (2008) Determination of genetic stability in surviving apple shoots following cryopreservation by vitrification. Cryo Letters, 29, 7–14.

Malik, S. K., Chaudhury, R. (2006) The cryopreservation of embryonic axes of two wild and endangered *Citrus* species. Plant Genetic Resources, 4, 204–209.

Martin, C., Cervera, M. T., Gonzalez-Benito, M. E. (2011) Genetic stability analysis of Chrysanthemum (*Chrysanthemum xmorifolium* Ramat.) after different stages of an encapsulation–dehydration cryopreservation Protocol. Journal of Plant Physiology, 168, 158–166.

Mitra, G. C. (1986) *In vitro* culture of orchid seeds for obtaining seedlings. In: S. P. Vij (Eds.)Biology, conservation and culture of orchids, Affiliated East-West Press Private Ltd, New Delhi, 401.

Noble, I. R., Dirzo, R. (1997) Forests as Human-Dominated Ecosystems. Science, 277, 522-525.

Popova, E., Kim, H. H., Paek, K. Y. (2010) Cryopreservation of coriander (*Coriandrum sativum* L.) somatic embryos using sucrose pre-culture and air desiccation. Scientia Horticulturae,124, 522–528.

Pritchard, H. W., Moat, J. F., Ferraz, J. B. S., Marks, T. R., Camargo, J. L. C., Nadarajan, J., Ferraz, I. D. K. (2014) Innovative approaches to the preservation of forest trees. Forest Ecology and Management, 333, 88–98

Rajasekharan, P. E., Ambika, S. R., Ganeshan, S. (2009) *In vitro* conservation of *Tylophora indica*: a threatened medicinal plant. IUP Journal of Genetics and Evolution, 11(3), 26–35.

Rawat, J. M., Rawat, B., Agnihotri, R. K., Chandra, C., Nautiyal, C. (2013) In vitro propagation, genetic and secondary metabolite analysis of *Aconitum violaceum* Jacq.: a threatened medicinal herb. Acta Physiologiae Plantarum, 35, 2589–2599

Rands, M. R., Adams, W. M., Bennun, L., Butchart, S. H., Clements, A., Coomes, D., Entwistle, A., Hodge, I., Kapos, V., Scharlemann, J. P., Sutherland, W. J., Vira, B. (2010) Biodiversity conservation: challenges beyond. Science, 329(5997), 1298-303.

Rogers, S. (2003) Tissue culture and wetland establishment of the freshwater monocots Carex, Juncus, Scirpus, and Typha. In Vitro Cellular and Developmental Biology – Plant, 39, 1–5.

Sakai, A., Engelmann, F. (2007) Vitrification, encapsulation-vitrification and droplet-vitrification: A review. Cryo Letters, 28, 151–172.

Sanchez, C., Martinez, M. T., Vidal, N., San-Jose, M. C., Valladares, S., Vieitez, A. M. (2008) Preservation of *Quercus robur* germplasm by cryostorage of embryogenic cultures derived from mature trees and RAPD analysis of genetic stability. Cryo Letters, 29, 493– 504.

Sarasan, V., Cripps, R., Ramsay, M. M., Atherton, C., McMichen, M., Prendergast,G., Rowntree, J. K. (2006) Conservation *in vitro* of threatened plants - progress in the past decade. In Vitro Cellular and Developmental Biology – Plant, 42, 206–214.

SCBD Aichi Biodiversity Targets (2017) retrieved fromhttp: //www.cbd.int/sp/ targets/ on 10/10, 2017.

Sharma, D. K., Sharma, T. (2013) Biotechnological approaches for biodiversity conservation. Indian Journal of Scientific Research, 4(1), 183-186.

Sharma, M., Rai, S. K., Purshottam, D. K., Jain, M., Chakrabarty, D., Awasthi, A., Nair, K. N., Sharma, A. K. (2009) In vitro clonal propagation of *Clerodendrum serratum* (Linn.) Moon (barangi): a rare and threatened medicinal plant. Acta Physiologiae Plantarum, 31, 379–383

Sharma, S., Thokchom, R. (2014) A review on endangered medicinal plants of India and their conservation. Journal of Crop and Weed, 10(2), 205-218.

Shibli, R. A., Shatnawi, M. A., Subaih, W. S., Ajlouni, M. M. (2006) *In vitro* conservation and cryopreservation of plant genetic resources: a review. World Journal of Agricultural Science, 2, 372-82.

Sinha, S. K., Sinha, L., Hedge, S. N. (1998) *In vitro* multiplication of *Aeridesrosea* Loddiges ex Paxt. Through asymbiotic seed germination. *Arunachal Forest News*. 16(1 and 2), 38-44.

Steward, F., Mapes, M., Mears, K. (1958) Growth and organized development of cultured cells. II. Organization in cultures grown from freely suspended cells. American Journal of Botany, 45, 705–8

Stewart, S. L., Kane, M. E. (2006)Asymbiotic seed germination and *in vitro* seedling development of *Habenaria macroceratitis* (Orchidaceae), a rare Florida terrestrial orchid. Plant Cell Tissue and Organ Culture, 86, 147-158.

Tscharntke, T., Batáry, P., Clough, Y., Kleijn, D., Scherber, C., Thies, C., Wanger, T. C., Westphal, C. (2012) Combining biodiversity conservation with agricultural intensification. In: D.Lindenmayer,S. Cunningham, A.Young (Eds.) Land use intensification - Effects on agriculture, biodiversity and ecological processes. CSIRO Publishing, Collingwood, Australia. 7-15.

Verma, P., Mathur, A. K., Jain, S. P., Mathu, A. (2012) *In vitro* Conservation of Twenty-Three Overexploited Medicinal Plants Belonging to the Indian Sub Continent. ScientificWorld Journal, Article ID 929650.

Vij, S. P. (2002) Orchids and tissue culture: Current status. In: S. K. Nandi, L. M. S. Palni, A. Kumar (Eds.) Role of plant tissue culture in biodiversity conservation and economic development, Gyanodaya Prakashan, Nainital, India. 491-502.

Yamamoto, S., Rafique, T., Priyantha, W.S., Fukui, K., Matsumoto, T., Niino, T. (2011) Development of a cryopreservation procedure using aluminium cryo-plates. Cryo Letters, 32, 256–265.

Chapter 15

Electroformation of Giant Unilamellar Vesicles from Bacterial Polar Lipid Extract using Copper Electrode

*Himadri Gourav Behuria, Gayatree Panda and Santosh Kumar Sahu**

Department of Biotechnology, North Orissa University, Baripada – 757 003, Odisha
e-mail: drsantoshnou@gmail.com

ABSTRACT

An easy, cheap and convenient protocol was developed to produce giant unilamellar vesicles (GUVs) from K. pneumoniae polar lipid extract using copper electrodes. The average size of vesicles varied from 50 to 100 μm when prepared for 3h at 30 V and 10 Hz in low salt solution. We successfully reproduced the procedure with several bacterial lipid extracts with a wide range of lipid composition. Formation of GUVs was confirmed by staining with membrane-staining dye NAO and visualizing under fluorescent microscope. The vesicles were spherical and unilamellar as indicated by the fluorescence imaging. By using these GUVs as a model membrane system, several membrane membrane-mediated events such as mechanism of drug targeting, mechanism of lipid translocators and lipid-protein interaction in bacterial membrane could be evaluated.

Keywords: *Liposome, Giant unilamellar vesicles, Phospholipids, Electroformation, NAO.*

Abbreviations

egg-PC, egg-phosphatidylcholine; GUV, giant unilamellar vesicle; LUV, Large unilamellar vesicle, LSS, low salt solution; NAO, 10 N-nonyl acridine orange. PE, phosphatidylethanolamine; PG, phosphatidylglycerol; PL, phospholipid; PMSF, phenylmethylsulfonylfluoride; SRB, sulforhodamine B, SUV, small unilamellar vesicle; Egg-PC, Egg yolk phosphatidylcholine; *Kp*PE, *K. pneumoniae* phosphatidylethanolamine; *KpPL, K. pneumoniae* polar lipids; *EcPL, E. coli* polar lipids

Introduction

Giant unilamellar vesicles (GUVs) constitute an ideal model membrane system for investigation of membrane-mediated events such as drug delivery, permeability to metabolites, diffusion of membrane proteins and drug resistance in bacteria. GUVs are composed up of synthetic or naturally derived lipid molecules organized into a single bilayer that enclose an aqueous interior. The vesicles are mostly round and usually range from 10 to 100 μm in diameter (Welde *et al.,* 2010). The lipid bilayer serves as a pure membrane system analogous to cellular membranes in its physico-chemical properties. GUVs prepared from bacterial polar lipid extract is an ideal model system to investigate plasma membrane mediated events such as host-bacteria interaction, mechanism of drug resistance and drug sensitivity (Carvalo *et al.,* 2008).

GUVs have been prepared by many workers following various methods such as electroformation (Oglecka *et al.,* 2014, Angelova *et al.,* 2007), hydration, sonication, lipid emulsification (Pautot *et al.,* 2003) and osmotic shock (Motta *et al.,* 2015). However, methods involving gentle hydration is time consuming and produces vesicles of variable size. Similarly, sonication method that involves swelling of dried lipid cakes using ultrasonic sound waves produces multilamellar vesicles those are to be less useful for investigation of membrane-mediated events. Electrofomation, the most common method of GUV preparation involves continuous swelling of dried lipid layers in an aqueous medium using an alternating electric current for a period of 1-3 h. Electroformation method that uses Indium tin oxide (ITO)-coated slides is the most common and ideal method to produce uniform sized GUVs from > 10 μm (Angelova *et al.,* 2007).

However, the currently used electroformation method requires sophisticated instrumentation facilities and expensive chemicals such as platinum electrodes or ITO coated slides. Cost analysis shows that these methods are too expensive to be used in average-furnished laboratories. It has been calculated that the average cost of GUVs preparation by electroformation method using platinum electrode in INR are "40, 000.00 and 25000.00 for platinum electrodes and ITO coated slides respectively. Further, reusability, washability and scalability are major issues when these electrodes are used in laboratories that requires frequent preparation of GUVs. Many workers have proposed the use of low expensive electrodes such as stain-less steel for successful preparation of GUVs (Table 15.1).

Table 15.1: Analysis of PL content of bacterial strains used in the study

Strain	*Total PL (nmol)*	*PE (Per cent)*	*PG (Per cent)*	*CL (Per cent)*	*Unknown*
E.coli ($DH_5\alpha$)	385.62±23.1	62.228 ±2.805	20.166±4.110	17.606±1.857	ND
Strain 1	361.82±6.47	81.59±2.29	14.74±2.51	ND	4.00±0.55
Strain 2	349.70±26.87	76.45±1.81	17.79±0.81	ND	5.75±1.38
Strain 3	240.03±64.31	77.84±1.15	22.16±1.45	3.2 ±1.45	ND
K. pneumonaie	325.67 ± 34.42	88 per cent ± 2.12	10.11 ± 1.13	2.1 ± 0.36	ND
Strain 4	325.63±18.31	63.90±2.81	16.01±2.71	14.82±1.25	5.27±1.78

To explore the use of low expensive electrodes for preparation of GUVs, we used copper electrodes to prepare GUVs from the polar lipid extract of different bacteria and evaluated its potential using various physico-chemical methods.

Materials and Method

Extraction, Purification and Identification of Lipids

Total lipid from bacteria was extracted using aqueous two phase method (Bligh and Dyer 1959) and total phospholipid (PL) content was quantitated by phosphate quantitation (Fiske and Subarrow, 1925). The PL composition of various lipid extract was determined by two-dimensional TLC. Quantitation of different PLs in the lipid extract was performed as described by Skipski *et al.,* from two dimensional thin layer chromatography (2-D TLC) (Skipski *et al.,* 1964). Polar lipids from total lipid extract containing 2 mg PL were separated from non-polar (*e.g.* neutral and storage) lipids using a 12 × 150 mm silica gel column containing silica beads of 60-120 μ mesh size using $CHCl_3$:CH_3OH 1:1 (v/v) as mobile phase (Rouseer *et al.,* 1970).

Electroformation of GUVs

1 mg polar lipid (corresponding to 1300 nmol PO_4) in chloroform was coated on a thin copper electrode (5cm x 0.5cm x 1mm) and dried at 45 °C for 1 h in a hot air oven. The dried lipid film was hydrated by placing inside a small (1.2cm x 5cm) round bottomed glass bottle containing 3 ml low salt solution (10 mM KCl and 100 mM Sucrose). The GUVs were formed by placing the electrodes 0.5 cm apart and passing an alternating current of 10 Hz and 30 V for 3 h at 25 °C.

Imaging of GUVs

Formation of GUVs were analyzed under an inverted fluorescent microscope (Nickon, Model-100, Japan) using 40X objective, visualize of both under bright field and fluorescent imaging. The vesicles were stained with 0.01 mM Nonyl acridine orange (NAO) (sigma) following the procedure of Fernandez (Fernandez *et al.*, 2003). Images were taken both under bright and fluorescent field. Fluorescence images were taken using green filter (excitation 440-480 nm and emission 510 - 560 nm). Sizes of GUVs were measured using NIS-Elements Basic Research software. All statistical analysis were performed using Graph pad prism (version 6). Result were found to be significant when $P < 0.05$.

Results and Discussion

Bacteria used in the present study contained two or three major PLs namely, phosphatidyl ethanolamine (PE), phosphatidyl glycerol (PG) and cardiolipin (CL). The PLs were identified by comparing them with standard PLs (Sigma) (Figure 15.1).

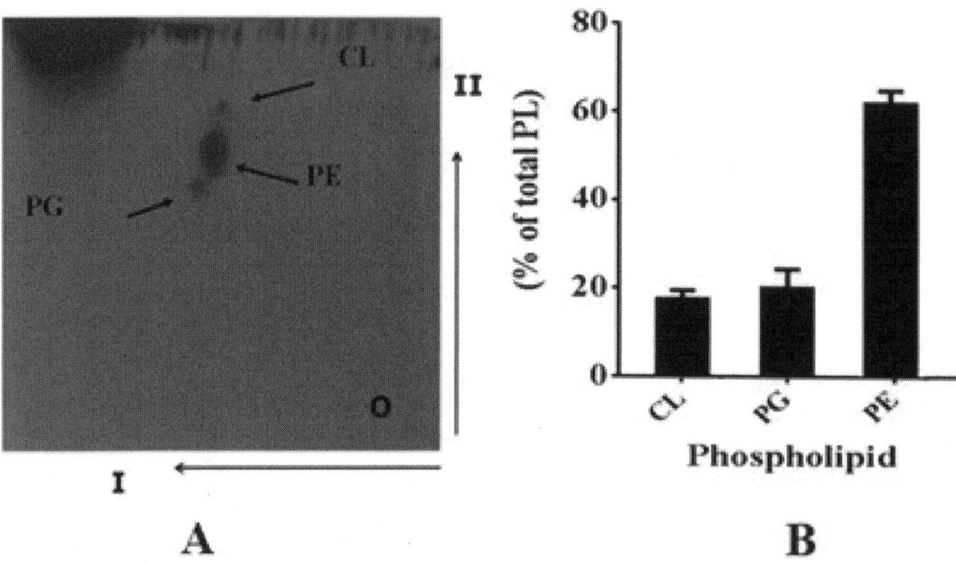

Figure 15.1: Extraction and estimation of lipids from bacteria: (A) 2-D TLC of total bacterial lipid showing phosphatidyl ethnolamine (PE), phosphatidyl glycerol (PG) and cardiolipin (CL). 'O' designates the origin where the lipid was loaded. I and II show the directions of solvents for first and second dimension respectively. (B) Percentage of each phospholipid component in *E. coli* extract as determined by phosphate estimation.

PE is the predominant PL in bacteria that varied from 60-80 per cent, followed by PG (15-20 per cent). CL was not observed in strains 1 and 2. However, *E.coli* ($DH_5\alpha$) and strain 4 contained ~ 15 per cent CL. PE is the predominant lipid present in all strains. However, in *K. pneumoniae*, 88 per cent of the lipids were PE. In strains 1, 2, 3 and 4, the phospholipid composition is shown in Table 15.2.

Table 15.2: A comparative analysis of cost and reusability of Electrodes Used in electroformation of GUVs

Sl. No.	*Metals*	*Cost (INR)*	*Dis-posable*	*Reusable*	*Availability*	*Expensive Accessories*	*Scalable*
1	Platinum	40,000	No	Yes	Commercial	Yes	Expensive
2	ITO-coated glass	25,000	Yes	Limited	Commercial	Yes	Limited
3	Stainless steel	2000	Yes	Yes	Limited	Yes	Yes
4	Copper	50	Yes	Yes	Locally made	No	Yes

The table 15.1 shows a wide range of PL composition from different bacteria. However, strains 2, 3 and 4 are isolated from mercury contaminated soil and are not yet identified.

Analysis of GUVs under Fluorescent Microscopy

Formation of GUVs were confirmed after their staining with NAO. The average diameter of the vesicles varied with time of swelling. Most vesicles (40-50 per cent of total count) were of 50-100 μ in diameter. The number of vesicles formed peaked at 2 h. Swelling for longer time led to formation of gigantic and distorted vesicles (Figure 15.2). Vesicle count was dependent upon PL composition. Presence of more anionic lipids (PG and CL) resulted in enhanced abundancy of the GUVs. Anionic PLs are known to enhance the stability of the vesicles. Further, these lipids are

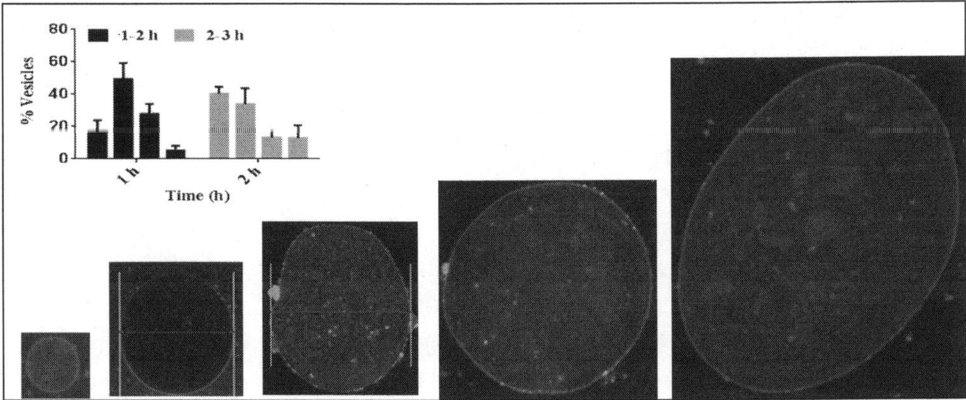

Figure 15.2: Giant unilamellar vesicles made from the polar lipid extract of *E. coli*. The images are obtained in 40X objective using green filter that optimally excites NAO. The vesicles shown from left to right are collected and imaged at 1, 2, 3, 4 and 5th hour of electroformation resoectively. The vesicles were counted manually and their size was measured using NIS-Element software provided along with the Nikkon fluorescence microscope.

cylindrical in shape in contrast to the zwitterionic and conical PE. Incubation for longer time (>4-5h) led to fusion of vesicles resulting in mega-sized membranous structures of irregular shape. Cost analysis shows much less preparation cost of GUVs (Table 15.2) compound to prevision methods.

Conclusion

In conclusion, we have developed an easy, cheap and convenient protocol to produce giant unilamellar vesicles from bacterial lipid extract. The average size of vesicles ranged from 50 to 100 μ depending upon the lipid composition and time of swelling. We have confirmed the procedure with multiple bacteria. By using these GUVs as a model membrane system, we can investigate the efficacy of membrane-targeted drugs, efficiency of lipid translocators and lipid-protein interaction in bacterial membrane. This will enable us to determine the mechanism of action for AMPs, mechanism of drug resistance in bacteria and the most efficient drug that effectively eliminates pathogens

REFERENCES

Bligh, E.G., Dyer, W.J., A rapid method for total lipid extraction and purification, *Can. J. Biochem. Physiol.* (1959) 37, 911-917.

Costa, B.A., Sanches, L., Gomide, A.B., Bizerra, F., Mas, C.D., Oliveira, E.B., Perez, K.R., Itri, R., Oguiura, N., Hayashi, M.A.F., Interaction of the rattlesnake toxin crotamine with model membranes, *J. Phys. Chem* (2014) 118, 5471–5479

Fiske, C., Subarrow, Y., The colorimetric determination of phosphorous, *J. Biol. Chem.* (1925) 66, 375-400.

Knol, J., Veenhoff, L., Liang, W., Henderson, P.J.F., Leblanci, G., Poolman, B., Unidirectional reconstitution into detergent-destabilized liposomes of the purified lactose transport system of Streptococcus thermophilus, *J. Biol. Chem* (1996) 271, 15358–15366

Leite, N.B., Aufderhorst-Roberts, A., Palma, M.S., Connell, S.D., Neto, J.R. Beales P.A., PE and PS lipids synergistically enhance membrane poration by a peptide with anticancer properties, *Biophys. J* (2015) 109, 936–947

Lira R.B., Dimova R., Riske, K.A., Giant unilamellar vesicles formed by hybrid films of agarose and lipids display altered mechanical properties, *Biophys J.* (2014) 107, 1609-19

MacDonald, R.C., MacDonald, R.I., Menco, B.Ph.M., Takeshita, K., Subbarao, N.K., Hu, L., Small-volume extrusion apparatus for preparation of large unilamellar vesicles, *Biochim. Biophys. Acta.* (1991) 1061, 297- 303

Moscho, A., Daniel, O.O., Chiu, T., Modi, B.P., Zare, R.N., Rapid preparation of giant unilamellar vesicles, *Proc. Natl. Acad. Sci.* USA (1996) 93, 11443-11447

Nagahama, M., Hara, H., Fernandez-Miyakawa, M., Itohayashi, Y., Sakurai, J., Oligomerization of clostridium perfringens ε toxin is dependent upon membrane fluidity in liposomes, *Biochemistry* (2006) 45, 296-302

Pereno, V., Carugo, D., Bau, L., Sezgin, E., de la Serna, J.B., Eggeling, C., Stride, E., Electroformation of giant unilamellar vesicles on stainless steel, electrodes, *ACS Omega* (2017) 2, 994–1002

Skipski, V.P., Peterson, R.F., Barclay, M., Quantitative analysis of phospholipids by thin-layer chromatography, *Biochem. J.* (1964) 90, 374-78

Weso³owska, O., Michalak, K., Maniewska, J., Hendrich, A.B., Giant unilamellar vesicles: a perfect tool to visualize phase separation and lipid rafts in model systems, *Acta Biochim. Pol* (2009), 56, 33-39

Chapter 16

Bioreduction of Toxic Selenite to Nontoxic Selenium using an Efficient Salt Tolerant *Bacillus* sp. Isolated from Mangrove Soil

R.R. Mishra[1], Priyambada Sarangi[1] and H.N. Thatoi[2]*

[1]Department of Biotechnology, MITS School of Biotechnology, Bhubaneswar – 751024, Odisha
[2]Department of Biotechnology, Sriram Chandra Bhanja Deo University Takatpur, Baripada – 757003, Odisha
**e-mail: rashmiranjan93@gmail.com*

ABSTRACT

In the present study, nine salt tolerant bacteria (BSB1-9) were isolated from mangrove soils collected from outside the National Park area of Bhitarkanika. The isolates were screened for their salt tolerance. Based on the high salt tolerance, five strains were selected for Minimum Inhibitory Concentration (MIC) towards high selenite concentrations. Out of the five bacterial strains, BSB4 could able to tolerate highest concentration of selenite (200 mM) and was further studied for bioremediation of selenite. In an unoptimised condition, at a fixed concentration of selenite (100 mM) the bacterium BSB4 could able to reduce 90 per cent of selenite in 60 h of incubation. Further, the reduction process was optimized in different pH, temperature and NaCl concentrations. It has been observed that at pH 9.0, temperature of 37°C and NaCl concentration of 4 % (w/v) was more effective for selenite reduction. The reduction of selenite to red elemental selenium was also from the evident of formation of red colour. With a view, to identify the most effective selenite reducing bacterial strain (BSB4), various morphological and biochemical tests were carried out. Based on various biochemical results the bacterial strain BSB4 was tentatively identified as Bacillus sp.

***Keywords**: Mangrove soil, Bioremediation, Selenite, Bacillus sp.*

Introduction

Environmental pollution by heavy metals is considered to be lethal for all forms of life. Trace amounts of selenium are important for humans, animals, and plants but high concentration results in detrimental effects (Fordyce, 2005). The difference between productive and noxious levels of selenium is pretty narrow making both selenium deficiency and selenium pollution a public health issue worldwide (Lin *et al.*, 2000). A number of microorganisms have the ability to reduce Se oxyanions (Theisen and Yee, 2014). They play a crucial role in recycling and transformation of selenium through redox and methylation reactions. The biomethylation of selenium in seleniferous environments is readily carried out by bacteria and plants and is thought to be an important detoxification mechanism in these organisms. Selenium is of considerable environmental importance as it is essential at low concentrations but toxic at high concentrations for animals and humans, with a relatively small difference between these values (Fordyce, 2005). Selenium occurs in different oxidation states as reduced form (selenide, Se^{2-}), least mobile elemental selenium (Se0) and water soluble selenite (Seo_3^{2-}) oxyanions. Listing in the sixth main group in the periodic table of elements, selenium is a crystalline "metalloid" and possesses similar chemical properties with sulfur. Selenium can occur in environment as various oxidation states such as: selenate (SeO_4^{2-}, Se (VI)), (SeO_3^{2-}, Se(IV), selenide (Se^{2-}) and elemental (Se0). However, Se (IV) and Se (VI) present in aqueous environment commonly. Besides, the Se (IV) is more toxic than Se (VI) (Awual *et al.*, 2015). As is known to all, selenium (Se) is an essential micronutrient for the creature and it is good for our health at low concentration. However, excessive Se in the environment causes serious pollution in soil, atmosphere and water, and brings severe risk and damage to human health (Sakamoto *et al.*, 2012). Industrial and agricultural activities are the dominant anthropogenic sources of Se pollution. Selenium absence and toxicity lead to contamination in drinking water and in waste water, which is a latent threat for water systems (Santos *et al.*, 2015b). Therefore, in recent years, selenium contamination has attracted tremendous interests and has been understood more and more deeply around the world (Sakamoto *et al.*, 2015). A wide variety of bacteria occurring in diverse terrestrial and aquatic environments, both selenium-rich and selenium free soils, have the ability to reduce aqueous selenite/selenate oxyions (Ghosh *et al.*, 2008). In particular, inorganic and organic selenium compounds have diûerent toxicity profiles, with the latter being much less toxic than the former (Mehdi *et al.*, 2013). The standards of public health for selenium should consider chemical form, because selenium exists in drinking water almost entirely with inorganic form. On the contrary, inorganic forms of selenium in foods are not generally found. In drinking waters, selenium is generally detected at relatively low concentrations (Mehdi *et al.*, 2013).

Microorganisms living in selenium-rich environments often equipped with certain mechanisms enable them to convert inorganic selenium compounds into volatile compounds. Those microorganisms were reported to secrete certain methyl transferases which stimulate the emission of DMSe but its mechanism is still unknown. The present study was conducted on bacterial strains capable of transforming toxic SeO_3^{-2} (IV) into elemental Se(0) so that they can be used in future for remediating toxic selenium oxyanions from soil. Many studies has been carried out for bioremediation of selenite using bacterial strains isolated from Se rich environment. However, the potential of salt tolerant bacteria for reduction of selenite or selenate and other anions in high salt concentrations has been relatively less explored. The mangrove is a typical costal biome located in the transition area in between land and sea. The periodic tidal flooding makes the salinity and nutrient content of mangrove highly variable and thus favours the growth of genetically diverse groups of aquatic and terrestrial microorganisms having potential for biotechnological applications (Kothari, 2002; Dias *et al.*, 2009). Both halophilic and halotolerant microorganisms of the saline environments are suitable for bioremediation or biomineralization in saline soils because they are capable to grow in high salt concentration (Ventosa, 2004). Among the different groups of bacteria, the Gram (+) bacteria is well represented in the saline habitat and the *Bacillus* spp. and *Micrococcus* spp. are mimetically dominant. It has also been reported that the halotolerant *Bacillus* species have greater biotechnological potential compared to other groups of non-tolerant bacteria (Oguntoyinbo, 2007). Keeping the above in view, the present work was aimed to isolate some of the predominant halotolerant bacterial strains with high Se (IV) resistance potential from one of the biologically diversified and less explored saline habitat, the mangrove soils of Bhitarkanika, Orissa, India and to use the most effective bacterial strain for bioremediation of selenite.

Materials and Methods

Reagents

All the chemicals, reagents and microbiological media used in the present study were of analytical grade. Nutrient agar (NA) and nutrient broth (NB) (Hi-media) were used as supplied. Stock solution (250 mg L^{-1}) of selenite was prepared by dissolving $Na_2SeO_32^- 4H_2O$ (E. Merck) in double distilled water, 2-mercaptoethanol (0.5 M) was prepared by diluting 3.5 mL (density, 1.14) of the original solution to 100 mL with double distilled water. Concentrated sulphuric and nitric acids (Merck, GR) were used as received.

Site Description and Sample Collection and Isolation of Bacteria

Mangroves of Bhitarkanika (20°30'–20°50'N latitude and 86°30'–87°6'E longitude) are situated in the Kendrapara district which extends about 139.39 km^2 in the Brahamani and Baitarani deltaic regions of Odisha. Soil samples were collected from the Bhitarkanika mangrove forest and brought to the laboratory in sealed sterile polythene packets. The bacteria were isolated in NA media (g L^{-1}: peptone, 5.0; NaCl, 3.0; beef extract 3.0; agar, 1.8) and at pH _7.0 containing 4–12 per cent NaCl (w/v) following the dilution plate technique method.

Physiological Characterization of the Isolated Bacterial Strains

Physiological characteristics of the organisms were checked following the standard methods for identification of the isolates. Physiological characteristics *viz.* growth at different salt (NaCl), temperature and pH was checked. Growth of the organisms on their respective media supplemented with different NaCl (5-15 per cent) and temperature (4-65°C) were checked. Highly diluted suspensions of the organisms were spotted on the plates, incubated at 30 ± 0.1°C for 72 h and growth was recorded.

Qualitative Screening of Bacterial Isolates and Bacterial Growth

A qualitative screening of selenite resistance was done to identify the most effective isolate for further study. For, NA plates supplemented with (50-200 mM) of sodium selenite at pH 7.5 were inoculated with different isolates. The plates were incubated for 48 h at 37°C. The intensity of orange–red/red colour around the bacterial colonies was taken as a qualitative efficiency for the reduction of selenite to elemental selenium. Among five bacterial isolates, one bacterial strain (BSB4) could tolerate up to 200 mM of selenite. Further, it developed the most intense colour were selected for further characterization and assessment of selenite reduction.

Determination of MIC

Minimal inhibitory concentration is best defined as the completely prevention of bacterial growth in modest concentration of metal. For this, dilution plate technique has been used. NA plates supplemented with different concentration of selenite (50-200 mM) were spot inoculated by different bacterial strains. Then the plates were incubated at BOD incubator at 37°C. After 48 hour the growth of bacterial strains were checked.

Bacterial Selenite Reduction Experiment

Aliquots of solution containing 0.8-15.4 μg/ml selenium were transferred into a series of 10ml calibrated flask. A volume of 1ml of 2 % potassium iodide solution was added followed by 1ml of 1M HCl and the mixture was gently shaken until the appearance of yellow colour indicating the liberation of iodine. A 0.5ml of 0.02 % safranine and 2 ml of acetate buffer solution of pH=4 were added in each flask and the reaction mixture was shaken for 2minuits. The content was diluted to 10ml with distilled water. The absorbance of resulting solutions was measured at 532 nm against a reagent blank. A blank was prepared by replacing the selenium solution with distilled water. The absorbance corresponding to the blank colour, which is in turn corresponds to the selenium concentration, was obtained by subtracting the absorbance of the blank solution from that of the test solution. The amount of selenium present in the volume taken was computed from the calibrate graph.

Effect of Various Temperatures on Selenite Reduction

In order to determine the effect of temperature on reduction of sodium selenite, three different temperatures (32, 37, and 42 °C) were selected as a temperature range on which these strains can grow. Nutrient broth was prepared in flasks and supplemented with 100 mM of sodium selenite. Three flasks were prepared for each strain for all the temperatures (32, 37, and 42 °C) and labeled properly. A negative control was also prepared. Afterward, inoculation was given in the same manner as already described. Flasks were incubated at 32, 37, and 42°C for 48 h and then checked the reduction in the selenite as described above.

Effect of Various pH on Selenite Reduction

Nutrient broth was prepared in 250 ml flasks. Three flasks for each strain were prepared and supplemented with 100 mM of sodium selenite. pH was adjusted to, 5 7, and 9 in different flasks and labeled properly. After inoculation, all the flasks were incubated at 37 °C for 48 h. Selenite content was estimated in the supernatant after 48 h of incubation. Optical densities were plotted on the calibration curve to measure the selenite concentration. Negative control (uninoculated nutrient broth) supplemented with 100 mM was also estimated for selenite content after 48 h.

Effect of NaCl on Selenite Reduction

Nutrient broth was prepared in 250 ml flasks. Three flasks for each strain were prepared and supplemented with 100 mM of sodium selenite. NaCl concentration was adjusted to 2, 4, 6 % in different flasks and labeled properly. After inoculation, all the flasks were incubated at 37 °C for 48 h. Selenite content was estimated in the supernatant after 48 h of incubation. Optical densities were plotted on the calibration curve to measure the selenite concentration. Negative control (uninoculated nutrient broth) supplemented with 100 mM was also estimated for selenite content after 48 h.

Biochemical Identification of the Bacterial Strain

In a view, to identify the most effective selenite reducing bacterial strain (BSB4) various biochemical tests such as catalase. MR, MRVP, indole test, citrate utilization test, hydrolysis of casein, gelatin, acid production from glucose *etc.* were carried out following the standard procedure of Mishra *et al.* (2012).

Results

Isolation of some Predominant Halotolerant Bacteria

In the present study 9 bacteria were isolated from Bhitarkanika mangrove soils using NA medium supplemented with 6 % NaCl and were designated as

Bhitarkanika soil bacteria (BSB1-BSB9). The colony characteristics of the isolated bacteria are given in Table 16.1. Majority of the isolates had irregular colonies with undulate margins. The bacterial isolates BSB1, BSB3 and BSB4 have entire margins. Colony of most of the bacterial isolates had flat surface. Colours of the isolates were white (BSB2, BSB5 and BSB7). The bacterial isolates showed variation in their size and they ranged from 0.1-4.0 mm. The colony characters of the isolates (BSB1-BSB7) were irregular, filamentous, punctiform or circular, yellowish-brown, brown, creamy white or pale brown. Range of colony size was 1.0-1.4 mm on nutrient agar which appeared gummy and sticky consistency (Table 16.1). Three out of nine isolates (BSB5, BSB7 and BSB9) were Gram (+)ve and the rest six isolates (BSB1, 2, 3, 4, 6 and 8) were Gram (-) ve (Table 16.1).

Table 16.1: Morphological Characteristics of the Halotolerant Bacteria Isolated from Mangrove Soils of Bhitarkanika

Bacterial Isolate	*Form*	*Colour*	*Elevation*	*Margin*	*Size (mm)*	*Consistency*	*Gram stain*
BSB 1	Circular	White	Flat	Entire	0.1-1.5	Gummy	-ve
BSB 2	Irregular	White	Flat	Undulate	0.1-1.8	Gummy	-ve
BSB 3	Circular	Creamy white	Flat	Entire	0.2-1.2	Gummy	-ve
BSB 4	Circular	Creamy white	Flat	Entire	0.2-1.5	Sticky	-ve
BSB 5	Irregular	White	Flat	Undulate	0.2-2.0	Gummy	+ve
BSB 6	Irregular	White	Flat	Undulate	0.1-1.5	Sticky	-ve
BSB 7	Irregular	White	Flat	Undulate	0.1-1.5	Gummy	+ve
BSB 8	Punctiform	White	Flat	Undulate	0.4-1.5	Sticky	-ve
BSB 9	Punctiform	Creamy white	Flat	Undulate	0.4-1.5	Sticky	+ve

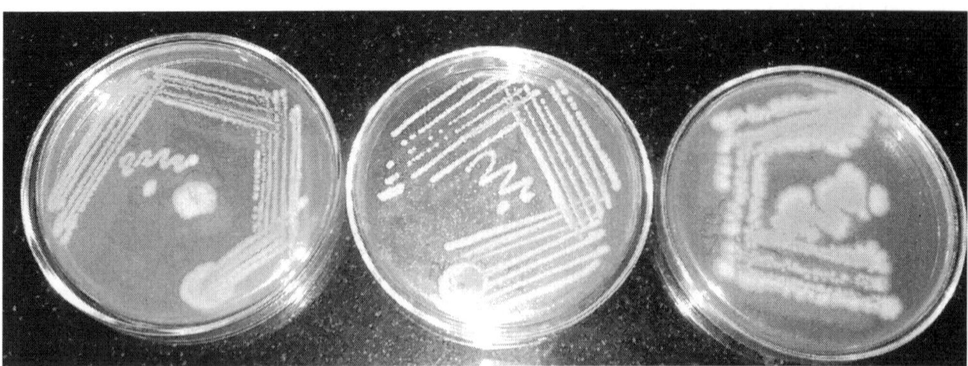

Figure 16.1: Pure Cultures of some Bacterial Isolates from Mangrove Soils of Bhitarkanika.

Screening for NaCl Tolerance

Nine bacterial isolates were screened for their salt tolerance in presence of 2-10 per cent of NaCl (w/v). All isolates (except BSB3 and BSB9) could grow in medium containing 10 % NaCl (w/v). Generally growth was better in the medium supplemented with NaCl. However, the optimum concentration of NaCl for growth of the isolates varied. Optimum NaCl concentration for growth of majority of the strains was 6 % NaCl. Besides salt tolerance, the bacterial isolates were also screened for pH and temperature tolerance. From the study, it has been observed that, better growth of the isolates occurred around pH 7.0 and temperature of 35 °C.

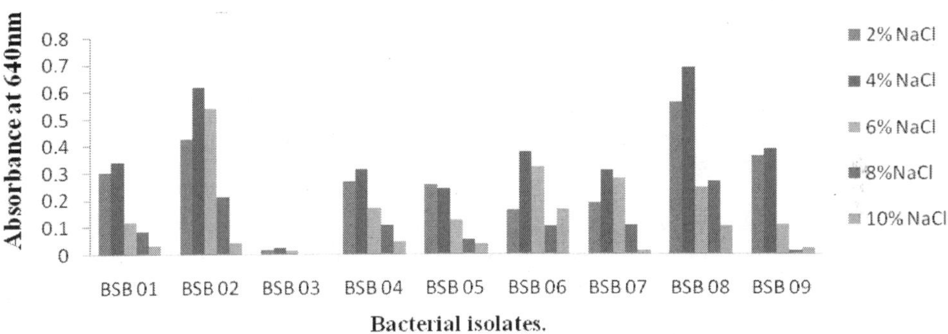

Figure 16.2: Growth of Bacteria at different Concentration of NaCl (w/v).

Determination of MIC

Based on salt tolerance, out of 9 bacterial isolates, 5 bacterial isolates could bale to grow in 10 per cent (NaCl). Hence, further selected for selenite screening through Minimum inhibitory concentration (MIC) test. MIC is the lowest concentration of metal that completely prevented bacterium growth. MIC towards selenite of all 5 isolates was listed below (Table 16.2 and Figure 16.3). Among five bacterial strains tested, BSB4 could able to tolerate highest selenite concentration (200 mM). Therefore, the strain BSB4 was selected for further optimization and reduction experiment.

Table 16.2: MIC Values of Bacterial Isolates against different Concentration of Selenite

Isolate No.	*MIC (mM)*
BSB 1	125
BSB 2	150
BSB 4	200
BSB 8	100
BSB 9	150

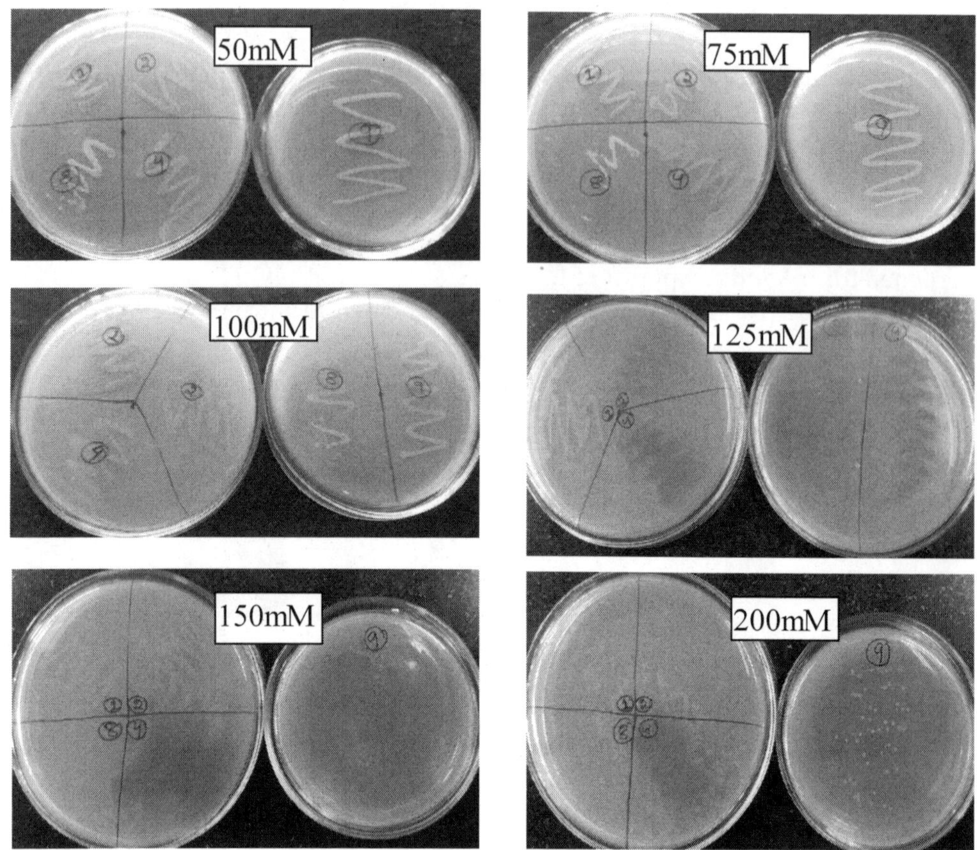

Figure 16.3: Detection of MIC of Bacterial Strains with Varied Concentration of Selenite (50- 200mM).

Reduction of Selenite

The potential of BSB4 for Se(IV) reduction was studied at fixed selenite concentration (100 mM). Under optimized growth conditions (*i.e.* at 37 °C and pH 7.5,) a representative time course of Se (IV) reduction along with control is shown in Fig. 4. There is practically no change of selenite concentration in the control during the incubation period. However, the bacterial strain BSB4 effectively reduced toxic Se (IV) to non toxic Se(0) within 60 h of incubation period. It is evident from the change in colour of the reduction medium to orange–red/red due to generation of allotropic form of elemental selenium during the exponential growth phase (Figures 16.4 and 16.5).

Effect of pH, Temperature and NaCl on Selenite Reduction

The reduction of selenite by the bacterial strain (BSB4) was carried out at different pH (5, 7, 9), temperature (32, 37 and 42 °C) and NaCl (2-6 per cent). At

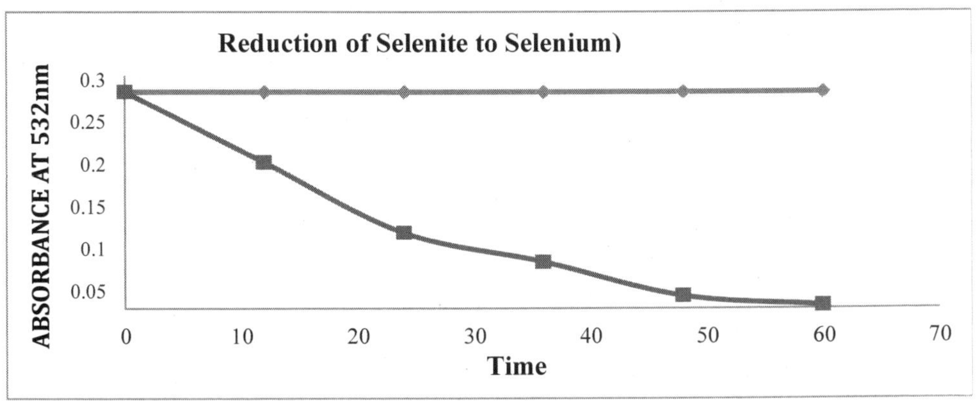

Figure 16.4: Time Course of Selenite Reduction by Bacterial Strain, BSB4.

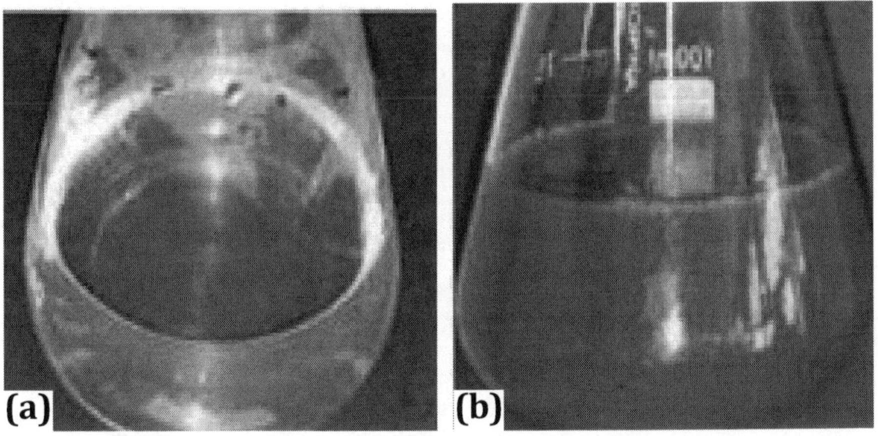

Figure 16.5: Change of Red Colour Indicates Reduction of Selenite against Control. (a) Control and (b) Change of colour of medium due to reduction.

pH 5, the bacterial strain BSB4 showed less selenite reduction (40 %). While, at pH 7 and 9, reduction percentage was increased. At a higher pH (pH 9.0) more selenite reduction (108 %) was observed by the bacteria. But at acidic pH, bacteria reduce selenium not so efficiently (Figure 16.6a). Formation of red precipitate is an indication that reduction has taken place. As the bacteria was isolated from saline habitat, it is very much essential to know it's reduction potential in presence of salt. For this purpose, the bacterial strain (BSB4) was inoculated in nutrient broth medium containing 100 mM of selenium but varied concentrations of salt (2-6 %). At different time interval the percentage of selenium reduction was estimated. The highest reduction of selenium (80 %) was observed in 4 per cent NaCl. However, less reduction (20 %) of selenium was observed in 2 per cent NaCl (Figure 16.6b). Similarly effect of temperature of selenite reduction was also studied at different temperature conditions (32, 37, 42 °C). It was observed that, the optimum temperature for selenite reduction was 37°C. However, at 42 °C 65 %

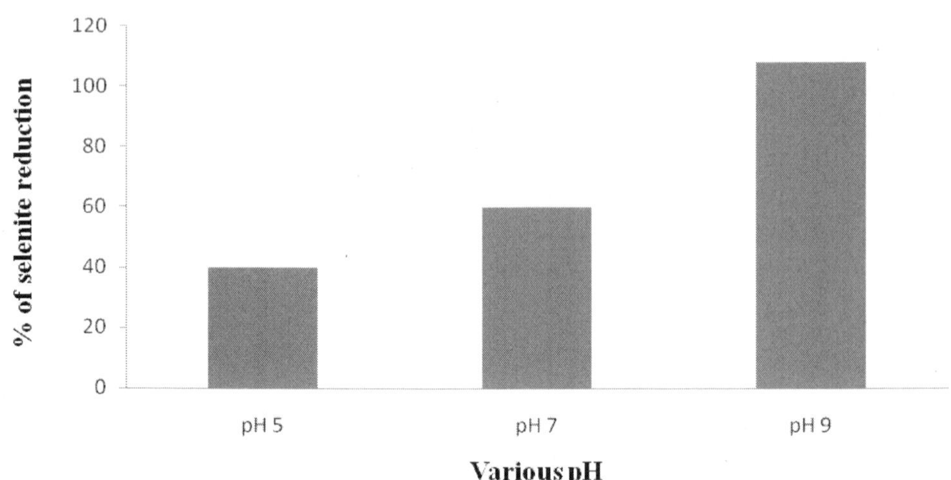

Figure 16.6a: Effect of pH on Selenite Reduction by BSB4.

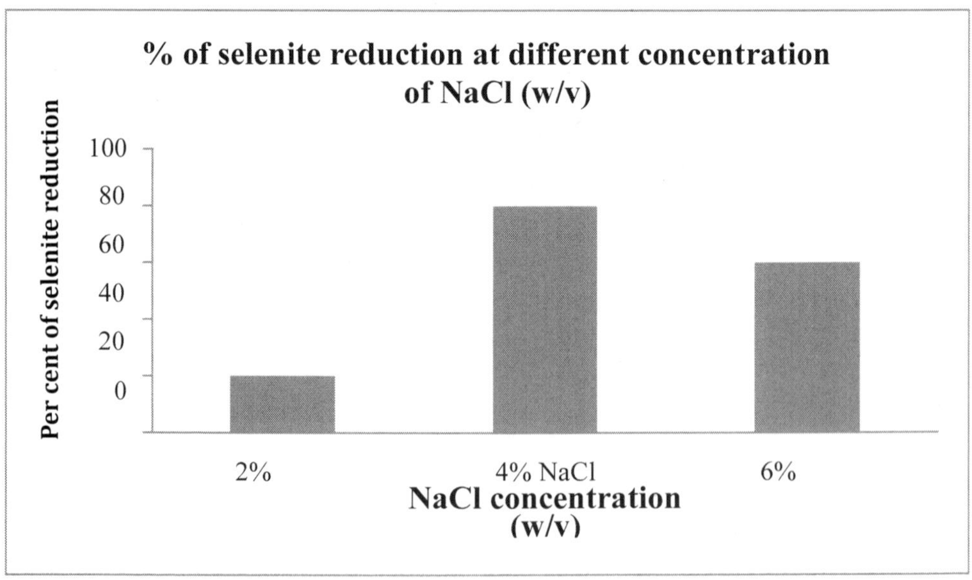

Figure 16.6b: Effect of NaCl on Selenite Reduction by BSB4.

of reduction of selenium was observed. Minimum reduction (60 %) was observed at 32°C (Figure 16.6c).

Biochemical Identification

In order to identify the most effective selenite reducing bacterial strain (BSB4), several biochemical tests were carried out. The bacterial strain (BSB4) was Gram-

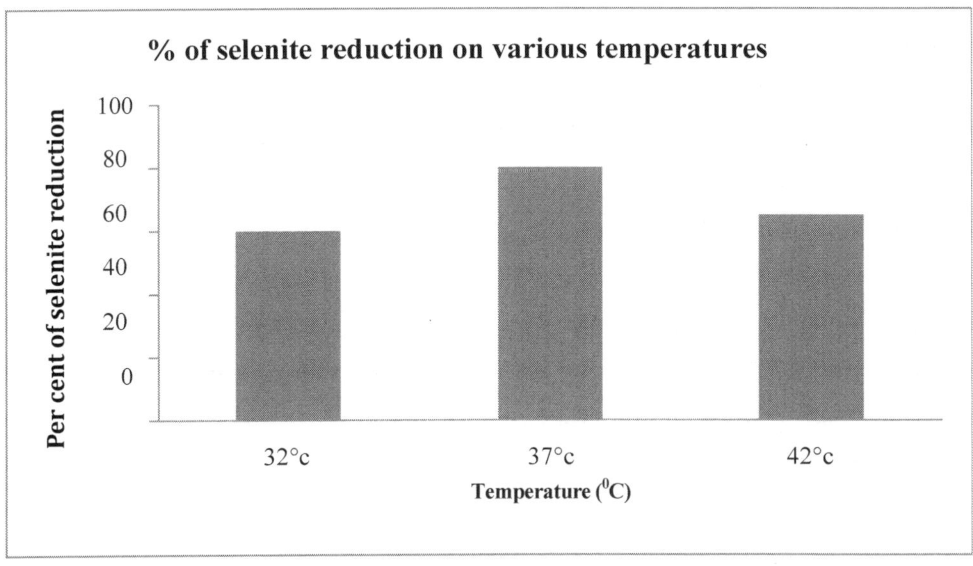

Figure 16.6c: Effect of Temperature on Selenite Reduction by BSB4.

positive, endospore forming motile rod. Biochemical tests such as catalase test, MR test, MRVP test, Indole test, citrate utilization test, carbohydrate utilization and fermentation test were carried out. The bacterial strain BSB4 was found negative for catalase test. The bacterial strain (BSB4) was found positive towards indole test and MR test and negative towards VP test. The bacterium found negative for citrate utilization test. The isolate showed positive result for acid production test while unable to hydrolyze gelatin and casein (Table 16.3). Based on the above biochemical data the bacterial strain was tentatively identified as *Bacillus* spp.

Table 16.3: Biochemical Identification of the Selenite Reducing Bacterial Strain (BSB4)

Characteristics	*BSB 4*	*Characteristics*	*BSB 4*
Gram staining	**+ve**	Spore forming	**+ ve**
Motile	**+ve**	Catalase test	+ve
MR test	+ve	VP test	-ve
Citrate utilization	+ve	Citrate	-ve
Indole production	+ve	Hydrolysis of casein	+ve
Hydrolysis of gelatin	-ve	Carbohydrate utilization	+ve
Carbohydrate fermentation	+ve		

Discussion

Selenium contamination due to various human activities is considered as one of the main environmental pollutions found almost everywhere in the earth. (David *et al.,* 2015). Selenium has widespread uses in the industrial and agricultural processes, which are responsible for its high toxic levels in the environment (Pierru *et al.,* 2006). The present research work focused on the bioremediation of selenite, the toxic form of selenium using a salt tolerant bacterial strain isolated from Bhitarkanika mangrove environment. Biotransformation of selenate and selenite to elemental selenium is known as dissimilatory selenate/selenite reduction (Narasingarao and Haggblom, 2004) and bacteria can play an important role in such biotransformation process. Bacteria showing tolerance/resistance to selenate or selenite would be considered as good candidate for bioremediation of selenium.

Minimum inhibitory concentration is the lowest concentration of metal that completely prevented bacterium growth. In the present study, MIC test was carried out for the selected 5 bacterial strains isolated from mangrove soils of Bhitarkanika. Among the five bacterial strains tested, BSB4 could able to tolerate highest selenite concentration (200 mM) and subjected to further studies. The potential of BSB4 for Se(IV) reduction was studied at a fixed selenite concentration (100 mM) using optimized growth conditions (*i.e.* at 37°C and pH 7.5). A representative time course of Se(IV) reduction along with control revealed that, there is practically no change of selenite concentration in the control during the incubation period. However, the bacterial strain BSB4 effectively reduced Se(IV) to Se(0) within 60h of incubation period. It is evident from the change in colour of the reduction medium to orange-red/red due to generation of allotropic form of elemental selenium during the exponential growth phase. In the contaminated environment, pH and temperature plays a very important role for reduction of toxic metals (Lortie *et al.,* 1992). Selenite reduction pattern was studied with respect to pH, temperature. In certain anaerobic microorganisms, selenium in the form of Se (VI) and Se (VI) can be used as a terminal electron acceptor, while few aerobic and phototrophic bacteria can catalyze the selenium oxyanions to insoluble Se(0) (Blum *et al.,* 1998; Mishra *et al.,* 2011). Effect of initial selenite concentration on selenite reduction potential of bacterial strains showed that more selenite is reduced at lower initial selenium concentration. The higher toxicity of the selenite ion may be responsible for the decrease in the rate of selenite reduction at higher selenium concentrations by BSB4. With the increase in incubation time, more selenite was reduced. Incubation time had a direct relation with selenite reduction. Greater the incubation time, greater will be the reduction potential. After 96 h of incubation on average, there was 92 % reduction, while at 48 and 24 h of incubation, it was 81 and 63 %, respectively. It has been shown in one of the previous works that Se(IV) reduction occurred only in late stationary phase under aerobic conditions in *Shewanella oneidensis* (Carpentier *et al.,* 2005). In the present study, the reduction of Se(IV) (100 mM) increased with an increase of pH and attained a maxima in between pH 7.0 and 8.0. Further increase of pH showed a marginal effect on Se(IV) reduction. It may be noted that the reduction of

Se(IV) to Se(0) at neutral to alkaline pH is similar to the observation of Lortie *et al.* (1992). Since the variation of salt concentration in the medium greatly affects the growth of bacterial strains, it was interesting to see the effect of salt concentration on Se(IV) reduction. The effect of NaCl concentrations in the range 4–6 % on Se(IV) (0.25 mM) reduction at pH 7.5 and 37 °C is shown in Figure 16.6b. It was seen that the reduction efficiency of the strain (BSB4) was marginally affected up to 6 % NaCl and thereafter sharply declined on further increase of salt concentration. The better reduction of selenite by BSB4 was observed at 4 per cent salt. The selenite reduction virtually stopped at higher concentration of salt in the media. The reduced activity is primarily due to detrimental effect of the ionic stress on the bacteria at higher salt concentrations which inhibits bacterial growth (Borg *et al.,* 2009).

Conclusion

Selenium exists in natural environment in trace quantity but the anthropogenic activities like mining and agricultural chemicals have increased the risk to the ecosystem. In the present study, one well characterised halotolerant bacteria *Bacillus sp.* (BSB4), isolated from mangrove soil of Bhitarkanika, Odisha without selenium contamination, were found capability of reducing Se(IV) to elemental selenium even in the presence of high salt concentrations. At fixed concentration of selenite (100 mM), the bacterial strain BSB4 could able to reduce 90 % of selenite in 60 h of incubation. Further the reduction process was optimized in the different pH, temperature, NaCl optimization. The reduction of selenite to red elemental selenium also from the evident of selenium reduction. Hence it is concluded that the bacterial strain BSB4 could be an efficient strain for bioremediation of selenite of industrial effluents containing relatively high amount of salt.

REFERENCES

Awual, M.R., Hasan, M.M. and Khaleque, M.A. (2015). Efficient selenium(IV) detection and removal from water by tailor-made novel conjugate adsorbent. Sens. Actuators B. Chem. 209: 194–202.

Borg, C., Friese, D. and Yin, T.T. (2009). Removal of Selenium in Contaminated Wastewater Streams. US patent 2009/0152194A1.

Blum, J.S., Bindi, A.B., Buzzelli, J., Stolz, J.F. and Oremland, R.S. (1998). *Bacillus arsenicoselenatis*, sp. nov., and *Bacillus selenitireducens*, sp. nov.: two haloalkaliphiles from Mono Lake, California that respire oxyanions of selenium and arsenic. *Archives of microbiology*, *171*(1): 19-30.

Carpentier, W., De Smet, L., Van Beeumen, J. and Brigé, A. (2005). Respiration and growth of Shewanella oneidensis MR-1 using vanadate as the sole electron acceptor. *Journal of bacteriology*. *187*(10): 3293-3301.

Dhanjal, S. and Cameotra, S.S. (2010). Aerobic biogenesis of selenium nanospheres by *Bacillus cereus* isolated from coalmine soil. Microb. Cell Fact. 9: 52–62.

Dias, A.C., Andreote, F.D., Dini-Andreote, F., Lacava, P.T., Sá, A.L., Melo, I.S., Azevedo, J.L. and Araújo, W.L. (2009). Diversity and biotechnological potential of culturable bacteria from Brazilian mangrove sediment. World J. of Microbiol. and Biotechnol. 25(7): 1305-1311.

David, N., Gluchowski, D.C., Leatherbarrow, J.E., Yee, D. and McKee, L.J. (2015). Estimation of Contaminant Loads from the Sacramento San Joaquin River Delta to San Francisco Bay. Water Environment Research, 87(4): 334-346.

Fordyce, F.M. (2005). Selenium deficiency and toxicity in the environment. In: Selinus, O., Alloway, B., Centeno, J.A., Finkelman, R.B., Fuge, R., Lindh, U., Smedley, P. (Eds.), Essentials of Medical Geology. Elsevier Academic Press, Amsterdam, Holland, pp. 373–416.

Ghosh, A., Mohod, K.M., Paknikar, M. and Jain, R.K. (2008). Isolation and characterization of selenite- and selenate-tolerant microorganisms from selenium contaminated sites. World J. Microbiol. Biotechnol. 24: 1607–1611.

Kothari, M.J. (2002). Mangrove diversity and its role for sustaining productivity of the North West coast in India. In: Proc. Natl. Sem. Creeks, Estuaries and Mangroves– pollution and conservation (Quadros Ged), Thane, India, pp. 226–233.

Lin, Z.Q., Schemenauer, R.S., Cervinka, V., Zayed, A., Lee, A., and Terry, N. (2000). Selenium volatilization from a soil—plant system for the remediation of contaminated water and soil in the San Joaquin Valley. J. of environmental quality. 29(4): 1048-1056.

Lortie, L., Gould, W.D., Rajan, S., McCready, R.G.L. and Cheng, K.J. (1992). Reduction of selenate and selenite to elemental selenium by a Pseudomonas stutzeri isolate. Appl. Environ. Microbiol. 58, 4042–4044.

Mehdi, Y., Hornick, J.L., Istasse, L. and Dufrasne, I. (2013). Selenium in the environment, metabolism and involvement in body functions. Molecules 18: 3292–3311.

Mishra, R.R., Prajapati, S., Das, J., Dangar, T.K., Das, N. and Thatoi, HN. (2011). Reduction of selenite to red elemental selenium by moderately halotolerant Bacillus megaterium strains isolated from Bhitarkanika mangrove soil and characterization of reduced product. Chemosphere 84: 1231-1237.

Narasingarao P. and Haggblom, M.M (2007). Identification of anaerobic selenate-respiring bacteria from aquatic sediments. Appl. Environ. Microbiol. 73: 3519–352.

Pierru, B., Grosse, S., Pignol, D. and Sabaty, M. (2006). Genetic and biochemical evidence for the involvement of a molybdenum-dependent enzyme in one of the selenite reduction pathways of Rhodobacter sphaeroides f. sp. denitrificans IL106. Appl. Environ. Microbiol. 72(5): 3147-3153.

Sakamoto, M., Chan, H.M., Domingo, J.L., Kubota, M. and Murata, K. (2012). Changes in body burden of mercury, lead, arsenic, cadmium and selenium in infants during early lactation in comparison with placental transfer. Ecotoxicol. Environ. Saf. 84: 179–184.

Sakamoto, M., Itai, T., Yasutake, A., Iwasaki, T., Yasunaga, G., Fujise, Y., Nakamura, M., Murata, K., Chan, H.M., Domingo, J.L. and Marumoto, M. (2015). Mercury speciation and selenium in toothed-whale muscles. Environmental research, 143: 55-61.

Santos, S., Ungureanu, G., Boaventura, R. and Botelho, C. (2015). Selenium contaminated waters: an overview of analytical methods, treatment options and recent advances in sorption methods. Science of the Total Environment, 521: .246-260.

Theisen, J., and Yee, N. (2014). The molecular basis for selenate reduction in *Citrobacter freundii*. Geomicrobiology Journal, 31(10): 875-883.

Ventosa, A. (2004). Halophilic microorganisms (p. 349pp). Berlin, Springer. New York.

Chapter 17

Control of Postharvest Fungal Decay of Root, Tuber and Bulb Crops Using Plant Products

Ashirbad Mohapatra[1], *Akhtari Khatoon*[2] *and Kunja Bihari Satapathy*[2]*

[1]*Department of Botany, Sri Jayadev College of Education and Technology, Bhubaneswar, Odisha*

[2]*Department of Botany, School of Applied Sciences, Centurion University of Technology and Management, Odisha*

**e-mail: kbs_bot@rediffmailmail.com*

ABSTRACT

An extensive survey undertaken in store houses and market places of Odisha could indicate that eight genera of fungi comprising of ten species were mostly responsible for causing postharvest decay in common root, tuber and bulb crops. The incidence of rotting in the samples collected was observed in the frequencies ranging from 9.74 per cent to 50.3 per cent of the total samples studied. Aspergillus, Geotrichum, Penicillium and Rhizopus were identified as the dominant genera of fungi for causing maximum loss of these vegetables in storage condition. Studies on the inhibition of mycelial growth of the ten fungal species under study by using aqueous, petroleum ether and methanolic leaf extracts of sixteen angiospermic plants confirmed that the petroleum ether and methanolic leaf extracts were more effective than their respective aqueous extracts besides showing severe to moderate impact upon the test pathogens compared to the mild and insignificant effect of the aqueous leaf extracts. Of the sixteen experimental plants, Ageratum conyzoides, Alstonia scholaris and Averrhoa carambola showed significant results in controlling the test pathogens than other plants tested. While comparing the efficacy of plant extracts (petroleum ether and methanolic extract of Ageratum conyzoides, Alstonia scholaris and Averrhoa carambola) with the commercial fungicides (Blitox-50, Dhanustin, Indofil and Macozeb) it was found that plant extracts were more efficient to inhibit the mycelial growth of ten test fungi. It was further interesting to observe that as a result of the combined effect of petroleum ether and methanolic extracts of Ageratum conyzoides, Alstonia scholaris and Averrhoa carambola could completely inhibit the mycelial growth of all the test fungi and could further be recommended as a suitable biofungicide.

Keywords: *Rooted crops, Postharvest management, Biofungicides.*

Introduction

Vegetables are an important source of human nutrition and are a rich and comparatively cheaper source of vitamins and minerals. They are important source for maintenance of good health and are also beneficial in protecting against some degenerative diseases. The vegetables grouped under root, tuber and bulb crops are most important as compared to other vegetables being non-perishable in nature. After harvest they can be stored and consumed throughout the year. Some important vegetables, included in this group are Potato (*Solanum tuberosum* L.), Sweet potato [*Ipomoea batatas* (L.) Lam.], Radish (*Raphanus sativus* L.), Carrot (*Daucus carota* L.), Beet (*Beta vulgaris* L.), Colocasia [*Colocasia esculenta* (L.) Schott.], Yam (*Dioscorea alata* L.), Elephant's foot yam [*Amorphophallus paenifolius* (Dennst.) Nicolson], Onion (*Allium cepa* L.) and Garlic (*Allium sativus* L.). After harvest, the produce is stored in locally prepared bins, in open store houses or in bamboo baskets for their consumption and marketing throughout the year.

The issue of postharvest disease management of these crops becomes an important concern due to inadequate agricultural storage practices for produce preservation from microbes-induced spoilages (Kana *et al.,* 2012). These crops are drought tolerant and provide a wide harvesting window which makes it act as a famine reserve. However, postharvest deteriorations caused by microbial invasion of the tubers are the most important causes of loss in its production and contribute hugely to the unsuccessful long term storage of the root tubers. The pathogenic organisms including fungi gain entry into tubers probably through the area where the tubers are separated from stems at harvest or the root tips which breaks during harvesting, or other natural openings and cracks on the tubers surfaces sustained during harvesting, transit or storage (Okigbo, 2009). Improper handling leaves off bruises on the produce during harvesting, transit or storage which may become contaminated from adherent soil or air-borne propagules which later colonies such abrasions to cause rots (Okigbo and Nmeka, 2005). In Odisha, the postharvest decay of these vegetables during storage pose serious problems and sustain heavy losses particularly for the warm and humid climate. According to various workers, rotting usually starts from the field and progresses in storage

(Okigbo and Ikediugwu, 2000). Unhealthy store conditions lead to the absorption of moisture by produce in storage as a result of defects in the storage facility, thus encouraging the development of hot-spots and moulds.

Antimicrobial chemicals such as benzimidazoles, aromatic hydrocarbons and sterol biosynthesis inhibitors are often used in the control of plant diseases in agriculture. However, there is a series of problems against the effective use of these chemicals in areas where the fungi have developed resistance (Brent and Hollomon, 1998). In order to overcome this problem, higher concentrations of the chemicals had been used, but this increased the risk of high-level toxic residues in the products which were not beneficial for human beings and environment. Thus, there has been a growing interest on the research of possible use of plant extracts for control of pest and diseases in agriculture which was less harmful to the human health and environment (Costa *et al.,* 2000; Duru *et al.,* 2003).

To avoid use of synthetic fungicides, it is imperative to find alternative sources from naturally occurring compounds that are easily biodegradable and of low mammalian toxicity for a safe control of fungal pathogens (Adongo *et al.,* 2012). Plants are the source of natural pesticides that make excellent leads for new pesticide developments (Shanmugavalli *et al.,* 2009). Recent studies on the use of plant extract have opened a new approach to control of plant diseases. These plant extract have been reported to be safe, non-toxic to man, but effective against plant pathogens (Shivpuri *et al.,* 1997). Plant diversity serves the humankind as renewable natural resources for a variety of biologically active chemicals. Active constituents of the medicinal and aromatic plants have been found to be less phytotoxic, more systemic and easily biodegradable (Fawcett and Spencer, 1970). Plants are the sources of natural pesticides that make excellent leads for new pesticide development (Gangadevi *et al.,* 2008; Satish *et al.,* 2008; Shanmugavalli *et al.,* 2009; Swarna Latha and Neelakanta Reddy, 2009).

In India, Dastur (1916) first studied the control measure for ripe rot of plantains. After a long period of gap, studies in this field gained rapid momentum and within two decades much knowledge was gained on fungi causing postharvest rot of fruits and vegetables from several parts of the country, particularly from Pune, Allahabad, Lucknow, Delhi, Kolkatta, Bhagalpur and Bangalore. Unfortunately, little work has been done so far in this field in Odisha (Das, 2012). In continuation of the previous work, the present topic was selected in view of the limited studies on postharvest fungal rots of root, tuber and bulb vegetables. The present article is, therefore, based upon the postharvest decay of root, tuber and bulb vegetables and their control by bio-fungicides. It is expected that the present investigations shall throw some light on the etiology and control of postharvest storage decay of vegetables such as potato, sweet potato, colocasia, yam, amorphophallus, beet, carrot, radish, onion and garlic caused by various species of fungi and their management by application of botanicals to avoid serious economic losses in the markets and store houses.

Methodology

Pathological Studies

Collection of Rotten Samples of Root, Tubers and Bulb Crops

Total one thousand samples of rotten potatoes, sweet potatoes, radish, carrot, beet, colocasia, yam, amorphophallus, onion and garlic were collected from various markets and store houses of Odisha during 2014-16. Rotten plant samples were collected in separate polythene bags with labels in respect of (1) date of collection (2) place of collection (3) extent of damage in sample lots (4) viable symptoms (smell, colour and mycelial growth) and (5) name of collector. All the samples were brought to the laboratory, surface sterilized with 0.1 per cent mercuric chloride solution for 2-3 minutes. Some specimens from each sample were examined for the presence of fungal infections by taking a small portion of the rotten tissues on a clean glass slides, macerated with lactophenol-cotton blue and examined under a microscope for the occurrence of fungal mycelium and fruiting bodies, if any. Specimens showing the signs of fungal infections were immediately processed for phytopathological analysis, after recording the visual symptoms of rotting.

Isolation of Fungi from Rotten Samples

Sterile moist chambers were prepared out of petridishes, 100 mm in diameter, by placing triple layer moist filter paper over which a platform was prepared by placing two pieces of clean glass slides over a pair of equal sized glass rods. It was sterilized at 15 PSI for 20 minutes in an autoclave.

Small pieces of tissues including both rotten and healthy areas were surface sterilized with 0.1 per cent mercuric chloride solution for 2-3 minutes, washed thoroughly three times with sterile distilled water and transferred aseptically into the sterile moist chamber at the rate of 5 pieces per chamber. Similar sterilized tissues were also transferred into petriplates containing potato dextrose agar (PDA) at the rate of 5 pieces per petri plate.

All the materials were incubated at room temperature (28° C ± 1° C) for four to five days. At the end of incubation, the sterile chambers as well as the petriplates were examined for the presence of fungi. Fungi of all kinds appearing on the incubated materials were transferred aseptically into PDA slants. Each fungus as obtained in PDA slants were subsequently purified by single spore/hyphal tip isolation technique and maintained by periodic sub culturing in PDA slants for further studies. The frequency of incidence of each fungus on different samples of rotten samples was estimated by the formula:

$$\text{Frequency of incidence} = X/Y \times 100$$

where,

X = The number of pieces showing the presence of different fungi and

Y = Total number of pieces investigated. The frequency of incidence was expressed in terms of percentage.

Identification of Fungi

The colony characteristics of each fungal isolates on PDA were recorded. The details of morphology of each isolate were studied under varying magnifications (× 100 to × 1000) of a microscope. The characteristics of mycelium and fructifications were recorded. The dimensions of each structure were measured with the help of an Ocular micrometer. Each fungal isolate was taxonomically identified with the help of available literatures and confirmed by the microbiology Department, OUAT, Bhubaneswar, and Odisha. The fungal isolate which found to be new their molecular identification was carried out by sequencing of 26S rRNA gene followed by submission of sequence to NCBI GenBank. Genomic DNA was extracted from these fungal isolates (Sambrook *et al.,* 1989) followed by amplification of 26S rRNA gene using universal primers DF (5'-ACCCGCTGAACTTAAGC-3') and DR (5'- GGTCCGTGTTTCAAGACGG -3') and sequencing at Xcelris Genomics, India. Raw sequences were analyzed by Bio-Edit software (v7.0.5.3) and identification of the isolates by BLASTN in NCBI database (www.ncbi.nlm.nih.gov/nucleotide). Multiple alignments of sequences were performed with the muscle (Edgar, 2004). The phylogenetic tree was constructed using the UPGMA distance algorithm (Wheeler and Kececioglu, 2007) using Mega software v5.1. The resultant tree topologies were evaluated by boot-strap analysis of muscle data sets based on 500 resamplings. The final sequences were submitted to NCBI GenBank to obtain the accession number of each isolates.

In vitro Evaluation of the Antifungal Efficacy of Plant Extracts

Selection of the Test Fungal Isolates

Total ten fungal species were found to be associated with postharvest decay of Potato, Sweet potato, Radish, Carrot, Beet, Taro, Yam, Amorphophallus, Onion and Garlic during the present investigations. All the fungal species were purified, identified and maintained in potato dextrose agar slants by periodically sub-culturing to use in the present studies.

Selection of Plants and Plant Parts

On the basis of local availability and medicinal values, the leaves of the 16 plant species such as *Abutilon indicum* (L.) Sweet (Malvaceae), *Ageratum conyzoides* L. (Asteraceae), *Alstonia scholaris* (L.) R.Br. (Apocynaceae), *Artocarpus heterophyllus* Lam. (Moraceae), *Averrhoa carambola* L. (Averrhoaceae), *Cassia fistula* L. (Caesalpiniaceae), *Centella asiatica* Urban. (Apiaceae), *Dillenia indica* L. (Dilleniaceae), *Eucalyptus globulus* Labill (Myrtaceae), *Haldinia cordifolia* (Roxb.) Rids (Rubiaceae), *Justicia adhatoda* L. (Acanthaceae), *Lawsonia inermis* L. (Lythraceae), *Murraya paniculata* Jack. (Rutaceae), *Pithecellobium dulce* (Roxb.) Benth. (Mimosaceae), *Pongamia pinnata* (L.) Panigrahi (Fabaceae) and *Tamarindus indica* L. (Caesalpiniaceae) were selected for the present investigations. Some of these plants have been reported to exhibit antifungal properties.

Collection, Processing and Extraction of Plant Material

Leaves of 16 angiospermic plants belonging to diverse plant families were collected from different parts of Odisha. Each material was collected in separate polythene bags and brought to the laboratory of Department of Botany, S.J. College of Education and Technology, Bhubaneswar, Odisha (India) for further studies.

Three solvent systems were used for leaf compound extraction *viz.* aqueous, petroleum-ether and methanol. For the preparation of aqueous extract (cold water extract), fresh plant leaves were washed thoroughly with sterile water, chopped into pieces and the extracts were obtained by macerating with 100 ml of sterile distilled water for each 20 g of fresh leaf material. The extracts thus, obtained were separately filtered through a three folds sterilized muslin cloth into sterilized conical flasks. Then the extracts were re-filtered with Whatman No. 1 filter paper using a power operated suction filter pump. These extracts formed the standard plant extract solution (20 per cent) and were stored at 5 °C in refrigerator for further use. For dried leaf extraction, Reflux method of solvent extraction was used for purification of different organic constituent. Solvent series for successive separation was as follows: petroleum-ether and methanol. First the suxhelt extraction unit was cleaned and rinsed with petroleum-ether then 30g of air dried and coarse powdered plant material were kept in it. Then this unit was kept on water bath and exhaustively extracted with petroleum-ether for 30 to 36 hrs. During extraction the round bottom flask (RBF) which contains the extracts should with porcelain chips for well siphonation. To confirm whether the extraction is complete or not, the extract from siphonating tube of suxhelt was taken in a watch glass. After petroleum-ether extraction the extracted plant material was then air dried, repacked in soxhelt apparatus and extracted with methanol for such time hour and maintain again. The crude extract and the fraction obtained at every step were filtered and distilled to evaporate the solvent from the extract. The liquid extracts were concentrated separately under vacuumed and the resulting dried extract was preserved in desiccators until further use. The diluted plant extracts were prepared by dissolving purified petroleum-ether and methanolic extract in dimethyl sulfoxide (DMSO) @ 20 mg plant extract in 1 ml DMSO.

Evaluation of Antifungal Efficacy of Plant Extracts

For the evaluation of antifungal effect of plant extracts on inhibition of mycelia growth of ten isolated fungal species namely, *Apergillus flavus, Apergillus niger, Fusarium oxysporum, Geotrichum candidum, Junghuhnia* sp. AK15 (NCBI Gene Bank accession number- KT 946990), *Mucor* sp. BOT15 (NCBI Gene Bank accession number- KT 946994), *Neoscytallidium* sp. AM15 (NCBI Gene Bank accession number- KT 946992), *Penicillium* sp. KBS15 (NCBI Gene Bank accession number- KT 946991), *Penicillium* sp. UU15 (NCBI Gene Bank accession number- KT 946993) and *Rhizopus oryzae* was adopted by poison food technique. For this purpose, Potato Dextrose Agar (PDA) medium was prepared and sterilized at 15 PSI for 20 minutes. After sterilization, 1 ml of each diluted plant extract solution

(20 mg extract/1 ml DMSO) were mix with 19 ml of PDA (Potato Dextrose Agar) and poured into each petriplate @ 1 mg/ml, mixed thoroughly and allowed to solidify. PDA medium without the plant extracts but substitute with 1 ml distilled water was served as control.

The mycelial disc of 5 mm in diameter was taken from the periphery of 2-3 day old colonies of the desired fungi with the help of a sterilized cork borer. These discs were transferred aseptically inside the laminar flow with the provision of ultraviolet light and inoculated at the centre of each petriplate containing PDA with plant extracts. Three replications were maintained for each fungus × plant extract combination. A separate set of petri plates containing only PDA was inoculated by each test fungus to serve as control. All the test petriplates were incubated at room temperature (28 ± 1 °C) and examined at regular intervals. Growth of the isolates was considered as optimum for recording the observations when it reached around 8 cm in any of the control sets. *Rhizopus oryzae* was incubated for 1 day, *Neoscytallidium* sp. AM15 for 3 days, *Junghuhnia* sp. AK15 and *Mucor* sp. BOT15 for 5 days, *Aspergillus flavus* for 7 days, *Aspergillus niger* for 8 days, *Penicillium* sp. UU15 and *Fusarium oxysporum* for 10 days *Geotrichum candidum* for 11 days and *Penicillium* sp. KBS15 for 15 days depending upon their complete growth (8 cm) on the petriplate. At the end of incubation period, the diameter of the colony in each petriplate was measured in two planes, one at the right angle to the other. Further, the data of each test plant extract was compared with that of the control in order to estimate the inhibition (positive) or stimulation (negative) effects of the plant extract tested. The formula used for this purpose was:

Per cent inhibition/Stimulation = X-Y/X × 100

where,

X = Colony diameter in control

Y = Colony diameter in treated plated.

For the sake of convenience of description of the results, it was classified by the method of Rath and Mohanthy (1978) with slight modifications as give below:

Per cent of Inhibition	***Rating***
0 or negative	No inhibition
1-10	Incipient inhibition
11-30	Mild inhibition
31-70	Moderate inhibition
71-100	Severe inhibition

While screening the data on the basis of the above noted scale, it was noted that most of the effective inhibition of the colony diameter were in the range of 31-70 per cent. Only in a few instances it exceeded 70 per cent.

Comparison of Plant Extracts and Estimation of Inhibition of Mycelial Growth by Selected Plant Extracts

For this purpose, 3 selected plants showing the mean maximum inhibition of mycelial growth of ten test fungal species namely *Aspergillus flavus*, *Aspergillus niger*, *Fusarium oxysporum*, *Geotrichum candidum*, *Junghuhnia* sp. AK15, *Mucor* sp. BOT15, *Neoscytallidium* sp. AM15, *Penicillium* sp. KBS15, *Penicillium* sp. UU15 and *Rhizopus oryzae*. The selected plant species were *Azeratum conyzoides*, *Alstonia scholaris* and *Averrhoa carambola*.

Comparison of the Efficacy of Synthetic Fungicides and Selected Plant Extracts

A separate experiment was conducted in order to compare the relative efficacy of four synthetic fungicides, *viz.*, Dhanustin, Blitox-50, Indofil-M-45 and Macozeb with the petroleum ether and methanolic leaf extracts of *Ageratum conyzoides*, *Alstonia scholaris* and *Averrhoa carambola*. The details of concentrations are given below:

Table 17.1: Comparison of Plants and Fungicides Used along with their Concentration

Treatment	*Solvent System*	*Concentration mg/ml*	*Chemical Constituent*
Blitox-50		0.1	Copper oxychloride
Dhanustin		0.1	
Indofil-M-45		0.1	Carbamate
Macozeb		0.1	Coordination product of zinc and magnesium ethylene bis dithiocarbamate
Ageratum conyzoides	*Aqueous*	4	Natural plant product
	Pet-ether	1	
	Methanol	1	
Alstonia scholaris	*Aqueous*	4	Natural plant product
	Pet-ether	1	
	Methanol	1	
Averrhoa carambola	*Aqueous*	4	Natural plant product
	Pet-ether	1	
	Methanol	1	
Control	-	-	No fungicide or plant product

Incorporation of the test fungicides/natural plant extracts was carried out as per the method described earlier. The required quantity of fungicides and plant extracts were incorporated after sterilization of the medium. The petriplates containing PDA and fungicides/plant extracts were inoculated by 10 test fungal isolates in three replicates and incubated for 1st to 15th days depending upon their respective complete growth as described earlier. At the end of incubation period, the colony diameter of each fungus-fungicide/plant extract combination was measured, transferred to percentage of mycelial inhibition as per the methods described earlier.

Formulation of Herbal Fungicides and Recommendation for Use

A combination of petroleum ether leaf extracts of *Ageratum conyzoides* (10 mg) and *Alstonia scholaris* (10 mg); *Ageratum conyzoides* (10 mg) and *Averrhoa carambola* (10 mg); *Alstonia scholaris* (10 mg) and *Averrhoa carambola* (10 mg) was tested against ten fungi tested. The concentration of the formulated combination was 20 mg/ml DMSO (10 mg extract of each plant).

Results and Discussion

The data on the incidence and taxonomic identity of fungi in root, tuber and bulb crops revealed that eight genera comprising of ten species of fungi were responsible for storage decay of ten root, tuber and bulb vegetables collected from different market places and store houses of Odisha in frequencies ranging from 9.74 per cent to 50.3 per cent of the cases studied. *Aspergillus, Geotrichum, Penicillium* and *Rhizopus* were found to be the dominant fungi causing maximum loss of those vegetables during storage due to rotting. It was also noted that these fungi are generally the soil inhabitants which come in contact with the tubers and fleshy roots of these vegetables during the crop stand as surface microflora. After harvest and during transport, transit and storage, they infected the tubers and roots through wounds arising out of harvest, transportation and storage handlings or by injuries caused by insects and other microorganisms when the environmental factors are favourable for their growth and disease development incurring heavy economic losses to both growers and consumers. Experiments conducted on the pathogenicity tests indicated that fungi isolated from rotten samples of potato, colocasia, amorphophallus, sweet potato, yam, carrot, radish, beet, onion and garlic were pathogenic to their respective hosts in an artificial inoculation test.

The data on the pathogenicity test of six fungi on potato tubers indicated that *Aspergillus niger* was the most pathogenic one and causing the complete rotting of tuber within 10 days of incubation. The pathogenicity test data of colocasia rotting caused by five fungi indicated that *Rhizopus oryzae* was the most pathogenic to corms than other isolates and complete rotting of the corm within 10 days of inoculation was recorded. Amorphophallus rotting was caused by three fungi and the most virulent was *Geotrichum candidum* which caused the rotting within 30 days of inoculation. Artificial rotting of this host took a long time. Among the two fungi, *Rhizopus oryzae* was the most virulent one causing the complete rotting of sweet potato within 15 days. Similarly among the five fungi causing storage rot in yam, *Neoscytalidium* sp. AM15 was the most virulent to cause complete rotting within 20 days. Rotting in carrot, radish and beet root was caused by three fungi and among them the most virulent fungus for carrot was *Rhizopus oryzae* while for radish and beet was *Aspergillus niger* but both the fungi cause complete rotting of their respective host within 10 days of inoculation. Rotting of onion and garlic was found to be caused by two and three fungi respectively. Among them for onion *Aspergillus niger* and for garlic *Penicillium* sp. UU15 was found to be more virulent than others. They cause the complete rotting of the respective host within 15 days of inoculation.

Studies on the inhibition of ten fungal species causing storage rot of root, tuber and bulb vegetables, by aqueous, petroleum ether and methanolic leaf extract of sixteen angiospermic plants revealed that petroleum ether and methanolic leaf extracts were more efficient than aqueous extracts (Tables 17.2–17.4 and Figures 17.1–17.3). Petroleum ether and methanolic extracts of all the sixteen test plants were severe to moderately efficient against the ten fungi tested while the aqueous leaf extract were mild to incipiently efficient to inhibit the mycelial growth. Out of the sixteen plants tested, extracts of *Ageratum conyzoides, Alstonia scholaris* and *Averrhoa carambola* were comparatively more efficient than extracts of other plants tested (Table 17.5 and Figure 17.4).

Table 17.2: Screening of Aqueous, Petroleum Ether and Methanolic Extracts of *Ageratum conyzoides* L. against Ten Test Fungi

Test Fungi	*Per cent of Inhibition of Mycelial Growth*			
	Aqueous Extract	*Petroleum Ether Extract*	*Methanolic Extract*	*Mean*
Aspergillus flavus	25.13 ± 0.56	62.90 ± 0.17	56.95 ± 0.68	48.33
Aspergillus niger	21.25 ± 0.73	62.54 ± 0.17	46.51 ± 0.63	43.43
Fusarium oxysporum	15.68 ±0.29	59.80 ± 0.76	57.41 ± 1.05	44.3
Geotrichum candidum	16.42 ± 0.63	89.03 ± 0.56	79.08 ± 0.65	61.51
Junghuhnia sp. AK15	33.12 ± o.51	98.13 ± 0.38	100	77.08
Mucor sp. BOT15	42.77 ± 0.73	99.9 ± 0.08	88.48 ± 1.19	77.05
Neoscytalidium sp. AM15	20.8 ± 0.80	86.86 ± 0.26	87.87 ± 0.67	65.17
Penicillium sp. KBS15	26.81 ± 0.66	70.29 ± 0.80	58.72 ± 1.04	51.94
Penicillium sp. UU15	14.79 ± 0.40	58.35 ± 0.36	59.5 ± 0.88	44.21
Rhizopus oryzae	11.78 ± 0.35	88.97 ± 1	88.88 ± 0.59	63.21
Mean	**22.85**	**77.68**	**72.34**	

Results expressed as mean ± S.E.M of three determinations.

Comparative efficacy of plant extracts (petroleum ether and methanolic extract of *Ageratum conyzoides, Alstonia scholaris* and *Averrhoa carambola*) with four commercial fungicides (Blitox-50, Dhanustin, Indofil and Macozeb) revealed that plant extracts were more efficient to inhibit the mycelial growth of ten fungi

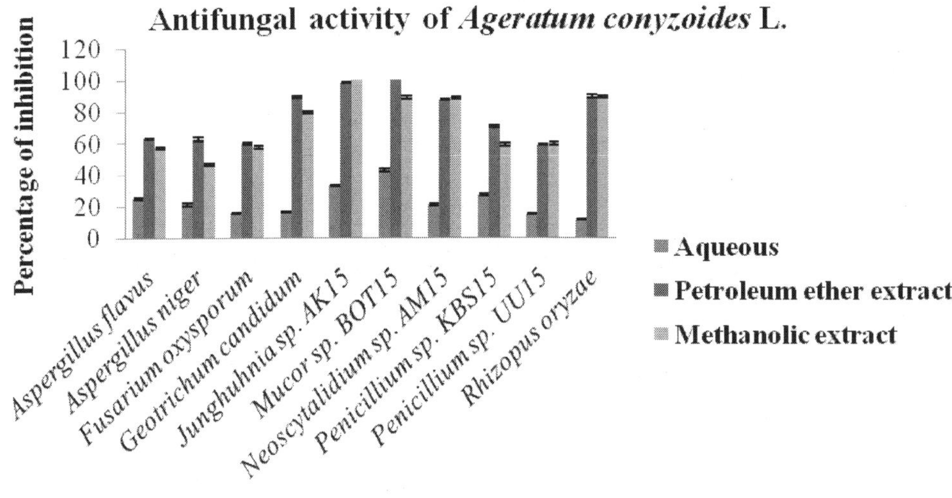

Figure 17.1: Antifungal Activity of *Ageratum conyzoides* L. against Ten Tested Fungi.

tested (Table 17.6 and Figure 17.5). Mishra and Dixit (1976) reported that the leaf extracts of *Clematis gouriana* was superior to Blitox-50, Dithane M-45, Dithane Z-78 and Ziram while Renu *et al.* (1985) reported that essential oil from leaf of *Aegle marmelos* possessed greater fungitoxicity than that of six commercial

Table 17.3: Screening of Aqueous, Petroleum Ether and Methanolic Extracts of *Alstonia scholaris* (L.) R.Br. against Ten Test Fungi

Test Fungi	*Per cent of Inhibition of Mycelial Growth*			
	Aqueous Extract	*Petroleum Ether Extract*	*Methanolic Extract*	*Mean*
Aspergillus flavus	16.98 ± 0.62	54.26 ± 0.82	45.66 ± 1.44	38.97
Aspergillus niger	10.88 ± 0.48	34.32 ± 0.36	54.72 ± 0.98	33.3
Fusarium oxysporum	5.58 ± 0.54	58.55 ± 0.22	57.33 ± 1.51	40.48
Geotrichum candidum	18.44 ± 1.11	73.41 ±1.87	80.14 ± 1.03	57.33
Junghuhnia sp. AK15	25.46 ± 0.65	100	99.46 ± 0.23	74.97
Mucor sp. BOT15	15.97 ± 0.99	100	74.48 ± 0.83	63.48
Neoscytalidium sp. AM15	8.99 ± 0.53	84.28 ± 0.75	87.25 ± 1.03	60.17
Penicillium sp. KBS15	10.5 ± 0.23	51.66 ± 0.98	59.91 ± 0.53	40.69
Penicillium sp. UU15	14.14 ± 0.65	63.1 ± 1.15	79.04 ± 1.05	52.09
Rhizopus oryzae	21.18 ± 0.93	86.29 ± 0.68	89.14 ± 0.65	65.54
Mean	**14.81**	**70.58**	**72.71**	

Results expressed as mean ± S.E.M of three determinations.

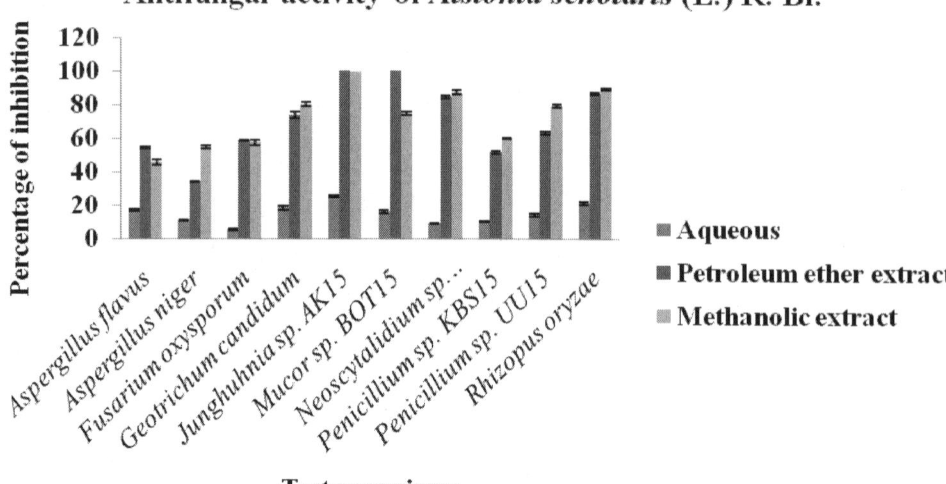

Figure 17.2: Antifungal Activity of *Alstonia scholaris* (L.) R. Br. against Ten Test Fungi.

fungicides. El-Shami *et al.* (1985) evaluated garlic extracts along with five different fungicides *in vitro* against *Fusarium oxysporum* sp. and found that the garlic extract inhibited spore germination and mycelia growth of the fungus in a similar manner to five different fungicides. Bandara *et al.* (1989) reported that the petroleum

Table 17.4: Screening of Aqueous, Petroleum Ether and Methanolic Extracts of *Averrhoa carambola* L. against Ten Test Fungi

Test Fungi	*Per cent of Inhibition of Mycelial Growth*			
	Aqueous Extract	*Petroleum Ether Extract*	*Methanolic Extract*	*Mean*
Aspergillus flavus	17.49 ± 0.38	57.39 ± 0.46	54.57 ± 0.86	43.15
Aspergillus niger	20.66 ± 0.64	27.8 ± 0.12	45.47 ± 0.75	31.31
Fusarium oxysporum	14.44 ± 0.72	73.5 ± 0.62	58.57 ± 0.44	48.83
Geotrichum candidum	14.03 ± 1.22	77.75 ± 0.65	70.46 ± 0.84	54.08
Junghuhnia sp. AK15	19.19 ± 0.45	98.93 ± 0.42	99.66 ± 0.27	72.59
Mucor sp. BOT15	9.3 ± 0.53	84.81 ± 0.53	79.83 ± 0.47	57.98
Neoscytalidium sp. AM15	22.86 ± 0.69	88.93 ± 0.67	90.21 ± 0.74	67.33
Penicillium sp. KBS15	15.93 ± 0.26	75.20 ± 0.73	73.08 ± 0.73	54.74
Penicillium sp. UU15	26.08 ± 0.58	67.38 ± 1.01	67.5 ± 0.7	53.65
Rhizopus oryzae	22.63 ± 0.65	87.94 ± 0.62	89.7 ± 1.56	66.75
Mean	**18.26**	**73.96**	**72.9**	

Results expressed as mean ± S.E.M of three determinations.

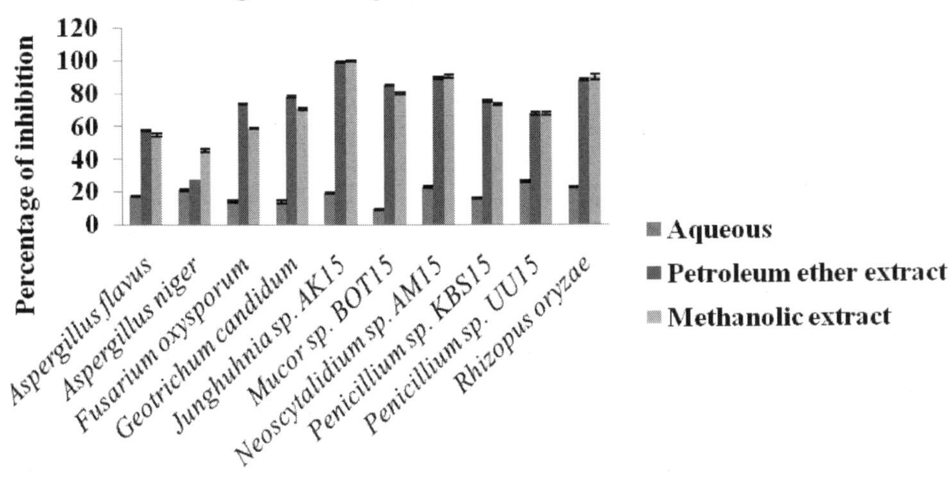

Figure 17.3: Antifungal Activity of *Averrhoa carambola* L. against Ten Tested Fungi.

and ethyl acetate extracts of *Butea monosperma* was fungitoxic to *Cladosporium cladosporoides.* During the present investigations, similar study on comparing the effect of four commercial fungicides with three promising plant extracts could reveal that the activity of a particular plant extract or fungicide effective against a particular test fungus was quite ineffective against others in inhibiting the mycelial growth. For inhibiting the mycelial growth of *Aspergillus flavus, Fusarium oxysporum* and *Penicillium* sp. KBS15, Dhanustin and Macozeb was found to be superior over the three test plants but the activity of plant extracts were found to be more efficient than Blitox-50 and Indofil. For growth retardation of *Aspergillus niger*, the plant extracts were found to be more efficient than the test commercial fungicides except Dhanustin. Against *Geotrichum candidum*, the fungitoxic potential of *Ageratum conyzoides* was found to be equal with Mancozeb while the effect of the rest of the plant extracts were more than rest of the three commercial fungicides. However, the antifungal activity of plant extracts was found to be superior over fungicides against *Mucor* sp. BOT15. Against *Junghuhnia* sp. AK15, all the test plant's extract and commercial fungicides showed significant inhibitory effect except Blitox-50. The effect of Dhanustin, which completely inhibited the mycelial growth of *Aspergillus flavus* and *Penicillium* sp. UU15, confirmed its superiority over all the test fungicides and three plant species. For the complete inhibition of mycelial growth in case of *Geotrichum candidum, Junghunia* sp. AK15, *Mucor* sp. BOT15, *Neoscytalidium* sp. AM15, *Penicillium* sp. KBS15, *Penicillium* sp. UU15 and *Rhizopus oryzae* all the three test plant species were significantly superior over all the four commercial fungicides.

Table 17.5: Comparative Efficacy of Antifungal Activity of Sixteen Angiospermic Plants

Test Plants	Growth Inhibition of Test Fungal Species (per cent)										
	Af	*An*	*Fo*	*Gc*	*Ja*	*Mb*	*Na*	*Pk*	*Pu*	*Ro*	*Mean*
Abutilon indicum	37.99 ± 19.24	26.4 ± 8.97	28.5 ± 4.74	43.75 ± 16.8	76.72 ± 28.59	51.86 ± 24.83	58.53 ± 24.42	28.59 ± 14.63	41.62 ± 11.63	52.2 ± 33.95	**44.61**
Ageratum conyzoides	48.33 ± 16.58	43.43 ± 16.99	44.3 ± 20.26	61.51 ± 32.14	77.08 ± 31.09	77.05 ± 24.68	65.17 ± 31.38	51.94 ± 18.38	44.21 ± 20.8	63.21 ± 36.36	**57.62**
Alstonia scholaris	38.97 ± 15.93	33.3 ± 17.91	40.48 ± 24.68	57.33 ± 27.63	74.97 ± 35.01	63.48 ± 35.17	60.17 ± 36.21	40.69 ± 21.61	52.09 ± 27.61	65.54 ± 31.38	**52.7**
Artocarpus heterophyllus	42.53 ± 25.42	31.03 ± 12.2	40.84 ± 17.07	61.77 ± 25.62	73.44 ± 34.5	63.85 ± 32.76	52.54 ± 28.29	41.33 ± 18.05	38.91 ± 23.46	64.42 ± 37.39	**51.06**
Averrhoa carambola	43.15 ± 18.18	31.31 ± 10.42	48.83 ± 25.07	54.08 ± 28.47	72.59 ± 37.76	57.98 ± 34.48	67.33 ± 31.45	54.74 ± 27.45	53.65 ± 19.49	66.75 ± 31.21	**55.04**
Cassia fistula	34.27 ± 20.27	35.14 ± 11.34	41.31 ± 14.25	54.07 ± 28.91	68.98 ± 33.4	63.01 ± 33.39	60.8 ± 36.34	35.94 ± 16.12	46.89 ± 22.98	62.94 ± 32,.26	**50.33**
Centella asiatica	38.56 ± 18.0	27± 13.78	22.08 ± 12.46	47.99 ± 19.95	68.76 ± 33.07	50.48 ± 23.06	57.16 ± 24.61	24.37 ± 15.74	40.96 ± 16.1	58.95 ± 24.92	**43.63**
Dillenia indica	34.76 ± 19.98	27.5 ± 18.7	33.62 ± 17.93	51.63 ± 25.85	69.07 ± 34.09	50.39 ± 19.7	63.01 ± 27.96	44.51 ± 22.13	51.51 ± 25.18	60.82 ± 26.06	**48.68**
Eucalyptus globulus	40.95 ± 16.88	37.19 ± 15.11	32.52 ± 19.36	57.6 ± 27.68	72.15 ± 32.4	63.85 ± 28.9	56.06 ± 27.8	39.75 ± 20.06	46.82 ± 22.4	55.63 ± 24.76	**50.25**
Haldina cordifolia	36.71 ± 17.23	32.09 ± 13.78	49.14 ± 23.26	53.07 ± 25.58	70.16 ± 30.99	59.31 ± 35.64	65.75 ± 30.3	46.64 ± 20.42	43.01 ± 24.2	55.2 ± 33.05	**51.1**
Justicia adhatoda	33.55 ± 17.24	31.08 ± 12.08	37.97 ± 23.38	56.01 ± 30.01	70.88 ± 33.16	59.78 ± 29.62	43 ± 25.08	29.95 ± 11.8	35.63 ± 15.95	59.67 ± 31.89	**45.75**
Lawsonia inermis	33.47 ± 21.8	37.68 ± 17.9	28.7 ± 18.71	55.62 ± 26.17	68.69 ± 34.41	52.08 ± 24.06	62.96 ± 32.67	36.06 ± 21.95	48.04 ± 23.28	74.28 ± 36.36	**49.75**

Test Plants	Growth Inhibition of Test Fungal Species (per cent)										
	Af	***An***	***Fo***	***Gc***	***Ja***	***Mb***	***Na***	***Pk***	***Pu***	***Ro***	***Mean***
Murraya paniculata	40.49 ± 22.33	26.81 ± 15.82	32.51 ± 12.89	48.84 ± 22.56	66.98 ± 38.56	52.66 ± 26.47	56.41 ± 25.05	27.56 ± 16.92	40.01 ± 22.7	48.68 ± 20.57	**44.09**
Pithecelobium dulce	36.74 ± 17.06	22.13 ± 14.93	27.41 ± 15.72	49.81 ± 26.88	68.93 ± 32.57	39.7 ± 22.15	57.77 ± 29.79	26.96 ± 19.58	45.71 ± 23.17	53.32 ± 24.93	**42.84**
Pongamia pinnata	37.26 ± 18.25	35.5 ± 18.5	36.52 ± 15.68	52.18 ± 28.56	74.77 ± 31.27	64.74 ± 28.64	54.92 ± 26.67	49.39 ± 20.67	40.35 ± 22.88	56.95 ± 28.60	**50.25**
Tamarindus indica	19.31 ± 14.32	16.02 ± 11.33	33.06 ± 18.13	54.5 ± 27.95	68.88 ± 36.15	53.51 ± 24.5	52.4 ± 23.92	35.41 ± 19.01	33.32 ± 17.66	65.51 ± 32.01	**43.19**
Mean	**37.31**	**30.85**	**36.11**	**53.73**	**71.44**	**57.73**	**58.37**	**38.36**	**43.92**	**60.25**	

Af: *Aspergillus flavus*; An: *Aspergillus niger*; Fo: *Fusarium oxysporum*; Gc: *Geotrichum candidum*; Ja: *Junghuhnia* sp. AK15; Mb: *Mucor* sp. BOT15; Na: *Neoscytalidium* sp. AM15; Pk: *Penicillium* sp. KBS15; Pu: *Penicillium* sp. UU15; Ro: *Rhizopus oryzae*.

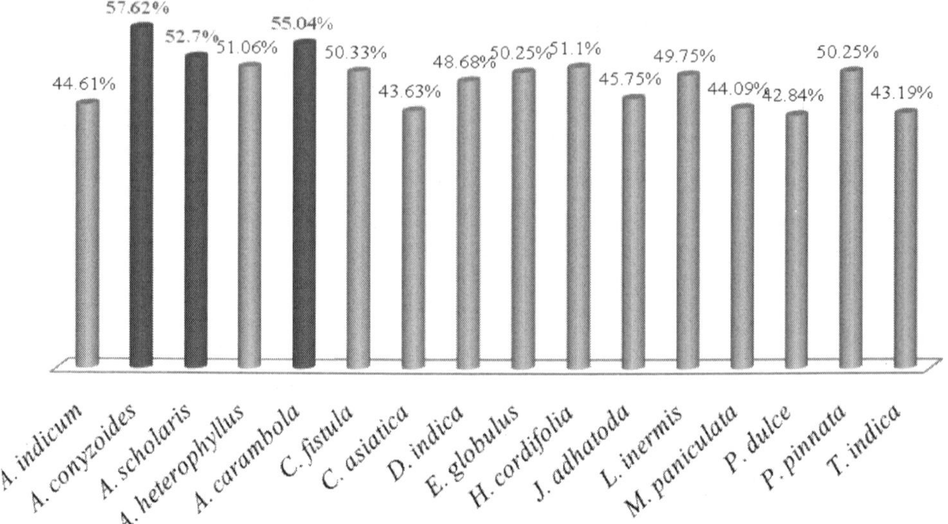

Figure 17.4: Comparative Efficacy of Antifungal Activity of Sixteen Medicinal Plants.

Table 17.6: Comparative Efficacy of Three Selected Plants and Four Commercial Fungicides on Mycelia Inhibition of Ten Fungal Species

Test Fungi	*Percentage of Inhibition of Mycelial Growth*							
	Ag	*As*	*Av*	*Blitox-50*	*Dhanu-stin (0.1mg/ml)*	*Indofil (0.1mg/ml)*	*Macozeb (0.1mg/ml)*	*Mean (0.1 mg/ml)*
A. flavus	48.33	38.97	43.15	21.74	100	19.63	82.5	**54.99**
A. niger	43.43	33.3	31.31	17.66	81.41	22.75	18.11	**39.10**
F. oxysporum	44.3	40.48	48.83	0	86.7	29.33	79.53	**62.81**
G. candidum	61.51	57.33	54.08	17.09	31.41	31.41	61.05	**53.12**
***J.* sp. AK15**	77.08	74.97	72.59	17.79	76.24	72.5	80.43	**78.024**
***M.* sp. BOT15**	77.05	63.48	57.98	24.8	11.83	33.17	30.75	**51.73**
***N.* sp. AM15**	65.17	60.17	67.33	58.33	67.13	77.28	73.2	**76.69**
***P.* sp. KBS15**	51.94	40.69	54.74	12.5	68.58	19.25	72.57	**51.78**
***P.* sp. UU15**	44.21	52.09	53.65	56.25	100	100	100	**78.57**
R. oryzae	63.21	65.54	66.75	100	6.48	31.75	37.16	**56.04**
Mean	**57.62**	**52.70**	**55.04**	**22.61**	**62.98**	**43.7**	**63.53**	

Ag: *Ageratum conyzoides*; As: *Alstonia scholaris*; Av: *Averrhoa carambola.*

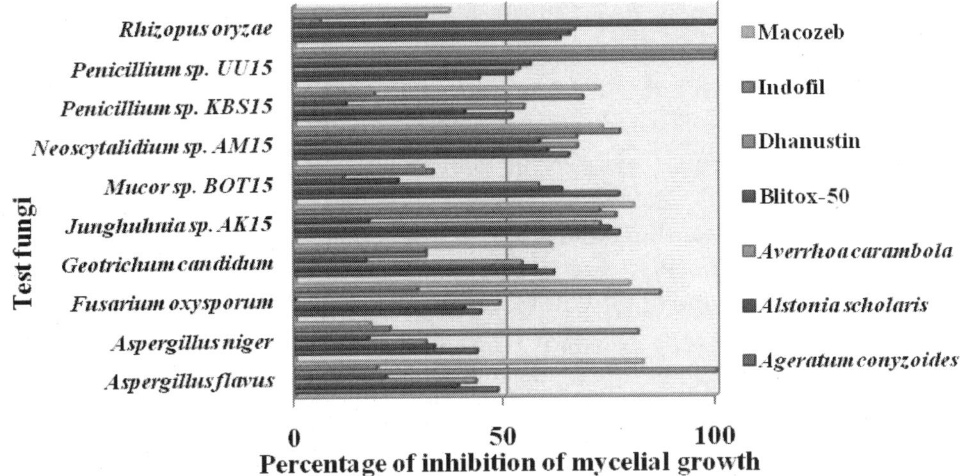

Figure 17.5: Comparative Efficacy of Three Selected Plants and Four Commercial Fungicides on Mycelial Inhibition of Ten Test Fungal Species.

Studies on the antifungal properties and potentials of plants and plant products has now been extended in several directions, namely identification of the potential microbiocidal products *in vitro*; their isolation, identification and characterization in the laboratory; comparison of plant products and extracts with commercial fungicides and testing the efficacy of plant products in the field, *in-vivo*. All these activities put together shall enable us not only locate and use the present herbal recommendations against postharvest management of root, tuber and bulb crops but also the application of biofungicides would certainly be an endeavour for sustainable agricultural practices. As the plant kingdom is a rich reservoir of useful materials, the future of plant products in plant diseases control seems to be very bright.

It is suggested that the extracts of effective experimental plants of the present study may be utilized as postharvest dips for the control of storage decay of root, tuber and bulb crops in order to avoid the side effects caused due to use of commercial fungicides. The findings of the present investigation will be of immense help to the agro-chemical industries to prepare and biofungicides for sustainable agricultural practices.

REFERENCES

Adongo, J.O., Njue, A.W., Omolo, J.O., Cheplogoi, P.K. and Otaye, D.O. (2012). *In vitro* inhibition of *Botrytis cinerea* - causative agent for grey mold by crude extracts of Basidiomycetes fungi. Sci. J. Biochem. 173-175.

Bandara, B.M.R., Fernando, H.I.S., Hewage, C.M., Karunarathe, V., Adikaram, N.K.B. and Wijsundara, D.S.A. (1989). Antifungal activity of some medicinal plants of Srilanka. Journal of the National Science Council of Srilanka, 17(1): 1-3.

Brent, K.J. and Hollomon, D.W. (1998). Fungicide resistance: the assessment of risk. FRAC, Global Crop Protection Federation, Brussels. Monograph, 2: 1-48.

Costa, T.R., Fernandes, F.L.F., Santos, S.C., Oliveria, C.M.A., Liao, L.M., Ferri, P.H., Paulo, J.R., Ferreira, H.D., Sales, B.H.N. and Silva, M.R.R. (2000). Antifungal activity of volatile constituents of *Eugenia dysenterica* leaf oil. J. Ethnopharm., 72(1-2): 111-117.

Das, S. (2012). Antimicrobial and Antioxidant activities of green and ripe fruits of *Averrhoa carambola* Linn. and *Zizyphus mauritiana* Lam. Asian Journal of Pharmaceutical and Clinical Research, 5(3): 102-105.

Dastur, J.F. (1916). Spraying for ripe rot of plantain fruits. Agric. J. India, 11: 142-149.

Duru, M.E., Cakir, A., Kordali, S., Zengin, H., Harmandar, M., Izumi, S. and Hirata, T. (2003). Antifungal activities of the leaves of three *Pistacia* species grown in Turkey. Fitoterapia, 74(1-2): 170-176.

EL-Shami, M.A., Fadl, F.A., Tawfick, K.A., Sirry, A.R. and Zayat, M.M. (1985). Antifungal property of garlic clove juice compared with fungicidal treatments against *Fusarium* wilt of watermelon. Egyptian Journal of Phytopathlology, 17(1): 55-62.

Fawcett, C.H. and Spancer, D.M. (1970). Plant Chemotherapy with natural products. Ann. Ref. Phytopath., 8: 403-418.

Gangadevi, V., Yogeswari, S., Kamalraj, S., Rani, G. and Muthumary, J. (2008). The antibacterial activity of *Acalypha indica* L. Indian J. Sci. Technol., 1(6): 1-5.

Kana, H.A., Aliyu, I.A. and Chammang, H.B. (2012). Review on neglected and underutilized root and tuber crops as food security in achieving the millennium development goals in Nigeria. J. Agric Vet Sci., 4: 27-33.

Mishra, S.B. and Dixit, S.N. (1976). Comparision of activity of the active principle of *Clematis gouriana* with commercial fungicides. Indian Phytopath., 29: 448.

Okigbo, R.N. (2009). Variation in phytochemical properties of selected fungicidal aqueous extract of some plant leaves in Kogi State, Nigeria. American-Eurasian J. of Sust. Agric., 3 (3): 407-409.

Okigbo, R.N. and Nmeka, I.A. (2005). Control of yam tuber rot with leaf extract of *Xylopia aethiopica* and *Zinguber officinale*. African Journal of Biotechnology, 4(8): 804 - 807.

Okigbo, R.N. and Ikediugwu, F.E.O. (2000). Studies on biological control of postharvest rot in yams (*Dioscorea* spp.) using *Trichoderma viride*. Journal of Phytopathology, 148: 351-335.

Rath, G.C. and Mohanty, G.N. (1978). Control of Macrophomina rot of stored garlic. Sci. Cult., 44: 467-468.

Renu, H.V., Dubey, N.K. and Dixit, S.N. (1985). Mycotoxic properties of the essential oil *Aegle marmelos*. Corr. Beitrage Zur Biologie der pflanzen, 60: 325-361.

Sambrook, J., Fritschi, E.F. and Maniatis, T. (1989). Molecular cloning: a laboratory manual, cold spring Harbor Laboratory Press, New York.

Satish, S., Raghavendra, M.P., Mohana, D.C. and Raveesha, K.A. (2008). Antifungal activity of a known medicinal plant *Mimusops elengi* L. against grain moulds. J. Agri. Technol., 4(1): 151-165.

Shanmugavalli, N., Umashankar, V. and Raheem. (2009). Anitmicrobial activity of *Vanilla planifolia*. Indian J. Sci. Technol., 2 (3): 37-40.

Shivpuri, A., Sharma, O.P., and Thamaria, S. (1997). Fungitoxic properties of plant extracts against pathogenic fungi. Journal of Mycology and Plant Pathology, 27(1): 29 - 31.

Swarna, L.L. and Neelakanta, R.P. (2009). Antimicrobial, antidiarrhoeal and analysis of phytochemical constituents of *Sphaeranthus amaranthoides*. Indian J. Sci. Technol., 2 (3): 45-48.

Wheeler, T.J. and Kececioglu, J.D. (2007). Multiple alignment by ligning alignments. Bioinformatics, 23(1): 559-568.

Chapter 18

Microbial Community Composition in the Skin of Pufferfish (*Tetraodon cutcutia*) Isolated from Mahanadi River

*Lipika Das, Sushanta Deb and Subrata K. Das**

Institute of Life Sciences,
Department of Biotechnology, Nalco Square, Bhubaneswar – 751 023, Odisha
**e-mail: subratkdas@hotmail.com/subrata@ils.res.in*

ABSTRACT

This study describes the culturable microbial community composition in the skin of pufferfish (Tetraodon cutcutia), a fresh water omnivorous fish species. The most dominant bacterial taxa represented by Acinetobacter and Exiguobacterium. Moreover, information on microbial community composition of fresh water pufferfish might help for the screening of potential tetrodotoxin (TTX) producing bacteria in future.

Keywords: *Pufferfish (Tetraodon cutcutia), Skin microbiota, Taxonomic composition, Functional attributes.*

Introduction

Pufferfish belongs to the order Tetraodontiformes with different local names such as toadfish, balloonfish and Fugu. Very little information are available on freshwater pufferfish (*Tetraodon cutcutia*) found in different parts of India *viz.*, Assam, Odisha and West Bengal (Chaudhry, 2010). Dwarf puffers used as an aquarium fish for recreation. Origin of tetrodotoxin (TTX) from pufferfish is widely reported fish poisoning in the coastal area of Asia. Toxin produced by pufferfish blocks the sodium channel of cell membrane leading to the dizziness, vomiting, rapid heart rate, decrease blood pressure and muscle paralysis (Zaman *et al.*, 1997; Chew *et al.*, 1983). Several deaths has been reported in Bangladesh, Singapore and Hong Kong due to pufferfish toxicity (Ahasan *et al.*, 2004). The pufferfish found in Indian fish market frequently mixed with other freshwater fishes. People are consuming this fish without having knowledge of its toxicity that leads to human casualty (Chew *et al.*, 1983). The reason behind the TTX acquisition in pufferfish includes food chain transmission and the existence of TTX producing bacteria (Lago *et al.* 2015). The TTX producing bacteria identified from pufferfish belonged to the genera *Aeromonas, Alteromonas, Bacillus, Pseudomonas, and Vibrio* (Lago *et al.*, 2015).

Knowledge on the microbes associated with pufferfish skin from Indian River is very limited. In this study, we have analyzed the skin microbiota of fresh water pufferfish (*Tetraodon cutcutia*), commonly found in Mahanadi River. Besides, we have described the culturable bacterial community of pufferfish in skin using 16S rRNA gene sequence. Finally, whole genome sequence of the strain *Vibrio cholerae* strain A1 has compared with the accessible genome sequences of closely related species of the genus *Vibrio* to enumerate the average nucleotide identity and functional variability.

Materials and Methods

Collection of Pufferfish and Isolation of Bacteria from Skin

The pufferfish (*Tetraodon cutcutia*) samples were collected from Mahanadi River near Mundali Bridge, Cuttack, India in the month of August 2018 and transported to the laboratory. For the isolation of bacteria from skin, mucus samples of five individual pufferfish were taken using sterile cotton swabs. Cotton swabs were pooled and suspended in 2 ml sterile phosphate buffer saline (PBS), pH 7.4 by vigorous vortexing and used as a master mix. (Badhai *et al.*, 2013). An aliquot (100 μl) of the skin mucus master mix were serially diluted using PBS and plated onto nutrient agar (BD, Difco) for the isolation of bacteria. Morphologically unique colonies appeared on nutrient agar plates were picked up, purified, and preserved in 15 per cent glycerol at -80°C for future use.

16S rRNA Sequencing and Phylogenetic Analysis of Bacteria

Genomic DNA was extracted from the bacterial strains following the methods of (Sambrook and Russel 2001). For amplification of 16S rRNA, the PCR was carried out using the bacterial universal primers 27F (52 -GAGTTTGATCCTGGCTCAG-32) and 1525R (52 -93 AAAGGAGGTGATCCA GCC-32) (Das *et al.*, 1996). The PCR product was purified using PCR cleanup kit (Promega, USA) and sequenced in an eight capillary DNA analyzer (Model number 3500, Applied Biosystems), following the manufacturer protocol. The nucleotide sequences were assembled using the sequence alignment editor program BioEdit (http://www.mbio.ncsu.edu/Bio-Edit/bioedit.html) (Hall 2007) and compared with those in GenBank after BLAST searches (Altschul 1997) and using the EzTaxon-e server (Kim *et al.* 2012). The phylogenetic tree was constructed using the neighbor-joining algorithm (Felsenstein 1985).

Whole Genome Sequencing and Annotation

Genomic DNA was isolated by Xcelgen Bacterial gDNA Kit. DNA sample was sheared to an average length of 15kb using Covaris system, as per the manufacturer's protocol (Covaris, Woburn, MA, USA). Fragmented DNA was used for SMRTbell library preparation as recommended by the manufacturer. Quantity and quality of the SMRTbell libraries were evaluated using the High Sensitivity dsDNA kit and Qubit fluorometer and DNA 12000 kit on the 2100 Bioanalyzer (Agilent, Santa Clara, CA, USA). The putative coding sequences (genes) were identified using Prodigal-2.6.3 gene prediction tool (https://github.com/hyattpd/prodigal/releases/).

Results and Discussion

Identification of Bacteria

By using culturable approach, number of bacterial strains were isolated from the pufferfish skin. 16S rRNA gene sequence analysis identified 11 different genera. Our data showed most dominant group of bacteria belongs to the class *Gammaproteobacteria* followed by *Firmicutes, Betaproteobacteria, Alphaproteobacteria* and *Flavobacteria*. 16S rRNA based neighbour-joining phylogenetic tree was similar in its topology to the tree reconstructed with the maximum parsimony and maximum likelihood algorithm (Figure 18.1). Culturable bacteria from the skin includes the taxa *viz., Shewanella, Bacillus, Vibrio, Exiguobacterium, Serratia, Delftia, Staphylococcus, Moraxella, Acinetobacter and Aeromonas.* Presence of *Bacillus, Vibrio* and *Aeromonas* in the skin further indicated the exogenous origin of TTX in pufferfish Lago *et al.* (2015). This study concludes the type of culturable microbes present in pufferfish skin.

Genomic Characterization

Reverse Position Specific-BLAST (RPS-BLAST) was used to find the Cluster of Orthologous Groups (COG) data of the vibrio genome. Total 19 COG functional category was identified. The number of genes associated with different COG functional category are energy production and conversion (182), cell cycle control, cell division, chromosome partitioning (42), amino acid transport and metabolism (284), nucleotide transport and metabolism (81), carbohydrate

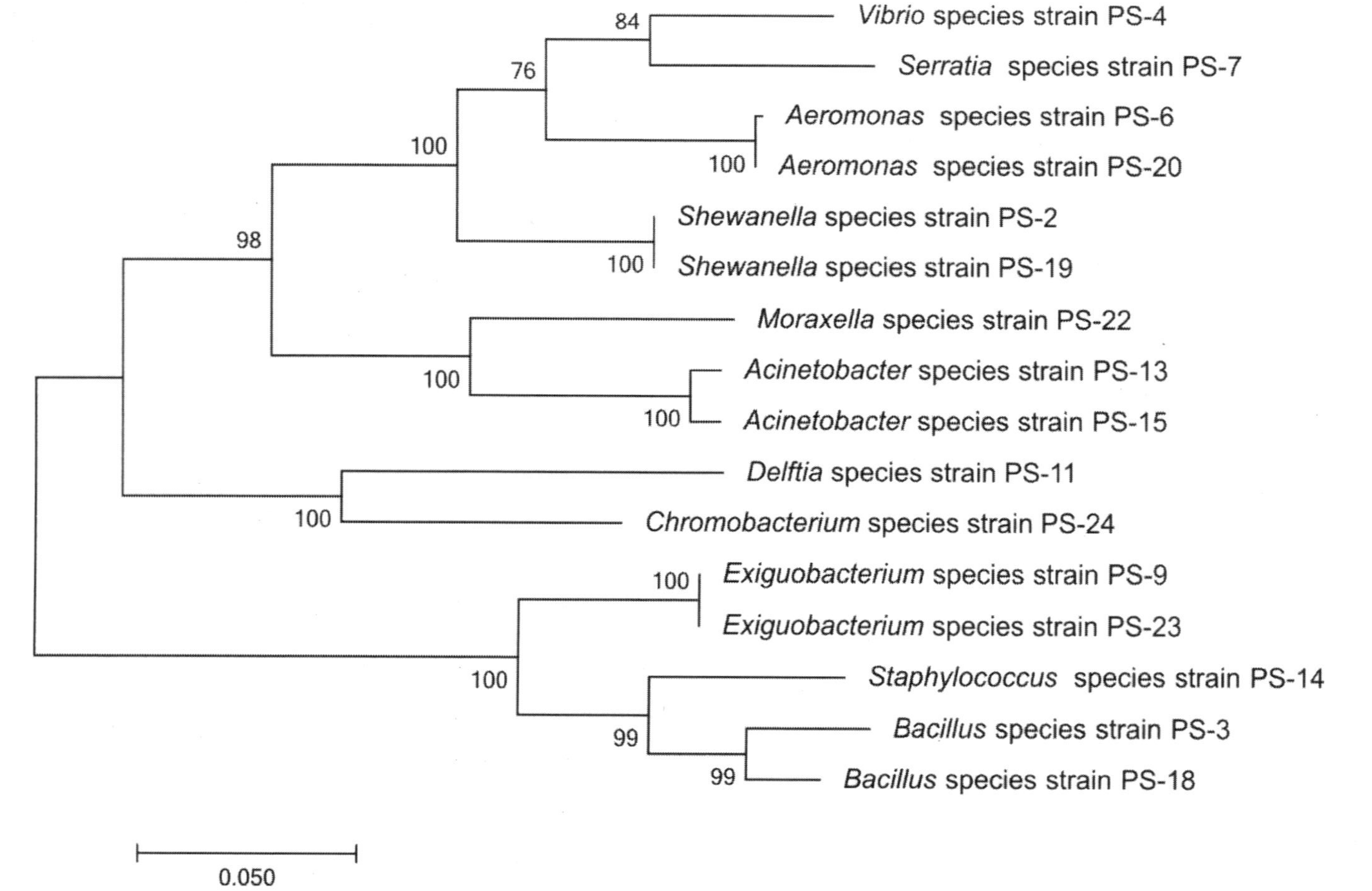

Figure 18..1: 16S rRNA Sequence Based Phylogenetic Tree. Bootstrap values expressed as percentages of 1000 replications are given at branch points. Bar, 0.050 substitutions per nucleotide positions.

transport and metabolism (201), coenzyme transport and metabolism (159), lipid transport and metabolism (85), translation, ribosomal structure and biogenesis (195), transcription (248), replication, recombination and repair (175), cell wall/membrane/envelope biogenesis (176), cell motility (149), posttranslational modification, protein turnover, chaperones (143), inorganic ion transport and metabolism (177), secondary metabolites biosynthesis transport and catabolism (58), general function prediction only (354), signal transduction mechanisms (289), intracellular trafficking, secretion, and vesicular transport (105) and defense mechanisms (54). A comparative COG functional analysis revealed strain A1 is functionally closely related with strain A2, A3, A4 and A5 (Figure 18.2). A circular representation of strain A1 genome depicting the position of tRNA and rRNA genes in the genome (Figure 18.3).

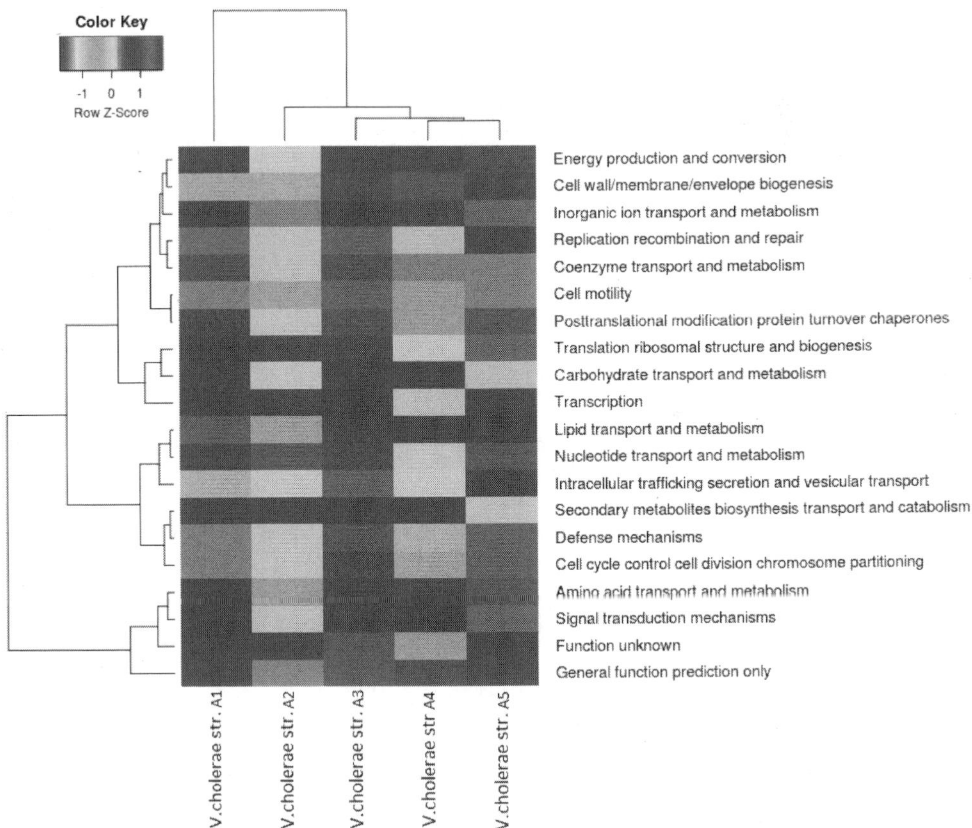

Figure 18.2: Heat Map Representing the different GOG Functional Category of Vibrio Strains.

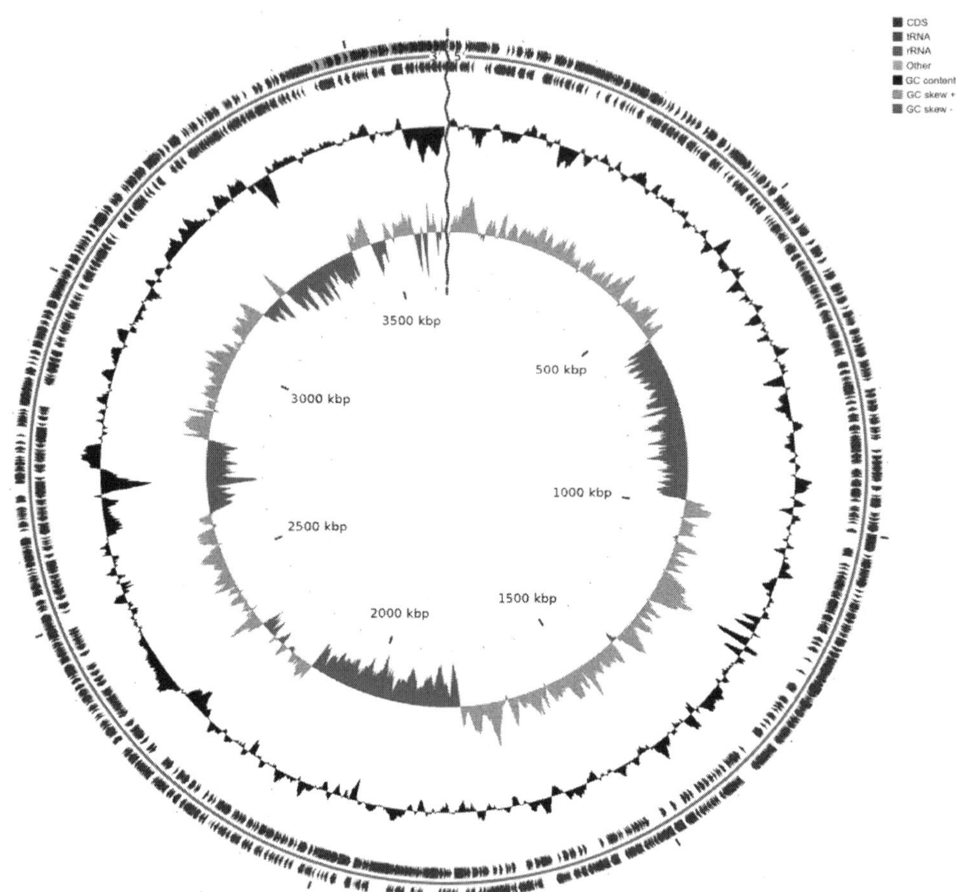

Figure 18.3: Circular Representation of Vibrio Strain A1 Genome.

REFERENCES

Ahasan HA, Mamun AA, Karim SR *et al.*, Paralytic complications of pufferfish (tetrodotoxin) poisoning. *Singapore Med J* 2004; **45:** 73-4.

Altschul SF, Madden TL, Schäffer AA *et al.*, Gapped BLAST and PSI-BLAST: a new generation of protein database search programs. *Nucleic Acids Res* 1997; **25:** 3389-402.

Badhai J, Kumari P, Krishnan P *et al.*, Presence of SXT integrating conjugative element in marine bacteria isolated from the mucus of the coral Fungia echinata from Andaman Sea. *FEMS Microbiol Lett* 2013; **338:** 118-23.

Chaudhry, S. Tetraodon cutcutia. The IUCN Red List of Threatened Species 2010: e.T166426A6206677.http: //dx.doi.org/10.2305/IUCN.UK.2010-4.RLTS.T166426A6206677.en (5th January 2019, date last accessed)

Chew SK, Goh CH, Wang KW *et al.,* Pufferfish (tetrodotoxin) poisoning: clinical report and role of anti-cholinesterase drugs in therapy. *Singapore Med J* 1983; **24:** 168-71.

Das SK, Mishra AK, Tindall BJ *et al.,* Oxidation of thiosulfate by a new bacterium, *Bosea thiooxidans* (strain BI-42) gen. nov., sp. nov.: analysis of phylogeny based on chemotaxonomy and 16S ribosomal DNA sequencing. *Int J Syst Bacteriol* 1996; **46:** 981-7.

Felsenstein J. Confidence limits on phylogenies: an approach using the bootstrap. *Evolution* 1985; **39:** 783-91.

Hall T. BioEdit. Biological Sequence Alignment Editor for Win95/98/NT/2K/XP. Carlsbad, CA: Ibis Biosciences. 2007. http: //www.mbio.ncsu.edu/BioEdit/bioedit.html

Kim OS, Cho YJ, Lee K *et al.,* Introducing EzTaxon-e: a prokaryotic 16S rRNA gene sequence database with phylotypes that represent uncultured species. *Int J Syst Evol Microbiol* 2012; **62:** 716-21.

Lago J, Rodríguez L, Blanco L *et al.,* Tetrodotoxin, an extremely potent marine neurotoxin: Distribution, toxicity, origin and therapeutical uses. *Mar Drugs* 2015; **13:** 6384-406.

Sambrook J, Russell DW, Russell DW. Molecular cloning: a laboratory manual (3-volume set). *Immunol* 2001; **49:** 895-909.

Zaman L, Arakawa O, Shimosu A *et al.,* Occurrence of paralytic shellfish poison in Bangladeshi freshwater puffers. *Toxicon* 1997; **35:** 423-31.

Chapter 19

Potential Antimicrobial Activity of Marine Cyanobacteria Isolated from Balasore Coast, Odisha

*Satyabrata Dash[1], Sabyasachy Parida[2], Bijayananda Sahoo[1], Tapan Kumar Panigrahi[1] and Biswajit Rath[1]**

[1]Department of Biotechnology, North Orissa University, Baripada – 757 003, Odisha
[2]Department of Microbiology, Maharaja Purnachandra College (Autonomous), Baripada, Odisha
**e-mail: brath_2000@yahoo.com*

ABSTRACT

Marine cyanobacteria are considered as an excellent source of novel bioactive compounds which has a broad range of biological activities. The present investigation was carried out to screen the bioactive substances from marine cyanobacteria. Two cyanobacteria (Phormidium molle and Lyngbya martensiana) were isolated from Talsari, Balasore, Northeast coast of Odisha and the extracts were prepared using three solvents (methanol, acetone and benzene) extracts of both the species were tested for antimicrobial activity against four human pathogenic bacteria (Bacillus subtilis, Escherichia coli, Staphylococcus aureus and Vibrio cholerae) and two pathogenic fungus (Aspergillus niger and Candida albicans) using agar well diffusion assay method. Among the two species, the acetone extract of Phormidium molle showed highest antibacterial activity against Bacillus subtilis (16±1mm) and the least inhibitory activity was observed in methanol extract of Lyngbya martensiana against

Staphylococcus aureus (8±1mm). The acetone extract of Lyngbya martensiana exhibited highest antifungal zone of inhibition against Aspergilus niger (18±1mm) whereas the acetone extract of Phormidium molle revealed least activity against Candida albicans (13±1mm). Thus the antimicrobial activity indicated the presence of active constituents in the extracts of cyanobacteria species which can be exploited for pharmaceutical applications through further investigation in this direction.

***Keywords**: Antimicrobial, Bioactive, Cyanobacteria, Marine, Phytochemicals,*

1. Introduction

Cyanobacteria have been screened extensively to isolate life saving drugs or biologically active substances all over the world. Selective utilization of marine microalgae and cyanobactria as a potential source of pharmaceutical agents has been increasing in recent years. Natural products are a major resource for drug development. A large number of plants, microbes and marine animals have been examined for bioactive secondary metabolites (Firakova *et al.,* 2007). Among different compounds with functional properties, antioxidants and antibacterial compounds are mostly widely studied. Bacterial infection causes high rate of mortality in human population and aquatic organisms (Kandhasamy and Arunachalam, 2008). The revolutionized therapy of infectious diseases by the use of antimicrobial drugs has certain limitations due to changing patterns of resistance in pathogens and the side effects they produced. These limitations demand for improved pharmacokinetic properties, which necessitate continued research for new antimicrobial compounds for the development of drugs (Al- Haj *et al.,* 2009). Pharmaceutical industries are giving importance to the compounds derived from traditional sources (soil and plants) and less traditional sources like marine organisms (Solomon and Santhi, 2008). Hence, the interest in marine organisms as a potential and a promising source of pharmaceutical agents has increased during recent years (Kim and Lee, 2008).Marine cyanobacteria are considered as diverse source of secondary metabolites having a broad spectrum of biological activities such as antimicrobial, antiviral, antifungal, anti-allergic, anticancer, anti-fouling and antioxidant that have been used to develop the pharmaceutical industries (Salvador *et al.,* 2007; Bouhlal *et al.,* 2011; Guedes *et al.,* 2012; Na *et al.,* 2005; Murugan and Iyer, 2013; Bhadury and Wright, 2004; Kumar *et al.,* 2008; Wijesekara *et al.,* 2010; Rath and Priyadarshani, 2013). They are assuming increasing importance in frontier areas of biotechnology (Pankratova, 1987).The antimicrobial substances involved may target kinds of microorganisms, prokaryotes as well as eukaryotes. The properties of secondary metabolites in nature are not completely understood (Metting and Pyne, 1986; Inderjit and Daskshini, 1994). Secondary metabolites influence other organisms in the vicinity and are thought to be of phylogenetic importance. Antimicrobial effects of Cyanobacterial aqueous and organic solvent extracts are visualized in bioassays using selected microorganisms as test organisms (Frankmoll *et.al.,* 1992). Despite the importance of medicinal properties and pharmaceutical applications of cyanobacteria, few research was undertaken in this region to detect novel therapeutic agents. The main objective of the present study was to examine the antimicrobial and antifungal effects of cyanobacterial extracts. For this reason, two species of cyanobacteria (*Phormidium molle and Lyngbya martensiana*) isolated from Balasore coast of Odisha were selected to evaluate antimicrobial activity using their different solvent extracts against four human pathogenic bacteria and two fungus species.

2. Materials and Methods

2.1 Isolation and Culture Conditions

Two species of Cyanobacteria (*Phormidium molle and Lyngbya martensiana*) were collected from Balasore, the north-east coast of Odisha (Figures 19.1a-b). The samples were cleaned and necrotic parts were removed in the laboratory and cultured in ASN III medium. Unialgal cultures were maintained under fluorescent tube light at a intensity of 2000-3000 lux with a photoperiod of 16hrs light and 8hrs dark and at 25 ± 1°C temperature.

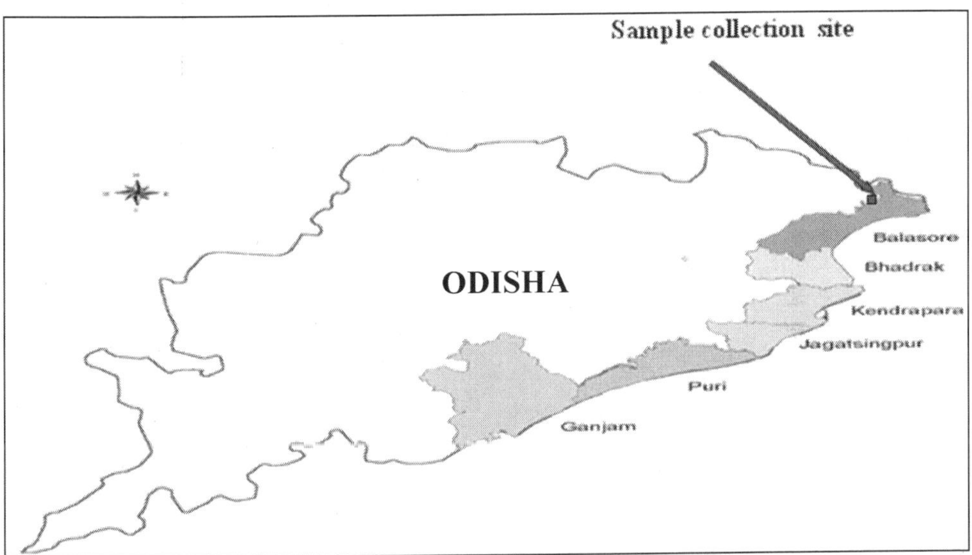

Figure 19.1a: Map Showing Sample Collection Site of Talasari, Balasore District of Odisha.

2.2 Preparation of Extracts

The test species of Cyanobacteria were harvested after 15-20 days of growth by centrifugation at 3500g for 10 minutes. Then, the cyanobacterial pellet were collected, weighed and used for extraction of antimicrobial agents. Aliquots of 0.5g of the powdered Cyanobacterial samples were added to 5 ml of solvents (methanol, acetone and benzene) and kept the mixture in a shaker for 24 hours. The extract

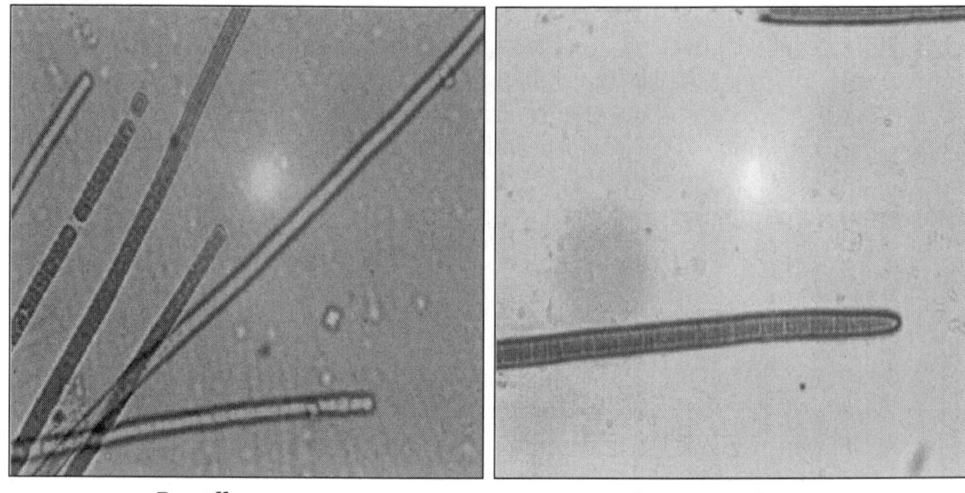

P. molle *L. martensiana*

Figure 19.1b: Microscopic Photograph Test Species of Cyanobacteria.

were filtered and poured in a petridish for air dry initially for 24 hours and then in the oven at 50°C for 2 hours The concentration was adjusted to 1mg/ml by using DMSO used for extraction was assayed for antimicrobial activity.

2.3 Antimicrobial Activity

The antimicrobial activity of the isolated Cyanobacteria was carried out by agar well diffusion method (Abedin and Taha, 2008). Ampicillin and Clotrimazole were used as positive control for bacteria and fungi respectively.

2.4 Phytochemical Tests

Cyanobacterial extracts were prepared using solvents (methanol, acetone and benzene). The extracts were analyzed for the presence of Phenolic compounds, Flavonoids, Saponins, Alkaloids, Glycosides, Tannins, Steroids, Anthrocyanin, Quinones, and Resins following the method of Dhanalakshmi *et al.*,2013 as mentioned below.

Phenolic Compounds

In extract (1 mL) few drops of Hydrochloric acid was added. Yellowish brown color indicated the presence of Phenolic compounds.

Flavonoid (Ferric chloride test): Few drops of Ferric chloride were added into the extracts and formation of brown precipitate, indicates presence of flavonoid.

Saponins: Each extract (3 mL) was taken in a tube. The suspension was vigorously shaken. The formation of stable foam was taken as an indication for the presence of saponins.

Alkaloids (Wagner's test): Few drops of Wagner's reagent was added at the side of the test tube. The formation of reddish-brown precipitate showed the presence of alkaloids

Glycoside test: 1 mL of concentrated sulphuric acid was added to extract and formation of reddish brown colour indicate a positive test for glycosides.

Tannins (Ferric chloride test): Few drops of ferric chloride solution was added to the extract. Brackish precipitate showed the presence of Tannins.

Steroids: To small amount of extact, few drops of acetic acid was added and subsequently few drops of conc. sulphuric acid was added. Appearance of reddish brown colour showed the presence of steroids.

Anthocyanin: A small amount of extract was treated with 2 ml of NaOH and observed. Blue green colour was formed which indicate the presence of anthocyanin.

Quinones: To 1 ml of the extract, Alcoholic KOH solution was added. Colour termed from red to blue showed the presence of quinones.

Resins: To small amount of extract, 2-3 ml of copper sulphate solution was added, and the contents were mixed for 2 minute and then solution was allowed to separate. Formation of green colour precipitate showed the presence of resin.

3. Results

3.1 Antimicrobial Activity

The acetone, methanol and benzene extracts of two marine cyanobacteria were tested for antimicrobial activity against four human bacterial pathogens and two fungal pathogens. The inhibition zones of extracts against the specific test organisms were measured. The extracts of two cyanobacterial species have significant inhibitory bioassay on the bacterial and fungal pathogens. Among the two species, the acetone extract of *Phormidium molle* showed highest antibacterial activity against *Bacillus subtilis* (16±1mm) and the least inhibition activity with methanol extract of *Lyngbya martensiana* (8+1mm) against *Staphylococcus aureus* (Figure 19.2). The acetone extract of *Lyngbya martensiana* exhibited highest antifungal activity showing zone of inhibition against *Aspergilus niger* as 18±1mm whereas the acetone extract of *Phormidium molle* revealed least antifungal activity (13±1mm) against *Candida albicans* (Figures 19.3-19.4).

3.2 Phytochemical Analysis

The study detected the presence of various phytoconstituents such as alkaloids, glycoside, tannin, flavonoids, steroid, saponins, resins, quinones, anthocyanin and phenols (Table 19.1). The acetone, methanol and benzene extract of *Phormidium molle* and *Lyngbya martensiana* indicated the presence of six major phytochemicals such as phenols, flavonoids, saponins, anthocyanin and resins, where as alkaloids was present only in *Lyngbya martensiana* and tannin was present in *Phormidium*

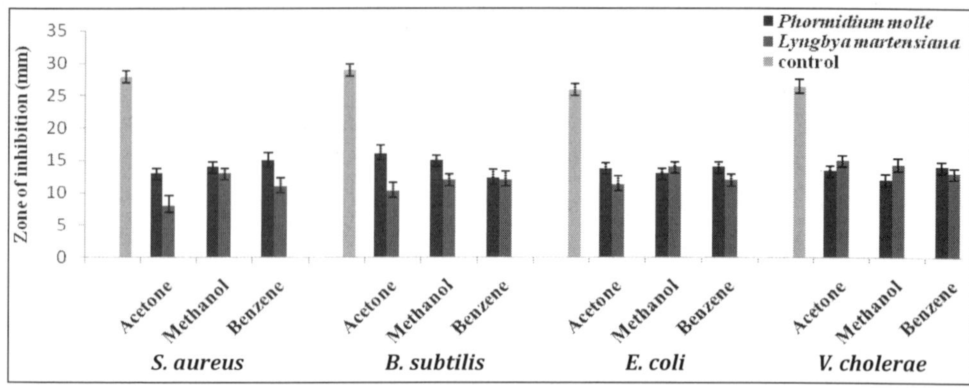

Figure 19.2: Antibacterial Activity of Two Species of Cyanobacteria.

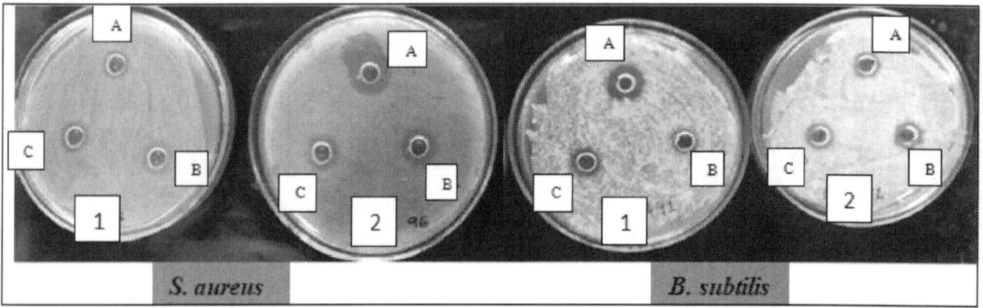

1: ***Phormidium molle*; 2: *Lyngbya martensiana;* A: Acetone; B: Methanol; B: Benzene.**

Figure 19.3: Antibacterial Activity of Two Species Cyanobacterial against *S. aureus* and *B. subtilis.*

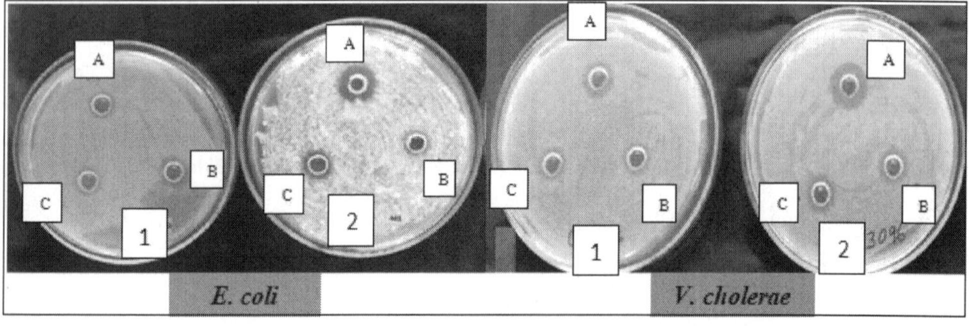

1: ***Phormidium molle*; 2: *Lyngbya martensiana;* A: Acetone; B: Methanol; B: Benzene.**

Figure 19.4: Antibacterial Activity of Two Cyanobacterial Apecies against *E. coli* and *V. cholerae.*

molle. Further,it was observed that glycoside, steroid and quinines were absent in boththe test species of cyanobacteria.

Table 19.1: Phytochemical Activity of Two Selected Cyanobacterial Extracts against Three Solvents

Species	*Solvents*	*Phenol*	*Flavonoid*	*Saponin*	*Alkaloid*	*Glycoside*	*Tannin*	*Steroid*	*Anthocyanin*	*Quinones*	*Resin*
Phormidium molle	Methanol	+	+	+	-	-	+	-	+	-	+
	Acetone	+	+	+	-	-	+	-	+	-	+
	Benzene	+	+	+	-	-	+	-	+	-	+
Lyngbya martensiana	Methanol	+	+	+	+	-	-	-	+	-	+
	Acetone	+	+	+	+	-	-	-	+	-	+
	Benzene	+	+	+	+	-	-	-	+	-	+

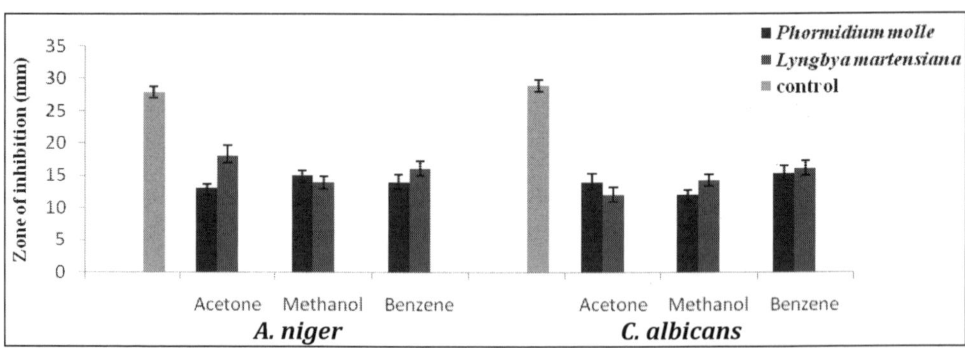

Figure 19.6: Antifungal Activity of Two Species of Cyanobacteria.

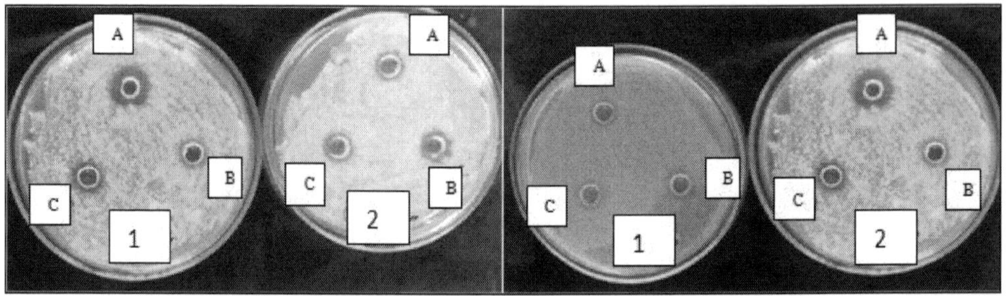

C. albicans *A. niger*

1: *Phormidium molle*; 2: *Lyngbya martensiana;* A: Acetone; B: Methanol; B: Benzene.

Figure 19.5: Antifungal Activity of Two Cyanobacterial Species against *A. niger* and *C. albicans.*

4. Discussion

The marine cyanobacterial biodiversity and associated chemical diversity constitute an unlimited reserve of bioactive substances in the field of bioactive products (Genovese *et al.,* 2012). Further, emerging resistance to antibiotics has raised serious concerns regarding the next source of new chemical entities that can meet the challenge of continually emerging drug resistance. Although considerable progress is being made within the fields of chemical synthesis and engineered biosynthesis of antimicrobial compounds, nature still remains the richest and the most versatile source for new antibiotics. Some recent studies reveal microalgae and cyanobacteria as a potential source of antimicrobial products (Alghazeer *et al.,* 2013; Mendes *et al.,* 2013; Al-Saif *et al.,* 2014). The antibacterial activities of *Phormidium molle* and *Lyngbya martensiana* extracts against selected pathogenic bacteria and fugus varied depending on the species and nature of solvents used for extraction. Acetone extract of *Phormidium molle* showed the highest inhibitory effect on *Bacillus subtilis* (16±1mm) which is in accordance with the work of Amaro *et al.,* 2011, Padmini *et al.,* 1986. So this study indicated that cyanobacteria should be extracted in different solvent systems in order to optimize their antimicrobial activity. Moreover, the used solvents were capable to extract diverse biological compounds from the cyanobacteria with antimicrobial activity.

Further marine cyanobacteria provide a rich source of structurally diverse secondary metabolite. Several studies have demonstrated that Cyanobacteria are an excellent source of components such as polysaccharides, flavonoids, phenolic acids, bromophenols, saponin, resin, and tannins have exhibited different biological activities (Rodriguez *et al.,* 2010; Bhacuni and Rawat, 2005; Priyadharshini *et al.,* 2011). These phytochemicals exhibited antimicrobial activity synergistically, the phenolic toxicity to microorganisms include enzyme inhibition by the oxidized compounds, possibly through reaction with sulfhydryl groups (Mason and wasserman, 1987).The activity of Flavonoids is probably due to their ability to complex with extracellular and soluble proteins and to complex with bacterial cell walls. More lipophilic flavonoids may also disrupt microbial membranes (Tsuchiya *et al.,* 1996). Hediat *et al.,*2010 reported that different solvents have been reported to have the capacity to extract different phytoconstituents depending on their solubility or polarity in the solvent which is in agreement to our findings.

5. Conclusion

Marine cyanobacteria are the substantial resources of natural bioactive substances with potential use in the pharmaceutical industry. Thus, it is worth mentioning that the extracts obtained from various solvents used in this study had antibacterial and antifungal activity. However, these cyanobacteria are potential sources of bioactive compounds and should be investigated for natural antibiotics. So these bioactive compounds will need further studies to identify the chemical structures of these active compounds and to examine their beneficial effect for inhibition of some pathogenic bacteria and fungus and subsequently will be useful for pharmaceutical applications.

6. References

Abedin RMA, Taha, HM(2008). Antibacterial and Antifungal Activity of Cyanobacteria and Green Microalgae. Gl Jr of Biotech and Biochem Res; 3, 22-31.

Al- Haj NA, Mashan NI and Shamudin MN. (2009). Antibacterial activity in marine algae Eucheuma denticulatum against Staphylococcus aureus and Streptococcus pyogenes. Res Jr. Biol Sci ; 4(4): 519-524.

Alghazeer R, Whida F, Abduelrhman E, Gammoudi F and Naili M. (2013) In vitro antibacterial activity of alkaloid extracts from green, red and brown macroalgae from western coast of Lybia. African Journal of Biotechnology.12: 7086-7091.

Al-Salif SSA, Abdel-Raouf N, El-Wazanani HA and Aref IA.(2014) Antibacterial substances from marine algae isolated from Jeddah coast of Red Sea, Saudi Arabia. Saudi Journal of Biological Sciences.21: 57–64.

Amaro HM, Guedes AC and Malcata FX. (2011) Science against microbial pathogens: communicating current research and technological advances A. Mendez-Vilas (Ed.), Antimicrobial activities of microalgae: an invited review. 1272–1280.

Bhacuni DS, Rawat DS (2005). Bioactive Marine Natural Products.Spinger/Anamaya Publishers; 400: 978.

Bhadury P, Wright CP. (2004)Exploitation of marine algae: biogenic compounds for potential antifouling applications. Planta. 219:561-78.

Bouhlal R, Haslin C, Cherman JC, Colliec-Jouault S, Sinquin C, Silmon G.(2011) Antiviral activities of sulfated polysaccharides isolated from Sphaerococcus coronopifolius (Rhodophytha, Gigartinales) and Boergeseniella thuyoides (Rhodophyta, Ceramiales). Mar Drugs. 9: 1187-209.

Firakova S, Sturdikova M and Muckova M. (2007). Bioactive secondary metabolites produced by microorganisms associated with plants. Biologia; 62(3): 251-257.

Frankmolle WP, Larsen LK, Caplan FR, Patterson GML and Knubel G. (1992). Antifungal cyclic peptides from the terrestrial blue-green alga Anabaena laxa. 1. Isolation and biological properties. Jr Antibiot; 45: 1451-1457.

Genovese G., Faggio C., Gugliandolo C., Torre A., Spano A., Morabito M., Maugeri T.L. (2012). In vitro evaluation of antibacterial activity of Asparagopsis taxiformis from the Straits of Messina against pathogens relevant in aquaculture. Mar Env Res; 73.1-6.

Guedes EA, Araujo MA, Souza AK, de Souza LI, de Barros LD, Maranhão FC. (2012). Antifungal activities of different extracts of marine macroalgae against dermatophytes and Candida species. Mycopathologia. 174: 223-32.

Hediat MH, Salama and Najat Marraiki. (2010). Antimicrobial activity and phytochemical analyses of Polygonum aviculare L. (Polygonaceae), naturally growing n Egypt. Saudi Jr of Bio Sc; 17: 57- 63.

Inderjit and Dakshini K.M.M.(1994). Algal allelopathy. Bot rev;60:182-196.

Kandhasamy M, Arunachalam KD. (2008). Evaluation of in vitro antibacterial property of seaweeds of southeast coast of India. Afr Jr Biotech; 7(12): 1958-1961.

Kim IH, Lee JH. (2008). Antimicrobial activities against methicillin resistant Staphylococcus aureus from macroalgae. Jr Ind Engg Chem; 14: 568-572.

Kumar Chandini S, Ganesan P, Bhaskar N.(2008) In vitro antioxidant activities of three selected brown seaweeds of India. Food Chem. 107: 707-13.

Mason,TL, and wasserman BP (1987).Inactivationof red beet beta-glucan synthase by native and oxidized phenolic compounds. Phytochem; 26:2197-2202.

Meeting B. and Pyne J.W.(1986). Biologically active compounds from microalgae, Enz Micro Tech;8: 388-394.

Mendes M, Pereira R, Sousa-Pinto I, Carvalho AP Gomes AM. (2013) Antimicrobial activity and lipid profile of seaweed extracts from the North Portuguese Coast. International Food Research Journal.20: 3337- 3345.

Murugan K, Iyer VV. (2013)Differential growth inhibition of cancer cell lines and antioxidant activity of extracts of red, brown, and green marine algae. In Vitro Cell Dev Biol Anim. 49: 324-34.

Na HJ, Moon PD, Lee HJ, Kim HR, Chae HJ, Shin T.(2005) Regulatory effect of atopic allergic reaction by Carpopeltis affinis. Jr Ethnopharmacol. 101: 43-8.

Padmini SR, Rao PS and Karmarkar SM. (1986)Antibacterial substances from brown algae II. Efficiency of solvents in the evaluation of antibacterial substances from Sargassum johnstonii Setchell et Gardner. Botanica Marina. 29: 503-507.

Pankaratova EM (1987). Participation of Cyanobacteria in the soil nitrogen cycle and formation of soil fertility. In Adv in microbiol, Nauka, Moscow; 21:212-242.

Priyadharshini S, Bragadeeswaran S. Prabhu K. and Ran SS (2011). Antimicrobial activity and hemolytic activity of seaweed extracts Ulva fasciata (Delile 1813) from Mandapam, Southeast coast of India. In Asian Pacific Journal of Trop Biomed;.1:S38-S39.

Rath B, Priyadarshani I.(2013.)Antibacterial and Antifungal activity of marine cyanobacteria from Odisha Coast"Int jrl of curr trnd in res. 2 (1): 248-251.

Rodriguez- Bernaldo de Quiros A., Large- Yusty M.A., Lopez- Hernadez J. (2010). Determination of phenolic compounds in macroalgae for human consumption. Food Chem ;121. 634-638.

Salvador N, Gomez Garreta A, Lavelli L, Ribera MA (2007). Antimicrobial activity of Iberian macroalgae. Sci Mar. 71: 101-14.

Solomon f and Santhi VS.(2008) Purification of bioactive natural product against human microbial pathogens from marine seaweed Dictyota acutiloba, W Jr Micro Biotech.24: 1747-1752.

Tsuchıya H., Sato M., Miyazaki T., Fujiwara S., Tanigaki S., Ohyama M., Tanaka T. and Iinuma M.(1996). Compparative study on the antibacterial activity of phytochemical flavanones against methicillin-resistant Staphylococcus aureus. Jr Ethnopharmacol. 50: 27-34

Wijesekara I, Kim SK. (2010). Angiotensin-I-converting enzyme (ACE) inhibitors from marine resources: Prospects in the pharmaceutical industry. Mar Drugs.8:1080-1093.

Chapter 20

Endosulfan Degradation by *Pseudomonas* sp. Isolated from Endosulfan Retreated Flooded Alluvial Soil

*Shuvashish Behera[1,2], Tapan Kumar Adhya[1,3] and Rama Chandra Mohanty[4]**

[1]Crop Production Division, National Rice Research Institute, Cuttack – 753 006, Odisha
[2]Department of Alcohol Technology and Biofuels, Vasantdada Sugar Institute, Manjari Budruk, Pune – 412 307, Maharashtra
[3]School of Biotechnology, Kalinga Institute of Industrial Technology, Patia, Bhubaneswar – 751 024, Odisha
[4]Department of Botany, Utkal University, Bhubaneswar – 751 004, Odisha
**e-mail: mohantyrc62@gmail.com*

ABSTRACT

In the present study, pure culture of Pseudomonas sp. isolated from endosulfan-retreated flooded alluvial soil used for rice growing was used to study the degradation of endosulfan isomers (α- and β-). The ability of the bacterium to degrade endosulfan was investigated by culturing it in mineral salt medium containing different concentrations of endosulfan. The isolated bacterium showed the ability to tolerate and degrade high concentrations of endosulfan and utilize it as the sole source of carbon and energy. Growth of Pseudomonas sp.

was found to be best at pH 7.5 and the cell number increased from 1.3 x 10^5 to 3650 x 10^8 cfu. ml^{-1}. Endosulfan biodegradation at concentrations beyond 50 µg.ml^{-1} was slow as it might be inhibitory to the growth of bacterial cells. The finding of this study will be useful for the development of bioremediation strategies for water logged environments contaminated with endosulfan.

Keywords: *Pesticide, Endosulfan, Pseudomonas sp., Biodegradation, Flooded soil.*

Introduction

Pesticides have made a tremendous impact on human health, production and preservation of foods, fiber and other cash crops by controlling pests and diseases and keeping in check many species of unwanted insects and pests (Akhtar *et al.*, 2009; Fianco *et al.*, 2011). Pesticides now constitute an integral part in the intensive agriculture involving modern agricultural practices wherein high-yielding crop cultivars are grown under precision management systems. However, being basically toxic compounds, unregulated and indiscriminate use of pesticides has often resulted in the contamination of environment as well as undesirable side effects like resurgence of pests (Damalas and Eleftherohorinos, 2011; Gill and Garg, 2014).

Endosulfan (6,7,8,9,10,10-hexachloro-1,5,5a,6,9,9a-hexahydro-6,9-methano-2,3,4- benzodioxathiepin-3-oxide, CAS No. 115-29-7), is a broad spectrum organochlorine pesticide and acaricide of the cyclodiene group. Endosulfan affects the central nervous system, kidney, liver, blood chemistry and parathyroid gland in human beings and has teratogenic and mutagenic effects (US Dept. of Health and Human Services, 2000). It is used extensively throughout the world as a broad spectrum insecticide on cotton, field crops such as paddy, sorghum, oil seeds and pulses as well as vegetables and fruit crops (Akhtar *et al.*, 2009; Tapsoba *et al.*, 2008). This broad-spectrum insecticide comprises of mixture of two stereo-isomers, α-endosulfan (64-67 per cent) and β-endosulfan (29-32 per cent) (Sutherland *et al.*, 2002).

Endosulfan is a persistent organic pollutant (POP) that enters the air, water and soil during its use and manufacture. Endosulfan has been ubiquitously detected in the atmosphere, soils, sediments, surface and ground waters and foodstuffs (Menezes *et al.*, 2017). When incubated with soil under aerobic conditions, endosuflan has been shown to undergo biotransformation and endosulfan sulfate was detected as the major metabolite (Rao and Murty, 1980). Endosulfan suflate is nearly as toxic as the parent chemical and does not undergo any further degradation (Coleman and Dolinger, 1982). Under different metabolic conditions, endosulfan diol, endosulfan sulfate, endosulfan ether, endosulfan hydroxyether, endosulfan lactone and endosulfan mono- and di-aldehyde have been reported as the major metabolites (Shivaramaiah and Kennedy, 2006; Pradeep and Subbaiah, 2016).

Detoxification of xenobiotic chemicals through biological means has received serious attention as an alternative to existing methods, such as incineration and landfill (Narwal *et al.*, 2017; Sang *et al.*, 2018). An essential step in the investigation of bioremediation technologies for endosulfan detoxification is the definitive identification of a biological source with endosulfan degrading activity (Kumar *et al.*, 2008; Li *et al.*, 2009). It has been observed that in some bioremediation process, heterotrophic microorganisms can break down substrates (hazardous compounds) to obtain chemical energy. Thus organic pollutants can serve as carbon, energy and nutrient sources for microbial growth (Bindhya and Thanga, 2010). Detoxification of endosulfan by a bacterial co-culture and mixed cultures has been reported by several researchers (Awasthi *et al.*, 1997; Sutherland *et al.*, 2000; Kafilzadeh *et al.*, 2015; Ozdal *et al.*, 2016). Certain fungi including *Trichoderma harzianum, Aspergillus niger, Trametes hirsuta* and *Mucor thermohyalospora* are also active in the biodegradation of endosulfan (Katayama and Matsumura, 1993; Mukherjee and Gopal, 1994; Shetty *et al.*, 2000; Kamei *et al.*, 2011). *Pseudomonas* sp. used in this study was earlier isolated from the flooded alluvial soil used to cultivate rice at National Rice Research Institute, Cuttack, Odisha. Endosulfan degradation was carried out using *Pseudomonas* sp. isolated from flooded alluvial soil with a view to develop strategy for bioremediation of flooded endosulfan contaminated soil.

Materials and Methods

Pesticide and Chemicals

The α- and β-endosulfan and endosulfan sulfate used in this study for their degradation by pure culture of bacteria were obtained from Dr. Ehrenstorfer GmbH, Germany. Commercial formulation of endosulfan (Thiodan EC, Hoechest, Mumbai) containing 50 per cent active ingredient was obtained from the local market.

Culture Medium

A mineral salts (MS) medium containing $(NH_4)_2SO_4$, 0.5 g; $MgSO_4$, $7H_2O$, 2g; $FeSO_4$, $7H_2O$, 0.01g; K_2HPO_4, 0.1g; $Ca(NO_3)_2$, 0.01g dissolved in distilled water (1 litre) and pH adjusted to 7.0 was used in the study. 10 ml portions of this medium were dispensed in 100 ml Erlenmeyer flasks and sterilized at 131.0 °C for 20 min in an autoclave. To the sterile medium, respective isomers of α- or β-endosulfan was added in acetone to a final concentration of endosulfan at 10 $\mu g.ml^{-1}$. Flask containing endosulfan in acetone was left overnight in a sterile laminar flow to allow evaporation of the acetone was used in subsequent cultural studies. For cometabolic studies, mineral salts medium was supplemented with glucose (1 per cent) and peptone (0.1 per cent) (MSGP medium) and endosulfan was added as mentioned earlier.

Bacterial Culture

One endosulfan-degrading bacterial isolate *Pseudomonas* sp. was used in this study. This bacterial culture was earlier isolated by selective enrichment of a flooded alluvial soil planted to rice at the local campus of National Rice Research Institute, Cuttack, Odisha and retreated with a commercial formulation of endosulfan and was identified. The isolated pure culture of the bacterium was effectively degrading both the isomers of endosulfan. The culture of endosulfan-degrading bacterium was maintained in MSGP slants fortified with endosulfan and sub-cultured at regular intervals.

Degradation of Endosulfan Isomers by Endosulfan-Degrading *Pseudomonas* sp.

In order to determine the degradation of endosulfan isomers (α- and β-) vis-a-vis its utilization as sole source of carbon and energy by *Pseudomonas* sp., 10 ml of sterilized mineral salts medium containing 10 $\mu g.ml^{-1}$ of α- or β-endosulfan were dispensed in 100 ml Ernelmeyer flasks and inoculated with 0.1 ml suspension of the individual bacterium separately from their respective dilutions in sterile distilled water. Uninoculated medium served as the control. Both inoculated and uninoculated medium were incubated at room (28 +2 °C) temperature. At periodic intervals, appropriate replicated flasks were removed and endosulfan residues were extracted with hexane and analyzed by gas liquid chromatography.

The cell proliferation of *Pseudomonas* sp. in the presence of endosulfan isomers as the sole source of carbon was followed by plating 0.1 ml portions of serial dilutions (10^{-1} to 10^{-8}) of the inoculated medium on to the mineral salts medium supplemented with glucose and peptone (MSGP) and 10 $\mu g.ml^{-1}$ of respective endosulfan isomer. Colony forming units (cfu) appearing on the agar plates were counted after 3-4 days of incubation.

Extent of Endosulfan Degradation by *Pseudomonas* sp.

The extent of endosulfan degradation by *Pseudomonas* sp. was studied by following the degradation in mineral salts medium supplemented with 10, 50 and 100 $\mu g.ml^{-1}$ of endosulfan isomers. Replicated sets of MS medium supplemented with appropriate concentration of α- or β-isomers of endosulfan in 100 ml Erlenmeyer flasks were inoculated with 10^6 $cells.ml^{-1}$ of *Pseudomonas* sp. and the degradation of endosulfan was monitored at regular intervals.

Influence of pH on the Growth of Endosulfan-Degrading Bacteria

To determine the growth of *Pseudomonas* sp. at different pH values, MSGP broth was supplemented with 10 $\mu g.ml^{-1}$ of endosulfan and inoculated with 0.1 ml of bacterial suspension from their respective dilutions and incubated. Uninoculated medium served as control. At periodic intervals, growth of the bacteria in MSGP medium was determined by measuring the O.D. values in a spectrophotometer at 610 nm.

Colorimetric Assay for Endosulfan Degradation

A colorimetric assay that detects haloalkane dehalogenase activity (Phillips *et al.,* 2001) was modified to detect degradation of endosulfan. Degradation is indicated by the colour change of phenol red from red to yellow, in a weakly buffered solution, as the solution becomes acidic due to release of SO_4 and the consequent formation of acidic radical. MSGP broth was used as the medium for the colorimetric assay. Other reagents required for the assay was 0.05M EDTA, 100 mM HEPES buffer, 1M Na_2SO_4, 0.85 per cent NaCl (saline), 12.5 µg/ml phenol red stock (take 50 mg phenol red and add in 4 ml Acetone), assay buffer (1mM HEPES+10 mM Na_2SO_4+0.5 mM EDTA), phenol red and assay buffer mixture (take 30 µl phenol red and dissolve in 29.970 ml assay buffer).

Bacterial cells were grown over night and centrifuged at 6000 rpm for 10 minutes. The pellet and the supernatant were collected in separate tubes. The pH of the supernatant was measured and pellets were washed twice with sterile saline. Pellets were dissolved in assay buffer and OD was taken. Then the cells were diluted with the assay buffer to make it 1 OD. From the OD cells amount to be taken was calculated. In the control, 1 ml of phenol red and assay buffer mixture was taken. In another tube buffer mixture and cell amount was taken and in the other tube same amount of cells, 50 µl endosulfan and rest buffer mixture was taken. Then it was kept for some hour for the colour development.

GC Analysis of Endosulfan and Preparation of the Standard Curve

Endosulfan residues extracted from microbial culture were analyzed by gas chromatography in a Varian gas chromatograph model 3400 equipped with a ^{63}Ni Electron Capture Detector (ECD) and a stainless steel column (2m length, 1/8 inch OD) containing 1.5 per cent OV17 + 1.95 per cent OV21 SS column (Mesh 80/100 of Chrompack). The column, injector and detector temperatures were maintained at 220, 240 and 350 °C, respectively. Ultrapure nitrogen was used as a carrier gas at a flow rate of 20 ml min^{-1}. The retention time of α-endosulfan under these conditions was 7.1 min and that of β-endosulfan was 12.20 min (Figure 20.1). Standard curve of endosulfan (α- and β-) was prepared by injecting different concentrations of the compound and plotting the area under the peak against the concentration on a graphical plot.

Results and Discussion

The impact of environmental contamination of pesticides is of great concern. Organochlorine pesticides are of particular interest because of their persistent nature, toxicity and bioaccumulation potential (Adhya *et al.*, 1996). Many developed and developing countries have already banned several of these toxic chemicals but a few of them are still in use in agriculture and public health. Endosulfan is one such chemical that is used on a large scale in agriculture and occupies the second position after chlorpyrifos in terms of use (Annual report of the Department of Chemicals, Ministry of Chemicals and Fertilizers, Govt. of India). However, recently the use of endosulfan in the dense forest of south India, where aerial spraying is regularly conducted on cashew plantations, has become controversial and has

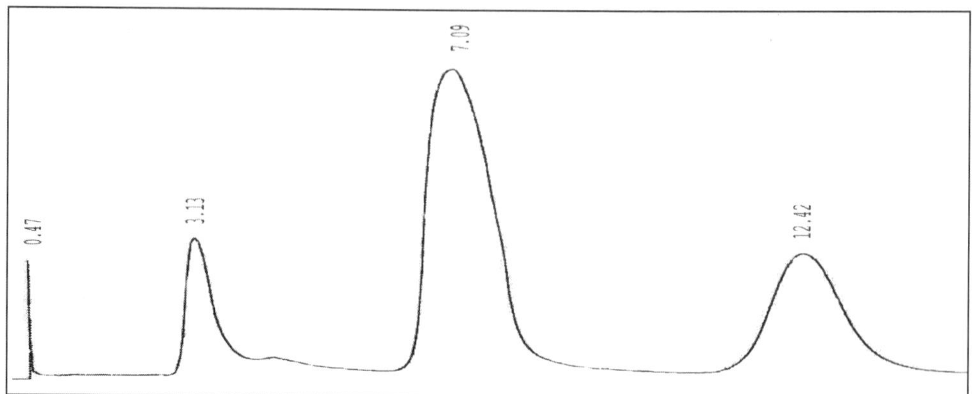

Figure 20.1: Normal Gas Chromatogram of Endosulfan Isomers and their Metabolite (Retention time 3.13 = Endosulfan diol, 7.09 = α endosulfan, 12.42 = β endosulfan).

led to restriction on its use by the Kerala State Government, India (Ramesh and Vijayalakshmi, 2002).

Microorganisms remain the most potent biological agents for the degradation of many pesticides in soil and water environment. If the environment is optimum for microbial growth, activity and survival and the contaminant is biodegradable, the last factor to consider is the population of catabolic microorganisms. While several microorganisms with defined catabolic potential have been isolated, it is now a common knowledge that only a small fraction of microorganisms can be cultivated. Biotechnological interventions with enriched microorganisms recruited for this purpose are feasible options but most efforts are still under developmental stages. However, isolation and characterization of effective degrading microorganisms is an essential primary step. In the present study, experiments were undertaken to characterize one endosulfan-degrading bacteria, isolated earlier from flooded alluvial soil planted to rice and retreated with endosulfan.

Degradation of Endosulfan (α- and β-) by *Pseudomonas* sp.

In this study, *Pseudomonas* sp. could effectively degrade α- and β-endosulfan. Thus, amount of α-endosulfan remaining in the medium after 8 days was a mere 1.63 $\mu g.ml^{-1}$ in medium inoculated with *Pseudomonas* sp. (Table 20.1). During the degradation of α-endosulfan, endosulfan diol was formed as a transitory metabolite during the first 2 days of incubation but disappeared subsequently. Narkhede *et al.* (2015) studied endosulfan degradation by a local isolate of *Pseudomonas aeruginosa* isolated from the soil of cotton field. The isolate could degrade 96 per cent of endosulfan after 288h under static condition with complete mineralization of endosulfan after 28 days. Endosulfan diol could not be quantified in the study due to lack of analytical standard and was only qualitatively identified. Degradation of α-endosulfan was associated with cell proliferation in the bacteria. For *Pseudomonas* sp., the cell number increased from 1.3×10^5 to 3650×10^8 $cfu.ml^{-1}$.

Table 20.1: Aerobic Degradation of Endosulfan at 10 μg.ml^{-1} by *Pseudomonas* sp.

Days after Incubation	*Endosulfan Recovered (μg.ml^{-1})*			
	α-Endosulfan		*β-Endosulfan*	
	Uninocualted	*Inoculated*	*Uninoculated*	*Inoculated*
0	9.75 ± 0.07*	8.87 ± 0.28	9.08 ± 0.65	8.74 ± 0.41
2	7.97 ±0.64	3.64 ± 0.53	8.27 ± 0.61	3.21 ± 0.20
4	5.89 ± 0.62	2.91 ± 0.27	7.23 ± 0.14	2.82 ± 0.35
6	4.23 ± 0.28	2.35 ± 0.02	6.72 ± 0.50	0.56 ± 0.38
8	4.06 ± 0	1.63 ± 0.04	5.57 ± 1.28	0.36 ± 0.17

* Average of two replicate observations + mean deviation.

Unlike α-isomer, β-endosulfan was degraded more extensively. Some reports suggested that β-endosulfan is more resistant to degradation (Rao and Murty, 1980; Singh *et al.*, 2000). However, in the present study, *Pseudomonas* sp. degraded β-endosulfan from 8.74 to 0.36 μg.ml^{-1} within 8 days of incubation (Table 20.1). Like α-endosulfan, degradation of β-endosulfan was associated with the proliferation of cells.

Concentration Dependent Degradation of Endosulfan Isomers

For effective bioremediation, the degrading bio-agent should have wide potential including capability of degrading the target xenobiotic at variable concentrations and tolerance to other environmental factors including changes in pH, salinity, temperature *etc.* In the present experiment, the tolerance of *Pseudomonas* sp. to varied concentrations of endosulfan (α- and β-) in affecting the degradation was quantified. *Pseudomonas* sp. was independently exposed to α- and β-endosulfan in mineral salts medium at concentrations of 10, 50 and 100 μg.ml^{-1} and the degradation potential was recorded.

The bacterial isolate was able to degrade endosulfan (α- and β-) at various concentrations. At a concentration of 10 μg.ml^{-1}, α- and β-endosulfan were degraded to negligible levels within 8 days of incubation by *Pseudomonas* sp. (Tables 20.1). The degradation also continued *albeit* at a slower rate when the concentration of endosulfan (α- and β-) was raised to 50 μg.ml^{-1} and it took more than 20 days to come down to insignificant levels (Tables 20.2). The picture was starkly contrasting when the concentration of endosulfan (α- and β-) was raised to 100 μg.ml^{-1} (Table 20.3). *Pseudomonas* sp. degraded only 25 per cent of added endosulfan in an incubation period of 20 days. In this study, the incubation was not continued beyond 20 days apprehending death/decay or possible contamination in keeping for longer incubation. Concentration of endosulfan beyond 50 μg.ml^{-1} might have been toxic for the degrading bacterial species. Similar result was found by Narkhede *et al.* (2015) in which *Pseudomonas aeruginosa* was able to degrade 96 per cent of endosulfan after 288 hours under static conditions. In another study, Zaffer *et al.*

(2018) isolated a *Pseudomonas* sp. from pesticide contaminated soil and done a batch degradation experiments of endosulfan with its different concentration *i.e.* 5, 25, 50, 75 and 100/mg/l. They found 70-80 per cent of degradation after three days of incubation and complete degradation after five days of incubation.

Table 20.2: Aerobic Degradation of Endosulfan at 50 μg.ml^{-1} by *Pseudomonas* sp.

Days after Incubation	*Endosulfan Recovered (μg.ml^{-1})*			
	α-Endosulfan		*β-Endosulfan*	
	Uninoculated	*Inoculated*	*Uninoculated*	*Inoculated*
0	41.55 ± 0.27*	36.89 ± 0.30	51.51 ± 5.46	39.73 ± 4.32
5	35.80 ±1.77	22.21 ± 2.10	48.07 ± 4.38	22.64 ± 2.62
10	29.19 ± 0.90	16.23 ± 0.01	37.53 ± 1.61	12.79 ± 3.05
15	28.00 ± 0.08	13.81 ± 1.55	36.01 ± 0.43	11.11 ± 0
20	25.58 ± 0.37	12.76 ± 0.78	31.79 ± 1.55	9.42 ± 1.22

* Average of two replicate observations + mean deviation.

Table 20.3: Aerobic Degradation of Endosulfan at 100 μg.ml^{-1} by *Pseudomonas* sp.

Days after Incubation	*Endosulfan Recovered (μg.ml^{-1})*			
	α-Endosulfan		*β-Endosulfan*	
	Uninoculated	*Inoculated*	*Uninoculated*	*Inoculated*
0	92.34 ± 13.23*	92.29 ± 9.87	89.86 ± 19.21	92.32 ± 0.06
5	89.01 ± 0.08	83.11 ± 8.53	87.33 ± 2.27	85.89 ± 3.36
10	87.81 ± 1.05	79.53 ± 1.57	90.19 ± 4.69	81.59 ± 1.60
15	74.75 ± 3.62	78.07 ± 0.57	80.80 ± 3.27	71.65 ± 5.75
20	75.41 ± 2.20	56.72 ± 1.92	81.38 ± 4.59	70.02 ± 0.14

* Average of two replicate observations + mean deviation.

Effect of pH on the Growth of Endosulfan-Degrading Bacterial Species

Endosulfan isomers are stable at low pH (Shivaramaiah *et al.*, 2005). However, increase in pH to near neutral to alkaline results into chemical conversion of endosulfan to endosulfan diol thereby reducing the toxicity levels of the insecticide (Sutherland *et al.*, 2000; Kumar and Philip, 2006). In the present experiment, growth of endosulfan-degrading *Pseudomonas* sp. was marginally inhibited at pH 6.0. Growth of the bacteria was best at pH 7.0 and was affected in a negative way again at pH 7.5 and 8.0. *Pseudomonas* sp. was isolated from flooded rice field wherein pH level usually hovers around near neutrality between 6.8 to 7.2. Microorganisms proliferating in such environment are expected to grow better at near neutral pH.

Development of a Colorimetric Assay for Endosulfan Degradation

In order to isolate potential endosulfan-degrading microorganisms from soil, the traditional approach is to inoculate mineral salts media containing endosulfan with soil suspensions from retreated soil. Presently, extent of endosulfan degradation is measured by gas chromatography. Degradation of endosulfan is usually followed by cleaving of the SO_3^- moiety and its release as SO_4. This results in the formation of H^+ and production of acidic reaction. In the colorimetric assay, this reaction causes a decrease in pH in a weakly buffered solution (Phillips *et al.,* 2001). In the presence of phenol red, a color change occurs from red to yellow, indicating release of SO_4 and degradation of endosulfan (Figure 20.2).

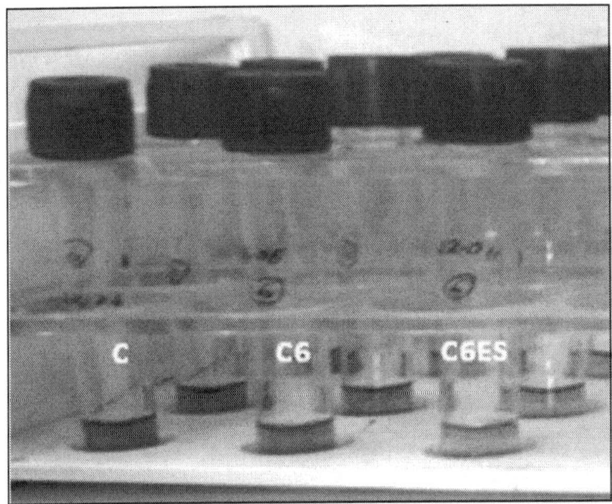

Figure 20.2: Colorimetric Assay for Endosulfan Degradation by Bacteria.

[C: Uninoculated control; C6: Inoculated with endosulfan degrading *Pseudomonas* sp. without endosulfan; C6ES: Inoculated with endosulfan degrading *Pseudomonas* sp. with endosulfan].

The phenol red indicator turned yellow when inoculated with known endosulfan-degrading *Pseudomonas* sp. (Figure 20.2). In both the negative (uninoculated medium with endosulfan) and positive controls (inoculated without endosulfan), color of phenol red did not turn yellow. This was also true for absolute control with endosulfan non-degrading *E. coli* wherein the color of phenol red did not change either with endosulfan or without it. Thus, the assay is effective as a qualitative test for degradation of endosulfan and may have potential as a screening tool.

The above results clearly demonstrate that *Pseudomonas* sp. isolated from flooded alluvial soil planted to rice and retreated with endosulfan, invariably exhibited catabolic potential in degrading endosulfan isomers. The bacteria *Pseudomonas* sp. could utilize endosulfan isomers as the sole source of carbon and energy. The degradation potential exhibited by the candidate bacteria would be an ideal source for field bioremediation studies in near future.

Conclusion

Bioremediation, a biodegradation process in which sites contaminated with xenobiotics are cleaned up by means of bacterial bio-degradation processes, preferably exploits the ability of microorganisms to reduce the concentration and/ or toxicity of a large number of pollutants. Microbial diversity offers an immense field of environment-friendly options for mineralization of contaminants or their transformation into less harmful non-hazardous compounds. Pure culture *Pseudomonas* sp. isolated from endosulfan-treated flooded alluvial soil planted with rice was used in the present study to characterize growth-related degradation of endosulfan isomers (α- and β-). Also, a qualitative test for degradation of endosulfan having potential as a screening tool has been developed. In summary, the bacterial strain capable of degrading endosulfan as a sole source of carbon and energy was characterized. Results suggest that the bacteria *Pseudomonas* sp. which degraded endosulfan as sole source of carbon and energy has wide potential in using them for field bioremediation studies. However, further studies are essential to evolve bioremediation strategies for field applications.

Acknowledgement

The article forms a part of the M.Sc. dissertation submitted to Utkal University by the senior author. Authors are thankful to the Director General, NRRI, Cuttack for providing necessary facilities to complete this work and constant motivation.

References

Adhya, T.K., Apte, S.K., Raghu, K., Sethunathan, N. and Murthy, N.B.K. (1996). Novel polypeptides induced by the insecticide lindane (γ-hexachlorocyclohexane) are required for its biodegradation by a *Sphingomonas paucimobilis* strain. Biochem. Biophys. Res. Commun. 221: 755-761.

Aktar, W., Sengupta, D. and Chowdhury, A. (2009). Impact of pesticides use in agriculture: their benefits and hazards. Interdisciplinary Toxicol. 2(1): 1-12.

Awasthi, N., Manickam, N. and Kumar, A. (1997). Biodegradation of endosulfan by a bacterial coculture. Bull. Environ. Contam. Toxicol. 59:928-934.

Bindhya, R. and Thanga, V.S.G. (2010). Differential utilization of organophosphorus pesticides as nutrient source by bacteria isolated from tropical soil. J. Biotechnol. 150: 230.

Coleman, P.F. and Dolinger, P.M. (1982). Environmental health evaluation of California restricted pesticides. (Endosulfan Monograph no. 4) PM Dolinger Associates, Menlo Park, California.

Cook, A.M., Daughton, C.G. and Alexander, M. (1978). Phosphorus-containing pesticide breakdown products: quantitative utilization as phosphorus sources by bacteria. Appl. Environ. Microbiol. 36(5): 668-672.

Damalas, C.A. and Eleftherohorinos, I.G. (2011). Pesticide exposure, safety issues, and risk assessment indicators. Int. J. Environmental Res. Public Health, 8(5): 1402-1419.

Fianko, J.R., Donkor, A., Lowor, S.T. and Yeboah, P.O. (2011). Agrochemicals and the Ghanaian environment, a review. J. Environmental Protection, 2(03): 221.

Gill, H.K. and Garg, H. (2014). Pesticide: environmental impacts and management strategies. Pesticides-toxic aspects, 8:187.

Kafilzadeh, F., Ebrahimnezhad, M. and Tahery, Y. (2015). Isolation and identification of endosulfan-degrading bacteria and evaluation of their bioremediation in Kor River, Iran. Osong public health and research perspectives, 6(1): 39-46.

Kamei, I., Takagi, K. and Kondo, R. (2011). Degradation of endosulfan and endosulfan sulfate by white-rot fungus *Trametes hirsuta*. J. Wood Sci. 57(4): 317-322.

Katayama, A. and Matsumura, F. (1993). Degradation of organochlorine pesticides, particularly endosulfan, by *Trichoderma harzianum*. Environ. Toxicol. Chem. 12(6): 1059-1065.

Kumar, M. and Philip, L. (2006). Bioremediation of endosulfan contaminated soil and water – optimization of operating condition in laboratory scale reactors. J. Hazard. Material. 136: 354-364.

Kumar, M., Lakshmi, C.V. and Khanna, S. (2008). Biodegradation and bioremediation of endosulfan contaminated soil. Bioresour. Technol. 99(8): 3116-3122.

Li, W., Dai, Y., Xue, B., Li, Y., Peng, X., Zhang, J. and Yan, Y. (2009). Biodegradation and detoxification of endosulfan in aqueous medium and soil by *Achromobacter xylosoxidans* strain CS5. J. Haz. Materials, 167 (1-3): 209-216.

Menezes, R.G., Qadir, T.F., Moin, A., Fatima, H., Hussain, S.A., Madadin, M., Pasha, S.B., Al Rubaish, F.A. and Senthilkumaran, S. (2017). Endosulfan poisoning: an overview. J. Forensic legal Med. 51: 27-33.

Mukherjee, I. and Gopal, M. (1994). Degradation of beta-endosulfan by *Aspergillus niger*. Toxicol. Environ. Chem. 46(4), pp.217-221.

Narkhede, C.P., Patil, A.R., Koli, S., Suryawanshi, R., Salunke, B.K. and Patil, S.V. (2015). Studies on endosulfan degradation by local isolate pseudomonas aeruginosa, Biocatalysis Agricul. Biotechnol. http://dx.doi.org/10.1016/j.bcab.2015.01.006.

Narwal, S.K. and Gupta, R. (2017). Biodegradation of xenobiotic compounds: an overview. In: Bhakta, J.N. (ed.) Handbook of research on inventive bioremediation techniques IGI Global., pp. 186-212).

Ozdal, O.G., Ozdal, M., Algur, O.F. and Sezen, A. (2016). Isolation and identification of α-Endosulfan degrading bacteria from insect microflora. Turkish J. Agric. Food Sci. Technol. 4(4): 248-254.

Phillips, T.M., Seech, A.G., Lee, H. and Trevors, J.T. (2001). Colorimetric assay for lindane dechlorination by bacteria. J. Microbiological Methods, 47(2): 181-188.

Pradeep, V. and Subbaiah, U.M. (2016). Use of Ca-alginate immobilized *Pseudomonas aeruginosa* for repeated batch and continuous degradation of Endosulfan. 3 Biotech, 6(2): 124.

Ramesh, A. and Vijayalakshmi, A. (2002). Environmental exposure to residues after aerial spraying of endosulfan: residues in cow milk, fish, water, soil and cashew leaf in Kasargode, Kerala, India. Pest Manage. Sci. 58:1048-1054.

Rao, D.M.R. and Murty, A.S. (1980). Persistence of endosulfan in soils. J. Agric. Food Chem. 28:1099-1101.

Sang, H., Hulvey, J.P., Green, R., Xu, H., Im, J., Chang, T. and Jung, G. (2018). A xenobiotic detoxification pathway through transcriptional regulation in filamentous fungi. MBio, 9(4): e00457-18.

Shetty, P.K., Mitra, J., Murthy, N.B.K., Namitha, K.K., Savitha, K.N. and Raghu, K. (2000). Biodegradation of cyclodiene insecticide endosulfan by *Mucor thermohyalospora* MTCC 1384. Current Sci. pp.1381-1383.

Shivaramaiah, H.M., Sanchez-Bayo, F., Al-Rifai, J. and Kennedy, I.R. (2005). The fate of endosulfan in water. J. Environ. Sci. Health, 40(5): 711-720.

Shivaramaiah, H.M. and Kennedy, I.R. (2006). Biodegradation of endosulfan by a soil bacterium. J. Environ. Sci. Health, 41(6): 895-905.

Singh, S., Dureja, P. and Kumar, S. (2000). Biodegradation of α and β isomers of endosulphan and endosulphan sulphate in Indian soils. J. Environ. Sci Health, 35(3): 337-346.

Sutherland, T.D., Home, I., Lacey, M.J., Harcourt, R.L., Russell, R.J. and Oakeshott, J.G. (2000). Enrichment of an endosulfan-degrading mixed bacterial culture. Appl. Environ. Microbiol. 66:2822-2828.

Sutherland, T.D., Weir, K.M., Lacey, M.J., Horne, I., Russell, R.J. and Oakeshott, J.G. (2002). Enrichment of a microbial culture capable of degrading endosulphate, the toxic metabolite of endosulfan. J. Appl. Microbiol. 92:541-548.

Tapsoba, K.H., Pare, S. and Bonzi-Coulibaly, L.Y. (2008). Endosulfan residues in soils from three cotton-growing areas in Burkina Faso. Science et technique, Sciences appliquees et Technologies, 2 (1): p.2.

US Department of Health and Human Services (2000). Toxicological Profile for Endosulfan (Update): Agency for Toxic Substances and Disease Registry (ATSDR), Atlanta, Georgia.

Zaffar, H., Ahmad, R., Pervez, A. and Naqvi, T.A. (2018). A newly isolated *Pseudomonas* sp. can degrade endosulfan via hydrolytic pathway. Pesticide Biochem. Physiology, 152: 69-75.

Chapter 21

Curcumin: Transforming the Spice to Herbal and Traditional Medicine

Laxmipreeya Behera, Kailash Chandra Samal and Sachin Subhash Kharade*

Department of Agricultural Biotechnology, College of Agriculture, Odisha University of Agriculture and Technology, Bhubaneswar – 751 003, Odisha
**e-mail: klaxmipreeya1@gmail.com*

ABSTRACT

The genus Curcuma (Family: Zingiberaceae) contains many species which have immense economic, medicinal, ornamental and cultural importance. There are 80 different species of curcuma genus out of which, Curcuma longa (spice turmeric), Curcuma amada (mango ginger), Curcuma caesia (black turmeric), Curcuma aromatica (wild/Kasthuri turmeric) and Curcuma angustifolia (Indian Arrow root) are widely cultivated in India and used as spices, medicines, dyes, foods, perfumes, tonics, and as tropical ornamentals. In this study full focuswas given on, morphological, biochemical and anti-microbial studies of five Curcuma species. All the five Curcuma species exhibited variation in different characters when they were evaluated for different qualitative traits viz. leaf and rhizome characters. In case of Curcuma longa and Curcuma aromatic much similarity found for most of the characters. With respect to rhizome colour, smell, colour of leaf sheath, and leaf midrib colourCurcuma caesia was found highly different than others and Curcuma amadashown different in aroma (raw mango) than other five species. Extraction of curcumin and curcuminoids from dry powders of rhizomes was done in methanol as solvent by soxhlet apparatus. The estimation of Curcumin, demethoxycurcumin and bis-demethoxycurcumin were done by with the help of HPTLC technique by comparing with standard sample. The highest curcumin content was observed in Curcuma aromatica (1.251 per cent) followed by Curcuma longa (0.845 per cent). The highest percentage of demethoxycurcumin (0.593 per cent) and bis- demethoxycurcumin (0.788 per cent) were recorded in Curcuma aromatica followed by Curcuma longa (0.437 per cent. 0.487 respectively). C. angustifolia exhibited the lowest curcumin (0.028 per cent), demethoxycurcumin(0.01 per cent),and bis- demethoxycurcumin(0.016 per cent) content. The antimicrobial activity was tested as per the zone of inhibition formed by the plant extract against the bacteria. Curcuma longa extract showed antimicrobial activity against E. coli (18.2 mm). Highest zone of inhibition against Pseudomonas fluorescence were measured by extract of C. longa(13.75 mm) and curcuma caesia (12.17 mm).

Introduction

The genus Curcuma belongs to the Family: Zingiberaceae and contains more than 80 different species which have immense economic, medicinal, ornamental and cultural importance and the center of diversity is in Southeast Asia, but some species extend to the Himalayas, Southern China, Australia and the Pacific Islands. Curcuma is commonly popularized by a number of common names *viz.* turmeric, Indian saffron and hidden lily.

As per the statement of the World Health Organization (WHO), half of the world's population relies on traditional medicine for their primary healthcare requirement. The development of traditional medicines and use of medicinal plants for the treatment of various diseases causes considerable economic benefits (Kapoor *et al.*, 2001). Most of the people especially rural origin are still forced to practice traditional medicines for their common day ailments due to less communication means, poverty and unavailability of modern health care facilities. In pharmaceuticals, cosmetics and for preparation of medicinal products Plant derived products are being promoted as primary source for the health care. Both developed and developing countries widely used herbal medicines in health-care which are complex chemical mixtures of medicinal plants. To standardize crude Ayurvedic drugs recently a number of laboratories are active, apart from a conventional objective to document the various aspects of identification of single crude drug, one of the underlying objectives of many advanced laboratories betrothed in detection of finger printing of marker components from the botanicals being employed and used traditionally for therapeutic use in many parts of the country.

Fingerprinting in essence is chemo-profiling which means establishing a characteristic chemical pattern for the plant material. This pattern is the line of divergence between the modern system of medicines based on synthetic components and the classical system of medicine like Ayurveda where one can make use of the chemical pattern of the secondary metabolites for the identification of plant material as our knowledge of chemistry has advanced sufficiently and through sophisticated analytical techniques. One can measure these compounds qualitatively and quantitatively. The foundation of the Ayurvedic drugs is to administer drugs as whole or processed extracts thereof. This means that the theory of synergy is very well highlighted and it is documented that it is not our pure compound but a set of compounds which are necessary for biological activity of a plant material. This shows that for a better assessment of the plant material one need to measure a set of compounds rather than one single compound. Among them *Curcuma longa* (spice turmeric), *Curcuma amada* (mango ginger), *Curcuma caesia*(black turmeric), *Curcuma aromatica* (wild/Kasthuri turmeric) and *Curcuma angustifolia* (Indian Arrow root) are widely cultivated in India and used as spices, medicines, dyes, foods, perfumes, tonics, and as tropical ornamentals. *Curcuma longa is* known as turmeric, safflower, or yellow ginger. The rhizomes are horizontal underground stems that send out shoots as well as roots. Curcumin is the active constituent of turmeric which has anti-inflammatory and anticancer activities having potential to prevent and treat the disease (Akram*et al.*, 2010). In addition other curcuminoids are found in turmeric which include demethoxycurcumin and bis-demethoxycurcumin. In addition to its use as a spice and pigment, turmeric has been used in India for medicinal purposes for centuries. The oleoresin component of turmeric is used for oil-containing products. Raw Turmeric juice is used to treat hyper acidity and indigestion. The juice of raw turmeric also acts as a blood purifier. Curcumin- an active component of turmeric has anti-oxidant properties and so turmeric is used in alternative medicine. Turmeric is used for cuts and burns as it is supposed to have antiseptic effects and promotes healing. Curcumin also has an anti-inflammatory effect by reducing histamine (hormone) levels. The flouride present in turmeric is essential for teeth. Turmeric also has a protective effect on the liver and also in atherosclerosis. Regular turmeric use is said to make the skin fair, soft and smooth. Turmeric is used for spots caused due to pigmentation or blotches and also for diseases like eczema.

Curcuma caesia Roxb. is known as black turmeric, It has deep bluish black or grayish black-colored rhizome having pungent smell and hot bitter taste and leaves with a deep red-violet patch, which runs throughout the whole lamina. The rhizome of black turmeric has a high economic importance owing to its putative medicinal properties. To get relief from migraine fresh rhizomes are crushed and applied as a paste on forehead and applied on the body for sprains and bruises.

Curcuma amada Roxb., prevalently famous as mango ginger, are perennials growing up to about 80 cm in height and having morphological resemblance with ginger but imparts a raw mango flavour. The plant species is endemic to South Asia and commonly distributed in many parts of northeast India and in the hills of south India. Mango ginger rhizome is used in the manufacture of pickles and culinary preparations. In Ayurveda and Unani mango ginger is considered as an appetizer, alexteric, and antipyretic, aphrodisiac, diuretic, emollient, expectorant and laxative and to cure itching, skin diseases, bronchitis, asthma, hiccough and inflammation due to injuries (Reshma *et al.,* 2013).

Curcuma aromatic Salisb. (Kasthuri/wild turmeric) is considered as medicinal and aromatic herb with strong antibiotic properties. Fresh rhizomes are yellowish with camphoraceous smell and produce clumps of about 1 m tall erect and unbranched leaf stems. The rhizomes are often used in cosmetic herbal medicines and as a culinary ingredient in limited quantities as a food flavor (Jangde*et al.*, 1998).

Curcuma angustifolia is also known as Indian Arrow root which is characterized by the presence of narrow, green, glabrous leaves with small inflorescences bearing yellow flowers with pink coma bracts, and its grains contain starch as chief carbohydrate. Starches are used as multifunctional excipient in the field of pharmaceutical sciences. The arrowroot starch has high digestibility and is commonly used as a thickener in many foods such as puddings and sauces, cookies and other baked goods. Arrowroot is bland, making it suitable for neutral diets, especially for people who are feeling nauseous (Kumalasari *et al.,* 2012).

Chemo-Profiling of *Curcuma*

Chemo profiling of curcuma done by High Performance Thin Layer Chromatography (HPTLC) which has been utilized to separate new promising pharmaceutical therapeutants which could be used in pharmaceutical industries (Maripandi *et al.*, 2010). It is also useful for phytochemical and biomedical analysis, herbal drug quantification, active ingredient quantification, fingerprinting of formulations, and check for adulterants in the formulations.

The active constituents of *Curcuma* are the flavonoid curcumin (diferuloylmethane), demethoxycurcumin (DMC), bisdemethoxycurcumin (BDMC) and various volatile oils, including tumerone, atlantone, and zingiberone (Li*et al.,*1985). Other constituents include sugars, proteins, and resins. Curcumin has been shown to possess a variety of pharmacological activities, including antioxidant, anti-inflammatory, cancer chemo-preventive, and neuro-protective activities (Jayaprakasha *et al.,* 2006). The structures of the curcumin and curcuminoids were determined in 1910 (Aggarwal, 2007).Curcumin, dimethyoxycurcumin and bis-demethoxycurcumin is a dietary photochemical obtained from dried rhizomes of *Curcuma.* Curcumin is a main coloring substance in *Curcuma longa* and two related compounds, demethoxycurcumin (DMC) and bis-demethoxycurcumin (BDMC) are altogether known as curcuminoid (Ahsan*et al.*, 1999). Curcumin is the pigment that provides the bright stunning yellow colour. It is the main biologically active phytochemical compound present in *Curcuma* and has proven disease preventing medicinal properties. From several studies Curcumin is found are a promising anticancer agent and the expected mechanisms of action include regulation of transcription factors, growth-regulatory molecules, and growth factor receptors, protein kinase, and tumor suppressor pathways. Phytochemical screening of the dye confirmed the presence of saponins, tannins, flavonoids and alkaloids. Among these compounds, tannins and flavonoids are the substances which can give the color. Flavonoids are water soluble compounds or chromophores responsible for colouration of rhizome and have reducing power and strong antioxidant and free radical scavenging activity. Tannins are the most important ingredients which are necessary for dyeing. Saponins are responsible for the reduction of surface tension and this property also enhances staining.

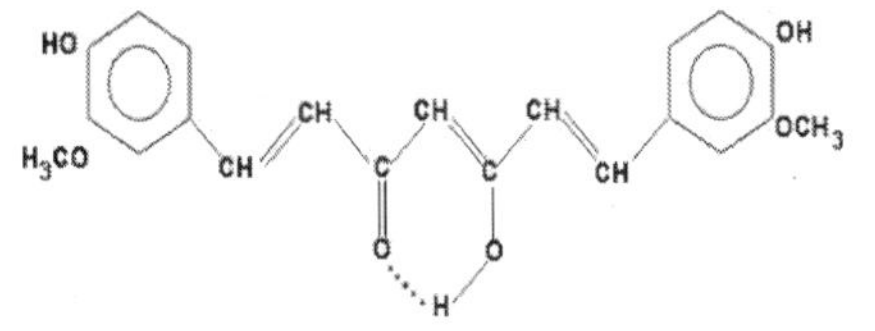

CURCUMIN

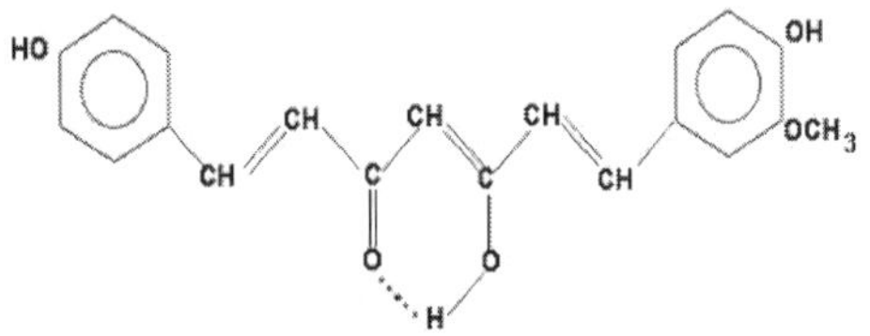

DEMETHOXYCURCUMIN

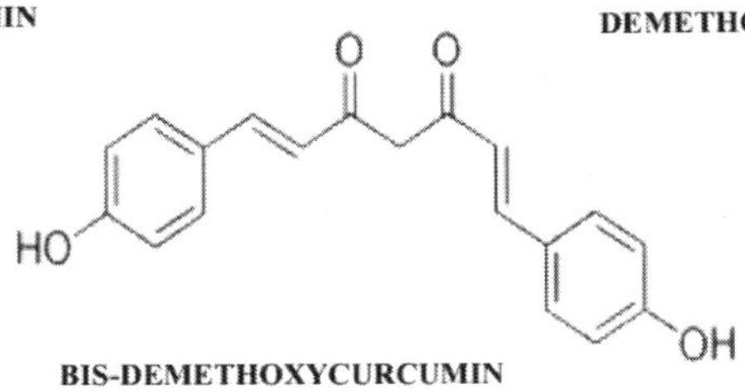

BIS-DEMETHOXYCURCUMIN

The present study aims for the determination of curcumin, demethoxycurcumin (DMC) and bisdemethoxycurcumin (BDMC) in *Curcuma longa, Curcuma amada, Curcuma caesia. Curcuma aromatica* and *Curcuma angustifolia* collected from Eastern regions of India and to study its antimicrobial potential (Mukherjee *et al.*, 1995).

Materials and Methods

Plant Material

In this present investigation five *Curcuma* species *viz. Curcuma longa, Curcuma amada, Curcuma caesia. Curcuma aromatica* and *Curcuma angustifolia* were collected from the germplasm garden of All India Coordinated Project on Medicinal Plants, Odisha University Agriculture and Technology (OUAT), Bhubaneswar, Odisha. These species were planted in the university experimental field for morphological and biochemical evaluation.

Chemo-Profiling Analysis of *Curcuma*

The rhizome of the plant samples were air dried and well macerated inelectronic mixer/grinder to obtain a fine powder. The powdered plant material (15gm) was loaded in the inner tube of the of Soxhlet apparatus and then fitted into around bottomed flask containing methanol. The solvent was boiled gently (64-65°C) over a heating mantle using the adjustable rheostat. The extraction was continued until complete extraction was effected (24-30 h) and the solvent was removed at the reduced pressure with the help of rotary vacuum evaporator to yield a viscous dark yellowish or brown residue (Kulkarni,*et al.,* 2012) The residue was dissolved in methanol and was further used for HPTLC analysis for presence of standard curcumin, demethoxycurcumin and bis-demethoxycurcumin purchased from Himedia and the extracts of five turmeric species and standard solutions were applied 1.5 cm away from the lower edge of the Aluminium plate pre-coated with silica gel 60 F254 TLC plates (20 × 10 cm) (Merck) with the help of micro-capillary tube or automatic TLC applicator (Paramasivam*et al.*, 2009). The solvent system used to identify curcumin, demethoxycurcumin and bis-demethoxycurcumin was toluene: ethyl acetate: methanol (18:1:1 v/v/v) under laboratory conditions (25 – 30°C and 40 – 50 per cent relative humidity). The loaded plates were then placed vertically in the chamber previously saturated with solvent system for 30 min. After the solvent front moved up to a distance of about 90 per cent of length, the plate was taken out, solvent front was marked and the plate was dried at room temperature. Developed plates were air dried and then immersed in a freshly prepared mixture of methanol and concentrated sulphuric acid (95:5 v/v) for dervitisation. After drying, the plates were heated at 110°C for 5 min to develop the colour of the spots. For quantitative determination, spots corresponding to standards were scanned using a Camag TLC Scanner 3 at 254 nm wavelength. The Rf values were calculated using formula: Rf = Distance travelled by solute/Distance travelled by solvent. The amount of curcumin and its derivatives present in each extract were calculated by comparing the peak area of standard and respective samples. The following formula was used to quantify the active constituent.

$$\textbf{\% Curcumin, DMC \& BDMC} = \frac{\text{Area of sample}}{\text{Area of standard}} \times \frac{\text{Conc of standard}}{\text{Conc. of sample}} \times 100$$

Phytochemical and Antimicrobial Analysis

The rhizome sample extracts of five important *Curcuma* species were prepared by extraction with methanol as a solvent using Soxhlet apparatus. From this, different concentrations were prepared *viz.* 20 mg/ml, 40 mg/ml, and 60 mg/ml. Bacterial cultures of *Escherichia coli* (E. coli), *Bacillus substalis, Psedomonas fluroscence* were obtained from the Department of Plant Pathology, College of Agriculture, Odisha University of Agriculture and Technology, India and were used for antimicrobial test organisms. The bacteria were maintained on nutrient agar at 37°C.The antibacterial tests of the rhizome extracts were tested on the test bacteria *i.e. E. coli, Bacillus substalis* and *Psedomonas fluroscence* using the paper disc diffusion method. The single bacterial colony inoculated in autoclaved nutrient broth (100 ml) and incubates for 24 hrs at 37°C. This overnight bacterial broth suspension used as inoculums for antimicrobial activity test. The autoclaved nutrient agar medium was poured in sterile petridishes in aseptic condition and allows to solidifying. Bacterial broth inoculums (50 µl) spread on solidified medium by "L-spreader" and the sterile paper discs (5mm) were soaked in the rhizome extract of different concentrations *viz.* 20 mg/ml, 40 mg/ml and 60 mg/ml for 2 hours. The paper discs containing the extracts were placed at different areas on the surface of each plate. The plates were incubated at 37°C for 24 h. A disc soaked in methanol was used as control. Antimicrobial activity of the extract against the test bacteria was indicated by growth -free "zone of inhibition" near the respective disc. The Minimum Inhibitory Concentrations (MIC) of the extracts was determined according to. The diameters of inhibitory zones were measured on the second day. The diameter of zone inhibited was measured in mm. The experiment was performed in triplicate and the average zone of inhibition was recorded.

Results and Discussion

All five species exhibited variation indifferent characters (Table 21.1). *Curcuma longa* and *Curcuma aromatica* were shown much similarity for most of the characters. *Curcuma caesia* was found highly different from others with respect to rhizome colour, smell, colour of leaf sheath, and leaf midrib colour. Curcuma amada (raw mango) was different in aroma (no aroma) than other five species.

Table 21.1: Morphological Characteristics of *Curcuma* Species

Sl. No.	*Species Name*	*Rhizome Characters*			*Aerial Characters*	
		Colour of Rhizome	*Aroma of Rhizome*	*Taste of Rhizome*	*Colour of Leaf Sheath*	*Leaf Midrib Colour*
1	***Curcuma longa*** (Cultivated turmeric)	Orange yellow	Mustardy and Compho-raceous	Slight bitter	Green	Green
2	***Curcuma aromatica*** (Wild turmeric)	Pale Yellow	Pleasant aroma	Bitter	Green	Green
3	***Curcuma caesia*** (Black turmeric)	Blue	Compho-raceous	Bitter	Purple Brown	Purple Brown
4	***Curcuma amada*** (Mango ginger)	Pale yellow white	Mango	Raw mango	Green	Green
5	***Curcuma angustifolia*** (Indian arrow root)	Creamy White	No such aroma	Slightly Bitter	Green	Green

Percentage of curcumin, demethoxycurcumin (DMC) and bis-demethoxycurcumin (BDMC) content were also estimated in five different *Curcuma* species (Table 21.2). *Curcuma aromatica* and *curcuma longa* showed the high percentage of curcumin, DMC and BDMC content as compared to other four species. The Rf values of the bands for reference standards were: Curcumin at Rf 0.37, de-methoxycurcumin at Rf 0.23 and bis-demethoxycurcumin at Rf 0.16. Constituents of the five sample extracts were recognized by comparison of bands in sample with reference standard on the same TLC plate. It was observed that curcumin was present in all five *Curcuma* species and percentage of active constituents was also estimated. The highest curcumin content was observed in *Curcuma aromatica* (1.251 per cent) followed by *Curcuma longa* (0.845 per cent). Other *Curcuma* species have lower curcumin content. With respect to curcminoids, similar trend was also recorded. The highest percentage of demethoxycurcumin (0.593 per cent) and bis- demethoxycurcumin (0.788 per cent) were recorded in *Curcuma aromatica* followed by *Curcuma longa* (0.437 per cent. 0.487 respectively). Among the five *Curcuma* species the lowest curcumin (0.028 per cent), demethoxycurcumin (0.01 per cent),and bis- demethoxycurcumin (0.016 per cent), were recorded in *C. angustifolia*

Table 21.2: Percentage of Curcumin, Demethoxycurcumin and Bis-demethoxycurcumin Content in different *curcuma* Species

Curcumin and Curcuminoides	*C. longa*	*C. aromatica*	*C. caesia*	*C. angustifolia*	*C. amada*
Curcumin (per cent)	0.845	1.251	0.137	0.028	0.212
Demethoxy-curcumin (per cent)	0.437	0.593	0.137	0.001	0.136
Bis-demethoxy-curcumin (per cent)	0.487	0.788	0.124	0.016	0.060

Antimicrobial Activity

The antimicrobial activity was evaluated as per the zone of inhibition formed by the plant extract against the bacteria (*Bacillus substillis, Pseudomonas fluorescence* and *E.coli*). The extracts of all the five species were tested against the pathogenic bacteria namely *Bacillus substillis, Pseudomonas fluorescence* and *E. coli* at different concentrations of extract *viz.* 20, 40, 60 mg/ml. The five species extracts exhibited antibacterial activity against these three bacteria and the zone of inhibition was recorded in millimetre (mm). The highest antimicrobial activity against *E. coli* was exhibited by *C. amada* (17.50mm) followed by *Curcuma longa* (13.43 mm). *Curcuma caesia* showed the highest zone of inhibition against *Escherichia coli* (13.20

Table 21.3: Antimicrobial Activity of Crude Extracts of *Curcuma* Species

Pathogenic Bacteria Tested	*Antimicrobial Activity of Curcuma Extract [Zone of inhibition (mean value) in millimeter]*				
	C. longa	*C. aromatica*	*C. caesia*	*C. angustifolia*	*C. amada*
Bacillus substillis	7.42	8.50	7.08	13.50	12.33
Pseudomonas fluorescence	13.75	11.92	12.17	5.75	2.92
Escherichia coli	13.43	12.87	13.20	12.53	17.50

mm). *Curcuma angustifolia* extracts showed antimicrobial activity against *Bacillus substalis* (13.50 mm) followed by *Curcuma amada* (12.33 mm).

Antimicrobial activity of crude extracts of *Curcuma* species were also studied by many workers. Hamdan (2011) reported that crude ethanolic extract of curcumin was tested against number of microorganisms; gram positive bacteria (*Staphylococcus aureus*), and gram negative bacteria (*E. coli, Pseudomonas, and Klebsiella*). The antibacterial activity of the extract was evaluated against bacteria using disc diffusion method and stated that the ethanolic extract showed a broad spectrum of antibacterial activity and the *Pseudomonas aeruginosa* was the most sensitive bacteria. Yasodamma *et al.* (2013) reported antibacterial activity of rhizome and leaf extracts of the *C. neilgherrensis,C. longa, C. zedoaria, C. malabarica, C. xanthorrhiza* on mouth and dental pathogens *S. aureus* and other human pathogens like *Klebsiella, Vibrio, Bacillus, Streptococcus, Azatobacter, Enterobacter, Pseudomonas* and *E. coli.* Antibacterial activity of rhizome cold water extracts of *C. neilgherrensis* was found very effective on *Staphylococcus aureus* at 10 mg/well than the standard drug Ampicillin.

However, suitable bacterial bioassays have been documented to recognize and quantify antibacterial effects of plant rhizome extracts. Further studies have to be made on fractionation and separation of crude extracts in order to find out the principle antibacterial compound.

Conclusion

In this above study it was observed that curcumin was found in all five *Curcuma* species and percentage of active ingredient was also estimated. The highest curcumin content was observed in *Curcuma aromatica* (1.251 per cent) followed by *Curcuma longa* (0.845 per cent). Other *Curcuma* species have lower curcumin content. With respect to curcminoids, similar trend was also recorded. The highest percentage of demethoxycurcumin (0.593 per cent) and bis- demethoxycurcumin (0.788 per cent) were recorded in *Curcuma aromatica* followed by *Curcuma longa* (0.437 per cent, 0.487 respectively). The lowest percentage of demethoxycurcumin (0.001 per cent) and bis- demethoxycurcumin (0.016 per cent) were recorded in *Curcuma angustifolia.* The antimicrobial activity was tested as per the zone of inhibition formed by the plant extract against the bacteria. The highest antimicrobial activity against *E. coli* was exhibited by *C. amada*(17.50) mm followed by *Curcuma longa* (13.43 mm). The highest zone of inhibition against *Pseudomonas fluorescence* were measured by extract of *Curcuma longa* (13.75 mm) and *curcuma caesia* (12.17 mm). *Curcuma angustifolia* and *Curcuma amada* extracts showed antimicrobial activity against *Bacillus substalis* (13.50 and 12.33 mm respectively).

Acknowledgement

We are highly thankful to DBT, Govt. of India, for providing financial assistance during course of research and also deep gratitude to All India Coordinated Research Medicinal and Aromatic Plants, Department of Floriculture OUAT, Bhubaneswar.

REFERENCES

Aggarwal BB, Sundaram C, Malani N, Ichikawa Haruyo. 2007. Curcumin the Indian Solid Gold, *Advances In Experimental Medicine and Biology* 595: 1-75

Ahsan H, Parveen N, Khan NU, Hadi SM. 1999. Pro-oxidant, antioxidant and cleavage activities on DNA of curcumin and its derivatives demethoxycurcumin and bis-demethoxycurcumin.*ChemBiol Interact*.121: 161-75.

Akram M, Shahab-Uddin, Afzal A, Khan Us, Abdul H, Mohiuddin E and Asif M. (2010). Curcuma Longa AndCurcumin: A Review Article. Rom. J. Biol.–Plant Biol. 55(2). 65–70

G.K. Jayaprakasha, L. JaganmohanRao *, K.K. Sakariah. 2006. Antioxidant activities of curcumin, demethoxycurcumin and bisdemethoxycurcumin G.K. Jayaprakasha, L. JaganmohanRao *, K.K. Sakariah. Food Chemistry 98 (2006) 720–724

Hamdan AA. 2011. Bioactivity of curcumin extract against of some pathogenic strains, *Eng. and Tech. Journal*. 29 (10) 2087-2092

Jangde, CR., Phadnaik, BS. and Bisen, VV. (1998).Anti-inflammatory activity of extracts of Curcuma aromaticaSalisb. Indian Veterinary Journal, 75, 76-77.

Kapoor LD, 2001. Handbook of Ayurvedic medicinal plants.Boca Raton, FL, CRC Press, p 216.

Kumalasari I D, Harmayani E, Arsanti LL, Raharjo S, Asmara W, Kosuke N and Takuya S. 2012. Evaluation of immune stimulatory effect of the arrowroot (*Marantaarundinacea*. L) *in vitro* and *in vivo*. *Cytotechnology*.64: 131–137

Li, C. Z. (1985). Anti-inflammatory effect of the volatile oil from Curcuma aromatica. Zhong Yao Tong Bao, 10, 38-40.

M. Paramasivam, R. Poi, H. Banerjee, A. Bandyopadhyay. 2009. High-performance thin layer chromatographic method for quantitative determination of curcuminoids in Curcuma longa germplasm. Food Chemistry 113 (2009) 640–644

Maripandi, A., L. Prakash, and Ali A. Al-Salamah (2010) HPLC Separation Of Antibacterial Compounds From PernaViridis And PortunusSanguinolentus And Its Activity Tested Against Common Bacterial Pathogens. *Advanced Biotech*.9: 24-28.

Reshma V. Lawand, Santosh V. Gandhi. 2013. Comparison of Curcuma caesiaRoxb. with other Commonly Used Curcuma Species by HPTLC, Journal of Pharmacognosy and Phytochemistry 2(4): 126-131.

S.J.Kulkarni, K.N. Maske, M.P.Budre and R.P.Mahajan. 2012. Extraction and purification of curcuminoids from Turmeric (curcuma longa L.) International Journal of Pharmacology and Pharmaceutical Technology (IJPPT), ISSN: 2277 – 3436, Volume-1, Issue- 2, 2012

Yasodamma N, Chaithra D and Alekhya C. 2013.Antibacterial activity of *Curcuma neilgherrensis* WT.A Wild *Curcuma* Species.*Int. J. Pharm Pharm Sci.* 5(3): 571-576

Mukherjee, P.K., P. Balasubramanian, K. Saha, B.P. Saha and M. Pal. 1995. Antibacterial efficiency of *Nelumbo nucifera* (Nymphaeaceae) rhizomes extract. *Indian Drugs*, 32: 274-276.

Chapter 22

In-silico Comparative Sequence and Structure Analysis of Fungal Tannase for Industrial Suitability: An Environment Friendly Enzyme Technology

*Debanjan Mitra, Ishita Biswas and Pradeep K. Das Mohapatra**

Department of Microbiology, Raiganj University, Raiganj – 733 134, Uttar Dinajpur, West Bengal
**e-mail: pkdmvu@gmail.com*

ABSTRACT

Tannase are inducible enzymes and are produced from many fungi, bacteria and yeast. Though fungal tannase have wide use in industries, but it is still unclear that which genus is beneficial for industries. Tannase, an enzyme that degrades tannin and produce glucose and gallic acid. This glucose is a powerful source of energy for all organisms. The present study helps to find out which fungal tannase is more efficient for industrial applications. In this work, a detailed analysis of sequences and structures of those fungal tannase have been performed. Physiochemical parameters from their sequences have calculated. Due to lack of fungal tannase structure (except Aspergillus sp.) we create a model structure of Penicillium stipitatum (strain ATCC 10500/CBS 375.48/QM 6759/NRRL 1006) where Aspergillus oryzae (PDB ID-6G21) used as a template. Result showed that Aspergillus sp. has high number of ordered residue and low disordered residues. Due to the presence of high number of charged residues, tannase of Aspergillus sp. has high number of salt bridge and intra-protein interactions which may helps in its stability than other fungal tannase. Due to high stability, it will able to perform enzymatic reactions at higher temperatures, high salt concentration and high pH. Thus tannase from Aspergillus sp. is better for industrial uses.

***Keywords**: Tannase, Salt bridge, Disordered residues, Intra-protein interactions, Stability.*

Introduction

"Green technology" is an umbrella term that describes the use of technology and science to create products that are environmentally friendly. The goal of green technology is to protect the environment and, in some cases, repair the damage resulted in the past. After the green revolution, huge number of pesticides was used to prevent the bacterial and insect attacks in crops which lead to the environmental pollution and soil contamination. But bio-based pesticides or natural products can be the best option of chemical-based antibacterial agents. Recently used tannin-based product reduced the development of tomato bacterial wilt has been reported (Vu *et al.,* 2017).Tannins are one of the important plant secondary metabolites and are the second most abundant group of water-soluble plant phenolic compound that forms complex with other macromolecules such as proteins, carbohydrates *etc.* (Hassanpour *et al.,* 2011).

Tannase or Tannin acyl hydrolase (E.C. 3.1.1.20) is an inducible enzyme that catalyzes the hydrolysis of the ester bonds present in the gallotannins, complex tannins, and gallic acid esters (Aguilar *et al.,* 2007) to form gallic acid and glucose. The enzyme tannase is produced in the presence of tannic acid by various fungi such as *Aspergillus* sp., *Penicillium* sp., *Rhizopus* sp., *Fusarium* sp. (Cruz *et al.,* 2013, Mukherjee *et al.,* 2004, 2006) *etc.* and other tannase producers including bacteria (Mondal and Pati, 2000, Mohapatra *et al.,* 2021) and yeast (Aoki *et al.,* 1976).

Tannase is commercially applicable in the production of instantaneous tea or acorn liquor and also used as a clarifying agent in some wines, beers, fruit juices, and in refreshing drinks (Lekha and Lonsane, 1997). Gallic acid, the byproduct of tannase mediated hydrolysis of tannic acid is an important intermediary compound in the synthesis of the antibacterial drug, trimethoprim. Therefore,gallic acid has enormous uses in the pharmaceutical industry and also in the food industry (Badhani *et al.,* 2015). Gallic acid is a substrate for the chemical or enzymatic synthesis of a potent antioxidant propyl gallate (Yao *et al.,* 2007). It is also reported to increase antioxidant property in green tea (Lu and Chan, 2008).

Till date it's not clear that which fungal tannase is better for industrial uses. *In silico* study of fungal tannases is very helpful to search those quarries. Comparison between fungal tannases at both sequence and structural level is helpful to find the stable and better tannase for industrial applications. An enzyme will be better for industries when it will show stability in high temperatures, high salt concentrations or high pH. Several factors affect the rate at which enzymatic reactions precede - temperature, pH, enzyme concentration, substrate concentration, and the presence of any inhibitors or activators. In previous study, researchers have discovered that salt bridge has an important role in pertaining the thermostability (Kumar and Nussinov, 1999), salt tolerance nature (Madern *et al.*, 2000; Dym *et al.*, 1995) and high pH activity (Gallivan and Dougherty, 2000) of an enzyme. So, by calculating the number of salt bridges and intra-protein interactions, we can understand the stability of an enzyme. Salt bridge of protein can remain stable within a high range of temperature (284 K to 348 K) (Belur and Mugeraya, 2011, Mitra *et al.*, 2021c). Hydrophilicity is also an important character for enzyme activity and stability (Sinha and Trehan, 2003).

Materials and Methods

Dataset

Detailed sequence and structure analysis of tannases (tannase and its family member feruloyl esterase only), obtained from *Aspergillus* sp. and other fungal sources have performed. All 32 *Aspergillus sp.* and 13 other fungal tannase sequences were retrieved from UniProtKB database (UniProt Consortium. 2019) in FASTA format (Pearson, 1990). Structure of *Aspergillus oryzae* (PDB ID- 3WMT, 6G21) was retrieved from Research Collaboratory for Structural Bioinformatics (RCSB) protein data bank (PDB) (Rose *et al.*, 2012). Due to lack of other fungal tannase structure it is necessary to create a model for compare the stability of the enzyme.

Table 22.1: Sequence Similarity (Per cent) between *Aspergillus oryzae* (PDB ID- 6G21) Tannase and Tannase from other Fungal sp.

Organism Name	*Maximum Similarity (Per cent)*
Penicillium sp.	70.00
*Emericella*sp.	67.60
Talaromyces sp.	62.30
Neosartorya sp.	57.50
Arthroderma sp.	29.20
Collimonas sp.	26.70
Macrophomina sp.	21.20

Construction of Model

Till date there are 2 structures of *Aspergillus* sp. tannase are available in database. Sequence similarity was calculated by Expasy SIM (Gasteiger *et al.*, 2003).Tannase from *Penicillium stipitatum* showed maximum sequence similarity with *Aspergillus oryzae* tannase, so it (*Penicillium stipitatum*) was considered as a target for the preparation of model structure. The model of *Penicillium stipitatum (strain ATCC 10500/CBS 375.48/QM 6759/NRRL 1006)* was developed against the template of *Aspergillus oryzae* (PDB ID- 6G21) by Modeller 9V11 (Fiser and Šali, 2003). The model structure was validated by Verify3D (Eisenberg *et al.,* 1997) and Procheck (Laskowski *et al.,* 1993).

Multiple Sequence Alignment

All sequences were subjected to multiple sequence alignment by Clustal Omega (Sievers *et al.,* 2011). The result of alignment was analyzed by JalView (Clamp *et al.,* 2004).

Sequence Analysis

Physiochemical analysis was done by ProtParam (Gasteiger *et al.,* 2005, Mitra *et al.,* 2021b) and ProtScale (Jones, 1975). Block of sequence was made manually from MSA result. Mean Kyte-Doolittle hydrophobicity was computed from the sequences and plotted their results. For compare the data MRA (Mean relative abundance) of amino acid composition was calculated from the mean value of *Aspergillus* sp. tannase relative to other fungal tannase.

$$MRA = \frac{Mean\ value\ of\ \ Aspergillus\ sp.tannase - Mean\ value\ of\ other\ fungal\ tannase}{Mean\ value\ of\ other\ fungal\ tannase}$$

Phylogenetic Tree Construction

The result of MSA was used for phylogenetic tree construction using Fig. Tree 1.4.3 (Rambaut, 2014). The phylogenetic tree was represented in cladogram format.

Structure and Stability Analysis

All structures were minimized by UCSF Chimera (Novoselov *et al.,* 2002). Salt bridges were extracted by WHAT IF server (Variend, 1990). Intra-protein interactions were calculated by PIC server (Tina *et al.,* 2007).

Results and Discussions

Evaluations of Physico-chemical Parameters

From analysis those blocks of *Aspergillus* sp. tannase and other fungal tannase, it was found that the percentage of acidic residue *i.e.* percentage of aspartic acid and glutamic acid were higher in *Aspergilllus* sp. tannase. The percentage of basic amino acid *i.e.* percentage of histidine, lysine and arginine were also high in *Aspergillus* sp. tannase. Most interestingly, the percentage of residues forming disorder (*i.e.* serine, proline, arginine, glutamic acid) was high in case of other fungal tannase, which means it may create toxicity or any disease in living cells (Dunker *et al.,* 2001; Radivojac *et al.,* 2007). The number of residues which are shown to be properly ordered was high in case of *Aspergillus* sp. tannase. The lower value of GRAVY (grand average of hydropathy) indicates *Aspergillus* sp. tannase was more hydrophilic in nature (Kyte and Doolittle, 1982; Chang and Yang, 2013).

Table 22.2: Block Analysis for the Tannase of *Asprgillus* sp. and other Fungal Tannase with Acidic Amino Acid Percentage (AC), Basic Amino Acid Percentage (BS), Disorder Forming Amino Acid Percentage (DFA), Order Forming Amino Acid Percentage (OFA), GRAVY, Isoelectric Point (pI) and Aliphatic Index (AI)

	SEQ ID	*AC*	*BS*	*DFA*	*OFA*	*GRAVY*	*pI*	*AI*
***Aspergillus* sp. Tannase**	P78581	10.9	6.8	21.1	13.4	-0.37	4.76	65.90
	Q8WZI8	10.1	9	19.1	14.4	-0.30	5.06	71.57
	Q2UP89	9.5	8.1	19.3	14.5	-0.23	4.99	66.15
	A2R0Z6	10.3	9.0	19.2	14.5	-0.30	5.00	71.00
	Q2UMX6	9.5	9.7	17.6	13.7	-0.29	5.61	70.30
	Q0CI21	9.3	9.7	20.9	14.5	-0.27	5.64	67.61
	A1CTK4	10.9	10.6	19.7	14.1	-0.30	5.44	69.39
	B8NPT0	10.9	9.6	21.3	14.1	-0.21	5.03	74.55
	B8NPA4	9.7	9.7	21.6	13.7	-0.29	5.52	70.63
	Q0CVS2	11.4	10.4	21.6	14.6	-0.42	5.18	64.91
	A0A397H1K1	9.8	7.2	19.6	13.5	-0.08	4.64	77.65
	A0A3F3R9P0	9.3	11.2	22.3	12.6	0.006	6.80	90.15
	A0A254TYE5	8.3	7.9	19.4	10.1	-0.31	5.29	69.26
	A0A3F3R5M7	11.2	11.3	19.2	15	-0.01	5.52	77.75
	A0A254U2N0	5.5	7.8	16.6	8.3	0.38	8.79	109.06
	A0A254UDF0	14	11	19.7	7.8	-0.02	4.84	101.6
	A0A505HN07	10.4	7.2	21.4	11.1	0.12	4.65	90.03
	A0A505IHH6	11	12.4	22.1	8.9	-0.23	5.77	80.35
	A0A505IFT8	4.3	5.7	19.9	14.2	0.762	7.50	107.87
	A0A505HZY5	9.5	14.5	23.1	8.9	-0.33	8.91	86.14
	A0A505I2T4	13.7	12.2	24.1	12.6	-0.43	5.14	76.64
	A0A505HMR2	9.4	12.5	23	11	-0.02	7.90	97.43
	A0A505IF96	10.8	11.2	20.8	8.8	0.35	5.79	75.18
	A0A505I3I3	9.8	10.4	23.7	12.8	-0.20	5.89	81.89
	A0A505HXY1	9.5	9.9	24.1	9.9	0.02	5.49	89.26
	A0A254UEC7	12.8	12.5	25.5	10.3	-0.15	5.45	84.48
	A0A254TWK0	9.9	14.6	22.4	14	-0.06	9.06	92.92
	A0A254U6L8	10.8	8.4	18.6	14.7	-0.32	4.76	63.44
	A0A254UEX5	11.7	15.1	22.6	10.2	-0.26	8.88	89.96
	A0A254TT94	14.7	14.4	22.9	12.8	-0.49	5.40	68.04
	A0A254U162	11	17.2	21.3	11.3	-0.28	8.76	85.75
	A0A254TZF1	12.4	9.0	12.9	10.2	-0.26	4.56	76.69
	Total	332.3	336.2	666.6	390.5	-5.49	192.02	2563.88
	Average	10.38	10.50	20.83	12.20	-0.17	6.00	80.11

	SEQ ID	*AC*	*BS*	*DFA*	*OFA*	*GRAVY*	*pI*	*AI*
Other fungal tannase	Q4WMR0	10.3	10.2	20.5	14.3	-0.29	5.57	70.99
	Q4W8Z9	8.7	9.5	19.7	14.4	-0.21	5.85	68.2
	A1DKV3	9.7	9	20	11.06	-0.29	5.50	64.69
	A1DMV3	8.9	10	20.1	14.4	-0.23	6.07	72.49
	B0Y7U1	10.3	10	20.8	14.4	-0.28	5.57	70.99
	B0YDW9	8.9	9	19.5	15.3	-0.22	5.76	68.20
	A0A2A9L7J7	8.7	7.7	18.4	10.8	-0.04	4.98	67.95
	A0A127QSE8	8.4	8.5	19.2	10.6	-0.08	5.75	77.54
	A0A127PZK5	8.4	8.5	19.2	10.7	-0.07	5.75	77.72
	K2RDB8	14.7	11.8	30.5	7.3	-0.34	5.10	83.36
	Q5BCF8	12	9	22.9	13.5	-0.29	4.73	74.57
	B8LV47	10.8	7.2	20.8	13.2	-0.23	4.81	71.92
	D4AS70	10.8	9.8	20.5	11.0	-0.11	5.56	81.73
	Total	130.60	99.90	251.60	157.00	-1.41	73.62	950.35
	Average	10.05	7.68	20.96	12.08	-0.11	5.66	73.10

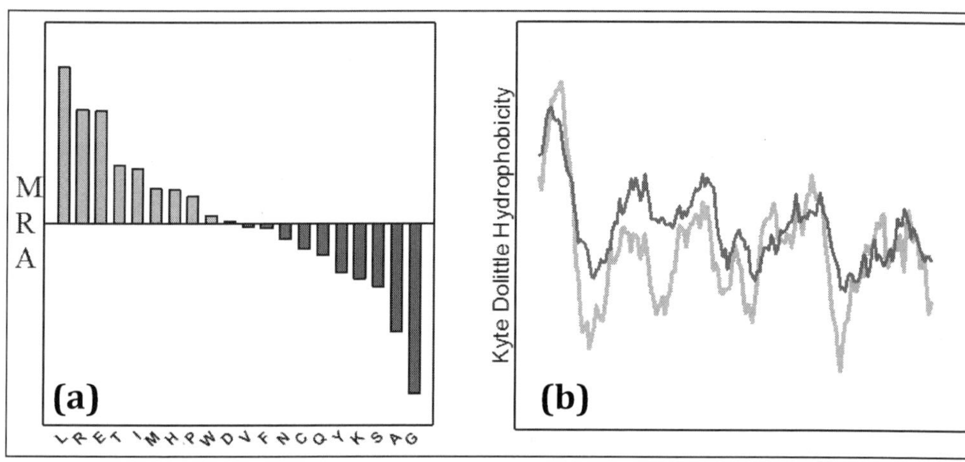

Figure 22.1: a: Mean Relative Abundance of Amino Acid Composition in Case of *Aspergillus* sp. Tannase; b: KyteDoolitle Hydrophobicity Scale of *Aspergillus* sp. Tannase (Green line) and other Fungal Tannase (Red line).

From Figure 22.1a it was clearly shown that the individual charged amino acid composition *i.e.* aspartic acid, glutamic acid, arginine, histidinehad higher abundance in *Aspergillus* sp. tannase. Leucine, threonine, isoleucine, methionine, proline and tryptophan also show positive abundance. Polar amino acid (except W, T) abundance was low in *Aspergillus* sp. tannase. KyteDoolitle Hydrophobicity (Figure 22.1b) shows that the *Aspergillus* sp. tannase (green line) was more hydrophilic in nature. That means it will more interact with water (Kyte and Doolittle, 1982; Chang and Yang, 2013).

Phylogenetic Relation Establishment

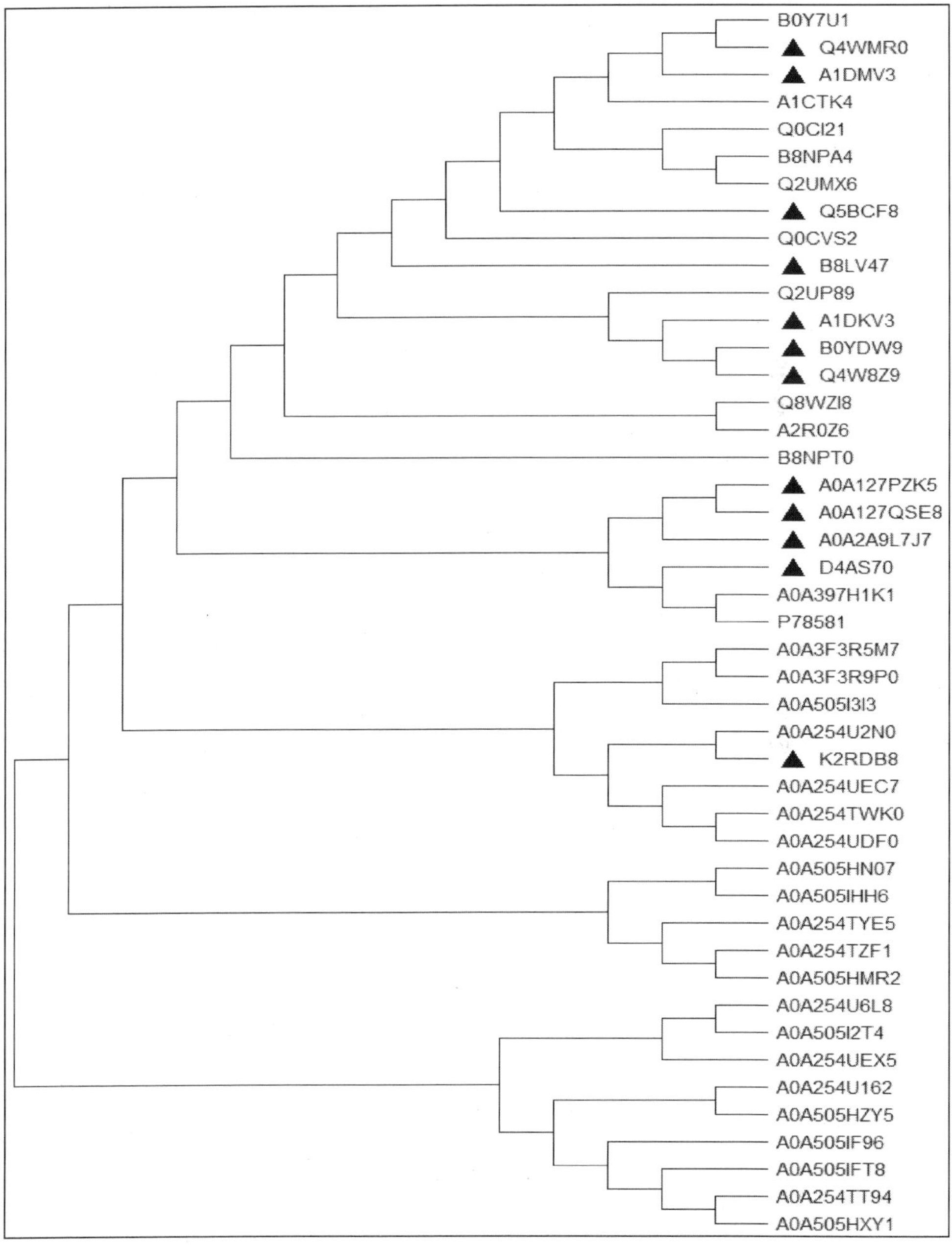

Figure 22.2: Phylgenetic Tree of *Aspergillus* sp. and other Fungal Species (Black Triangle).

Phylogenetic tree shows a close relation between *Aspergillus* sp. and other fungus. Blue line indicates *Aspergillus* sp. and red line shows other fungi that produce tannase. The other fungal tannase does not form any large separate clade, they forming small clade with those *Aspergillus* sp. clades of phylogenetic tree showing their close relation with *Aspergillus* sp. tannase.

Evaluation of Model

Sequence alignment showed that the *Penicillium stipitatum* tannase has 70 per cent sequence similarity with the template sequence of *Aspergillus oryzae* tannase. Global Model Quality Estimation (GMQE) reported a value of 0.83 which indicate that the modeled structure has higher reliability. Qualitative Model Energy Analysis (QMEAN) showed a value -0.56 that indicate the modeled structure is almost similar with the template structure. In case of Ramachandran plot, both template and modeled structure showed a similar pattern in the plot region.

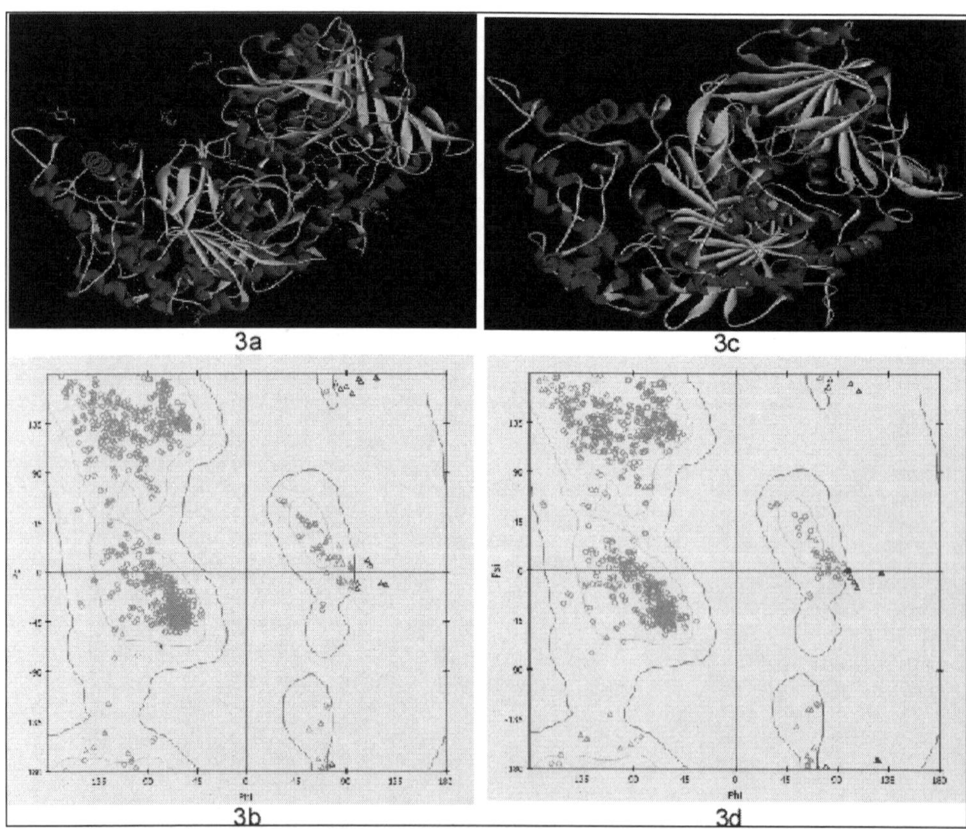

Figure 22.3: (a) Shows the structure of *Aspergillus oryzae* Tannase (PDB ID- 6G21). (b) Shows the Ramachandran Plot of 6G21. (c) Shows the Model Structure of *Penicillium stipitatum*. (d) Shows the Ramachandran plot of that Model Structure.

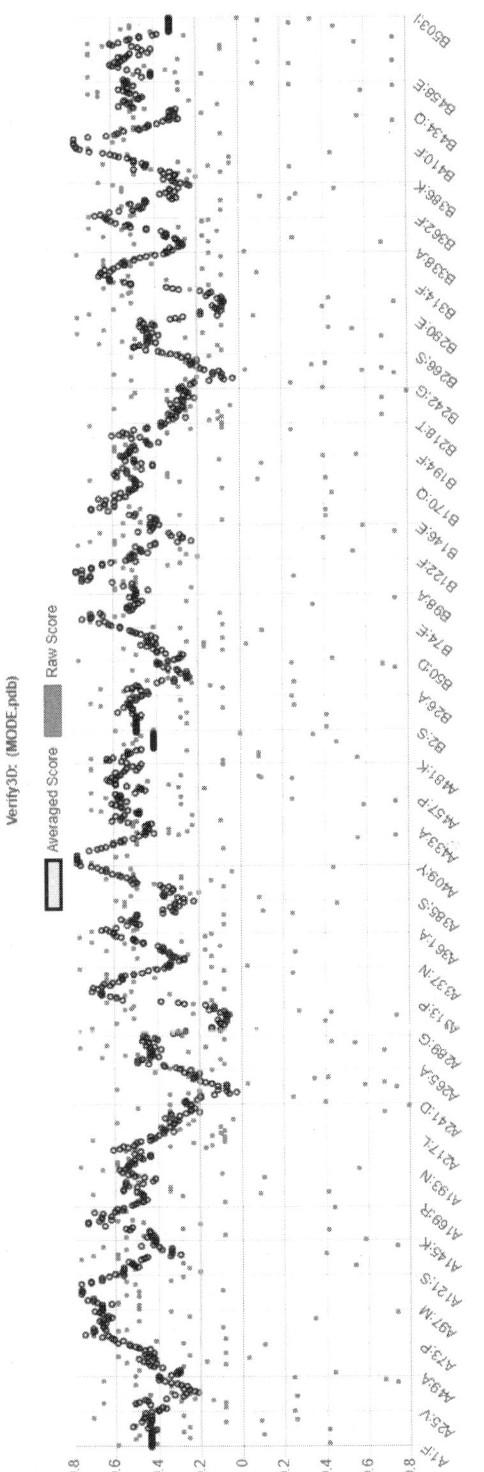

Figure 22.4: Average 3D-1D Score of that Model Structure of *Penicillium stipitatum*.

Study of Ramachandran plot (Table 22.3) showed most residues of both *Aspergillus oryzae* and *Penicillium stipitatum* structures are in favored region. *Aspergillus oryzae* showed maximum deviation 6.1 and *Penicillium stipitatum* showed maximum deviation 5.6. Overall G- factors showed a negative value which indicates that it is a good model.

Table 22.3: Properties of Ramachandran Plot of *Aspergillus oryzae* and *Penicillium stipitatum*

	Properties	*Aspergillus oryzae (PDB ID-6G21)*	*Penicillium stipitatum (Model)*
Plot area	Residues in favored region	86.80 per cent	89.70 per cent
	Residues in allowed region	11.60 per cent	9.10 per cent
	Residues in general region	1.30 per cent	1.20 per cent
	Residues in disallowed region	0.20 per cent	0.00 per cent
Residue properties	Max. deviation	6.1	5.6
	Bad contacts	0	0
G-factors	Dihedrals	-0.35	-0.24
	Covalent	-0.01	0.01
	Overall	-0.2	-0.13

The average 3D-1D score of that model structure of *Penicillium stipitatum* shows that 94.14 per cent of the total residues had an average 3D (atomic model)-1D (amino acid sequence) score of >0.2, and this results was favorable (Tong *et al.,* 2016).

Proof of Higher Stability

It is known that the salt bridge of protein helps to stabilize its three-dimensional structure in normal or extreme conditions (high temperature, high pH and high salt concentration). Both type of salt bridges *i.e.* isolated and network salt bridges were calculated. Network salt bridges were identified manually from isolated salt bridges. Intra-protein interactions also calculated to check the stability of those proteins.

Table 22.4: No. of Salt Bridges (Isolated and network), Solvation Free Energy, Cation-pi, Aromatic-Aromatic Interactions of *Aspergillus* sp. Tannase Structure (PDB ID- 6G21) and *Penicillium stipitatum tannase* (Model)

Properties	*6G21*	*model*
Isolated salt bridge	11	10
Network salt bridge	5	4
Cation-pi interactions	24	20
Aromatic-aromatic interactions	74	70

In case of *Aspergillus* sp. tannase structure (PDB ID- 6G21), it had 11 isolated salt bridges and 5 network salt bridges (Table 3) whereas the model structure of *Penicillium stipitatum* had 10 isolated salt bridges and 4 network salt bridges. It means that *Aspergillus* sp. tannase has higher stability (Kumar and Nussinov, 1999;Madern *et al.,* 2000; Mitra *et al.,* 2021d) because it is highly stabilized by its salt bridges. Intra-protein interactions like cation-pi and aromatic-aromatic interactions are high in *Aspergillus* sp.(PDB ID- 6G21). Those interactions are helpful to increase the protein stability (Serrano *et al.,* 1991; Chakravarty and Varadarajan, 2002; Prajapati *et al.,* 2006; Mitra *et al.,* 2021a).

Conclusion

Comparative study of sequences and structures between *Aspergillus* sp. and other fungal group showed that *Aspergillus* sp. has low disorder forming residue. So, it can't create any toxicity or disease in human. But other fungal tannase shown to retain high number of disorder forming residue, which is not applicable for industrial purposes. Also, *Aspergillus* sp. tannase has higher stability than other fungal tannase based on the number of salt bridges and intra-protein interactions. So the *Aspergillus* sp. tannase can be stable in high temperature, high pH and high salt concentrations. Stable enzymes would allow carrying out enzymatic reactions at higher temperatures, and therefore, increasing reaction rates and substrate solubility and reducing the risk of microbial growth and the viscosity of the reaction medium. Thus, *Aspergillus* sp. tannase is much better than any other fungal tannase in the field of industrial application.

References

Aguilar, C.N., Rodríguez R., Gutiérrez-Sánchez, G., Augur, C., Favela-Torres, E., Prado-Barragan, L.A., Ramírez-Coronel, A., Contreras-Esquivel, J.C. (2007). Microbial tannases: advances and perspectives. Appl. Microbiol. Biot.76, 47-59.

Aoki, K., Shinke, R., Nishira, H. (1976). Purification and some properties of yeast tannase. Agr. Biol. Chem. 40, 79-85.

Badhani, B., Sharma, N., and Kakkar, R. (2015). Gallic acid: a versatile antioxidant with promising therapeutic and industrial applications. Rsc Advances, 5(35), 27540-27557.

Belur P.D., Mugeraya, G. (2011). Microbial production of tannase: state of art. Research Journal of Microbiology, 6: 25-40.

Chakravarty, S., and Varadarajan, R. (2002). Elucidation of factors responsible for enhanced thermal stability of proteins: a structural genomics based study. Biochemistry, 41, 8152-8161.

Chang, K. Y., and Yang, J. R. (2013). Analysis and prediction of highly effective antiviral peptides based on random forests. PloS one, 8(8), e70166.

Clamp, M., Cuff, J., Searle, S. M., and Barton, G. J. (2004). The jalview java alignment editor. Bioinformatics, 20(3), 426-427.

Cruz, R., de Lima, J.S., Fonseca, J.C., dos Santos Fernandes, M.J., Lima, D.M., Duda, G.P., Moreira, K.A., de Souza Motta, C.M. (2013). Diversity of filamentous fungi of area from Brazilian Caatinga and high-level tannase production using mango (Mangifera indica L.) and surinam cherry (Eugenia uniflora L.) leaves under SSF. Adv. Microbiol. 3, 52-60.

Dunker, A. K., Lawson, J. D., Brown, C. J., Williams, R. M., Romero, P., Oh, J. S., Ausio, J. (2001). Intrinsically disordered protein. J. Mol. Graph. Model. 19(1), 26-59.

Dym, O., Mevarech, M., and Sussman, J. L. (1995). Structural features that stabilize halophilic malate dehydrogenase from an archaebacterium. Science, 267, 1344-1346.

Eisenberg D, Lüthy R, Bowie JU. (1997). VERIFY3D: assessment of protein models with three-dimensional profiles. InMethods in enzymology 1997 Jan 1 (Vol. 277, pp. 396-404). Academic Press.

Fiser, A., Šali, A. (2003). Modeller: generation and refinement of homology-based protein structure models. Method Enzymol. 374, 461-491.

Gasteiger, E., Gattiker, A., Hoogland, C., Ivanyi, I., Appel, R.D., Bairoch, A. (2003). ExPASy: the proteomics server for in-depth protein knowledge and analysis. Nucleic acids res. 31, 3784-3788.

Gasteiger, E., Hoogland, C., Gattiker, A., Wilkins, M. R., Appel, R. D., and Bairoch, A. (2005). Protein identification and analysis tools on the ExPASy server. In The proteomics protocols handbook (pp. 571-607). Humana press.

Hassanpour, S., MaheriSis, N., and Eshratkhah, B. (2011). Plants and secondary metabolites (Tannins): A Review. 1(1), 47-53.

Jones, D. D. (1975). ProtScale Tool: Amino acid scale: Refractivity. J. Theor. Biol, 50, 167-184.

Kumar, S., Nussinov, R. (1999). Salt bridge stability in monomeric proteins. Journal of molecular biology. 293, 1241-1255.

Kyte, J., and Doolittle, R. F. (1982). A simple method for displaying the hydropathic character of a protein. J. Mol. Biol. 157(1), 105-132.

Laskowski, R. A., MacArthur, M. W., Moss, D. S., and Thornton, J. M. (1993). PROCHECK: a program to check the stereochemical quality of protein structures. J. Appl. Crystallogr. 26(2), 283-291.

Lekha P.K., Lonsane B.K. (1997), Production and application of tannin acyl hydrolase: state of the art. Adv. Appl. Microbiol., 44: 215–260.

Lu, M.J., Chen, C. (2008). Enzymatic modification by tannase increases the antioxidant activity of green tea. Food Res. 41:130-137.

Madern, D., Ebel, C., and Zaccai, G. (2000). Halophilic adaptation of enzymes. Extremophiles, 4, 91-98.

Mitra, D., Das Mohapatra, P. K. (2021a). Cold adaptation strategy of psychrophilic bacteria: an in-silico analysis of isocitrate dehydrogenase. Systems Microbiology and Biomanufacturing, 1(4), 483-493.

Mitra, D., Dey, A., Biswas, I., Das Mohapatra, P. K. (2021b). Bioactive compounds as a potential inhibitor of colorectal cancer; an insilico study of Gallic acid and Pyrogallol. Annals of Colorectal Research, 9(1), 32-39.

Mitra, D., Mohapatra, P. K. D. (2021c). Discovery of Novel Cyclic Salt Bridge in Thermophilic Bacterial Protease and Study of its Sequence and Structure. Applied Biochemistry and Biotechnology, 193(6), 1688-1700.

Mitra, D., Pal, A. K., Mohapatra, P. K. D. (2020d). In-silico study of SARS-CoV-2 and SARS with special reference to intra-protein interactions, A plausible explanation for stability, divergency and severity of SARS-CoV-2.

Mohapatra, P. K. D., Biswas, I., Mondal, K. C., Pati, B. R. (2020). Concomitant yield optimization of tannase and gallic acid by Bacillus licheniformis KBR6 through submerged fermentation: An industrial approach. Acta BiologicaSzegediensis, 64(2).

Mondal, K.C., Pati, B.R. (2000). Studies on the extracellular tannase from newly isolated Bacillus licheniformis KBR 6. J Basic Microbio. 40: 223-232.

Mukherjee, G., Banerjee, R. (2004). Biosynthesis of tannase and gallic acid from tannin rich substrates by Rhizopus oryzae and Aspergillus foetidus. J Basic Microbio. 44, 42-48.

Mukherjee, G., Banerjee, R. (2006). Effects of temperature, pH and additives on the activity of tannase produced by a co-culture of Rhizopus oryzae and Aspergillus foetidus. World J MicrobBiot. 22, 207-212.

Novoselov, K. P., Shirabaikin, D. B., Umanskii, S. Y., Vladimirov, A. S., Minushev, A. K., and Korkin, A. A. (2002). CHIMERA: a software tool for reaction rate calculations and kinetics and thermodynamics analysis. J. comp. chem., 23, 1375-1389.

Pearson, W.R. (1990). Rapid and sensitive sequence comparison with FASTP and FASTA. Methods in Enzymology. 183, 63-98.

Prajapati, R. S., Sirajuddin, M., Durani, V., Sreeramulu, S., and Varadarajan, R. (2006). Contribution of cation-π interactions to protein stability. Biochemistry, 45, 15000-15010.

Radivojac, P., Iakoucheva, L. M., Oldfield, C. J., Obradovic, Z., Uversky, V. N., Dunker, A. K. (2007). Intrinsic disorder and functional proteomics. Biophys. J. 92(5), 1439-1456.

Rambaut, A. (2014). Figtree, a graphical viewer of phylogenetic trees [Internet]. Doi: http://tree. bio. ed. ac. uk/software/figtree.

Rose, P.W., Bi, C., Bluhm, W.F., Christie, C.H., Dimitropoulos, D., Dutta, S., Green, R.K., Goodsell, D.S., Prliæ, A., Quesada, M., Quinn, G.B. (2012). The RCSB Protein Data Bank: new resources for research and education. Nucleic acids res. 41, D475-D482.

Serrano, L., Bycroft, M., and Fersht, A. R. (1991). Aromatic-aromatic interactions and protein stability: investigation by double-mutant cycles. J. mol. bio. 218, 465-475.

Sievers, F., Wilm, A., Dineen, D., Gibson, T. J., Karplus, K., Thompson, J. D. (2011). Fast, scalable generation of high quality protein multiple sequence alignments using Clustal Omega. Mol. Syst. Biol. 7(1), 539.

Sinha, V.R., Trehan, A. (2003). Biodegradable microspheres for protein delivery. J Control. Release. 90, 261-280.

Tina, K. G., Bhadra, R., and Srinivasan, N. (2007). PIC: protein interactions calculator. Nucleic Acid. Res. 35, W473-W476.

Tong, X., Guo, M., Jin, M., Chen, H., Li, Y., Wei, J. F. (2016). In silico epitope prediction, expression and functional analysis of Per a 10 allergen from the American cockroach. Int. J. Mol. Med. 38(6), 1806-1814.

UniProt Consortium. (2019). UniProt: a hub for protein information. Nucleic acids res. 43: D204-D212.

Vriend, G. (1990). WHAT IF: a molecular modeling and drug design program. J. mol. graph. 8, 52-56.

Vu, T. T., Kim, H., Tran, V. K., Vu, H. D., Hoang, T. X., Han, J. W., Kim, J. C. (2017). Antibacterial activity of tannins isolated from Sapiumbaccatum extract and use for control of tomato bacterial wilt. PloS one, 12(7), e0181499.

Yao, J., Guo, G.S., Ren, G.H., Liu, Y.H. (2014). Production, characterization and applications of tannase. J. Mol. Catal. B Enzym. 101, 137-147.

Chapter 23

Metagenomics Analysis of the Bacterial Community Structure and Function in the Sukinda Chromite Mine Soil

Sukanta Kumar Pradhan[1] *and Hrudayanath Thatoi*[2]*

[1]*BIF Centre, Department of Bioinformatics, Odisha University of Agriculture and Technology, Bhubaneswar – 751 003, Odisha*

[2]*Director, Centre for Similipal Studies & Coordinator(DBT- Bioinformatics Infrastructure Facility) Maharaja Sriram Chandra Bhanjdeo University (formerly North Orissa University), Sriram Chandra Vihar,Takatpur, Baripada-757003,Odisha, India*

**e-mail: hn_thatoi@rediffmail.com*

Introduction

Excessive mining operations lead to accumulations of heavy metals in the terrestrial environment, which has become an increasingly serious environmental problem posing threat to mining sites. The presence of high concentration of toxic heavy metals in mining soils affects both the structure and function of microbial communities present in that ecosystem. Several studies conducted on mining area reported that, toxic heavy metals have detrimental effect on soil microbial biomass (Brookes and McGrath 1984; Chander *et al.,* 1995), thereby leading to a decrease in their functional diversity (Kandeler *et al.,* 1996) and shifts in microbial community structure (Frostegård *et al.,* 1993; Gadd 1993; Elsgaard *et al.,* 2001; Sandaa *et al.,* 2001; Akmal *et al.,* 2005). Heavy metals such as Chromium (Cr),

Aluminum (Al), Lead (Pb), Cadmium (Cd), Mercury (Hg), Silver (Ag), and Gold (Au) are usually toxic to living organisms including microorganisms (Siddiquee *et al.*, 2015; Lakherwal *et al.*, 2014, McLean and Beveridge, 2001; Turpeinen *et al.*, 2002). For example, Cr and Cd are capable of inducing oxidative damage and denaturation of microorganisms (Bernard *et al.*, 2018). Chromium contamination found to decrease the taxonomic species of microbes in soil and change the soil microbial composition (Pei *et al.*, 2018). Copper causes injury to cytoplasmic molecules, DNA, lipids and other proteins (Giner *et al.*, 2014 and Osman and Cavet, 2008) and Aluminum stabilizes superoxide radicals and is responsible for DNA damage (Booth *et al.*, 2015). Similarly, Lead exhibit damage in cell membrane and DNA structure (Olaniran *et al.*, 2013). However, there are evidences that, microorganisms also exhibit resistance and can able to survive under heavy metal contamination (Bruins *et al.*, 2000). The study conducted by De Jayasankar and his co-workers identified some bacterial species from metal contaminated environment that can confer resistance towards toxic heavy metals (Hg, Cd, and Pb) (Ramaiah and Vardanyan, 2008). Similarly, other researchers have also isolated bacteria from different heavy metal contaminated environment which exhibit resistance towards Cr, Zinc (Zn), Cd, Pb, Copper (Cu), Nickle (Ni) and, Co (Abioye *et al.*, 2018; Ashruta *et al.*, 2014 and Joo *et al.*, 2015). Thus, knowledge of the bacterial diversity and their activities in chromate contaminated soil are important for understanding function of chromite mine soil ecosystems (Miao *et al.*, 2015; Sheik *et al.*, 2012).

The chromite ore belt at Sukinda valley spreading over an area of approximately 200 sq. km. in the Jajpur district of Odisha (India) is a well-known chromite hub in the world (Das *et al.*, 2013). It is estimated that, approximately 183 million tonnes of chromite deposit is located in Sukinda valley (Das *et al.*, 2013). Due to its economic reason, the extraction of chromite ore has been taken up expeditiously by the State Government through different public and private agencies. The extensive mining of chromite in Sukinda valley over the past several years has resulted in serious heavy metal pollution of soil and water environments which has posed threat to all living organisms including microorganisms in and around the mining sites.

Chromium is a transition metal with atomic number 24 and is known to be a toxic in nature. The oxidation state of chromium determines its level of toxicity inside the medium. Chromium exists in several oxidation states (-2 to -6) of which Cr (VI) and Cr (III) are found in stable forms in the natural environment. Cr (VI) is highly soluble, thus mobile, toxic and strong oxidizing in nature (Das *et al.*, 2014; Chandra and Singh, 2014). Whereas, its lower oxidation state Cr (III) is comparatively less toxic and impermeable to the biological membrane by forming insoluble oxides and hydroxides above pH 5.0 (Das *et al.*, 2014; Chandra and Singh, 2014; Pal and Paul, 2004; Megharaj *et al.*, 2003). In chromite mining areas, Cr (VI) contamination is a serious environmental problem due to the leaching of heavy metals. Micro-organisms that can tolerate such toxic metals can able to survive in that environment. It is also reported that, geochemical cycling of chromium is possible through microbes under both aerobic (Campos *et al.*, 1995) and anaerobic (Marsh *et al.*, 2001) conditions. Microbes observed in such a contaminated environment showed resistance not only towards Cr but also well to other heavy metals (Bruins *et al.*, 2000). Assessment of those heavy metal resistant bacterial communities available under such a hostile environment is important from bioremediation point of view, which can be made by culture-dependent and culture-independent methods (Schmitt *et al.*, 2012; Andreote *et al.*, 2012). Several studies had been undertaken by different researchers who have identified a number of chromium resistant bacteria from the Sukinda chromite mine environment through the culture-dependent method and evaluated the bioremediation potential of few selected bacteria. Some of the bacteria isolated from Sukinda chromite mine soil and water samples and evaluated for their Cr(VI) resistance include *Bacillus amyloliquefaciens* (Das *et al.*, 2013), *Bacillus* sp. (Dhal *et al.*, 2013), *Arthrobacter* sp. and *Pseudomonas* sp. (Dey *et al.*, 2013), *Bacillus subtilis* VITSUKMW1, *Acinetobacter junii* VITSUKMW2, and *Escherichia coli* VITSUKMW3 (Samuel *et al.*, 2012), *Enterobacter cloacae* SUKCr1D (Harish *et al.*, 2012), *Lactobacillus acidophilus, Staphylococcus lactis, Lactobacillus sp.* and *Propionibacterium acnes* (Acharya *et al.*, 1998), *Brevibacterium* sp. (Das and Mishra, 2010), *Staphylococcus aureus* (Mishra *et al.*, 2009) *etc.* However, no attempt has been made so far to assess the bacterial diversity through a culture-independent method in the Sukinda chromite mine area although it is estimated that only 0.1-1 per cent of soil bacteria are readily cultivable (Hugenholtz *et al.*, 1998; Handelsman *et al.*, 2004), while the rest are uncultivable. Thus, the wealth of information obtained from the deep metagenomic sequencing makes it possible to capture the genomic information of low-abundance populations and to reveal the multiple activities in soil. In recent years, the metagenomics study has become crucial for understanding the structural and functional activities of microbial communities in different environmental conditions (Simon and Daniel, 2011). Several methods of metagenomics studies are Functional/Phenotypic Screens, SIP, RFLP, DGGE, and TGGE, Probe-based Technologies, Gene and Protein Expression Analysis, Substrate-Induced Gene Expression (SIGEX), Single-Cell Analysis and NGS (Next Generation Sequencing) are used to study the unculturable microorganisms. Technological progress in high-throughput sequencing technologies and the development of robust bioinformatics analysis tools enable us to generate metagenome from the soil. This knowledge discovery is based primarily on recent developments in DNA sequencing technology, which have made it possible to generate millions of sequence reads quickly and economically using NextGen sequencing and have revolutionized the study of microbial diversity, adaptation, and evolution (Handelsman *et al.*, 2007; He *et al.*, 2007; Riesenfeld *et al.*, 2004). The results of high throughput sequencing have clearly indicated that the assessment of microbial diversity through the high-throughput techniques is 100 times more as compared to the traditional culture-dependent methods (Schmitt *et al.*, 2012; Godoy *et al.*, 2012 and Sogin *et al.*, 2006). The 16S rRNA gene is used for phylogenetic studies as it is highly conserved between different species of bacteria and archaea. It is suggested that 16S rRNA gene can be used as a reliable molecular clock because 16S rRNA sequences from distantly related bacterial lineages are shown to have similar functionalities. So, among the various sequencing technologies, the 16S rRNA hypervariable region-specific amplicon studies based on NGS (Next Generation Sequencing) techniques have the potential to unfold the microbial diversity present in the environment (Gilbert *et al.,* 2011 and Andreote *et al.,* 2012). Studies of microbial communities from several environmental samples like acid-mine drainage (Tyson *et al.,* 2004), marine water and sediments (DeLong *et al.,* 2006), human gut (Gill *et al.,* 2006; Turnbaugh *et al.,* 2007; Jovel *et al.,* 2016) and soils (Smets and Barkay, 2005; Voget *et al.,* 2006 and Gandhi *et al.,* 2015) have yielded novel insights on community structure, function, evolution, gene discovery and metabolism. More particularly study had also been conducted to analyze the bacterial structure in heavy metal (Ni, Co, Cd, Cu, Zn, and Fe) contaminated soils (Harichova *et al.,* 2012; Bajkiæ *et al.,* 2013 and Go³êbiewski *et al.,* 2014). In addition to study the taxonomy and diversity of microorganisms, the 16S rRNA amplicon sequencing approach can also be useful to explore the potential metabolic pathways and community structures present in the environment.

Keeping these in view, the present study highlights a detailed insights of the bacterial community structure and function of Sukinda chromite mine of Odisha, India using Illumina HiSeq based sequencing study.

Materials and Methods

Sample Collection and Soil Physico-chemical Analysis

Soil samples were collected from different sites of *in situ* mine area of Sukinda chromite mine of state of Odisha, India. The sampling areas were not coming under protected area and did not need any special permission for the study. The sampling sites represent high level of chromium contamination. Soils were collected from a depth of 10-20 cm without any vegetation and then individual sample was mixed to homogeneity mixture to generate a representative composite soil sample for metagenomics. Then the prepared composite sample was transferred to laboratory in sterilized sealed plastic bags on ice (4°C) and stored at -80°C for metagenome isolation, soil physico-chemical properties characterization and heavy metal concentration detection. The physico-chemical properties of the soil samples were estimated at the beginning of the experiment in order to have a basic idea of the quality of the soil samples prevailing in the chromite mine area. Soil parameters were analysed based on different methods for heavy metal concentration (PinAAcle 900F, Perkin Elmer, USA), pH, EC, organic carbon (Flash, 2000; Thermo Scientific, USA), available N, P, and K (Subbiah and Asija, 1956; Bray and Kurtz., 1945; Piper., 1966).

DNA Extraction from Contaminated Soil Sample

DNA from the *in situ* soil sample was isolated using Power Soil® DNA Isolation Kit (MO-BIO, USA). Then the equal concentration (= 200 μg) of environmental DNA from the sample, 10 composite pool were mixed to form a composite genetic pool representing total DNA composition for the site. The purity and quantity of nucleic acids present in the extracted DNA was checked using NanoDropTM 2000

Spectrophotometer (Thermo Scientific) by taking aqueous solutions of nucleic acids (1 μl each) against Nuclease Free Water (1 μl) as blank. Further, the exact concentration of dsDNA present in the extracted DNA was checked using Qubit® 2.0 Fluorometer (Thermo Fisher Scientific). This instrument neglects the single nucleotides and quantifies only long DNA strands. Isolated total DNA was stored at -20 °C until further processing.

Preparation of 2x250 HiSeq Libraries

The NEBNext Ultra DNA Library Prep Kit of New England Biolabs, Inc. (NEB) for Illumina contains enzymes and buffers such as NEBNext End Repair Reaction Buffer (10X), NEBNext End Prep Enzyme Mix, Blunt/TA Ligase Master Mix, NEBNext Ligation Enhancer and NEBNext Q5 Hot Start HiFi PCR Master Mix were used to convert a small amount of DNA input into indexed libraries for next-generation sequencing on the Illumina platform (Illumina, Inc). The reagents for each step in the library preparation workflow have been reformulated to enable high yield preparation of high-quality libraries from 500 picograms to 1 microgram of input DNA. This new generation of NEBNext reagents uses a fast, streamlined, automatable workflow and enables use of fewer PCR cycles while also improving GC coverage. The DNA is fragmented, end repaired and A tailed and adapters are ligated, after gel selection an enrichment is performed with an indexed primer. The quality and quantity of the library is validated and sequenced in HiSeq 1000.

Read Processing and Assembly

The sequencing reads were checked for base quality score distribution, sequence quality score distribution, average base content per read, GC distribution in the reads, PCR amplification issue, check for over represented sequences, adapter contamination. Then the sequence reads were trimmed using Cutadapt (version - 1.8.1) to retain only high-quality sequence for further analysis and the low-quality sequence reads were excluded from the analysis. The *de novo* assembly of the adapter trimmed fastq files was carried out using Ray Meta assembler (v2.3.1). Ray Meta is a massively distributed metagenome assembler that is coupled with Ray Communities in which profiles microbiomes based on uniquely-colored k-mers. Further contigs were linked by the assembly algorithm to create scaffolds. N50 contigs were kept as good assembly. Bad or mis assemblies were removed from the result. Assembly was performed with default k-mer length (31-size) using de-bruijn graph method. In-house PERL and Python code were used to parse the fastq files for the downstream analysis.

Taxonomic Annotation

The taxonomic classification was carried out through BLASTN analysis against SILVA, SSUref and LSUref databases with an e-value of 1e–5 (Urich *et al.,* 2008) followed by annotation of BLAST output files using MEGAN (Huson *et al.,* 2007). The lowest common ancestor algorithm in MEGAN assigns rDNA or rRNA sequences for their lowest common ancestor in the taxonomy from a subset of best scoring matches in the BLAST result (cut-off: BLAST bit score 86, relative cut-off: 10 per cent of the top hit) (Urich *et al.,* 2008). As because random sequence reads exhibit very different levels of evolutionary conservation so all ranks of the NCBI taxonomy, placing more conserved sequences higher up in the taxonomy (*i.e.* closer to the root) and more distinct sequences onto nodes that are more specific (*i.e.* closer to the leaves, which represent species and strains). The taxonomy tree was generated based on neighbour-joining method using MEGAN software.

ORF Prediction and Functional Assignment

The contigs obtained from the assembly were used as input to Meta Gene Annotator (MGA) for the prediction of open reading frames (ORFs). The gene-finding program MGA has identified the coding regions and distinguished them from non-coding DNA. The predicted ORFs were taken for the functional annotation and taxonomic classification. Then the ORFs were searched against the non-redundant (NCBI-nr) database using DIAMOND (v0.7.9.58). DIAMOND is a new high-throughput program for aligning a file of short reads against a protein reference database such as NR, at 20,000 times the speed of BLASTX, with high sensitivity. The alignment file along with the putative ORFs were used for functional annotation using MEGAN5 (MEtaGenome ANalyzer) software. MEGAN5 is a fast and easy tool for metagenomic data analysis and can be used to analyze and compare metagenomic and metatranscriptomic data, both at taxonomically and functional level.

Statistical Analysis

Fisher's exact test (Fisher, 1922) tool at Cenargen Bioinformatics platform was used to compare and find out the levels and significance of contig expression between generated libraries which had passed through quality control.

Results

Sample Characterization

Sample collected from active chromite mine soil of Sukinda contains various heavy metals apparently at a higher concentration such as Mn (5444 mg kg^{-1}), Fe (5418 mg kg^{-1}), Pb (4713 mg kg^{-1}), Cr (2993 mg kg^{-1}), Ni (2278 mg kg^{-1}), and Co (1548 mg kg^{-1}). Other soil physico-chemical parameters analyzed include pH (7.32), EC (0.32 ds/m), organic carbon (0.19 per cent), available N (81.53 mg ha^{-1}), available P(62.35 mg ha^{-1}) and available K (27.05 mg ha^{-1}) (Table 23.1).

Overview of Metagenome Sequencing

The library was sequenced (2 x 250 bp chemistry) using Illumina Hiseq 1000 with mean size of 550 bp. The metagenome dataset consists of 10,349,864 paired end reads which corresponded to 2.1 GB data consisting of 5.2G bases. 36.21 per cent of the total reads were having phred score >=30. There were total 907,765 numbers of contigs generated after assembly of reads. After appropriate filtration and removable of bad/mis assemblies, total of 780,534 nos of filtered contigs were generated with different contig lengths such as 239,701(100-200 bps), 452,040(201-300 bps), 52,530(301-400 bps), 19,509(401-500 bps) and 16,754(>=500 bps).Then the processed reads were submitted to NCBIs' Sequence Read Archive (SRA) database for further references with SRA accession SRP148083 (https://www.ncbi.nlm.nih.gov/sra/?term=SRP148083).

Table 23.1: Physico-chemical Analyses of Chromite Mine Soil, Sukinda, Odisha

Parameters	Value
pH	7.32±0.11a
EC (dS/m)	0.32±0.03a
Organic carbon (per cent)	0.19±0.01c
Avail. N (kg ha^{-1})	81.53±3.23c
Avail. P (kg ha^{-1})	62.35±4.21a
Avail. K (kg ha^{-1})	27.05±2.54c
Fe (mg kg-^{1})	5418.02±34.54a
Cr (mg kg-^{1})	2993.68±45.83a
Co (mg kg-^{1})	1548.50±23.65c
Ni (mg kg-^{1})	2278.01±21.22a
Cu (mg kg-^{1})	ND
Zn (mg kg-^{1})	225.79±10.23a
Pb (mg kg-^{1})	4713.01±31.21a
Cd (mg kg-^{1})	5.15±0.23a
Mn (mg kg-^{1})	5444.01±23.32b

Values represent means ± standard error (n=3). Different letters in the rows indicate significant difference among the treatments at Tukey HSD ($p<=0.05$).

Bacteria Community Structure

Taxon abundance and their assignment to different functional categories are considered to be critical in whole metagenomic study. The taxonomic profiling of the metagenomic sample provided an insight into bacterial community structure of the chromium contaminated mine soil collected from Sukinda chromite mine. The taxonomic tree at phylum level and species level showed the hierarchy of comparative taxonomic abundance in the sample based on contig abundance (Figure 23.1). Bacteria were the most abundant among the three domains and accounted for more than 90 per cent of the microbial population of the metagenome. From the Figure 23.1 it can be inferred that *Proteobacteria*, *Actinobacteria* and *Bacteroidetes* were found to be highly enriched. Phylum level taxonomic hits distribution of the mining residue sample showed that *Proteobacteria* were the most abundant (66.45 per cent) followed by *Actinobacteria* (17.32 per cent), *Bacteroidetes* (4.65 per cent),

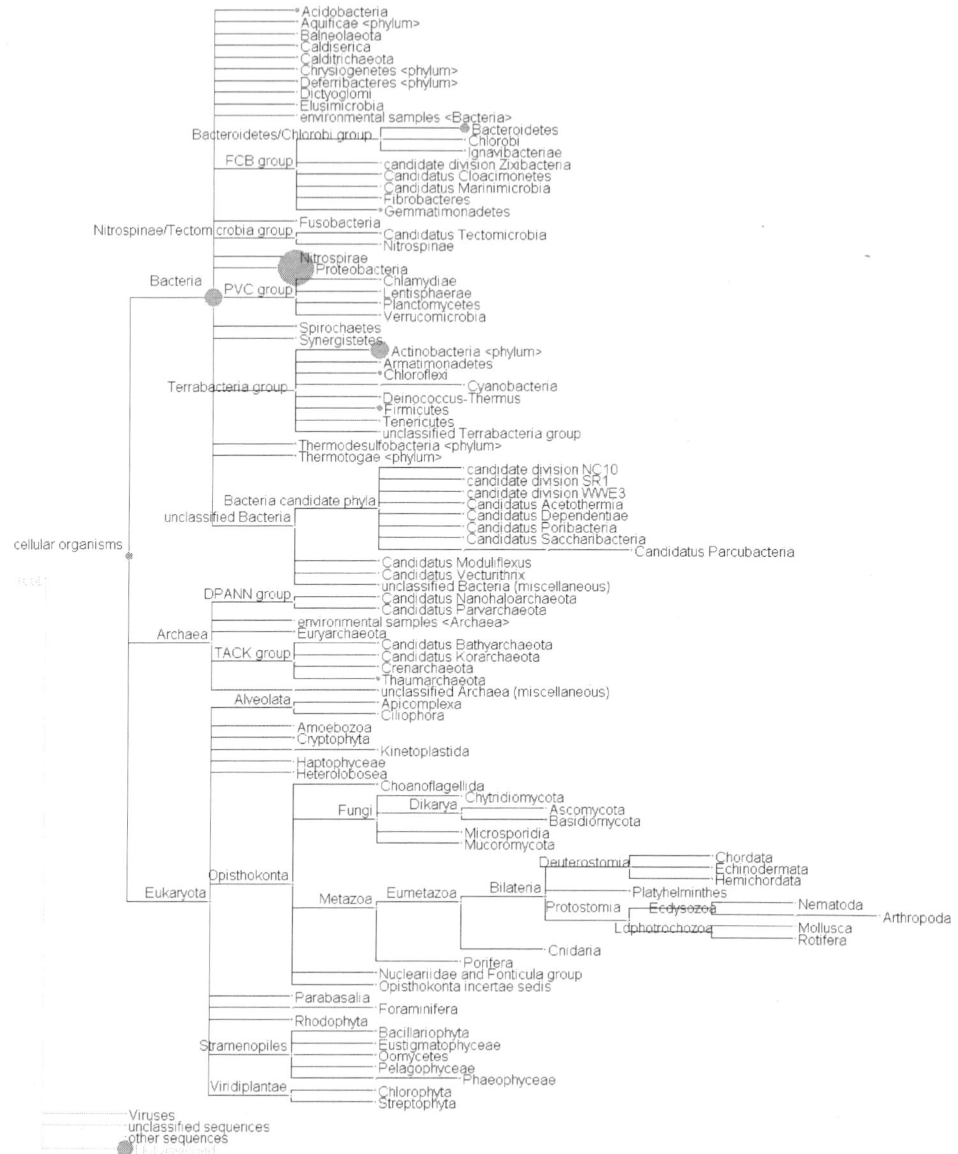

Figure 22.1: Phylogenetic Tree at Phylum Level Constructed for *in situ* Soil.

Firmicutes (1.78 per cent), *Nitrospirae* (1.47 per cent) and *Acidobacteria* (1.375 per cent) (Figure 23.2). Other phyla like *Chloroflexi, Gemmatimonadetes, Planctomycetes, Thaumarchaeota, Basidiomycota* and *Deinococcus-Thermus* constituted ~4 per cent of the whole metagenome (Figure 23.2). Among the *Proteobacteria* phyla, *Alpha proteobacteria* (25.47 per cent) was found to be most abundant class followed by *Betaproteobacteria* (23.61 per cent), *Gammaproteobacteria* (9.71 per cent) and

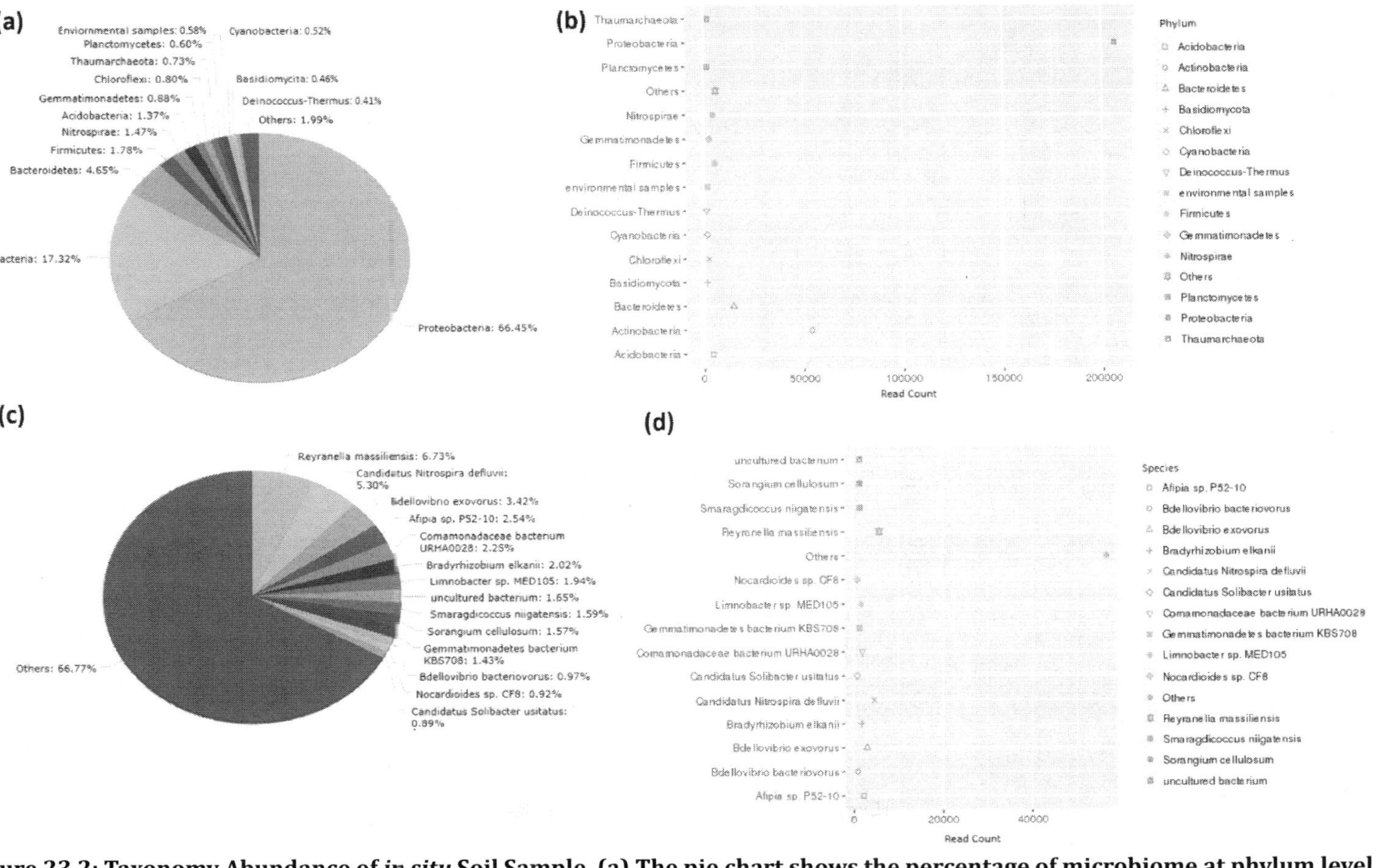

Figure 23.2: Taxonomy Abundance of *in situ* Soil Sample. (a) The pie chart shows the percentage of microbiome at phylum level (b) The scatter plot represents the abundance level of phylum with respect to read count (c) The pie chart shows the percentage of various species reported in the sample (d) The scatter plot shows the read abundance at species level.

Deltaproteobacteria (7.72 per cent). *Bradyrhizobium* was found to be the most predominant genus with 15796 hits followed by *Bdellovibrio* (9809), *Acinetobacter* (6457), *Reyranella* (5695), *Burkholderia* (5388) and *Pseudomonas* (5240). Bacteria like *Nitrospira, Sediminibacterium, Nocardioides, Propionibacterium etc.* were the other members as revealed from the metagenome of the chromite mine soil.

Rarefaction Curve

Rarefaction curve was generated by comparing the species abundance based on number of leaves in the taxonomy and number of sequences occurred. The curves represent the average number of different species annotations for the complete dataset. The curve is made for all taxa include Bacteria, Archaea, Eukaryote, Viruses, unclassified and other sequences. The plot (Figure 23.3) showed the rarefaction curve of annotated species richness in the metagenome. The steep slope on the left of the curve represented a large fraction of the species diversity existed and the curve pointing towards the right meant that a reasonable number of individuals were sampled; more intensive sampling is likely to yield few additional species. Initially, the sampling curve rose very rapidly but gradually leveled off indicating fewer new species per unit of individuals collected.

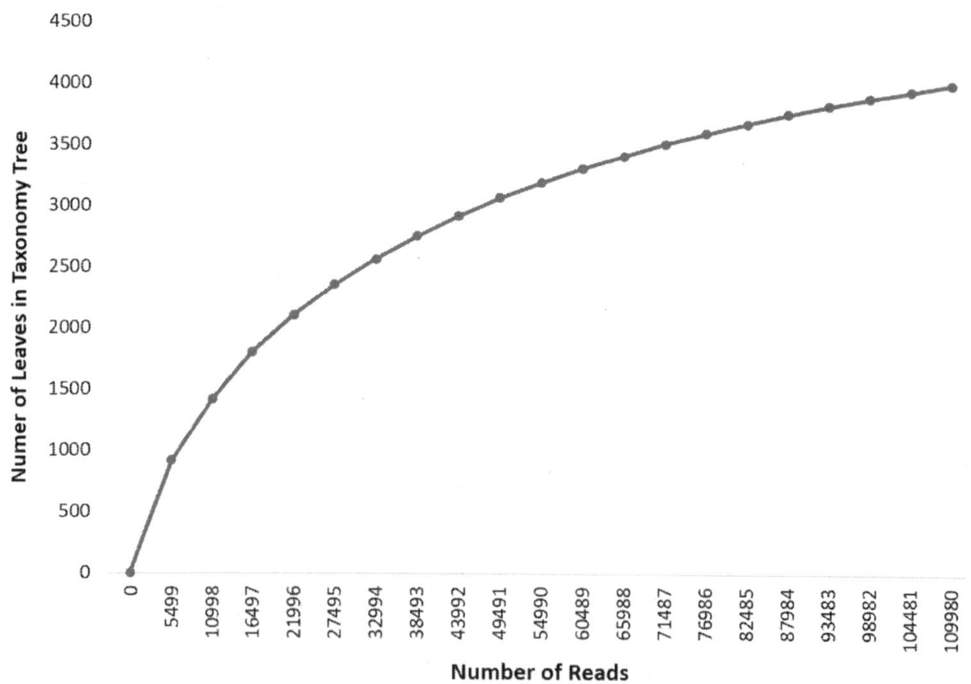

Figure 23.3: The Alpha Rare Faction Studies. The curve plot shows the observed

taxonomic unit of bacterial diversity distributed in *in situ* soil sample.

Figure 23.4: Microbial Attributes of *in situ* Soil Sample.

KEGG and SEED Functional Category Hits Distribution and Analysis

KEGG pathway analysis was performed for each contig sequences by assigning KEGG Orthology (KO) numbers obtained from the known reference hits. Further, enzyme and pathway information were assigned to contigs based on KO. The pathway classifications were represented as bubbles in bubble plot shown in Figure 23.5 (b). Each bubble from the plot illustrated the distribution of different functional categories at the highest level supported by the functional hierarchies. The protein regulating the metabolic functions were higher in number which comprises about 20 per cent of the whole metagenome (88470 reads) followed by the proteins regulating the genetic information processing which were 5.21 per cent of the metagenome with 23036 hits and environment information processing 5.13 per cent with 22671 hits. Further, proteins responsible for cellular processes (7277), human diseases (5106) and organismal systems (3831) constituted ~4 per cent of the whole metagenome. The SEED functional category has 112542 predicted metabolic functions. Out of which 11.73 per cent are involved in carbohydrate metabolism with 13202 hits followed by 10.57 per cent Amino Acids and Derivatives with 11899 hits, 8.24 per cent protein metabolism with 9282 hits, 6.88 per cent

DNA metabolism with 7746 hits and 5.94 per cent Cofactors, Vitamins, Prosthetic Groups, Pigments with 6687 hits are shown in (Figure 23.5a).

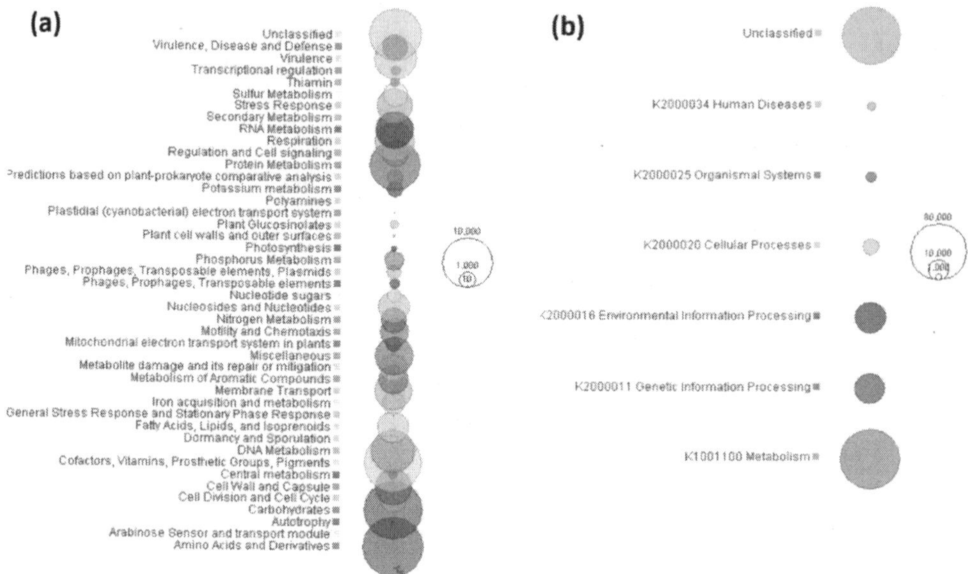

Figure 23.5: Bubble Plot Showing Functional Analysis of Predicted ORFs Obtained after Metagenome Assembly. Larger bubble size indicates more numbers of ORF assigned to the functional groups. The size of the circle is scaled logarithmically to represent the number of ORFs assigned directly to the functional categories. (a) Bubble plot shows the SEED based gene ontology (b) Primary pathway summary as a result of KEGG annotation.

Discussion

Metagenomics has the potential to improve our knowledge by study the genetic components of uncultured microorganisms (Singh A. H. *et al.,* 2009; Mirete *et al.,* 2015; Chauhan *et al.,* 2017; Yadav *et al.,* 2017). The techniques of metagenomics have been used to explore soil microbes for its composition and understanding its functional mechanism at molecular level that allowing successful adaption of microbes in mining environment. Mining activity has greater impact on environment because its process releases toxic metal such as Cr, Fe, Mn and Ni in soil. The microbial community plays an important role in the soil system, particularly in extreme ecosystems where they are responsible for the majority of organic matter decomposition. The dataset presented in this study is a taxonomic and functional characterization of the microbial community in the largest chromite mine area using whole metagenomic approach.

The *in situ* mine soil having high concentration of various heavy metals such as Mn, Fe, Cr, Cu and low concentration of organic carbon, available nitrogen, phosphorous and potassium. The bacterial consortia dwelling in the mine soil sample exhibited resistant to Cr along with other heavy metals. The maximum percentage (35.96 per cent) of Cr(VI) removal was occurred in case of 250 mg/l of initial Cr(VI) concentration as compared to higher (>250 mg/l) Cr(VI) concentration. This metal tolerant ability of these bacterial consortia may be due to the presence of inherent mechanisms evolved while growing in such metal contaminated environment. This study is supported by earlier research conducted by different workers (Das *et al.*, 2013; Dhal *et al.*, 2013; Dey *et al.*, 2013; Samuel *et al.*, 2012; Kumar *et al.*, 2014). Taxonomic analyses revealed that bacteria were the most abundant among the three domains accounting more than 90 per cent of the microbial population of the metagenome. Among the bacterial community Proteobacteria groups were the most abundant division comprising 66.45 per cent followed by *Actionbacteria* 17.32 per cent. Among the *Proteobacteria* phyla, *Alpha proteobacteria* was found to be most abundant class followed by *Betaproteobacteria*, *Gammaproteobacteria* and *Deltaproteobacteria. Bradyrhizobium* was found to be the most predominant genus followed by *Bdellovibrio, Acinetobacter, Reyranella, Burkholderia* and *Pseudomonas*. Bacteria like *Nitrospira, Sediminibacterium, Nocardioides,* and *Propionibacterium etc.* were the other members as revealed from the metagenome of the chromite mine soil. This finding is in accordance with other studies on bacterial community soil (Pradhan *et al.*, 2016; Reis *et al.*, 2013; Gremion *et al.*, 2003; Vishnivetskaya *et al.*, 2011) in which the most dominant phylum found to be *Proteobacteria*. Among the *Proteobacteria*, *α-proteobacteria, β-proteobacteria* found to be abundant class. Study suggested that the abundant of the *Proteobacteria* and *Actinobacteria* phyla is significantly correlated with metal contaminated environment. The predictive functional profiling of the bacterial communities of *in situ* chromite mine soils showed that "metabolism" was the most dominant category, of which "carbohydrates metabolism", "amino acid metabolism", and" DNA metabolism" were the top three abundant pathways. "Carbohydrate metabolism" involves a complex series of enzymatic steps to convert external substrate into metabolic precursors, such as acetyl-CoA, pyruvate, and d-fructose-6-phosphate, and it involves a hydrolysis reaction in which polymers are hydrolysed to monomers (Guo *et al.*, 2015; Gao *et al.*, 2016). In the "amino acid metabolism" pathway, which is commonly used by bacteria for osmoregulation, soils have been exposed to frequent moisture stress, mainly in mine area (Fierer *et al.*, 2012; Monard *et al.*, 2012). DNA metabolism in which cellular deoxyribonucleic acid (DNA) is maintained, and includes both DNA synthesis and degradation reactions involved in DNA replication and repair of damaged DNA due metal toxicity (Majka and Burgers, 2004; Moldovan *et al.*, 2007).

Conclusion

Environment plays a key role in regulating the microbial community structure and their adaptations in the contaminated mining areas. Metagenomic analysis of the bacterial diversity and their functional annotation in the chromite mine deposits was investigated in order to assess their role in showing resistance to toxic Cr (VI). The metagenome sequence based analysis revealed the dominance of bacteria over other domains in the contaminated site. Taxonomic analyses revealed that bacteria were the most abundant among the three domains and among the bacterial community *Proteobacteria* were the most abundant phylum in the chromite mine soil. Functional profiling of the bacterial communities of *in situ* chromite mine soils showed "carbohydrate metabolism", "amino acid metabolism", and "DNA metabolism" pathways carried out and confer resistance to chromate. The results reported here expand the current knowledge of the bacteriall taxonomic in chromite mining areas; data may be a relevant contribution to the exploration of natural bioremediation strategies in mining and surrounding areas.

Chapter 24

Effects of Three different Types of Auxins (IAA, NAA and IBA) on Rooting in Linseed

Sadashiba Gochhait and Umaballava Mohapatra*

P.G. Department of Botany,
North Orissa University, Baripada – 757 003, Odisha
**e-mail: sadashiba.nou@gmail.com*

ABSTRACT

Linum usitatissimum L. is one of the valuable crops, having immense value in industries, medicine and others. Being an important crop, its genetic upgradation is essential and for this in vitro propagation is a prerequisite. In vitro callogenesis, shooting and rhizogenesis were achieved from hypocotylexplants. The explants were inoculated on MS medium supplemented with 1.0mg/l of IAA and 1.5mg/l of BAP. Auxin is known to involve in the rooting process. Both concentration and type of auxin markedly influenced the number of root formation per stem cutting. After six weeks of initial culture, regenerated shoots were excised and aseptically transferred on to MS medium supplemented with three different auxins (IAA, NAA and IBA) at four different concentrations (0.5mg/l, 1.0mg/l, 1.5mg/l and 2.0mg/l). Their effects on rooting were evaluated individually and a comparison was made among these. It was found that all the three types of auxins have potential for rooting but IBA at concentration 1.5mg/l was found to be the most effective for rooting from the regenerated shoots of six different linseed cultivars.

Keywords: *Linseed, Rooting, In vitro, Auxin.*

Introduction

Linseed (*Linum usitatissimum* L.) is an erect annual herbaceous plant 30-120cm. in height with slender glabrous, grayish green stem. The linseed types are grown for extraction of oil from the seeds. The flax types are grown for fibre extraction from the stems. The flax types of *L. usitatissimum* L. are relatively taller (80-120 cm) in height with straight culms, less number of secondary branches towards the top of the stem (Gill, 1987). *L. usitatissimum* L. occupies an important position in world market because of its multiple trade use. Different industries like fibre, textile, paint, soap, paper, *etc.* use the raw materials obtained from linseed. After extraction of fibre from stalk, the woody core of the stem and short fibres are used as raw pulp for making high grade paper. Chemically linseed contains moisture, protein, fatty oil, carbohydrate, fibre, calcium, phosphorus, iron, carotene, thiamin, riboflavin, niacin, pantothenic acid, choline and vitamin E. The flowers are considered as cardiotonic. Raw linseed oil can be used in pharmaceuticals as an emollient, demulcent, expectorant and diuretic. The whole seed is used as laxative. The extracted mucilage is often used in cosmetics and pharmaceutical industries as ademulcent. The oil has analgesic, anti-estrogenic, anti-inflammatory, purgative properties. The oil cake is used for cattle feed. Linseed oil is suitable for human consumption and is used as a nutritional supplement. It is rich in omega-3 fattyacid, especially alpha-linolenic acid that appears to be beneficial for heart diseases, inflammatory bowel disease, arthritis and a variety of other health conditions (Soni *et al.,* 2016). Development of a standard protocol for its multiplication is a pre-requisite for genetic modification. Keeping this in view, the present investigation is aimed to develop a standard protocol for maximum root regeneration.

Materials and Methods

Seeds of six flax cultivars namely “OL-98-7-3”, “OL-98-15-2”, “Padmini”, “Sharada”, “Indira-Alsi-32” and “V1K2-99-57” were collected from the Regional Research and Technology Transfer Substation, Jashipur, Mayurbhanj, Odisha. The healthy seeds were then selected and stored in the seed container to carry out the investigation. For *in vitro* shoot regeneration and rooting of *Linum usitatissimum* L. different constituents of MS medium (Murashige and Skoog, 1962), sucrose and the solidifying agent agar were procured from Himedia Pvt Ltd., Mumbai. The surface sterilizing agent used for surface sterilization of seeds as well as the explants used during *in vitro* propagation was procured from Merck Specialties Pvt. Ltd., Mumbai. Different glassware like Petri dishes, conical flasks, beaker, test tubes of different length and sizes used for this experiment were made up of Borosil Glass Work Ltd., Mumbai.

The seeds of six linseed cultivars after surface sterilization with 2 per cent v/v of sodium hypochlorite for 2 minutes and germinated *in vitro* to get the hypocotyls explants which were cultured on MS medium containing 1.0mg/l of IAA and 1.5mg/l of BAP. After successful shoot regeneration, *in vitro* regenerated shoots were then excised with the help of a sterile scalpel. The excised shoot were then transferred to tubes containing MS medium along with different concentrations (0.5, 1.0, 1.5 and 2.0mg/l) of auxin (IAA, NAA and IBA) alone to induce roots. The test tubes were then incubated at temperature 22±2°C under cool white fluorescence light with 16h light/8h dark photoperiod inside the culture room and tubes were under observation for root induction. Each experiment was repeated thrice and data were analyzed by using ANOVA.

Results and Discussion

Auxin is known to involve in the rooting process. The success of *in vitro* propagation is mostly depends upon the production of good quality of adventitious roots. Both the concentration and type of auxins used markedly influence the number of root formation per explants. Data pertaining to response of different concentration of different auxins towards rooting in stem cuttings for plantlet regeneration was subjected to statistical analysis by using three factor analysis. Six numbers of linseed cultivars were considered as factor A, three numbers of hormones were considered as factor B and four numbers of concentrations were considered as factor C, respectively. However, the analysis result so obtained has been presented in Table 24.1a–f; Figure 24.1 and Plate 24.1a,b.

Plate 24.1: Rooting from Stem Cuttings of *Linum usitatissimum* L. on MS Medium Supplemented with 1.5 mg/l of IBA.

It is evident from the table that all the factors were statistically significant ($p<0.01$) indicating that all the 6 numbers of cultivars, 3 numbers of hormones and 4 numbers of concentrations are different from each other. The CD value for factor was found to be 0.298 and it can be inferred here that, the mean difference between any two cultivars which is greater than 0.298 is significantly different from each other and the mean difference between any two cultivars is less than

Table 24.1: Comparison of Three different Types of Auxins (IAA, NAA and IBA) on Rooting

(a): A X B X C Mean Table

	A1	**A2**	**A3**	**A4**	**A5**	**A6**												
	B1	B2	B3	B1	B2	B3	B1	B2	B3	B1	B2	B3	B1	B2	B3	B1	B2	B3
C1	5.333	6.000	5.333	6.333	7.000	5.667	5.667	5.667	6.000	5.333	5.333	5.667	5.667	5.667	5.000	5.667	6.000	5.667
C2	6.000	6.667	6.333	7.000	7.000	7.333	6.333	7.333	7.333	6.667	7.000	7.000	6.000	6.667	7.000	5.333	7.000	7.333
C3	9.333	9.000	11.667	9.667	10.667	12.000	11.667	11.000	13.667	11.000	11.000	13.000	10.000	11.000	13.000	10.333	10.333	12.000
C4	6.333	6.667	7.667	6.667	7.000	8.000	6.000	6.667	8.000	5.667	6.667	7.333	6.000	6.667	7.000	6.333	6.667	7.333

A1: OL-98-15-2; A2: OL-98-7-3; A3: Sharada; A4: Padmini; A5: Indira-Alsi-32; A6: V1K2-99-7.

B1: IAA; B2: NAA; B3: IBA; C1: 0.5mg/l; C2: 1.0mg/l; C3: 1.5mg/l; C4: 2.0mg/l.

(b): A × B Mean Table

	B1	***B2***	***B3***	***Mean A***
A1	6.750	7.083	7.750	7.194
A2	7.417	7.917	8.250	7.861
A3	7.417	7.667	8.750	7.944
A4	7.167	7.500	8.250	7.639
A5	6.917	7.500	8.000	7.472
A6	6.917	7.500	8.083	7.500
Mean B	7.097	7.528	8.181	

(c): A× C Mean Table

	C1	*C2*	*C3*	*C4*	*Mean A*
A1	5.556	6.333	10.000	6.889	7.194
A2	6.333	7.111	10.778	7.222	7.861
A3	5.778	7.000	12.111	6.889	7.944
A4	5.444	6.889	11.667	6.556	7.639
A5	5.444	6.556	11.333	6.556	7.472
A6	5.778	6.556	10.889	6.778	7.500
Mean C	5.722	6.741	11.130	6.815	

(d): B ×C Mean Table

Factors	*C.D.*	*SE(d)*	*SE(m)*
Factor (A)	0.297	0.150	0.106
Factor (B)	0.210	0.106	0.075
Intraction A × B	N/A	0.261	0.184
Factor (C)	0.243	0.123	0.087
Intraction A × C	0.595	0.301	0.213
Intraction B × C	0.421	0.213	0.150
Intraction A × B × C	N/A	0.521	0.369

(e): Table of SEM, SED AND C.D.

	C1	*C2*	*C3*	*C4*	*Mean B*
B1	5.667	6.222	10.333	6.167	7.097
B2	5.944	6.944	10.500	6.722	7.528
B3	5.556	7.056	12.556	7.556	8.181
Mean C	5.722	6.741	11.130	6.815	

A: Linseed cultivar; B: Type of Auxin; C: Concentration

A1: OL-98-15-2; A2: OL-98-7-3; A3: Sharada; A4: Padmini; A5: Indira-Alsi-32; A6: V1K2-99-7

B1: IAA; B2: NAA; B3: IBA; C1: 0.5mg/l; C2: 1.0mg/l; C3: 1.5mg/l; C4: 2.0mg/l

0.298 is at par though their magnitude is different. Likewise the CD values for hormones and concentrations were found to be 0.210 and 0.243 respectively and interpretation may be made like the cultivars. As per the mean table placed above, it is observed that, linseed cultivar Sharada (A3) exhibited highest root numbers (13.667) per stem cutting, followed by Padmini (A4), Indira-Alsi-32 (A5), V1K2-99-7 (A6), OL-98-7-3 (A2), and least rooting was observed in OL- 98-15-2 (A1).

(f): Analysis of Variance Table

Source of Variation	*DF*	*Sum of Squares*	*Mean Squares*	*F-Calculated*	*Significance*
Factor A	5	13.648	2.730	6.700	0.00001
Factor B	2	42.843	21.421	52.580	0.00000
Int A X B	10	2.269	0.227	0.557	0.84663
Factor C	3	936.315	312.105	766.076	0.00000
Int A X C	15	23.130	1.542	3.785	0.00001
Int B X C	6	38.713	6.452	15.837	0.00000
Int A X B X C	30	12.176	0.406	0.996	0.48053
Error	144	58.667	0.407		
Total	215	1,127.759			

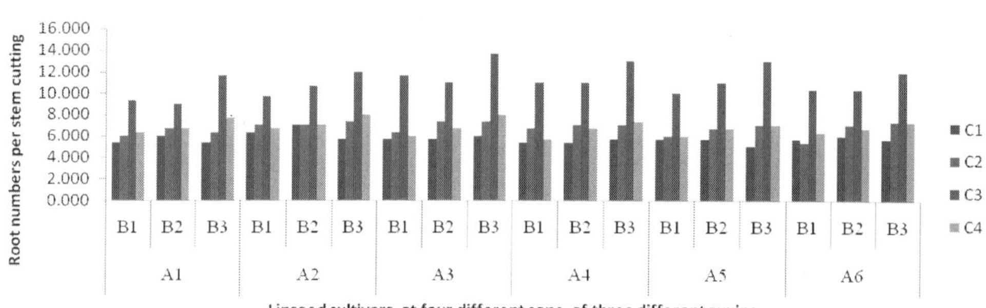

Figure 24.1: Graph Showing a Comparison between Three Types of Auxins on *in vitro* Rooting.

Among the three different types of auxins IBA (B3) was found to be best with hormone followed by NAA (B2) and least root number produces by IAA (B1). The best concentration of auxin concentration producing highest root number is 1.5mg/l. This repeats the findings of Benelli *et al.,* 2001 and Tanimoto, 2005 that IBA is the most effective auxin in Olive rhizogenesis. Highest number of roots 5.33/ shoot have been produced in Sharada followed by 5.00/shoot in Padmini, OL-98-15-2, OL-98-7-3 and 4.33/shoot in V1K2-99-57, Indira-Alsi-32. Thus, rooting in the *in vitro* propagated shoots depends upon the type and concentration of auxins which is in accordance to Sabatini *et al.* (1999); Blakesley and Chaldecott (1997). On increasing the concentration of IBA to 2.0mg/l there was a sharp decrease in the root numbers which was in accordance with Barcelo *et al.,* 1999 and Wynne and McDonald, 2002. They reported that IBA enhanced the cell division of first root initials in Avocado. It is also clear from the results that rooting percentage showed the tendency to increase with increasing concentration of IBA, however, the higher concentration (>1.5 mg/l) than the optimal exerted its inhibitory effect on the number of rooted shoots. In comparison to IAA, NAA exerted more efforts towards rooting. The result corresponds to the findings of Ansar *et al.,* 2009 while studying the effect of concentration of auxins *in vitro* rooting of Olive, reported that same concentration of NAA produces more number of roots in comparison to IAA.

REFERENCES

Ansar, A., Ahmad, T., Abbasi, N. A. and Hafiz, I. A., 2009. Effect of differentconcentrations of auxins on *in vitro* rooting of Olive cultivar 'M oraiolo'.Pak. J. of Bot. 41 (3): 1223-1231.

Barcelo, A. M., Encina, C. L. and Perez, E.S., 1999. Micropropagation of adult olive. Pl. Cell, Tissue and Organ Cult. 36 (3): 321-326.

Benelli, C., Fabbri, A., Grassi, S., Lambardi, M. and Rugini, E., 2001. Histology ofsomatic embryogenesis in mature tissue of Olive. The J. of Hort.Sc. and Biotechnol. 76 (1): 112-119.

Blakesley, D. and Chaldecott, M. A., 1997. The role of endogenous auxins in rootinitiation. Plant Growth Regul. 13 (1): 77-84.

Gill, K.S. 1987. Linseed. Indian Council of Agricultural Research, New Delhi.

Murashige T, Skoog F. 1962. A revised medium for rapid growth and bioassays withtobacco tissue cultures. Physiol. Plant. 15: 431-497.

Sabatini, S., Beis, D. and Wolkenfelt, H., 1999. An auxin dependent distal organizerof pattern and polarity in the *Arabidopsis* root. Scientia Horticult.99: 463-472.

Soni R. P., Katoch M., Kumar A. and Verma P., 2016. Flaxseed- composition and its health benefits. Research in Env. and Life Sci. 9(3): 310-316.

Tanimoto, E., 2005. Regulation of root growth by plant hormones: Roles for auxinand gibberellins. Critic. Reviews in Plant Sci. 24 (4): 249-265.

Wynne, J. and McDonald, M. S., 2002. Adventitious root formation in woody planttissue: The influence of light and indole-3-butyric acid. *In vitro* Cellular andDevelop. Biol.-Plant. 38 (2): 210-212.

Chapter 25

Nanotechnological Interventions Through Application of Ecofriendly Nanoparticles for Sustainable Aquaculture

Sukanta Kumar Nayak and Dipsha Nayak*

Department of Biotechnology,
North Orissa University, Takatpur – 757 003,
Baripada, Mayurbhanj, Odisha
**e-mail: sukantanayak@rediffmail.com*

ABSTRACT

Aquaculture is the fastest growing food producing sector and its increasing contribution to global food security, livelihoods, employment/income generation has been duly recognised by the developing countries around the world. However, aquaculture sector has to play an important role in providing livelihoods and nutritional security to billion people. Therefore, semi-intensive, intensive and now super intensive farming practices with high stocking densities and supplementation of nutritionally balanced feeds has been adopted to increase production. However, such types of aquaculture practices need to be practiced in a responsible and sustainable manner. Recently the sustainability of this sector is at stake basically due to factors like poor culture practices, deteriorated water quality, use of low grade/contaminated feed, outbreak of diseases etc. Therefore there is a concern to adopt new alternative approaches to maintain the culture environment at optimum level, to develop high quality feed as well as to prevent diseases in order to prevent severe economic loss. Looking into the potential and applications in living system, nanotechnology has ample of opportunities in aquaculture sector. Herein in this chapter emphasis has been given on the applications of nanoparticles especially green synthesized nanoparticles for sustainable aquaculture practices.

Keywords: *Aquaculture, Nanoparticles, Feed, Pathogens, Water quality.*

Introduction

Aquaculture, the farming of aquatic organisms including plants, is highly diverse in terms of cultured species, environment, scale, intensity, management and marketing. It has been practiced in different types of aquatic environments throughout world. Historically, fish are cultured for human consumption in ponds as early as 1100 B.C., while oyster farming has been recorded as early as Han Dynasty (Hishamunda and Subasinghe, 2003). Aquaculture which remained substantially a low level, small scale farming activity until major experimental husbandry practices for different fish and shellfish species were established, progressed and developed into industry in the mid 20th century (Bondad- Reantaso *et al.,* 2005).

This sector has the potency not only to provide livelihoods but also to ensure food security to global production that will reach 9 billion by 2050. Nowadays it has already been the fastest growing food producing sector world-wide, with an average annual growth rate of 8.9 per cent (FAO, 2018). However, this industry needs to be practiced responsibly and sustainably. The word sustainability from an aquaculture perspective can be defined as " the wise and productive methods of culturing aquatic animals and plants using natural resources in a manner that is environmentally non-degrading, technically appropriate, economically viable and socially acceptable; ensuring the attainment and continued satisfaction of critical human needs for the present and future generations" (ADB/NACA, 1998). The word "sustainability" became very crucial for modern day aquaculture practices as it is constantly threatened by increasing environmental problems especially deteriorated pond environment, lack of quality seed and feed, serious outbreaks of diseases at all stages of culture period *etc.*

A true sustainable aquaculture is a major concern today on the management point of view. According to Food and Agriculture Organization (FAO), United Nations, nearly 50 per cent of the world's recognized fish stocks are fully exploited and 32 per cent are over exploited or depleted. On the contrary, phenomenal changes have been made in the aquaculture practices with a shift from extensive to intensive and super intensive type since past few decades. To achieve a successful and sustainable aquaculture, three major factors *i.e.,* a good pond environmental condition, proper but nutritionally balanced feed supplementation and a healthy pathogen free/reduced conditions need to be achieved. Various stress factors like high stocking density, non-optimal water quality parameters, poor nutrition and diseases due to infectious agents can impact the sustainability (Figure 25.1). Water quality determines to a great extent the success of a aquaculture practices

(Piper *et al.,* 1982).The water quality enables propagation, growth and survival of the cultured aquatic species which ultimately determine the production. High stocking density of fish/shellfish usually linked with the deterioration of cultured environment. The wastes generated during culture are mostly due to faeces and unconsumed feed. Such wastes start to settle/accumulate in the bottom sediment, as a consequence of organic wastes and metabolites of degraded organic matter.

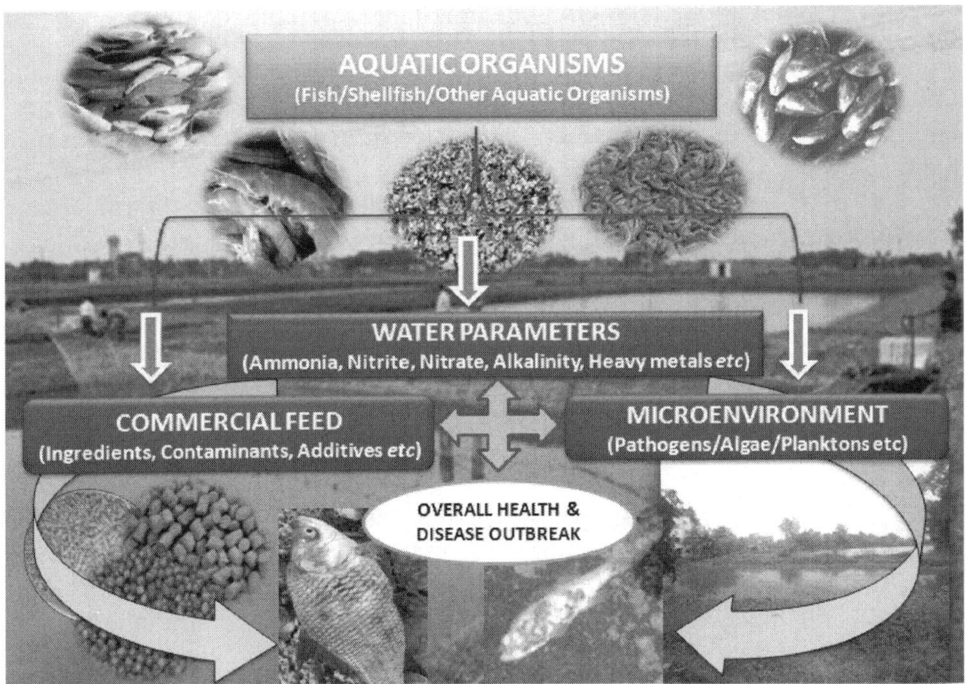

Figure 25.1: Major Factors Affecting the Growth and Production of Cultured Species in Aquaculture Practice.

Another factor that affects the growth and production of aquatic species is feed. The production has increased significantly with semi-intensive and intensive farming systems beyond any doubt. But such practices require many types of supplementation including the most important component *i.e.,* nutritionally balanced feeds. In intensive culture practices, the nutritional requirement of the cultured species is obtained from artificial feed. Aquafeed consisting of nutritionally balanced mixture of ingredients to support the maintenance, growth, reproductive performance and health of the animals. However, poor quality feed can have adverse influence on the sustainability of this sector.

Modern culture practices often give opportunities to infectious agents to affect fish irrespective of stages of growth. The cultured fish species are affected by all types of microbial pathogens but the degree of incidence may vary from species to species. Further, like any other living system, aquaculture system which is very diverse often faces serious setbacks due to emergence and re-emergence of infectious diseases for which sustainability of this sector is at stake.

Nanotechnology, Nanoparticles and their Applications in Aquaculture

Nanotechnology is defined as the understanding and control of matter at the nanoscale, at dimensions between approximately 1 and 100 nanometers, where unique phenomena enable novel applications (National Nanotechnology Initiative 2013, USA). Nanoparticles (From Greek nanos – dwarf) are the organic or inorganic solid particles in between 1 to 100 nanometers in size. The dimension of nanoparticles is defined in a uniform manner. a- particles in the sub micron range (< 1 μm), b- materials science: < 100 nm (nano scaled particles), c-pharmaceutics: < 500 nm, < 1000 nm = 1μm.

In the recent years, nanoparticles have gained tremendous response owing to their unique properties such as such as depression of melting temperature, conductivity, magnetism and light absorption different from their bulk counterparts. Nanomaterials can be synthesized by various processes including chemical, physical, and biological methods. In chemical and physical methods, nanoparticles are synthesized by electrodeposition, sol-gel process, chemical solution deposition, chemical vapour deposition, soft chemical method and wet chemical method. On the contrary, Green synthesis of nanoparticles is environmental friendly, more stability, uniform dimensions, non toxic and safe reagents and is synthesized using biological technique or green technology. Former methods often use high radiation and concentrated reluctant and stabilizing agents that are harmful for the environment and human health. Hence, green synthesis of nanoparticles is a stable and eco friendly, which is better and fruitful than other methods of synthesis. A range of nanoparticles, now been synthesized through green synthetic methods appear to be safer and more environment friendly while retaining the efficacy of the formed nanoparticles (Kalishwaralal *et al.,* 2008; Jayaseelan *et al.,* 2012).

Nanotechnology with new tools like water remediation, rapid disease diagnosis, enhancing the ability of fish to absorb drugs like hormones, vaccines and nutrients rapidly has the potency to revolutionize the fisheries and aquaculture industry. Nano materials, nano sensors, nanoparticles based products have potentials to improve fish nutrition, production, reproduction, prevention and control of diseases. In aquaculture application of nanotechnology has already practiced for some time and includes the disinfection of ponds, water/waste treatment and recycling, detection and control of pathogens, efficient delivery of nutrients and improvement of aquafeed (Bhattacharyya *et al.,* 2015, Huang *et al.,* 2015). Some of the applications of nanoparticles especially green synthesized nanoparticles in major aquaculture practices are as follows:

Water Quality Management

Water quality is one of the critical factors that regulate the overall health status of fish and shellfish. The quality of water takes into consideration physical, chemical and biological parameters that interact individually or collectively to influence the productivity. Deteriorated water quality leads to poor performance, injury to the organisms and a retarded growth rate (Boyd and Tucker 2012). In order to maximize the production, effective culture management system need to be implemented in aquaculture. Such management includes ensuring optimum water quality parameters (*e.g.* dissolved oxygen levels, temperature, salinity) for the cultured species, removal/elimination/minimise of wastes and toxic components as well as elimination of pathogens (Handy and Poxton, 1993). The cconventional aquaculture wastewater treatment technologies like re-circulatory systems are mostly high energy intensive and hence avoidable especially in the present scenario of climate change and energy conservation. On the contrary, the nanotechnology has immense scope for improving aquaculture production through water quality management. The quality of the effluents has been improved dramatically by using nanotechnological system. These systems include basic separation processes such as reverse osmosis and nanofiltration. Certain metal based nanoparticles like calcium, magnesium, selenium, silver, gold, zinc oxide, titanium oxide, iron oxide *etc.,* help to reduce wastes, pollution or eutrophication and ultimately promote safe aquaculture practices. Among different types of nanotechnological interventions, green synthesized metal nanoparticles are found to play a role in the management of water quality in aquaculture by removing various heavy metals/pollutant/ organic wastes.

Silver Nanoparticles

Silver nanoparticles (Ag NPs) have been used in aquaculture waste water treatment. The integration of Ag NPs with filtration membranes or filters has been considered useful for the disinfection of water due to their effective antibacterial abilities and cost effectiveness (Quang *et al.,* 2003). Besides, Ag NPs also exhibited effective antibacterial properties towards several microbial pathogens while filtering through the sheets (Madhura *et al.,* 2018). According to World Health Organization and the Environment Protection Agency, the concentration of Ag ions generated from Ag NPs is lower than the standards of the silver ions in the drinking water (Dankovich and Gray 2011). Thus the membrane filters prepared with Ag NPs could be an effective tool or the treatment of waste water contaminated by bacteria. Ag NPs are highly toxic to microorganisms resulting in effective microbial defense against fungi (Krishnaraj *et al.,* 2012), bacteria (Kalhapure et.al, 2015) and viruses (Borrego *et al.,* 2016). Due to this activity, Ag NPs have been extensively used for the disinfection of water. It was reported that Ag NPs interact with bacterial cell wall and penetrate the cell resulting in a change in the cell wall structure and consequently enhancing permeable capacity (Sondi and Salopek 2004).

Iron Nanoparticles

Iron nanoparticles (Fe NPs) have a moderate standard reduction potential to exhibit good reduction abilities when compared to various redox contaminants. The Fe NPs have much advantage over other metal nanoparticles like Ag and Zn due to a weaker reduction capacity. This property allows researchers to use Fe NPs extensively in water treatment processes, owing to its outstanding adsorption capacity, its cost- effectiveness, oxidation and precipitation. Because of the stronger reduction ability, small size and large surface area, Fe NPs have been widely used metal nanoparticles in waste water treatment (Fu *et al.*, 2013). Different separation methods such as precipitation, adsorption, oxidation and reduction were adopted by Fe NPs for the removal of wide range of pollutants, including radioactive elements, heavy toxic metals *etc.* (Markova *et al.*, 2013; Ling *et al.*, 2015).

Zinc Nanoparticles

Zinc nanoparticles (Zn NPs) possess high pollutants degrading activity in waste water and are considered as an alternative for the Ag and Fe NPs (Bokare *et al.*, 2013). With extra negative standard reduction ability, Fe NPs are a weak reductant as compared to Zn NPs. Therefore the pollutant degradation ability of Zn NPs is faster than that of other metal nanoparticles (Madhura et,al, 2018). Recent studies have specified that the reduction of carbon tetrachloride with Zn NPs were significantly affected by physico-chemical properties than by size of the particle or surface morphology. Further, it was found that Zn NPs could degrade carbon tetrachloride faster than Fe NPs under optimal experimental conditions (Tratnyek *et al.*, 2010). Though various reports have validated that degradation of pollutants with Zn NPs like octachlorodibenzo-*p*-toxin could be fruitful, the applications of Zn NPs in degrading organic pollutants are limited only to carbon tetrachloride (Bokare *et al.*, 2013).

Iron Oxide Nanoparticles

Iron oxide nanoparticles (Fe_2O_3 NPs) are used to eliminate toxic heavy metals due to their unique physic-chemical properties such as large surface area, superior adsorbent property. There are several forms of Fe_2O_3 NPs but maghemite, magnetite and haematite are commonly used nanoadsorbents. While the recovery and separation of other nanoadsorbents are challenging tasks after wastewater treatment due to their small size, maghemite and magnetite nanoparticles are easily separated after treatment. Therefore, these are commonly used as promising nano-adsorbents in the removal of a wide range of toxic heavy metals from wastewaters (Lei *et al.*, 2014; Ngomsik *et al.*, 2012; Tan *et al.*, 2014). On the other hand, nanohaematite is promising cost-effective nanomaterial for environmental applications and also used as sensors (Liang *et al.*, 2014). Over the years, it is proved to be a powerful adsorbent for the elimination of heavy metal ions from the waste water samples (Shipley *et al.*, 2013). The easy existence of iron and its facile synthesis methods make Fe_2O_3 NPs cost effective material for the adsorption and removal of toxic metals. Nonetheless, these nanoparticles are eco-friendly and can be applied effectively to treat wastewaters with no/minimal secondary toxic contaminations (Li *et al.,* 2006).

Zinc Oxide Nanoparticles

Zinc oxide nanoparticles (ZnO NPs) are very promising and effective nanomaterials for the wastewater management. ZnO NPs are considered as excellent materials for waste treatment due to their high oxidation potency and excellent photolytic property (Chen *et al.,* 1998; Reynolds *et al.,* 1999; Janotti and Van de Walle 2009). Moreover, ZnO NPs especially green synthesized are eco- friendly as they are coherent with organisms, making them appropriate for the treatment of waste waters (Schmidt- Mende and MacManus-Driscoll, 2007). Furthermore, they are cost-effective, adsorb solar light from a wider spectrum and have more light quanta as compared to other semiconducting metal oxides (Behnajady *et al.,* 2006).

Table 25.1: Effects of Various Metal/Metal Oxide Based Nanoparticles on Reducing/Eliminating Contaminants, Pollutants, Heavy Metals and other Water Quality Parameters in Aquaculture

Sl.No.	*Nanoparticles*	*Effective in Removing/ Minimizing/Eliminating*	*References*
1	Silver	Ammonia, Nitrite	Jeanette Brown 2017
2	Gold	Mercury	Kamarudin and Mohammad 2010
3	Copper	Nitrous oxide	Chen *et al.,* 2012
4	Zinc	Carbon tetrachloride	Tratnayek *et al.,* 2010
5	Titanium oxide	Oxytetracycline	Antonio *et al.,* 2017
6	Magnesium oxide	Lead, Cadmium	Li *et al.,* 2003
7	Manganese oxide	Arsenic	Wang *et al.,* 2011
8	Zinc oxide	Mercury, Cadmium, Nickel	Sing *et al.,* 2013, Dai *et al.,* 2014
9	Iron oxide	Lead	White *et al.,* 2009

Feed Quality Management

The production of aquaculture has been greatly regulated by the diets (Tom and Van-Nostrand, 1989). In sustainable aquaculture, nutrition represents more than just feeding. As a consequence, aquafeed production is currently one of the fastest expanding agricultural sectors in the world with 59.9 million tonnes production (Gonçalves and Kovalsky, 2015). In order to maintain the current rate of feed production which accounts for over 60 per cent of the operating costs (FAO, 2014), the supply of feed inputs need to grow at equal pace (Shepherd and Bachis, 2014). Dietary nutrients set the necessary levels for energy, protein, amino acids, lipids, vitamins and minerals (Prabu *et al.,* 2017). However, uunlike other animal feeds, aquafeeds require higher protein and lower carbohydrate content and traditionally depend on fishmeal and oil, apart from other additives in optimum proportions to fulfil the nutritional requirements of the cultured species. Deficiencies or excess of the nutrition can lead to reduced growth or disease susceptibility.

The improvement of nutritional interventions supports the aquaculture industry sustainable, economical and production (Robinson *et al.,* 1998). Further, the major challenge in a sustainable aquaculture is to develop relatively cheap but high quality feed. Traditionally, feed is formulated based on the daily nutritional requirements fish/shellfish containing fats, proteins, carbohydrates, minerals, vitamins *etc.* Nanoparticles in aquafeed has been found to increase the nutrient absorption in the gut while passing through the gut tissue rather than directly passing to the digestive system unused. According to the studies nanoparticles like selenium, iron, zinc *etc.* in diet are found to improve the growth of fish (Albrecht *et al.,* 2006; Wang *et al.,* 2007). Further the nano-mineral based diet can significantly improve the overall health of aquatic species including their growth and development (Chric and Agarwal, 2018).

Zinc Nanoparticles

Zinc nanoparticles (Zn NPs) have been used as an alternative to the conventional zinc in livestock feeding for better performances (Swain *et al.,* 2016; Ramachandraiah *et al.,* 2018; Tawfik *et al.,* 2017). Zinc is one of the most important trace minerals required for good health status of an animals. In recent years promising results have been recorded on the beneficial effects of ZnO NPs supplementation in fish and shellfish (Swain *et al.,* 2016). Supplementation of Zn NPs in aquafeed leads to weight gain alongwith certain biochemical compositions of fish. This is basically due to the fact that Zn NPs form has higher bioavailability and more readily absorbed in the gastrointestinal tract of different fishes in comparison to the bulk form.

Selenium Nanoparticles

Selenium Nanoparticles (Se NPs) in diet enhances the growth performance of aquatic species. Further it also improves antioxidant defense system (Ashouri *et al.,* 2015). Selenium is an essential trace element for human and animal health and also constitutes an integral part of the glutathione peroxidase enzyme (Rotruck *et al.,* 1973). Further, it protects the cell membrane from oxidative stress (Watanabe *et al.,* 1997). Administration of dietary selenium enhanced the growth performance and immunological function in various fish species (Jaramilo and Gatlin, 2004; Zhou *et al.,* 2009). Various forms of selenium are found to be supplemented in different fish nd shellfish species. It has been demonstrated that the organic form Se is more digestible better accumulated and more biologically active then inorganic Se (Zhou *et al.,* 2009).

Gold Nanoparticles

Gold nanoparticles (Au NPs) have a wider range of applications in aquaculture. Further, the recent development and implementation of new technologies have led the era of nano-revolution which unfolds the role of green synthesized Au NPs by synthesizing stable nanoparticles (Kavitha *et al.,* 2003). Further, they are less toxic in comparison to their chemically synthesized counter parts (Shine *et al.,* 2018). Oxidative stress plays important roles in cellular signaling, inflammatory and genotoxic proliferative responses leading to pathological consequences and diseases. Chemically synthesized Au NPs are found to induce significant oxidative stress in the fish tissues. Green synthesized Au NPs can reduce oxidative stress, enhance antioxidant defense mechanism and immunity in fish (Knaapen, *et al.,* 2004; Zhong *et al.,* 1997).

Iron Nanoparticles

Iron nanoparticles (Fe NPs) by virtue of their unique physico-chemical properties have a great potential as fish feed additives and around 40mg per kg is an optimum amount for achieving better growth in fishes (Akter *et al.,* 2018). Is an established fact that iron is an essential micronutrient involved in oxygen transport and cellular respiration through oxidation reduction activity. It's deficiency causes hypochromic microcytic anemia in several fish like brook trout, and common carp, higher levels can reduce growth, diarrhea and histopathological damage to liver cells and even mortality. As per Russian Academy of Sciences fishes grow faster when they are fed with Fe NPs (ETC, 2003). However, the requirement of Fe NPs needs to be determined for the optimum growth and physiology of individual fish.

Table 25.2: Effects of Dietary Supplementation of Various Metal/ Metal Oxide Based Nanoparticles in different Fish Species

Sl.No.	*Nanoparticle*	*Beneficial Effect(s)*	*Reference(s)*
1	Zinc	Growth, anti oxidative stress	Kumar *et al.,* 2009
2	Selenium	Weight gain, antioxidant defense mechanism	Ashouri *et al.,* 2015; Kohrle *et al.,* 2000
3	Copper	Growth, Immunity	Bausim *et al.,* 2016
4	Silver	Growth, Immunity, Stress, Survivability	Kumar *et al.,* 2018
5	Gold	Anti oxidative stress	Geffroy *et al.,* 2012
6	Titanium dioxide	Weight gain	Christopher *et al.,* 2009

Disease Management

Diseases due to microbial pathogens are the major stumbling block in the sustainable growth of the aquaculture (Huang *et al.,* 2010). While infectious diseases continue to be responsible for severe mortality and morbidity in all cultured species, the incidence rate and severity of infectious diseases is very often dependant on several factors. The outbreak of diseases is closely related to environmental deterioration. Poor water quality has been recognized as one of the main factors responsible for occurrence of diseases. Some water quality factors *viz.* dissolved oxygen, temperature, ammonia *etc.* are more likely to be involved with production losses: On the contrary, parameters like pH, alkalinity, hardness, clarity usually are not directly toxic but they can influence other water quality factor to affect fish and shellfish in interlinked complex ways.

Nowadays chemotherapeutic and prophylactic agents have also negative impact on this sector. Different chemical disinfectants, antibiotics, and drugs are frequently used in large quantities to control diseases in this sector. The cost, efficiency and side effects of the used agents are always an issue (Yang 2003). Further, unregulated and unrestricted use of antibiotics can lead to the emergence of resistant bacteria (Aarestrup *et al.*, 2005: Cabello *et al.*, 2013). Further, emerging incidence of drug resistance among pathogens due to over-use of antibiotics have made the search for new prophylactic measures in aquculture (Atyah *et al.*2010; Son *et al.*, 1997).

The complexity of the aquaculture sector and increasing needs for interdisciplinary efforts to ensure its effective drug development and management, recently much attention has been paid towards the use of nanoparticles as an alternative to antibiotics (Gunalan *et al.*, 2012). The utilization of nanoparticles as alternative antimicrobials to combat emergence of microbial resistance to antibiotics and other diseases in aquaculture has been recorded due to their distinct advantages over conventional antimicrobial agents (Jayaseelan *et al.*, 2012; Swain *et al.*, 2014). Among different types of nanoparticles, metals and metaloxides have been proved to be the best due to their broad spectrum of antimicrobial activity, durability, high resistance, selectivity and specificity (Singh *et al.*, 2012). Therefore, several antimicrobial metal nanoparticles are now been applied to combat microbial resistance in aquaculture.

Silver Nanoparticles

Silver nanoparticles (Ag NPs) are the most investigated and effective antibacterial agent due to their multiple inhibitory and killing mechanisms (Cesare *et al.*, 2012). Among different types of nanoparticles, Ag NPs have proved to be the most effective showing good antimicrobial property against bacteria, virus and other eukaryotic micro-organisms (Ahmad *et al.*, 2005). The release of Ag ions is a mechanism in which the Ag^+ binds to the bacterial cell membrane proteins causing disruption of the membrane, leading to the death of the bacterial cells (Knetsch *et al.*, 2011; Lara *et al.*, 2016). Intracellularly Ag^+ binds to cytochrome and nucleic acids, damage them and inhibiting cell division (Huang *et al.*, 2011). Further, it has been demonstrated that Ag NPs demonstrate high antibacterial efficacy against multi-drug resistant pathogens (Pakash *et al.*, 2013).

The bactericidal effect of Ag NPs has also been reported against methicillin resistant *S. aureus* and *E. tarda* and anti cyanobacterial activity towards *Anabaena* and *Oscillatoria* species (Swain *et al.*, 2014). Green synthesized Ag NPs from *Rhizophora mucronata* showed antimicrobial effect against *P. fluorescens*, *Proteus* species and *Flavobacterium* species (Umashankari *et al.*, 2012). In a study Vaseeharan *et al.* (2010) demonstrated that Ag NPs can help to reduce mortalities of juvenile shrimp *Feberopenaeus indicus* by 71 per cent infected with *Vibrio harveyi*. Ag NPs also exhibited high antifungal activity and their inhibitory effect against *Candida* species was found to be similar to the commercial antifungal Amphotericin B (Sanjenbam 2014; Mallmann *et al.*, 2015). Green synthesized Ag NPs and also Ag NPs -chitosan composite also found to exhibit anti-viral properties (Elechiguerra *et al.*, 2015, Mallmann *et al.*, 2015).

Gold Nanoparticles

Gold nanoparticles (Au NPs), being less toxic to eukaryotic cells (Mehrbod 2009) are now explored as antimicrobial agents for aquaculture. Au NPs can interact with bacterial proteins, lipopolysaccharides *etc.* to alter their biological function. The bactericidal effect against *Escherichia coli* and *Salmonella typhi* has been reported (Lima *et al.*, 2013). Further functionalized Au NPs are also effective against multi drug resistance bacterial pathogens (Li *et al.*, 2014). Green synthesized Au NPs show antibacterial activity against various aquatic pathogens (Velmurugan *et al.*, 2014). Au NPs also show fungicidal activity in a size dependent manner (Ahmad *et al.*, 2013 Wani *et al.*, 2013).

Table 25.3: Effects of Various Metal/Metal Oxide Based Nanoparticles on different Aquatic Pathogens Associated with Diseases of Fish and Shellfish Species

Sl.No.	*Nanoparticles*	*Antibacterial Activity against*	*References*
1	Silver	*Aerommonas hydrophila, Edwardseilla tarda, Flavobacteium branchiophilum, Citrobacter* spp., *Staphylococcus aureus, Vibrio* spp., *Bacillus cereus, Pseudomonas* *a***a***eruginosa*	Kumar *et al.*, 2014
2	Copper	*S. aureus, Vibrio* spp., *B. cereus, P. aeruginosa*	Meneses *et al.*, 2018
3	Zinc	*A. hydrophila, E. tarda, F. branchiophilum*	Meneses *et al.*, 2018
4	Titanium dioxide	*Streptococcus iniae, E. tarda, Photobacterium damselae,*	Mohamed *et al.*, 2015
5	Zinc oxide	*P. aeruginosa, Escherichia coli, Enterococcus faecalis, Aspergillus flavus, Candida albicans*	Jayaseelan *et al.*, 2012
6	Magnesium oxide	*Bacillus subtilis, Serratia* spp	Ravikumar *et al.*, 2012
7	Copper oxide	*Bacillus cereus, Vibrio* spp	Swain *et al.*, 2013

Zinc Oxide Nanoparticles

Zinc oxide nanoparticles (ZnO NPs) have drawn much attention due to their antibacterial and antifungal effects (Gunalan *et al.*, 2014; Son *et al.*, 1997). This is basically due to the fact that many times ZnO NPs are found to be better antibacterial agents as compared to their bulk counterparts (Ramamoorthy *et al.*, 2013). Literatures indicate that green synthesized ZnO NPs can inhibit the growth of *A. hydrophila, E. tarda, F. branchiophilum, Citrobacter* spp., *S. aureus, Vibrio* spp., *Bacillus cereus* and *P. aeruginosa* (Liu *et al.,* 2009). In another interesting study biologically synthesized using *Aeromonas hydrophila,* ZnO NPs showed antibacterial activity against the same bacteria as well as other pathogens like *P. aeruginosa, E. coli, Enterococcus faecalis, A. flavus* and *Candida albicans* (Jayaseelan *et al.,* 2012).

Titanium Oxide Nanoparticles

Titanium oxide nanoparticles ($Ti\ O_2$ NPs) have a bactericidal effect against *S. iniae, E. tarda* and *P. damselae,* when doped with Fe_3O_4 NPs (Cheng *et al.,* 2009, 2011). These particles can be used as disinfecting agent, as the fish pathogens bind with the nanoparticles, which can be easily extracted from water using magnet. However it is also concluded that the titanium oxide nanoparticles affect the fish immune system by inhibiting the activity of neutrophils.

Conclusion

Nanotechnology has emerged as a boon to the aquaculture sector with immense potential in varied area applications. It enhances the quality and availability of products which are of a great use in aquaculture. This article provides a summary of the implications of the growth of nanotechnology in the aquaculture and fisheries. The use of nanoscale materials is growing exponentially in aquaculture sector but there are also concerns about the environmental hazard to aquatic biota. The best way to minimize the potential environmental effects of manufactured nanomaterials is to create and use nanomaterials in products that are known to cause the least damage to potential targets. It is found that green synthesized nanoparticles are far better than the chemically synthesized ones are they are less toxic and more effective as they are extracted from the environment. However, there exist concerns due to the fact that nanoparticles can make their way into the aquatic environment where water chemistry will influence their fate and transport. As nanotechnology is still in its infancy, more research and development in this sector could potentially redirect nanotech efforts towards the useful goals for aquaculture and fisheries.

REFERENCES

Antony, J.J., Nivedheetha, M., Siva, D., Pradeepha, G., Kokilavani P., Kalaiselvi S., *et al.* (2013). Antimicrobial activity of *Leucas aspera* engineered silver nanoparticles against *Aeromonas hydrophila* in linfected *Catla catla*. Colloids Surf B Biointerfaces, 109: 20-24.

Atyah MAS, Zamri-Saad M, Siti-Zahrah A.(2010) First report of methicillin-resistant *Staphylococcus aureus* from cage-cultured tilapia (*Oreochromis niloticus*). Vet Microbiol; **144**(3): 502-504.

Ayala-Núñez NV, Lara HH, Turrent LDCI, Padilla CR. (2009) Silver nanoparticles toxicity and bactericidal effect against Methicillin- resistant *Staphylococcus aureus*: Nanoscale does matter. Nanobiotechnol; **5**: 2–9.

Behnajady MA, Modirshahla N, Hamzavi R (2006). Kinetic study on photocatalytic degradation of C.I. Acid Yellow 23 by ZnO photocatalyst. J Hazard Mater133(1-3): 226-32.

Bhattacharya A, Gupta A, Kaur A, Malik Darshan (2015) Simultaneous bioremediation of phenol and Cr (VI) from tannery waste water using bacterial consortium. International Journal of Applied Sciences and Biotechnology 3(1), 50-55.

Bokare V, Jung J, Chang YY, Chang YS (2013) Reductive dechlorination of octachlorodibenzo-p-dioxin by nanosized zero-valent zinc: Modeling of rate kinetics and congener profile. Journal of hazardous materials 250, 397-402.

Bondad-Reantaso MG, Subasinghe RP, Arthur JR, Ogawa K, Chinabut S, Adlard R, Tan Z, Shariff M. (2005) Disease and health management in Asian aquaculture. Vet parasitol; 132(3-4): 249-72.

Boyd CE and Tucker CS (2012) Pond aquaculture and water quality management. Springer and Science Business Media 700p.

Cabello FC, Godfrey HP, Tomova A, Ivanova L, Dolz H, Millanao A *et al.* (2013) Antimicrobial use in aquaculture re examined: its relevance to antimicrobial resistance and to animal and human health. Environ Microbiol, **15**(7): 1917-1942.

Chen Y, Bangnall DM, Koh H, Park K, Hiraga k, Zhu Z, Yao T,(1998) Plasma assisted molecular beam epitaxy of ZnO on c-plane sapphire: Growth and Characterization. Journal of Applied Physics 1998; 84(7): 3912-3918.

Cheng TC, Yao KS, Yeh N, Chang CI, Hsu HC, Chien YT, *et al.* (2009) Visible light activated bactericidal effect of TiO_2/Fe_3O_4 magnetic particles on fish pathogens. Surf Coat Technol; **204**(6): 1141- 1144.

Cheng TC, Yao KS, Yeh N, Chang CI, Hsu HC, Gonzalez F, *et al.* (2011) Bactericidal effect of blue LED light irradiated TiO_2/Fe_3O_4 particles on fish pathogen in seawater. Thin Solid Films; **519**(15): 5002- 5006.

Dankovich TA, and Gray DG (2011) Bactericidal paper impregnated with silver nanoparticles for point of use water treatment. Environmental Science and Technology 45(5): 1992-8.

Di Cesare A, Luna GM, Vignaroli C, Pasquaroli S, Tota S, Paroncini, P *et al.*(2013) Aquaculture can promote the presence and spread of antibiotic-resistant Enterococci in marine sediments. PloS One, **8**(4): e62838.

Di Cesare A, Vignaroli C, Luna GM, Pasquaroli S, Biavasco F. (2012). Antibiotic-resistant enterococci in seawater and sediments from a coastal fish farm. Microb Drug Resist, **18**(5): 502-509.

FAO, National aquaculture overview: India. Country profile fact sheets. FAO Fisheries and Aquaculture Department, FAO, Rome. http: //www.fao.org/fishery/factsheets/en, 2017 (Accessed 1 December 2017).

Frankic A and Hershner C (2003) Sustainable aquaculture: developing the promiseof aquaculture. Aquaculture International 11; 517-530.

Gehrke I, Geiser A, Schulz SS (2015) Innovations in nanotechnology for water treatment. Nanotechnology,science and applications 8,1.

Gunalan S, Sivaraj R, Rajendran V. (2012) Green synthesized ZnO nanoparticles against bacterial and fungal pathogens. Prog Nat Sci Mater Int; **22**(6): 693-700.

Handy RD and poxton MG (1993). Nitrogen pollution in mariculture: toxicity and excretion of nitrogenous compounds by marine fish. Reviews in Fish Biology and Fisheries, 3(3), 205-241.

Hindi KM, Ditto AJ, Panzner MJ, Medvetz DA, Han DS, Hovis CE,*et al.* (2009) The antimicrobial efficacy of sustained release silver–carbine complex-loaded L-tyrosine polyphosphate nanoparticles: Characterization, in vitro and in vivo studies. Biomaterials; **30**(22): 3771-3779.

Hishamunda N and Subasinge RP,(2003) Aquaculture development in China: the role of public sector policies. FAO Fisheries Technical Paper. No. 427. 64p.

Huang DL, Wang RZ, Liu YG, Zeng GM, Lai C, Xu P, Lu AB, Xu JJ, Wang C, Huang C (2015) Application of molecularly imprinted polymers in waste water treatment: a review. Environmental Science and Pollution Research 22(2), 963-977.

Janotti A and Van de Walle CG (2009) Fundamentals of zinc oxide as a semiconductor. Reports on progress in Physics 72; 12p.

Jaroenram W, Arunrut N, Kiatpathomchai W. (2012) Rapid and sensitive detection of shrimp yellow head virus using loop-mediated isothermal amplification and a colorogenic nanogold hybridization probe. J Virol Methods; **186**(1): 36-42.

Jayaseelan C, Rahuman AA, Kirthi AV, Marimuthu S, Santhoshkumar T, Bagavan A, *et al.* (2012) Novel microbial route to synthesize ZnO nanoparticles using *Aeromonas hydrophila* and their activity against pathogenic bacteria and fungi. Spectrochim Acta A; **90**: 78-84.

Kalhapure RS, Suleman N, Mocktar C, Seedat N, Govender T (2015) Nanoengineered drug delivery system for enhancing antibiotic therapy. Journal of pharmaceuticals sciences. 104(3) 872-905.

Kalishwaralal K, Deepak V, Ramkumarpadian, Nellaiah H, Sangiliyandi G (2008) Extracellular biosynthesis of silver nanoparticle by the culture supernatant of Bacillus licheniformis. Materials letters 62 (29), 4411-4413.

Kalishwaralal K, Jeyabharathi S, Sundar K, Muthukumaran A (2016) A novel one pot green synthesis of selenium nanoparticles and evaluation of its toxicity in zebrafish embryos. Artificial cells, nanomedicine and biotechnology 44 (2), 471-477.

Knetsch ML, Koole LH. (2011) New strategies in the development of antimicrobial coatings: the example of increasing usage of silver and silver nanoparticles. Polymers; **3**(1): 340-366.

Krishnaraj C, Ramachandran R, Mohan K, Kalaichelvan PT(2012) Optimization for rapid synthesis of silver nanoparticles and its effect on phytopathogenic fungi. SpectrochimicaActa Part A 93; 95-99.

Kuan GC, Sheng LP, Rijiravanich P, Marimuthu K, Ravichandran M, Yin LS, *et al.* (2013) Gold-nanoparticle based electrochemical DNA sensor for the detection of fish pathogen *Aphanomyces invadans*. Talanta; **117**: 312-317.

Lei Y, Chen F, Luo Y, Zhang L, (2014) Three dimensional magnetic graphene oxide foam/Fe_3O_4 nanocomposite as an efficient absorbent for Cr(VI) removal, J.Mater.Sci.49,4236-4245.

Li MH, Robinson EH (2006) Use of cottonseed meal in aquatic animal feed diets: a review.North Am.J.Aquaculture, 68(1): 14-22.

Liang CY, Keller SM, Sepulveda AE, Bur A, Sun WY, Wetzlar K, Carman GP(2014) Modeling of magnetoelastic nanostructures with a fully coupled mechanical-micromagnetic model. Nanotechnology 25(43),435701.

Ling D, Lee N, Hyeon T(2015) chemical synthesis and assembly of uniformly sized iron oxide nanoparticles for medical applications. Acc Chem Res. 48(5): 1276-85.

Madhura L, Singh S, Kanchi S, Sabela M, Biestty K, Innamudin, (2018) Nanotechnology based water quality management of waste water. Environmental Chemistry Letters.

Markova Z, Siskova KM, Filip J, Cuda J, Kolar M, Safarova K, Medrik I, Zboril R (2013) Air stable Fe-Ag nanoparticles for advanced antimicrobial treatment and phosphoris removal. Environmental science and technology 47(10), 5285-5293.

Mozafari MR, Johnson C, Hatziantoniou S and Demetzos C (2008) Nanoliposomes and their applications in food nanotechnology. Journal of Liposome Research, 18: 309-327.

Ngomsik A, Bee A, Talbot D, Cote G,(2012) Magnetic solid-liquid extraction of Eu(III), La(II), Ni(II) and Co(II) with maghemite nanoparticles.Sep.Purif. Technol.86,1-8.

Pelgrift RY, and Friedman AJ (2013) Nanotechnology as a therapeutic tool to combat microbial resistance. Adv Drug Deliv Rev; **65**(13): 1803- 1815.

Piper, R. G., I. B. McElwain, L. E. Orme, J. P. McCraren, L. G. Flower, and J. R. Leonard. 1982. Fish hatchery management. U. S. Fish and Wildlife Service, Washington, D. C.

Prabhu S, and Poulose EK. (2012)Silver nanoparticles: mechanism of antimicrobial action, synthesis, medical applications, and toxicity effects. Int Nano Lett; **2**: 32.

Prabu E, Felix S,Felix N, Abhilan B, Ruby P (2017). An overview on significance of fish nutrition in aquaculture industry. International Journal of Fisheries and Aquatic Studies; 5(6): 349-355.

Prakash P, Gnanaprakasam P, Emmanuel R, Arokiyaraj S, Saravanan M. (2013) Green synthesis of silver nanoparticles from leaf extract of *Mimusops elengi*, Linn. for enhanced antibacterial activity against multi drug resistant clinical isolates. Colloids Surf B Biointerfaces; **108**: 255-259.

Reynolds DC, Look DC, Joga B, Litton CW, Cantwell G, Harsch WC(1999) Valence band ordering in ZnO, Phys.Rev.B 60(4), 2340-2344.

Saleh M, and El-Matbouli M. (2015) Rapid detection of Cyprinid herpesvirus-3 (CyHV-3) using a gold nanoparticle-based hybridization assay. J Virol Methods; **217**: 50-54.

Saleh M, Soliman H, El-Matbouli M. (2015) Gold Nanoparticles as a Potential Tool for Diagnosis of Fish Diseases. Methods Mol Biol.; 1247: 245-252. doi: 10.1007/978-1-4939-2004-4_19.

Saleh M, Soliman H, Haenen O, El-Matbouli M. (2011) Antibody- coated gold nanoparticles immunoassay for direct detection of *Aeromonas salmonicida* in fish tissues. J Fish Dis; **34**(11): 845- 852.

Sankar R, Karthik A, Prabu, A, Karthik S, Shivashangari KS, Ravikumar V. (2013) *Origanum vulgare* mediated biosynthesis of silver nanoparticles for its antibacterial and anticancer activity. Colloids Surf B Biointerfaces; **108**: 80-84.

Schmidt ML and MacManus DJL(2007) ZnO nanostructures, defects and devices. Mater Today 10, 40-48.

Seetang-Nun Y, Jaroenram W, Sriurairatana S, Suebsing R, Kiatpathomchai W. (2013) Visual detection of white spot syndrome virus using DNA-functionalized gold nanoparticles as probes combined with loop-mediated isothermal amplification. Mol Cell Probes; **27**(2): 71-79.

Seil JT, Webster TJ. (2012) Antimicrobial applications of nanotechnology: methods and literature. Int J Nanomedicine; **7**: 2767-2781.

Shipley HJ, Engates KE, Grover VA(2013) Removal of Pb(II), Cd(II), Cu(II), and Zn (II) by hematite nanoparticles: effect of sorbent concentration, pH, temperature and exhaustion, Environmental Science and Pollution research 20(3), 1727-1736.

Son R, Rusul G, Sahilah AM, Zainuri A, Raha AR, Salmah I. (1997) Antibiotic resistance and plasmid profile of *Aeromonas hydrophila* isolates from cultured fish, Telapia (*Telapia mossambica*). Lett Appl Microbiol; **24**(6): 479-482.

Sondi I and Salopek B (2004) Silver nanoparticles as antimicrobial agent: A case study on E.coli as amodel for gram negative bacteria. Journal of Colloid and Interface Science 275(1): 177-82.

Sorum H, Lie O. (2008) Antibiotic resistance associated with veterinary drug use in fish farms. Improving farmed fish quality and safety, 157-182.

Swain P, Nayak SK, Sasmal A, Behera T, Barik SK, Swain SK, *et al.* (2014) Antimicrobial activity of metal based nanoparticles against microbes associated with diseases in aquaculture. World J Microbiol Biotechnol; **30**(9): 2491-2502.

Tan L, Huang C, Peng R, Tang Y, Weiming L,(2014) Development of hybrid organic-inorganic surface imprinted Mn-doped Zn and their application as a sensing material for target proteins. Biosensors amnd Bioelectronics 61, 506-511.

Tom L and Van Nostrand R (1989). Nutrition and feeding of fish. New York NY; 978-1-4757-1174-5.

Toubanaki DK, Margaroni M, Karagouni E. (2015) Nanoparticle- based lateral flow biosensor for visual detection of fish nervous necrosis virus amplification products. Mol Cell Probes; (3): 158- 166

Tratnyek PG, Salter AJ, Nurmi JT, Sarathy V (2010) Environmental applications of zerovalent metals: Iron vs zinc. Nanoscale Materials in Chemistry: Environmental Applications, 165-178.

Tusevljak N, Dutil L, Rajic A, Uhland FC, McClure C, St Hilaire S *et al.* (2013) Antimicrobial Use and Resistance in Aquaculture: Findings of a Globally Administered Survey of Aquaculture Allied Professionals. Zoonoses Public Health, **60**(6): 426-436.

Vaseeharan B, Ramasamy P, Chen JC. (2010) Antibacterial activity of silver nanoparticles (AgNps) synthesized by tea leaf extracts against pathogenic *Vibrio harveyi* and its protective efficacy on juvenile *Feneropenaeus indicus.* Lett Appl. Microbiol; **50**(4): 352-356.

Yang SY, Wu JL, Tso CH, Ngou FH, Chou HY, Nan FH, *et al.* (2012) A novel quantitative immune magnetic reduction assay for Nervous necrosis virus. J Vet Diagn Invest; **24**(5): 911-917.

Zhao L, Seth A, Wibowo N, Zhao CX, Mitter N, Yu C. *et al.* (2014) Nanoparticle vaccines. Vaccine, **32**(3): 327-337.

Zhou X, Wang Y, Gu Q, Li W, (2009) Effects of different Dietary selenium sources on growth performance, muscle composition and glutathionine peroxidise enzyme activity of crusian carp; Aquaculture 291: 78-81.

Chapter 26

Photosystem I: A Revisit of Structure and Function

P. K. Mohapatra[1], Debasmita Panda[2] and Manasweta Mishra[1]*

[1]Department of Botany, Ravenshaw University, Cuttack – 753 003, Odisha
[2]Department of Life Science, Presidency University, Kolkata – 700 083, West Bengal
**e-mail: pradiptamoha@yahoo.com*

ABSTRACT

In plants, algae and cyanobacteria, the primary step of oxygenic photosynthesis is light induced charge separation, driven by the photosystem I (PSI) and photosystem II (PSII). PSI is one of the most fascinating, multisubunit protein complex located in the stromal membrane of thylakoid in plants, algae and cyanobacteria. The PSI proteins are highly conserved from cyanobacteria to higher plants with very few group specific exceptions. However, the structural complexity and polymerization of PSI monomers differs in different photosynthetic organisms. PSI catalyzes the light driven electron flow from plastocyanin/cytochrome on the lumenal side of the membrane to the ferrodoxin/flavodoxin at the stromal side with the help of electron carriers associated with proteins and pigments. Recent approches in understanding of structural details of PSI has opened up an opportunity for comparison of available information on the structure of PSI core and the antenna complex and the function of the components for effective energy transfer. This review presents the structural and functional organization of PSI and its diversity among photosynthetic organisms.

Keywords: *Photosystem I, Protein complex, LHC I, Electron transport, Chlorophyll α, Carotenoids cofactor.*

Introduction

Photosystem I (PSI; plastocyanin-ferredoxin oxidoreductase; EC: 1.97.1.12) is the multiprotein complex that generates electron flow to reduce ferredoxin and oxidize plastocyanin. The concept of two different plant photosystems, PSI being the younger, emerged in 1960s, with the path breaking contribution of Rabinowitch. Robert Emerson's discovery of the Red Drop in the action spectrum of photosynthesis (Emerson and Lewis, 1943), two-light effect in photosynthesis (Emerson *et al.,* 1957; Govindjee and Rabinowitch, 1960), Kok's two light effect on the reaction center P700 (Kok, 1959), and the most important experiment of Duysens *et al.* (1961) were the significant contributions for the origin of the concept of PSI.

PSI generates the most negative redox potential in nature and is extremely efficient in its utilization of light for electron transport leading to reduction of $NADP^+$. Abstractly a PSI can be explained to have a core complex with chlorophylls *a* (Chls *a*) and all the cofactors of the electron transport chain and in most cases of an outer antenna system that increases the light-harvesting capacity (Gobets and van Grondelle, 2001). The core complex is conserved in all organisms performing oxygenic photosynthesis, while the outer antenna varies with the species. In plants, it is composed of Chl *a* and *b* binding proteins (Lhc*a*'s) belonging to the light-harvesting complex (Lhc) multigenic family and together they are called LHCI. In total, the PSI-LHCI complex of higher, plants coordinates around both Chl molecules and carotenoid molecules having high precision to absorb light energy and performs electron flow.

Recent breakthrough in understanding of structural details of PSI has opened up an opportunity for comparison of the available information on the structure of PSI core and the antenna complex. This review discusses the principles of structure and functional organization of PSI complex from photosynthetic organisms. The presentation is largely motivated by the search for the structural implications of the functional features of the PSI complexes predicted and confirmed through research evidences.

The Photosystem I Structure

PSI exists in both monomeric and trimeric forms, but the trimeric forms are primarily abundant in cyanobacteria (Figure 26.1). There also occurrence of miniPSI, dimer of dimmers and oligomeric complexes in cyanobacteria (Watanabe *et al.,* 2011; Li *et al.,* 2014; Zheng *et al.,* 2019; Perez-Boerema *et al.,* 2020). The stability of these oligomeric states varies with the environmental conditions such as light intensity and nutrient supply. The trimeric form is much more stable in thermophilic than in mesophilic cyanobacteria (Chitnis *et al.,* 1993; Malavath *et al.,* 2018). These two forms contain same reaction centers, but differ in their long wavelength antenna chlorophylls. The monomer present in the trimer leads to a large amount of long wavelength chlorophylls and results a higher photochemical activity of the trimer under red and far-red illuminations. This also enables the cyanobacteria to show effective photosynthesis in such illuminations (Gobets and van Grondelle, 2001; Schlodder *et al.,* 2007).

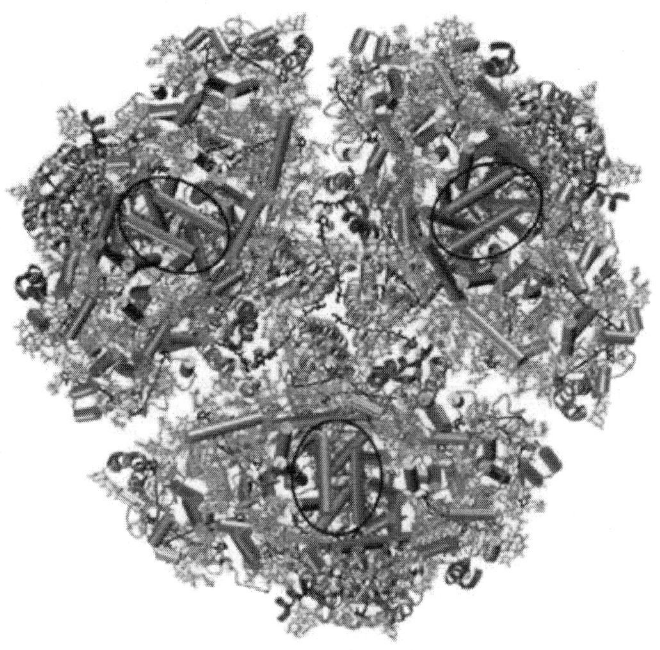

Figure 26.1: Structure of Cyanobacterial PSI Trimer. (PsaA-silver, PsaB-orange, PsaC-yellow, PsaD-cyan, PsaE-lime, PsaF-ice blue, PsaI-purple, PsaJ-blue, PsaK-tan, PsaL-pink, PsaM-mauve, PsaX-ochre; Fromme and Grotjohann, 2006).

In *Synechocystis* sp. PCC6803 it has been observed that the trimeric PSI forms a clover-leaf structure with a diameter of 220Å and molecular weight of 1068 KDa with its protein and cofactors composition. Out of the total mass of PSI, the large contents of cofactors provides more than 30 per cent of mass (Grotjohann and Fromme, 2005). Each trimer is composed of 33 protein subunits, 72 carotenoids, 285 chlorophyll *a* molecules, 51 lipids, 9 iron–sulphur clusters, 6 plastoquinones, 6 putative calcium ions, and over 870 water molecules (Malavath *et al.,* 2018; Netzer-El *et al.,* 2019; Kladawska *et al.,* 2019). Each monomer consists of 12 proteins and 127 cofactors, performing light capturing and electron transport. These cofactors per PSI monomer contain 96 chlorophylls, 22 carotenoids, two phylloquinone, three iron-sulfur (4Fe-4S) clusters and lipids. The 12 different protein subunits are PsaA, PsaB, PsaC, PsaD, PsaE, PsaF, PsaI, PsaJ, PsaK, PsaL, PsaM and PsaX. It is also known that most of the main proteins, including all cofactors binding sites, are well conserved between plants and cyanobacteria, whereas PsaM is a unique protein in cyanobacteria and PsaX has only been identified in thermophilic cyanobecteria (Table 26.1).

Table 26.1: Subunit Composition and Oligomeric State of PSI in different Photosynthetic Producers. The table is compiled from the mentioned references (Jordan *et al.*, 2001, Kargul *et al.*, 2003, Amunts *et al.*, 2007, Peng *et al.*, 2009, Busch *et al.*, 2013, Li *et al.*, 2014).

Structural Components	*T. elongates BP-1*	*Chroococci-diopsis sp. TS-821*	*C. reinhardtii*	*P. patens*	*P. sativum*
Core proteins	PsaA-F, PsaI-L, M, X	PsaA-F, PsaI-L, M, X	PsaA-F, PsaI-L, G, H, N, O	PsaA-F, PsaI-L, G, H, M, O	PsaA-F, PsaI-L, G, H, N, O
Light-harvesting proteins	Phycobili-protein, linker protein	Phycobili-protein, linker protein	Lhca1–9	Lhca1–3, Lhca5	Lhca1–4, Lhca 5, 6
Oligomeric state	Trimer	Tetramer, dimer	Monomer	Monomer	Monomer

Plant PSI is composed of a reaction center up to 15 subunits (PsaA to PsaL, PsaN to PsaP) and antenna complex (LHC I) that consists of 6 Lhca proteins that captures light and transfer the absorbed energy to the reaction center (Table 2). It contains approximately 200 cofactors, which is much larger than cyanobacterial cofactors (Figure 26.2). The structural model of PSI based on X-ray crystallography diffraction reveals the location and structure of 12 of the 15 core subunit within the complex. The main difference between cyanobacterial and plants PSI is monomeric both in vitro and in vivo and the extra subunits of plant PSI are PsaG, PsaH, PsaN, PsaO and PsaP (Haldrup *et al.*, 2006).

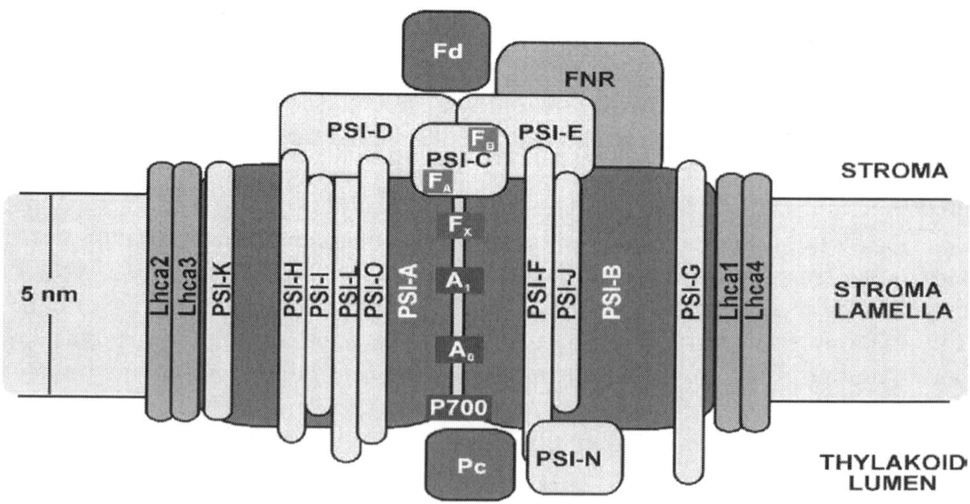

Figure 26.2: Structure of a Monomeric Form of Plant PSI showing the Location of different Protein Subunits in the Thylakoid Membrane. Also shown the path of electron movement from the PSI core to FNR (Haldrup *et al.*, 2006).

Zheng *et al.* (2019) reported the 2.37-Å structure of a tetrameric photosystem I complex from a heterocyst-forming cyanobacterium *Anabaena* sp. PCC 7120. Unlike other photosynthetic organisms, the PSI of the cyanobacterium was a dimer of dimer, form two distinct interfaces that are largely mediated by specifically orientated polar lipids, such as sulfoquinovosyl diacylglycerol. The structure depicts a more closely connected network of chlorophylls across monomer interfaces than those seen in trimeric PSI, possibly allowing a more efficient energy transfer between monomers. The authors also observed a functional link of PSI oligomerization to cyclic electron flow and thylakoid membrane organization. Perez-Boerema *et al.* (2020) determined a simple structure of PSI–LHCI (smallest known PSI) from *Dunaliella salina* with 7 subunits, compared to to a PSI with 14–16 subunits in plants and the algae. This indicated that PSI of *D. salina* represents a different type of the molecular organization that provides important information for reconstructing the plasticity and evolution of PSI.

In the red alga *Cyanidioschyzon merolae* two different froms of PSI were reported, one with three Lhcr subunits attached to the side, similar to that of higher plants, and the other with two additional Lhcr subunits attached to the opposite side, indicating an ancient form of PSI-LHCI (Pi *et al.,* 2018). Such a structure showed features of both cyanobacterial and higher plant PSI core suggesting an intermediate type during evolution from prokaryotes to eukaryotes. There was a unique energy transfer pathways in this alga, which was different from those of higher plant PSI-LHCI, indicating its adaptation to the changing environment, and the possible changes of PSI-LHCI during evolution from prokaryotes to eukaryotes.

PSI Proteins

The Membrane Intrinsic Central Core Proteins

PsaA and PsaB constitute the central core of PSI. These two proteins contain 11 transmembrane helices each (A/B-a to A/B-k) that construct the central part of PSI (Golbeck, 1994). The majority of the chlorophylls is coordinated by PsaA and PsaB and 6 Chls are involved in photosynthetic electron transfer [two Chls (Chla and Chla') constitute P700 reaction centre as the site of charge separation (Käss *et al.,* 2001; Webber and Lubitz, 2001), two Chls are accessory molecules (A) and two Chls are primary electron acceptors (A_0) (Dashdorj *et al.,* 2005). Further electron acceptors are two phylloquinones (A_1) (Kamlowski *et al.,* 1998) and three iron–sulphur clusters of which one is coordinated by conserved cysteines in both PsaA and PsaB proteins (F_x) (Vassiliev *et al.,* 1999) and the other two are coordinated by PsaC stromal subunit (F_A and F_B) (Lakshmi *et al.,* 1999).

The subunits of PsaA/PsaB have two domains: i) the five C-terminal α-helices surrounding the electron transport chain and forms reaction center domain and ii) six N-terminal helices flanking the central region on both sides, arranged as "trimers of helix pairs" where the helix pairs are formed by transmembrane helices a/b, c/d and e/f. (Sun *et al.,* 1997). The C-terminal α-helices separate electron transfer

Table 26.2: The Photosystem I Subunits of *Arabidopsis thaliana*. The table has been compiled from the mentioned references (Amunts *et al.*, 2007; Jensen *et al.*, 2007).

Gene	*Gene Location*	*Accession*	*Protein*	*Mol. mass (kDa)*	*Cofactors*	*Function*	*Mutation*	*Antibody*
PsaA	C	AtCG00350	PsaA	83.2	≈ 79 Chl a, β-car,	Light-harvesting		AS06 172
PsaB	C	AtCG00340	PsaB	82.5	P700, A0, A1, FX	Charge separation Electron transport		AS06 166
PsaC	C	AtCG01060	PsaC	8.9	FA, FB	Electron transport		AS04 042
PsaD	N (2)	At4g02770 At1g03130	PsaD	17.9/17.7		Binding of redoxin Binding of PsaC	SUP and KO	AS04 046
PsaE	N (2)	At4g28750 At2g20260	PsaE	10.4/10.5		Binding of Fd and FNRInvolved in cyclicelectron transport	KO	AS04 047
PsaF	N	At1g31330	PsaF	17.3		Binding of plastocyaninBinding of Lhca1/4	SUP and KO	AS06 104
PsaG	N	At1g55670	PsaG	11.0	1 Chl a 1–2 β-car	Binding of Lhca1/4 Regulation of PSI	SUP and KO	AS04 048
PsaH	N (2)	At3g16140 At1g52230	PsaH	10.4/10.4	1 Chl a	Binding of LHCII (state transitions) Stabilization of PsaD	SUP and KO	AS06 143
PsaI	C	AtCG00510	PsaI	4.1		Stabilization of PsaL		
PsaJ	C	AtCG00630	PsaJ	5.0	2 Chl a	Stabilization of PsaF		
PsaK	N A	t1g30380	PsaK	8.5	2 Chl a	Binding of Lhca2/3	SUP and KO	AS04 049
PsaL	N	At4g12800	PsaL	18.0	3 Chl a	Stabilization of PsaH and PsaO	SUP	AS06 108
PsaN	N	At5g64040	PsaN	9.7		Docking of plastocyanin	SUP and KO	AS06 109
PsaO	N	At1g08380	PsaO	10.1	Chl a ?	Binding of LHCII (state transitions)	SUP and KO	AS04 050
PsaP	N		PsaP					
Lhca1	N	At3g54890	Lhca1	21.5	13 Chl3 car	Light-harvesting	KO	AS01 005

Gene	Gene Location	Accession	Protein	Mol. mass (kDa)	Cofactors	Function	Mutation	Antibody
Lhca2	N	At3g61470 (At5g28450)	Lhca2	23.2	13 Chl2 car	Light-harvesting	SUP	AS01 006
Lhca3	N	At1g61520	Lhca3	24.9	13 Chl3 car	Light-harvesting	SUP	AS01 007
Lhca4	N	At3g47470	Lhca4	22.3	13 Chl2 car	Light-harvesting	KO/SUP	AS01 008
Lhca5	N	At1g45474	Lhca5	24.3	13 Chl2 car	Light-harvesting	KO	AS05 082

system from antenna pigment. The N-terminal α-helix pairs, A/B-e and A/B-f are attached to the C-terminal α-helices. Between α-helices A/B-e to A/B-k hydrophilic interaction are prominent, whereas α-helices A/B-a to A/B-d are separated from the core by approximately 17Å wide elliptical belt containing chlorophylls and carotenoids surrounding this inner core.

The C-terminal region of PsaA/PsaB is called the reaction center domain (Fromme *et al.,* 1994) and shows structural homology to the reaction center core subunits L and M in the PbRC as well as to the reaction center core subunits D_1 and D_2 in PSII (Schubert *et al.,* 1998; Zouni *et al.,* 2001). The C-terminal domain of PsaA and PsaB not only bind the cofactors of electron transfer chain, but also allows the coordination of 25 chlorophylls of the antenna system (12 in PsaA and 13 in PsaB). The N terminal domain of PsaA and PsaB coordinate with 54 chlorophylls (28 in PsaA and 26 in PsaB).

For the docking of soluble electron donor (cytochrome c_6/plastocyanin) of PSI, PsaA and PsaB, also play an essential role. The docking site for plastocyanin or cytochrome c_6 is located at the lumenal side of PSI. The binding pocket of PsaA and PsaB is located at the site that is mainly hydrophobic, forming two surface helices in the loop between α-helices i and j (Sommer *et al.,* 2004). The segment of these alpha helices contains IleA646 to SerA664 in PsaA and TyrB621 to AsnB639 in PsaB. Two tryptophan residues TrpA655 and TrpB63, are located in these alpha helices partially exposed to the aqueous phase are important features of the docking site (Fromme *et al.,* 2003). The aromatic side chains of tryptophan suggest possible role for the interaction of PSI with Cytc_6 and plastocyanin.

There is a high degree of structural symmetry between transmembrane helices of PsaA and PsaB, but symmetry is much weaker in the loop regions, except for the stromal loop that coordinates with F_X and PsaC, PsaD, and PsaE (Antonkine *et al.,* 2002). F_x is rare inter-protein iron-sulfur cluster that coordinates with the loop between transmembrane helices h and i. The loop region of A/B-hi is very important for the function of PSI, which has conserved the cysteine residues (CysA578, CysA587 and CysB565, CysB574) in the motif (FPCDGPGRGGTCXXSAWDH). These conserved cysteines coordinate with the iron sulfur center F_x. This loop is 98 percent conserved between PSI from all species indicating that this structural conservation is the basis for uniform function of PSI in all species. There is another loop region A/B-jk, which provides the binding site of phylloquinone with conserved TrpA697 and TrpB677.

Membrane Extrinsic Proteins

The extrinsic site or the stromal side of PSI contains three subunits *viz.,* PsaC, PsaD and PsaE. PsaC carries two terminal 4Fe-4S clusters, F_A and F_B coordinated by cysteines (Xu *et al.,* 2003). The conserved 4Fe-4S clusters binding motif consists of the sequence C(I)XXC(II)XXC(III)XXXC(IV)P, where C is cysteine, P is proline and X are other noncysteine aminoacids (Golbeck and Bryant, 1991). Conserved cysteines I, II and III of the first binding motif ligate the first (4Fe-4S) clusters, while cysteine IV of the first binding motif ligate the second (4Fe-4S) cluster. On the other hand, cysteines I', II'and III' on the second binding motif ligate the second (4Fe-4S) clusters, while cysteine IV' of the second binding site ligate the first (4Fe-4S) cluster (Zhao *et al.,* 1992).

PsaD is located at the stromal hump coordinating the transfer of electron from PSI to Fd (Chitnis *et al.,* 1997; Barth *et al.,* 1998). It consists of a large antiparallel four standard β-sheet that is connected to one α-helix by a short loop and this helix forms ionic interaction both with PsaC and PsaA (Lagoutte *et al.,* 2001). It forms a clamp on stromal side and performs important function for holding PsaC in its correct orientation and act as a stabilization factor of the electron acceptor sites in PSI. The segment D124-129 forms many hydrogen bonding to PsaB. The C-terminal region of PsaD contacts the N-terminal region of PsaL (Janson *et al.,* 1996). A N-terminal extension of PsaD in plants stabilizes the interaction of PsaC with the rest core of PSI making the stromal side of PSI in plant more stable than in cyanobacteria (Ihnatowicz *et al.,* 2004).

PsaE is with 5 antiparallel strand β-sheet and is located at the distal site of the stromal hump (Jordan *et al.,* 2001) It assembles with PSI core without involvement of any assembly factors (Lushy *et al.,* 2002). Relatively weak interactions are found in PsaE with PsaC and PsaD. Therefore, elimination of PsaE from the complex does not cause notable change in the stromal structure formed by PsaC and PsaD (Barth *et al.,* 1998). However, PsaE is involved in the docking of flavodoxin and ferredoxin (Meimberg *et al.,* 1998). Further it is also involved in cyclic electron transport (Zhao *et al.,* 1993).

Distal Proteins

Four hydrophobic protein subunits, which are membrane exposed such as PsaF, PsaJ, PsaK and PsaX form the distal protein region of PSI. However, PsaF, PsaJ and PsaK are conserved form cyanobacteria to plant but PsaX is found in thermophilic cyanobacteria only. PsaF and PsaJ are variously attached with PsaA, PsaB and PsaF but PsaK is located at the periphery of PsaA. All three subunits can stabilize the core antenna system of PSI. Another important role of these membrane exposed subunits is to interact with membrane intrinsic peripheral antenna system, the IsiA ring and the phycobilisomes (in cyanobacteria).

PsaF contains 3 domains- i) the terminal lumenal domain, ii) a transmembrane domain with one transmembrane helix, and iii) two short helical pieces in a V-shaped arrangement. The hydrophilic α-helices F-c and F-d are the most prominent N-terminal domains. They run parallel to the membrane plane and terminate at a distance of 15Å form the putative docking site of cytochrome c_6 or plastocyanin in cyanobacterial PSI. In plants the lumenal domain is responsible for the formation of a tight complex between plastocyanin and PSI (Hippler *et al.,* 1998). PsaF does not axially coordinate with chlorophylls but forms hydrophobic contacts with chlorophylls and carotenoids. Two function of PsaF are i) shielding of the carotenoids and of chlorophylls form the lipid phase and ii) transfer of excitation energy from external antenna system to PSI core.

PsaJ is a small hydrophobic subunit (6 KDa) with a single transmembrane α -helix in neighbourhood to PsaF (Schubert *et al.,* 1997). It binds three chlorophyll in case of cyanobacteria, whereas two chlorophylls in case of photosynthetic eukaryotes (Ben-Shem *et al.,* 2003). The N-terminus of PsaJ is located in the stroma and C-terminus in the lumen. PsaJ also plays an important role in the stabilization of PsaF. These three chlorophylls play an important role in the transfer of excitation energy from the IsiA ring to the PSI core (Fischer *et al.,* 1999). PsaK with two trans-membrane α-helices is located at the periphery of PSI complex (Mühlenhoff *et al.,* 1993). It is not in contact with any other membrane intrinsic proteins but is in the neighbourhood of PsaA. It coordinates with carotenoids and two chlorophylls. PsaG is subunit unique to plants, with sequence similarity to PsaK (Jensen *et al.,* 2007). In plants PsaK has been shown to interact with the LHCI proteins with a role in state transition (Varotto *et al.,* 2002; Ben-shem *et al.,* 2003).

The Low Molecular Weight Proteins

The PsaL and PsaI are present in the monomer-monomer interface in trimeric cyanobacterial complex, whereas in plant their work is different. PsaL subunit provides the sites for interaction for the three PSI monomers (Chitnis and Chitnis, 1993; Chitnis *et al.,* 1993) and the presence of phosphatidylglycerol is essential for trimer formation (Domonkos *et al.,* 2004). It coordinates with three antenna Chl *a* and hydrophobic contacts with three carotenoids for the excitation energy transfer between monomers. The structure of PsaL contains three transmembrane helices, named L-d, L-e and L-g. The first two (L-d, L-e) form hydrophobic contacts between the monomers which gives a tight trimetric structure with ability for energy shaving (Grotjohann and Fromme, 2005). Most of the further contact side between the monomers in the trimerization domain are provided by hydrogen bond and electrostatic interaction within the loop region of PsaL. The N-terminal loop contains three β-strand and one α-helix. This loop connects with PsaA and PsaD. There are metal binding sites, in which the metal ion Ca^{2+} is coordinated by amino acid side chains of PsaL in two adjacent PSI monomers and by PsaA giving stability to PSI trimer (Grotjohann and Fromme, 2005). The complex is additionally stabilized by PsaI and PsaM proteins (K^{3}odawska *et al.,* 2015, 2019). PsaI is one of the very small subunit of PSI, which contains only one transmembrane helix. This helix does not bind with Chl *a* but with carotenoid forms hydrophobic interaction. PsaI is located between PsaL and PsaM and influences the stability of these two proteins.

The Unique Protein Subunits of Cyanobacterial PSI

Psa M and PsaX are considered as the earliest subunit of cyanobecterial PSI and are unique to this group. Psa M is the smallest subunit of PSI containing one transmemberance α-helix neighbourhood of PsaI and PsaB, although an open reading frame for PsaM was found in the liverwort chloroplast genome. It forms hydrophobic contacts with carotenoid molecules and coordinate with one Chl *a*. The chlorophyll which bound with PsaM plays an important role in excitation energy transfer between monomers. It is functionally coupled with the chlorophylls of neighbouring moromer as well (Sener *et al.,* 2004), so that the side chain of PsaM does not make direct contact between monomers. Psa X (present only in thermophilic cynobecteria) contains a single trans-membrane helix nearer to PsaF with its N terminal amino acids present on the stroma (Grotjohann and Fromme 2005). It coordinates with one Chl and hydrophobic interaction with several carotenoid molecules and one phospholipid. It is assumed that PsaX gives stability to the PSI at high temperature (Grotjohann and Fromme 2005).

The Unique Proteins of Plant PSI

There are five proteins *viz.,* PsaG, PsaH, PsaN, PsaO and PsaP unique to PSI of algae and/or higher plants. PsaG is a small membrane extrinsic protein (ca 10 KDa) containing two transmebrane α-helices connected by a stromal exposed loop (Mant *et al.,* 2001; Rosgaard *et al.,* 2005) and bound to PsaB. It has a role in stabilizing the peripheral antenna but is not needed for binding and function of the peripheral antenna. PsaH is also a small protein (ca 10 KDa) containing one transmembrane α helix and a 20Å long helix lying parallel to the membrane. It has a single transmembrane helix adjacent to PsaL and is involved in phosphorylation-dephosphorylation cycles (state transition; Lunde *et al.,* 2000; Ben-shem *et al.,* 2003).

PsaN is the only subunit located entirely in the thylakoid lumen. It can be dissociated from the PSI complex at high ionic strength and therefore has been concluded to be located externally in the complex and involved in PC oxidation (He and Malkin, 1992; Haldrup *et al.,* 1999). The PsaO (ca 10 KDa) has two transmembrane helices connected by a 29 amino acid long lumenal loop, which is chemically crosslinked with PsaL and PsaH (Knoetzel *et al.,* 2002). Pi *et al.* (2018) found that PsaO binds three chlorophylls *a* and may be important in harvesting energy and in mediating energy transfer from LHCII to the PSI core under state-2 conditions. PsaP is a 14 KDa thylakoid membrane phosphoprotein with unknown function (Hansson and Vener, 2003).

The Components of Electron Transport Chain

The electron transfer chain, which is located in the center of PSI complex, performs the trans-membrane charge separation. The primary electron donor of PSI is P700 (a diamer of Chl a molecules) when charge separation is initiated and other carriers are A_0 (A Chl *a* molecule), A_1 (a phylloquinone) and a three (4Fe-4S) cluster-F_X, F_A and F_B. The chlorophyll and quinone (phylloqinone) cofactors of the electron transfer chain are arranged in a similar way in PSI as in PSII with coordinated oxidation –reduction states (Schubert *et al.,* 1998, Klukas *et al.,* 1999). At 2.5 Å resolution, it was observed that six chlorophyll molecules, two phylloquinones and three (4Fe-4S) clusters are present in electron transfer from PSI core to $NADP^+$. Out of them organic cofactors (Chlorophylls and phylloquinone) are arranged into two branches, named A and B, because most cofactors of the A and B branch are coordinated by PsaA and PsaB. The A-branch contains chlorophyll eC-A1, eC-B1, eC-A3 and phylloquinone molecule Qk-A and the B-branch contains chlorophylls eC-B2, eC-A2, eC-B3 and the phylloquinone Qk-B. It was concluded that cofactors presenting is one branch of PSI are not bound exclusively to one subunit of PSI. The amino acid environment of the reaction center chlorophylls is given in Table 26.3.

Table 26.3: Amino Acids Environment of the Reaction Center Chlotophyll of PSI (P_A, P_B, A_0, Chl_A (A branch, Chl_A (B branch). The table has been compiled from Jordan *et al.* (2001) and Loll *et al.* (2003).

RC Chls	*Amino Acids within H-bonding Distance, Ligands*	*Amino Acids within van der Waals Distance*
PSI PA	A-His680A-Tyr603(?)A-Tyr735 (?)	A-Phe611; A-Trp683; A-Thr742; A-Thr743; A-Phe746, and some more
PSI PB	B-His660B-Tyr727	A-Leu654; B-Phe656; B-Trp663; B-Tyr723
PSI Chl AA (Chl A on A branch; Nr 1012)	-	A-Leu 677; A-Ala684; B-Leu531; B-Trp588; B-Asn591; B-Trp663
PSI Chl AB (ChlA on B branch; Nr. 1022)	-	A-Phe456; A-Phe544; A-Trp601; A-Trp683; B-His660; B-Ala664
PSI A0A (Ao on A branch; Nr. 1013)	A-Met 688; A-Tyr 696; B-Ser429; B-Trp 588	A-Phe681; A-Ala 684; A-Trp697; B-Ser426; B-Phe587
PSI A0B (Ao on B branch; Nr 1023)	B-Ala664; B-Threo665; B-Met668	B-Trp677; A-Asp446; A-Val548; A-Ile 552; A-Phe600

P700

A pair of two cholorophylls is located close to the lumenal surface of PSI. When P700 is excited to its lowest excited state $P700^+$, an electron is very rapidly (Within 1-3 ps) transferred to A_0 (Fromme *et al.,* 2001). P700 has a highly negative redox potential (-1300 mv) and is hence a very strong reductant. The free energy change of the reduction-oxidation of the initial charge separation step was estimated to be 250 mev (Brettel, 1997) indicating that nearest electron accepter of $P700^+$ is located at around -1100mv redox potential (Figure 26.3).

The two chlorophylls of P700 are chemically different; the chlorophyll on the B-branch is the "common" Chl *a* molecule chemically identical to all other 95 Chls is PSI, whereas the chlorophyll at the A-branch in Chl a', the epimer at the C13 position of the chlorine ring system. Both the chlorophylls are axially coordinated to the histidine residues (A680 and B660) but their hydrogen bonding pattern is asymmetric. There is no hydrogen bonding between surrounding proteins and

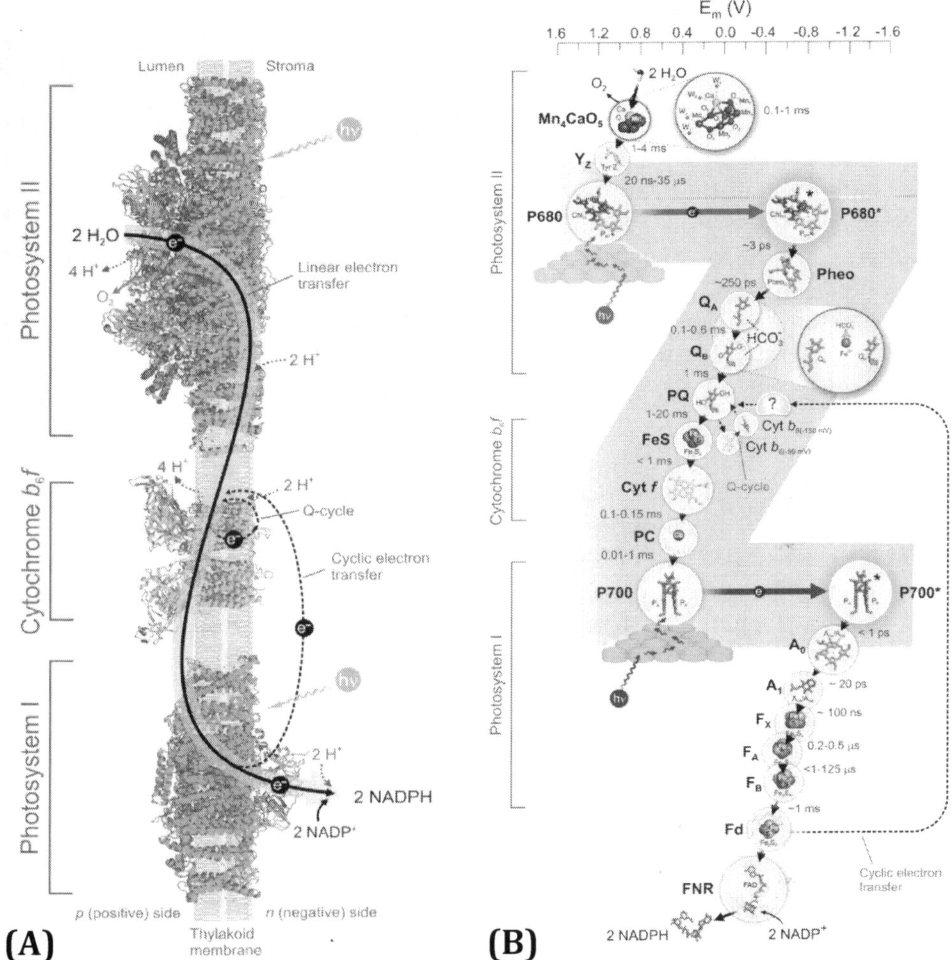

Figure 26.3: The Z Scheme of Electron Transport in Photosynthesis Showing (A) The location of photosystems in the thylakoid membrane and (B) The redox potential and life time of electron transport among the components (Govindjee *et al.*, 2017).

the chlorine head group of eC-B_1, whereas chlorophyll on the A-branch of P700 (eC-A_1) is the Chl *a* molecule that forms three hydrogen bonds to the side chains of transmembrane α-helices A-i and A-k and water molecule. The distance between the Mg^{2+} ions of the two chlorophylls is 6.3Å, which is much shorter than the corresponding distance between the bacteria chlorophylls in the special pair of PbRC. The planes of eC-A_1, and eC-B_1, are oriented perpendicularly to the membrane plane and parallel to each other with an inter planner distance of 3.6Å.

The Accessory Chlorophyll A

The second pair of chlorophylls, eCA_2 and eCB_2 is located between P700 and eCA_3 and it is the initial electron acceptor (Jordan *et al.,* 2001). The second pair of chlorophyll is located at the middle of the membrane. The center to center distance of the second pair of chlorophyll is estimated to be 12Å. For both the branches water molecule is fifth ligand to the central Mg^{2+} ion of eCA_2 and eCB_2. The axial ligand of both these chlorophylls is water molecule that is hydrogen bonded to an asparagine. A branch Chl *a*, coordinates with water molecule is hydrogen bonded to Asn B591 and B branch Chl *a* water molecule is hydrogen bonded to Asn A604 so that the second pair of chlorophyll (eCA_2 for B branch and eCB_2 for A branch) participate in the electron transfer. Studies suggest that charge separation may start from the accessory chlorophyll at B branch instead of P700 (Müller *et al.,* 2003).

The Primary Acceptor A_0

The chlorophylls named eCA_3 and eCB_3 are located in the middle of the membrane similar to the position of pheophytin (pheo) in PSII (Ferreira *et al.,* 2004). There is a parallel orientation of the chlorine rings of eCB_2<= >eCA_3 and eCA_2< =>eCB_3. These two chlorophylls, together termed as A_0, are the first electron acceptors. Both chlorophylls are hydrogen bonded with the tyrosine A696 and B676, which are neighbourhood to the Trp residues W697 of PsaA and W677 of PsaB (Figure 26.4). It stacks with the quinine electron acceptor (Fromme *et al.,* 2001; Jordan *et al.,* 2001). The electron flow primarily operates in the B branch but is some cyanobacteria electron flow in A branch has been reported (Cohen *et al.,* 2004; Fromme and Grotjohann, 2006).

The Phylloquinone A_1

There are two phylloquinone molecules (Q_KA and Q_kB) present in PSI. They are both bound tightly but in case of PSII the quinone Q_A is tightly bound and Q_B is mobile, leaving to the binding pocket, after double reduction, of two protons (forming PQH_2). In PSI, the phyloquinone is singlely reduced and involved in electron transfer to the first iron sulphur center F_X. Out of the two phylloquinones only one is actively involved in electron transfer and the other is in readiness to function, when necessary.

The phylloquinones are located at the stromal side and the protein ligandations of both phylloquinones are identical on A and B branch, but differ from all other quinone binding pockets found in proteins. Both quinones are pi-stacked with a tryptophan residue and show asymmetric hydrogen bonding: only one of the two oxygen atoms forms an H bond to NH backbone group of a serine, whereas the other oxygen atom is not hydrogen bonded (Xu *et al.,* 2003). This asymmetric ligandation may cause very negative redox potential (-770mv) of A1. When the phylloquenone synthesis is blocked, PSI incorporates with plastoquinone and they show photodamage at low light intensity (Semenov *et al.,* 2000).

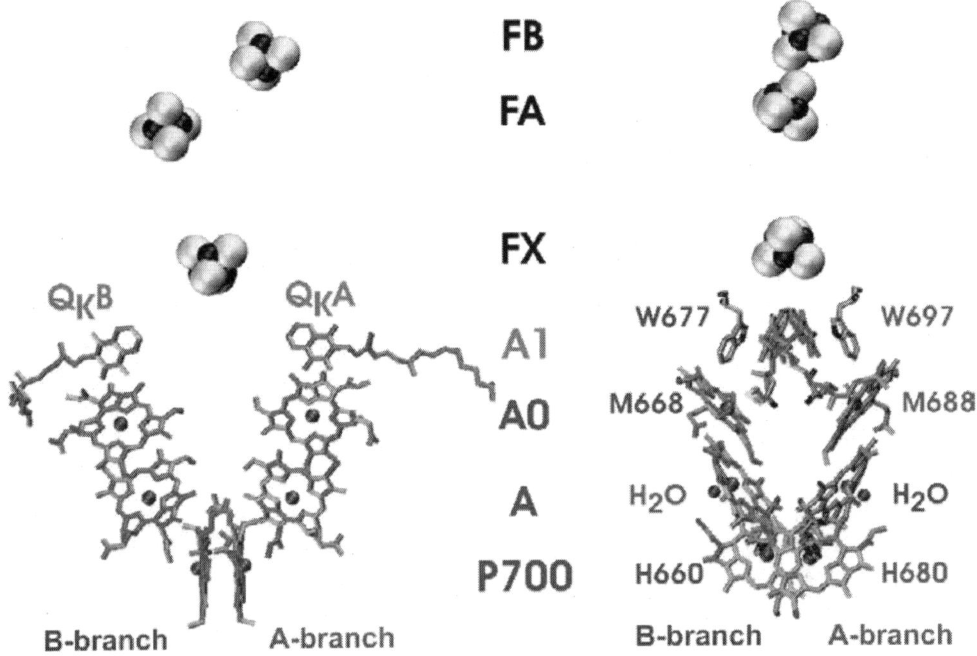

Figure 26.4: The Path of Electron Transfer from P700 through Phylloquinone and (4Fe-4S) Clusters. The phylloquinones, and Fe/S clusters of the electron transfer chain, the coordinating amino acids and water molecules are indicated (Fromme and Grotjohann, 2006).

The electron transfer from Q_KA to F_X is slow ($t_{1/2}$=200ns) because a significant activation energy barrier exists, whereas the electron transfer from Q_KB to F_X is very fast ($t_{1/2}$=15-25ns) (Bittl *et al.*, 1997). The structure shows that the distance from Q_KB and Q_KA to F_X are identical leading to the conclusion that ΔG' must be lower in the B branch than in the A branch thus preferring B branch, over A branch for electron transfer (Akhmatskaya *et al.*, 1999).

The First Fe-S Cluster F_X

F_X is the first of the 4Fe-4S clusters that play an important structural and functional role in PSI. It transfers the electron from the quinones to the terminal FeS clusters F_A and F_B. F_X is ligated by Cys578 and Cys587 of PsaA and Cys565 and Cys574 of PsaB and provides the histidine ligands to the iron atoms of the subsequent FeS clusters. The ligands are located in the loop connecting transmembrane helices h and i (Jordan *et al.*, 2001; Fromme *et al.*, 2001). F_X also plays an important role for stabilization and assembly of PSI complex. There is an assembly protein called rubredoxin (rebA), which is necessary for the assembly of F_X with PsaC, PsaD and PsaE (Shen *et al.*, 2002).

The Terminal Fes Clusters F_A and F_B

The geometric structure of a generic (4Fe-4S) cluster is a protein environment with a distorted cube, which can be depicted as a tetrahedron of four iron atoms interpreting a large tetrahedron of four sulphur atom. The two terminal FeS clusters of PSI are bound to PsaC. The proximal cluster with a center to center distance of 12A° to F_X is F_A and the terminal FeS cluster with a distance of 22A° to F_X is F_B (Yu *et al.,* 1995). PsaD and PsaE play an important role for docking of PsaC, so that PsaC leads to equivalent reduction of F_A and F_B (Li *et al.,* 1991). In the electron transfer process from F_X to $F_{B,}$ F_A, more than 80 per cent electron is present on F_A because it has more negative redox potential than F_B, so electron transfer from F_A to F_B is uphill. The F_B is solvent exposed and F_A is deeply buried within the stomal hump. This arrangement may reduce the wasteful and dangerous electron transfer from PSI to oxygen (Badger *et al.,* 2000; Lima-Melo *et al.,* 2019).

Four types of lipids, *viz.,* phosphatidyl glycerol (PG), monogalactosyldiacyl glycerol (MGDG), digalactosyldiacyl glycerol (DGDG) and sulfoquinovosyldiacyl glycerol (SQDG), are present in thylakoid membranes, and a variety of their functions in PSI have been suggested, such as trimerization, light protection, biosynthesis and activity regulation (Sato, 2004; Mizusawa *et al.,* 2009).

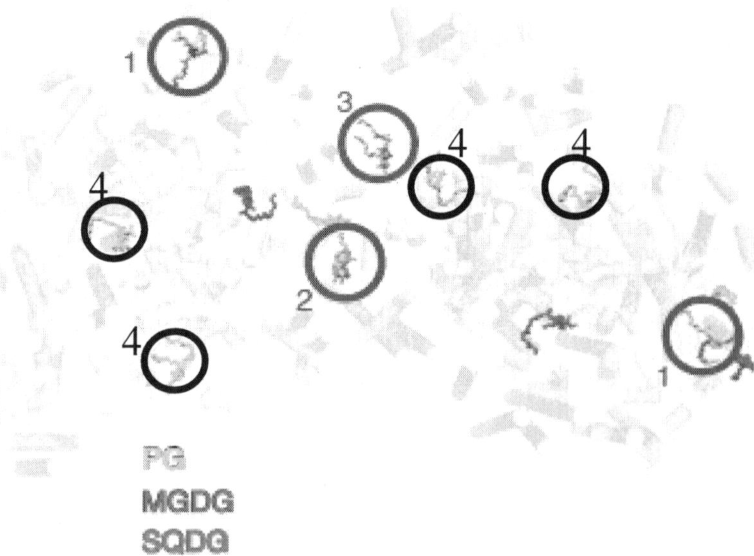

Figure 26.5: The Distribution and Interaction of Lipids in a PSI Dimer. The circles indicated are MGDG (1), SQDG (2,3) and PG (4) (Zheng *et al.,* 2019).

The crystal structure shows that PSI contains three molecule of phosphatidylglycerole (PG) and one molecule of monogalactosyldiacylglycerole (MGDG). Immunological investigation of *S. elongatus* shows these two lipids in the structure of PSI. One of the PGs and the MGDG are located close to *etc.* and function for the balancing of the electron transfer chain. Their head groups completely cover by the loops of PsaA and PsaB and three stromal subunits PsaC, psaD and PsaE. Studies have shown that no PSI can be assembled in mutants lacking PG (Domonkos *et al.,* 2004).

In a PSI dimer, Zheng *et al.* (2019) could model 13 lipid molecules of three types, including six PG, three MGDG and four SQDG (Figure 26.5). One interesting observation is that their head groups are all pointing towards the stromal side. While no SQDG was identified in the PSI trimer from *T. elongatus* (Jordan *et al.,* 2001), seven unsymmetrically distributed SQDG molecules were reported in the PSI trimer from *Synechocystis* 6803 (Malavath *et al.,* 2018). In contrast, Zheng *et al.* (2019) found eight SQDG in the PSI tetramer, only one of which (position 3 in Figure 26.5) is in a nearly identical position in the *Synechocystis* PSI trimer. In the tetramer, four SQDG are located between PsaB and PsaX of each monomer, occupying the equivalent position of a PG in the PSI trimer (Jordan *et al.,* 2001; Malavath *et al.,* 2018). These four lipid molecules play a role in the assembly and stabilization of PsaX on a PSI monomer. The side chain of Arg414 of PsaB 'grabs' this SQDG through a pair of hydrogen bonds with the head group. The sulphate group of this SQDG is embedded into the pocket formed by an N-terminal helix of PsaX, and extensive hydrogen bonds could form between the suphate group and three main chain nitrogens of PsaX, Thr19, Phe20, and Arg21, and the side chain of Arg21. The rest of SQDG in the PSI tetramer in interface I is capable of forming extensive hydrogen bonds with surrounding residues from the two monomers.

The lipids play specific role not only in the assembly of PSI monomers to stabilize subunit interactions within a monomer, but also in the organization of monomers into oligomeric PSI. The distinct distribution of lipids within the PSI tetramer and trimer indicate that the regulation of lipid biosynthesis may influence the oligomeric state of PSI. Given the presence of a large amount of lipids in the monomer interface, it is seen that lipid–PSI interactions are vital for proper functioning of electron transport in PSI. Van der Waals interactions play a major role in the assembly process, and the contribution from such interactions between lipids and PSI is about 50-fold higher than the interactions among PSI monomers. When the potential energy is fractioned into lipid classes, it is seen the largest contribution is from SQDG and MGDG (Zheng *et al.,* 2019). These results suggest that the establishment of specific lipid interactions within the PSI monomer as well as among the monomers in an oligomer is a factor for monomer-monomer coordination. Further, lipid partition might be another factor in regulating the PSI assembly.

Assembly Factors of the Subunit and Cofactors of PSI

The biogenesis of thylakoid membrane is a complex process including the synthesis and maintainance of protein subunit and cofactors. Scientific researches have identified that many factors are found for protein sorting, insertion and assembly of photosynthetic complexes (Rochaix, 2011, Chi *et al.,* 2012). For the assembly of PSI, a set of facilitating factors has been found to help the proper folding of newly synthesized PSI subunits, integration of many cofactors and light harvesting antenna (Table 26.4).

Assembly Factors of PSI Proteins

A large family of assembly factors is involved in proper integration and orientation of PSI proteins. Those studies so far are Alb3, VIPP1, Ycf3, Ycf4, Ycf37, Pyg7, Y3IP1, PPD1 and Psa2. Alb3 (Albino3) belongs to the membrane insertases YidC-Oxal-Alb3 family and highly conserved in bacteria (YidC), mitochondria (Oxal) and chloroplasts (Wang and Dalbey 2011, Saller *et al.,* 2012; Dalbey *et al.,* 2014). Alb_3 was first found as a component of the machinery for post-translational insertion of the LHC proteins into the thylakoid membrane (Moore *et al.,* 2000). In addition to that Alb3 may mediate PSI assembly by assisting the insertion of the reaction center protein, PSaA and PSaB. Alb3.1 and Alb3.2 were mainly responsible for the assembly of the LHC complex and reaction center complex respectively (Pasch *et al.,* 2005).

VIPP1 (Vesicle- inducing protein in plastid 1) is an important assembly factor in thylakoid biogenesis from cyanobacteria to plant although the role differs from species to species. In eukaryotic species like *C. reinhardtii,* VIPP1 is involved in the formation of thylakoid membrane mediated by Alb3 (Gohre *et al.,* 2006; Nordhues *et al.,* 2012). In case of *Arabidopsis,* VIPP1 is an essential factor for thylakoid maintenance by vesicle transport to target membrane (Kroll *et al.,* 2001). Ycf3 interacts with PSaA and PSaD and weakly associated with stromal ridge (Albus *et al.,* 2010). Ycf4 forms a large intermediate complex of PSI from PsaA to PsaF (Ozawa *et al.,* 2009) in *C. reinhardtii* but it acts as non-essential factor for PSI assembly (Krech *et al.,* 2012). It is assumed that the role of Ycf4 has been minimized during evolution. Ycf37 plays important role in the formation of PSI trimers in cyanobacteria (Wilde *et al.,* 2001).

Y3IP1 (Ycf3-interacting protein1) is a newly evolved nuclear encoded protein consesved in photosynthetic eukaryotes and interact with Ycf3 (Albus *et al.,* 2010). In addition to that, Y3IP1 is involved in the formation of stromal ridge during PSI assembly. PPD1 (PSBP-domain protein1) has been identified as a first thylakoid lumenal protein that is essential for PSI assembly. Loss of PPD1 results in destability of PSI complex (Liu *et al.,* 2012). PPD1 interacts with PsaA and PsaB in the luminal sides of the thylakoid membrane. Psa2, a DnaJ-type zinc finger protein, is a thylakoid luminal protein interacting with PsaG-containing subcomplex in Maize and *Arabidopsis* (Fristedt *et al.,* 2014).

Assembly Factors for PSI Cofactors

The assembly factors for PSI cofactors are CnfU/Nfu2 and Hcf101. RubA (rubredoxin) is required for the insertion of iron sulfur cluster F_x into the PsaA/B heterodimer and is essential for assembly in *Synechocystis sp.* PCC7002 (Shen *et*

Table 26.4: PSI Assembly Factors of Oxygenic Organisms. The information have been compiled from the source indicated in the reference column.

Assembly	*Syn*	*C. r.*	*P. p.*	*A. t.*	*Locali-zation*	*Proposed Functions*	*Reference*
Sub unit assembly factor							
Alb3	Slr1471	CRFT_187295	PP_13532	AT2G28800	Intrinsic, PDM, TM	Integration, folding, and/or assembly of PsaA/B, pD1 and LHC	Pasch *et al.*, 2005; Gohre *et al.*, 2006
VIPP1	sll0617	CRFT _134824	PPFT_117800	AT1G65260	TM	PSI assembly, thylakoid biogenesis, chloroplast envelope maintenance	Nordhues *et al.*, 2012; Kroll *et al.*, 2001
Ycf3	Slr0823	ChreCp037	PhpapaCp038	ATCG00360	TM	PSI assembly and/or stromal ridge formation	Ruf *et al.*,1997; Albus *et al.*, 2010
Ycf4	Sll0226	ChreCp038	PhpapaCp028	ATCG00520	TM	Scaffold for PSI assembly and/or stromal ridge formation	Ozawa *et al.*, 2009; Boudreau *et al.*, 1997
Ycf37/ Pyg7	Slr0171	CRFT _184916	PPFT _148653	AT1G22700	TM	Formation of PSI trimers, later steps of PSI assembly	Wilde *et al.*, 2001
Y3IP1	No	CRFT _158401	PPFT _142249	AT5G44650	TM	PSI assembly by interacting with Ycf3	Albus *et al.*, 2010
PPD1	No	CRFT _205916	PPFT _131430	AT4G15510	Extrinsic,L	Early steps of PSI assembly by integrating with specific domains of PsaA/B	Liu *et al.*, 2012
Psa2	No	CRFT _160591	PPFT _167917	AT2G34860	L	Early and/or later steps of PSI assembly	Fristedt *et al.*, 2014
RubA	Slr2033	CRFT _182463	PPFT _163749	AT1g54500	TM	Integration of Fx in PSI assembly, PSII accumulation	Shen *et al.*, 2002
Hcf101	Slr0067	CRFT _102098	PPFT _227402	AT3G24430	S	Scaffold for [4Fe–4S] cluster assembly	Schwenkert *et al.*, 2010
CnfU	Sll2667	CRFT _104310	PPFT _26756	AT4G01940 AT5G49940	S	Scaffolds for iron–sulfur cluster assembly and delivery	Touraine *et al.*, 2004

Syn: Synechocystis; C.r.: Chlamydomonas reinhardtii; P.p..: Physcomitrella patens; A.t..: Arabidopsis thalina; CRFT.: CHLREDRAFT; PPFT.: PHYPADRAFT; TM.: Thylakoid membrane, L.: lumen, S-Stroma. PDM.: Part A-defiened membrane.

al., 2002). Studies of different oxygenic organisms show that RubA has no effect on the PSI functions. In *Arabidopsis,* CnfU-V/Nfu2 mutant exhibit a dwarf and faint pale-green phenotype with reduced protein of PSI and ferredoxin. It may act as an iron sulfur cluster scaffold protein that is required for the biogenesis of PSI and ferredoxin (Touraine *et al.,* 2004). Hcf101 belongs to the 4Fe- 4S cluster containing P-loop NTPase superfamily and acts as the mediator in biogenesis of iron sulfur clusters (Schwenkert *et al.,* 2010). It functions as a scaffold for 4Fe-4S cluster assembly (Stockel and Oelmuller, 2004).

The Electron Flow in PSI

The lumenal surface of PSI contains a pair of cholorophylls that acts as the PSI reaction center and is designated as p700. When P700 is excited there is a rapid movement of electron to a phylloquinone (A_0) which acts as the primary electron acceptor like pheophytin of PSII (Fromme *et al.,* 2001; Ferreira *et al.,* 2004). Because of its highly negative redox potential (-1300 mv), P700 functions as a strong reductant. Both chlorophylls are hydrogen bonded with the tyrosine A696 and B676, which are neighbourhood to the Trp residues W697 of PsaA and W677 of PsaB (Figure 26.3). The two phylloquinone molecules (Q_KA and Q_kB) bound tightly to PSI are singlely reduced and involved in ET to the first iron sulphur center F_X. Out of the two phylloquinone molecules only one is actively involved in electron transfer and the other is in readiness to function, when necessary. The electron from the reduced phylloquinone is transferred to the first 4Fe-4S cluster (F_x), which transfers the electron to the terminal FeS clusters F_A and F_B. F_X is ligated by Cys578 and Cys587 of PsaA and Cys565 and Cys574 of PsaB and provide the histidine ligands to the iron atoms of the subsequent FeS clusters (F_A and F_B). From F_B, electron is donated to ferredoxin (Fd). The electrons from Fd are used to produce NADPH by ferredoxin-NADP+ reductase (FNR) thus compleing the linear electron flow from water vis photosystems to NADP+.

In cyanobacteria thylakoid membranes are the site for both photosynthesis and respiration. Therefore, some of the components of electron transport chains are shared between these two networks. This greatly facilitates the occurrence of the crosstalk between the two processes (Liu, 2016). It is not entirely clear how the fate of an electron is regulated in either of the transfer chains. Depending on the cellular requirement and environmental regulation, two modes of photosynthetic electron transfer: a non-cyclic from PSII via PSI to $NADP^+$ and a cyclic around PSI take place. In *Synechocystis,* there are potentially three pathways by which electrons might turn back to PQ pool from the acceptor side of PSI for ATP formation (cyclic electron flow): i) via NAD(P)H dehydrogenase type 1 complex (NDH-1) (He *et al.,* 2015), ii) via type 2 NAD(P)H dehydrogenase (NDH-2) (Howitt *et al.,* 1999) and iii) via Fd-PQ reductase (FQR) (Yeremenko *et al.,* 2005). There is a number of other potential electron sinks from the classical photosynthetic electron transport chain (terminal oxidases, flavodiiron proteins (Flv1–4), cytochromes *c*M and *c*6B/*c*6C, ferredoxins *etc.*) that may play protective roles by tunnelling excess electrons away thus preventing reactive oxygen species generation (Lea-Smith *et al.,* 2016).

The balance between the electron pressure in the photosynthetic electron transport chain and the capacity of electron acceptors in sink pathways is crucial to maintain the photosynthetic activity. Excess photosynthetic electron transport might be dangerous due to the risk of molecular oxygen reduction, generating reactive oxygen species (ROS) over-accumulation. In photoautotrophs, the rate of electron transport in thylakoid membranes should be balanced with the capacity of electron acceptor pool to reduce an excess ROS production (Lima-Melo *et al.,* 2019).

Conclusion

A good amount of knowledge has been gained regarding the structure of PSI during the last two decades. The structure of subunits of PSI, the discovery of new subunits, the structure of LHCI and the spectral characteristics of the pigments are the most important parts, which are responsible for excitation energy transfer, light harvesting, charge separation and electron transport. Their structure, organization and function are on the control of nuclear as well as chloroplast genome. Detailed dissection of molecular organization of PSI in cyanobacteria to plants has provided the important insights into the evolution of this complex. However, many issues regarding their structure and function have still remained unsolved. This is mainly due to high complexity of the system (the core protein around 100 Chls) which still represents a great challenge for modeling. The antenna system of PSI varing in each species and it was studied that variation play an important role for adaptation.

REFERENCES

Akhmatskaya, E.V., Cooper, M.D., Burton, N.A., Masters, A.J. and Hillier, I.H. (1999). Monte Carlo simulations of water clusters on a parallel computer using an ab initio potential. Int. J. Quant. Chem. 74: 709–719.

Albus, C.A., Ruf, S., Schöttler, M.A., Lein, W., Kehr, J. and Bock, R. (2010). Y3IP1, a nucleus-encoded thylakoid protein, cooperates with the plastid-encoded Ycf3 protein in photosystem I assembly of tobacco and *Arabidopsis.* Plant Cell 22: 2838–2855.

Amunts, A., Drory, O., and Nelson, N. (2007).The structure of a plant photosystem I supercomplex at 3.4 angstrom resolution. Nature 447(7140): 58–63.

Antonkine, M.L., Liu, G.H., Bentrop, D., Bryant, D.A., Bertini, I., Luchinat, C., Golbeck, J.H. and Stehlik, D. (2002). Solution structure of the unbound, oxidized Photosystem I subunit PsaC, containing [4Fe-4S] clusters FA and FB: a conformational change occursupon binding to Photosystem I. J Biol.Inorg. Chem. 7: 461–472.

Badger, M.R., Caemmerer, S. von., Ruuska, S. and Nakano, H. (2000). Electron flow to oxygen in higher plants and algae: rates and control of directphotoreduction (Mehler reaction) and rubisco oxygenase. Phil. Trans. R. Soc.Lond. B. Biol. Sci. 355: 1433–1446.

Barth, P., Lagoutte, B. and Se´tif, P. (1998). Ferredoxin reduction by Photosystem I from *Synechocystis sp.* PCC 6803: toward anunderstanding of the respective roles of subunits PsaD and PsaE in ferredoxin binding. Biochemistry 37: 16233–16241.

Ben-Shem, A., Frolow, F. and Nelson, N. (2003). Crystal structure of plant Photosyștem I. Nature 426: 630–635.

Bittl, R., Zech, S.G., Fromme, P., Witt, H.T. and Lubitz, W. (1997). *Pulsed EPR structure analysis of photosystem I single crystals: Localization of the phylloquinone acceptor.* Biochemistry 36: 12001-12004.

Boudreau, E., Takahashi, Y., Lemieux, C., Turmel, M. and Rochaix, J.D. (1997). The chloroplast ycf3 and ycf4 open reading frames of *Chlamydomonas reinhardtii* are required for the accumulation of the photosystem I complex. EMBO. J. 16: 6095–6104.

Brettel, K. (1997). Electron transfer and arrangement of the redox cofactors in Photosystem I. Biochim.Biophys. Acta. 1318: 322–373.

Busch, A., Petersen, J., Webber-Birungi, M.T., Powikrowska, M., Lassen, L.M., Naumann-Busch, B., Nielsen,A.Z., Ye, J.,Boekema, E.J., Jensen, O.N., Lunde, C. and Jensen, P.E.(2013). Composition and structure of photosystem I in the moss Physcomitrella patens. J. Exp. Bot. 64: 2689–2699.

Chi, W., Ma, J. and Zhang, L. (2012). Regulatory factors for the assembly of thylakoid membrane protein complexes. Philos. Trans. R. Soc. Lond. B Biol. Sci. 367: 3420–3429.

Chitnis, V.P., Ke, A. and Chitnis, P.R. (1997). The PsaD subunit of Photosystem I. Mutations in the basic domain reduce the levelof PsaD in the membranes. Plant Physiol. 115: 1699–1705.

Chitnis, V.P., Xu, Q., Yu, L., Golbeck, J.H., Nakamoto, H., Xie, D.L. and Chitnis, P.R. (1993). Targeted inactivation of the gene psaL encoding a subunit of photosystem I of the cyanobacterium Synechocystis sp. PCC 6803. J. Biol. Chem. 268: 11678–11684.

Cohen, R.O. Shen, G., Golbeck, J.H., Xu, W., Chitnis, P.R., Valieva, A.I., Est Avan der, Pushkar, Y. and Stehlik, D. (2004).Evidence.for asymmetric electron transfer in cyanobacterial Photosystem I: analysis of a methionine-to-leucine mutation of the ligand to the primary electron acceptor A0. Biochemistry 43: 4741–4754.

Dalbey, R.E., Kuhn, A., Zhu, L. and Kiefer, D. (2014). The membrane insertaseYidC. Biochim. Biophys. Acta. 1843: 1489–1496.

Dashdorj, N., Xu, W. and Cohen, R.O., Golbeck, J.H. and Savikhin, S. (2005). Asymmetric electron transfer in cyanobacterial photosystem I: charge separation and secondary electron transfer dynamics of mutations near the primary electron acceptor A_0. Biophys. J. 88: 1238–1249.

Domonkos, I., Malec, P., Sallai, A., Kovacs, L., Itoh, K., Shen, G., Ughy, B., Bogos, B, Sakurai, I., Kis, M., Strzalka, K., Wada, H., Itoh, S., Farkas, T. and Gombos, Z. (2004). Phosphatidylglycerol is essential for oligomerization of Photosystem I reaction center. Plant Physiol. 134: 1471–1478.

Duysens, L.N.M., Amesz, J., and Kamp, B.M. (1961). Two photochemical systems in photosynthesis. Nature 190: 510–511.

Emerson, R. and Lewis, C.M. (1943). The dependence of the quantum yield of Chlorella photosynthesis on wavelength of light. Am J. Bot. 30: 165–178.

Emerson, R., Chalmers, R. and Cederstrand, C.N. (1957). Some factors influencing the long-wave limit of photosynthesis. Proc. Natl. Acad. Sci. U.S.A 43: 133–143.

Ferreira, K.N., Iverson, T.M., Maghlaoui, K., Barber, J. and Iwata, S. (2004). Architecture of the Photosynthetic oxygen-evolving center. Science 303: 1831–1838.

Fischer, N., Boudreau, E., Hippler, M., Drepper, F., Haehnel, W. and Rochaix, J.D. (1999). A large fraction of PsaF is nonfunctional in photosystem I complexes lacking the PsaJ subunit. Biochemistry 38: 5546–5552.

Fristedt, R., Williams-Carrier, R., Merchant, S.S. and Barkan, A. (2014). A thylakoid membrane protein harboring a DnaJ-type zinc finger domain is required for Photosystem I accumulation in plants. J. Biol. Chem. 289: 30657–30667.

Fromme, P. and Grotjohann, I. (2006). Structural analysis of cyanobacterial photosystem I. In. Golbeck, J.H. (ed): Photosystem I: The Light-Driven Plastocyanin: Ferredoxin Oxidoreductase, Springer, pp. 47–69.

Fromme, P., Jordan, P. and Krauß, N. (2001). Structure of Photosystem I. Biochim. Biophys. Acta.1507: 5–31.

Fromme, P., Melkozernov, A., Jordan, P. and Krauß, N. (2003). Structure and function of Photosystem I: interaction with its soluble electron carriers and external antenna systems. F.E.B.S. Lett. 555: 40–44.

Fromme, P., Schubert, W.D. and Krauss, N. (1994). Structure of photosystem I: suggestions on the docking sites for plastocyanin, ferredoxin and the coordination of P700.Biochim. Biophys. Acta. 1187: 99-105.

Gobets, B. and van Grondelle, R. (2001). Energy transfer and trapping in photosystem I. Biochim. Biophys. Acta. 1507: 80–99.

Göhre, V., Ossenbühl, F., Crèvecoeur, M., Eichacker, L.A. and Rochaix, J.D. (2006). One of two ALB3 proteins is essential for the assembly of the photosystems and for cell survival in *Chlamydomonas*, Plant Cell 18: 1454–1466.

Golbeck, J.H. (1994). Photosystem I in cyanobacteria. In: Bryant DA (ed) Advances in Photosynthesis: The Molecular Biology of Cyanobacteria, Vol. 1: 319–360. Kluwer Academics, Dordrecht, The Netherlands.

Golbeck, J.H. and Bryant, D.A. (1991) Photosystem I. In: Lee CP (ed) Current Topics in Bioenergetics, Vol 16, pp 83–177. Academic Press, Inc., San Diego.

Govindjee and Rabinowitch, E. (1960). Two forms of chlorophyll a in vivo with distinct hotochemical functions. Science 132: 355–356.

Govindjee, Shevela, D. and Björn, L.O. (2017). Evolution of the Z-scheme of photosynthesis. Photosynth. Res. 133: 5-15.

Grotjohann, I. and Fromme, P. (2005). Structure of cyanobacterial photosystem I. Photosynth. Res. 85: 51–72.

Haldrup, A., Jensen, P.E., Scheller, H.V. (2006). The lower molecular subunits of higher plants Photosystem I In: John H. Golbeck (ed): Photosystem I: The Light-Driven Plastocyanin: Ferredoxin Oxidoreductase. Springer: 139–154.

Haldrup, A., Naver, H. and Scheller, H.V. (1999). The interaction between plastocyanin and photosystem I is inefficient in transgenic *Arabidopsis* plants lacking the PSI-N subunit of photosystem I. Plant J. 17: 689–698.

Hansson, M. and Vener, A.V. (2003). Identification of three previously unknown in vivo phosphorylation sites in thylakoid membranes of Arabidopsis thaliana. Mol. Cell. Proteomics 2: 550–559.

He, W.Z. and Malkin, R. (1992). Specific release of a 9-kDa extrinsic polypeptide of photosystem-I from spinach-chloroplasts by salt washing. F.E.B.S. Lett. 308: 298–300.

He, Z., Zheng, F., Wu, Y., Li, Q., Lv, J., Fu, P. and Mi, H. (2015). NDH-1L interacts with ferredoxin via the subunit NdhS in *Thermosynechococcus elongatus*. Photosynth. Res. 126: 341–349.

Hippler,M., Drepper, F., Haehnel, W. and Rochaix, J.D. (1998). The N-terminal domain of PsaF: precise recognition site for binding and fast electron transfer from cytochrome c6 and plastocyanin to Photosystem I of Chlamydomonas reinhardtii. Proc. Natl. Acad. Sci. U.S.A 95: 7339–7344.

Howitt, C.A., Udall, P.K. and Vermaas, W.F (1999). Type 2 NADH dehydrogenases in the cyanobacterium *Synechocystis* sp. strain PCC 6803 are involved in regulation rather than respiration. J. Bacteriol. 181: 3994–4003.

Ihnatowicz, A., Pesaresi, P.,Varotto, C., Richly, E., Schenider, A., Jahns, P., Salamini, F. and Leister, D. (2004). Mutants for photosystem I subunit D of *Arabidopsis thaliana*: effects on photosynthesis, photosystem I stability and expression of nuclear genes for chloroplast functions. Plant J. 37: 839–852.

Janson, S., Andersen, B. and Scheller, H.V. (1996). Nearest-neighbor analysis of higher-plant Photosystem I holocomplex. Plant Physiol. 112: 409–420.

Jensen, P.E., Bassi, R., Boekema, E.J., Dekker, J.P., Jansson, S., Leister, D., Robinson, C. and Scheller, H.V. (2007). Structure, function and regulation of plant photosystem I. Biochim.Biophys. Acta.Bioenerg. 1767(5): 335–352.

Jordan, P., Fromme, P., Witt, H.T., Klukas, O., Saenger, W. and Krauss, N. (2001). Three-dimensional structure of cyanobacterial photosystem I at 2.5 A ° resolution. Nature 411: 909–917.

Kamlowski, A., Altenberg-Greulich, B., Van Der Est, A. Zech, S.G., Bittl, R., Fromme, P., Lubitz, W. and Stehlik, D. (1998). The quinone acceptor A1 in photosystem I: binding site, and comparison to QA in purple bacteria reaction centers. J. Phys. Chem. B 102: 8278–8287.

Kargul, J., Nield, J. and Barber, J. (2003). Three-dimensional reconstruction of a light-harvesting complex I – Photosystem I (LHCI-PSI) supercomplex from the green alga *Chlamydomonas reinhardtii*. Insights into light harvesting for PSI. J. Biol. Chem. 278: 16135–16141.

Käss, H., Fromme, P., Witt, H.T. and Lubitz, W. (2001). Orientation and electronic structure of the primary donor radical cation P-700(+center dot) in Photosystem I: a single crystals EPR and ENDOR study. J. Phys. Chem. B. 105: 1225–1239.

K^{3}odawska, K., Kovács, L., Várkonyi, Z., Kis, M., Sozer, Ö., Laczkó-Dobos, H., Kóbori, O., Domonkos, I., Strza^{3}ka, K., Gombos, Z. and Malec, P. (2015). Elevated growth temperature can enhance photosystem I trimer formation and affects xanthophyll biosynthesis in cyanobacterium *Synechocystis* sp. PCC 6803 cells. Plant Cell Physiol. 56: 558–571.

K^{3}odawska, K., Kovács, L., Vladkova, R., Rzaska1, A., Gombos, Z., Laczko-Dobos, H. and Malec, P. (2019). Trimeric organization of photosystem I is required to maintain the balanced photosynthetic electron flow in cyanobacterium *Synechocystis* sp. PCC 6803. Photosynth. Res. 143(3): 251-262.

Klukas, O., Schubert, W.D., Jordan, P., Krauß, N., Fromme, P., Witt, H.T. and Saenger, W. (1999). Localization of two phylloquinones, QK and QK', in an improved electron density map of Photosystem I at 4 A ° resolution. J Biol Chem 274: 7361–7367.

Kok, B. (1959). Light induced absorption changes in photosynthetic organisms. II. A split beam difference spectrophotometer. Plant Physiol. l34: 184–192.

Krech, K., Ruf, S., Masduki, F.F., Thiele, W., Bednarczyk, D., Albus, C.A., Tiller, N., Hasse, C., Schottler, M.A. and Bock, R. (2012). The plastid genome-encoded Ycf4 protein functions as a non-essential assembly factor for photosystem I in higher plants. Plant Physiol.159: 579–591.

Kroll, D., Meierhoff, K., Bechtold, N., Kinoshita, M., Westphal, S., Vothknecht, U.C., Soll, J. and Westhoff, P. (2001). VIPP1, a nuclear gene of Arabidopsis thaliana essential for thylakoid membrane formation. Proc. Natl. Acad. Sci. U. S. A. 98: 4238–4242.

Lagoutte, B., Hanley, J. and Bottin, H. (2001). Multiple functions for the C terminus of the PsaD subunit in the cyanobacterialphotosystem I complex. Plant Physiol. 126: 307–316.

Lagoutte, B., Hanley, J. and Bottin, H. (2001.) Multiple functions for the C terminus of the PsaD subunit in the cyanobacterial photosystem I complex. Plant Physiol. 126: 307–316.

Lakshmi, K., Jung, Y. and Golbeck, J. (1999). The iron–sulfur clusters F_A and F_B in photosystem I: an electron paramagnetic resonance. Biochemistry 38: 13210–13215.

Lea-Smith, D.J., Bombelli, P., Vasudevan, R. and Howe, C.J. (2016). Photosynthetic, respiratory and extracellular electron transport pathways in cyanobacteria. Biochim. Biophys. Acta 1857: 247–255.

Li, M., Dmitry, A.S., Egbert, J.B. and Barry, D.B. (2014). Characterization and evolution of tetrameric photosystem I from the thermophilic cyanobacterium *Chroococcidiopsis sp*TS-821. Plant Cell 26: 1230–1245.

Li, N., Zhao, J.D., Warren, P.V., Warden, J.T., Bryant, D.A. and Golbeck, J.H. (1991).PsaD is required for the stable binding of PsaC to the Photosystem I core protein of *Synechococcus sp.* PCC 6301. Biochemistry 30: 7863–7872.

Lima-Melo, Y., Alencar, V.T.C.B., Lobo, A.K.M., Sousa, R.H.V., Tikkanen, M., Aro, E-M., Silveira, J.A.G. and Gollan, P.J. (2019). Photoinhibition of photosystem I provides oxidative protection during imbalanced photosynthetic electron transport in *Arabidopsis thaliana*. Front. Plant Sci. 10: 1–13.

Liu, J., Yang, H., Lu, Q., Wen, X., Chen, F., Peng, L., Zhang, L., Lu, C. (2012).PsbP-domain protein 1, a nuclear-encoded thylakoid lumenal protein, is essential for photosystem I assembly in *Arabidopsis*. Plant Cell 24: 4992–5006.

Liu, L.N. (2016). Distribution and dynamics of electron transport complexesin cyanobacterial thylakoid membranes. Biochim. Biophys. Acta 1857: 256–265.

Loll, B., Raszewski, G., Saenger, W. and Biesiadka, J. (2003). Functional role of Cα-H_O hydrogen bonds between transmembrane α-helices in photosystem I. J. Mol. Biol. 328: 737–747.

Lunde, C., Jensen, P.E., Haldrup, A., Knoetzel, J. and Scheller, H.V. (2000). The PSI-H subunit of photosystem I is essential for state transition in plants. Nature 408: 613–615.

Lushy, A., Verchovsky, L. and Nechushtai, R. (2002). The stable assembly of newly synthesized PsaE into the photosystem I complex occurring via the exchange mechanism is facilitatedby electrostatic interactions. Biochemistry 41: 11192–11199.

Malavath, T., Caspy, I., Netzer-El, S.Y., Klaiman, D. and Nelson, N. (2018). Structure and function of wild-type and subunit-depleted photosystem I in *Synechocystis*. Biochim. Biophys. Acta 1859: 645–654.

Mant, A., Woolhead, C.A., Moore, M., Henry, R. and Robinson, C. (2001). Insertion of PsaK into the thylakoid membrane in a "Horseshoe" conformation occurs in the absence of signal recognition particle, nucleoside triphosphates, or functional albino3. J. Biol. Chem. 276: 36200–36206.

Meimberg, K., Lagoutte, B., Bottin, H. and Mu¨hlenhoff, U. (1998). The PsaE subunit is required for complex formation between Photosystem I and flavodoxin from the cyanobacterium *Synechocystis sp.* PCC 6803. Biochemistry 37: 9759–9767.

Mizusawa, N., Sakata, S., Sakurai, I., Sato, N. and Wada, H. (2009). Involvement of digalactosyldiacylglycerol in cellular thermotolerance in *Synechocystis* sp. PCC 6803. Arch. Microbiol. 191: 595–601.

Moore, M., Harrison, M.S., Peterson, E.C. and Henry, R. (2000). Chloroplast Oxa1 homolog albino3 is required for post-translational integration of the light harvesting chlorophyll-binding protein into thylakoid membranes. J. Biol. Chem. 275: 1529–1532.

Mühlenhoff, U., Haehnel, W., Witt, H.T. and Herrmann, R.G. (1993). Genes encoding eleven subunits of Photosystem I from the thermophilic cyanobacterium *Synechococcus*sp. Gene 127: 71–78.

Müller, M.G., Niklas, J., Lubitz, W. and Holzwarth, A.R. (2003). Ultrafast transient absorption studies on Photosystem I reaction centers from Chlamydomonas reinhardtii. 1. A new interpretation of the energy trapping and early electron transfer stePSIn Photosystem I. Biophys. J. 85: 3899–3922.

Netzer-El, S.Y., Caspy, I. and Nelson, N, (2019). Crystal structure of photosystem I monomer from *Synechocystis* PCC 6803. Front. Plant Sci. 9: 1–7.

Nordhues, A., Schöttler, M.A., Unger, A.K., Geimer, S., Schönfelder, S., Schmollinger, S., Rütgers, M., Finazzi, G., Soppa, B., Sommer, F., Mühlhaus, T., Roach, T., Krieger-Liszkay, A., Lokstein, H., Crespo, J.L. and Schroda, M. (2012). Evidence for a role of VIPP1 in the structural organization of the photosynthetic apparatus in *Chlamydomonas.* Plant Cell 24: 637–659.

Ozawa, S., Nield, J., Terao, A., Stauber, E.J., Hippler, M., Koike, H., Rochaix, J.D. and Takahashi, Y. (2009). Biochemical and structural studies of the large Ycf4-photosystem I assembly complex of the green alga *Chlamydomonas reinhardtii.* Plant Cell 21: 2424–2442.

Pasch, J.C., Nickelsen, J. and Schünemann, D. (2005). The yeast split-ubiquitin system to study chloroplast membrane protein interactions. Appl. Microbiol. Biotechnol. 69: 440–447.

Peng, L.W., Fukao, Y., Fujiwara, M., Takami, T. and Shikanai, T. (2009). Efficient operation of NAD(P)H dehydrogenase requires supercomplex formation with photosystem I via minor LHCI in arabidopsis. Plant Cell 21(11): 3623–3640.

Perez-Boerema A., Klaiman D., Caspy, I., Netzer-El, S. Y., Amunts A. and Nelson N. (2020). Structure of a minimal photosystem I from the green alga *Dunaliella salina.* Nature Plants 6(3): 321-327.

Pi, X., Tian, L., Dai, H. -E., Qin, X., Cheng, L., Kuang,T., Suia, S. -F. and Shen J.-R. (2018). Unique organization of photosystem I–light-harvesting supercomplex revealed by cryo-EM from a red alga. PNAS 115(17): 4423-4428.

Rochaix, J.D. (2011). Assembly of the photosynthetic apparatus. Plant Physiol. 155: 1493–1500.

Rosgaard, L., Zygadlo, A., Scheller, H.V., Mant, A. and Jensen, P.E. (2005). Insertion of the plant photosystem I subunit G into the thyklakoid membrane: *in vitro* and *in vivo* studies of wild-type and tagged versions of the protein. F.E.B.S. J. 272: 4002–4010.

Ruf, S., Kössel, H., Bock, R. (1997). Targeted inactivation of a tobacco intron-containing open reading frame reveals a novel chloroplast-encoded photosystem I-related gene. J. Cell Biol. 139: 95–102.

Saller, M.J., Wu, Z.C., de Keyzer, J. and Driessen, A.J. (2012). The YidC/Oxa1/Alb3 protein family: common principles and distinct features. Biol. Chem. 393: 1279–1290.

Sato, N. (2004). Roles of the acidic lipids sulfoquinovosyl diacylglycerol and phosphatidylglycerol in photosynthesis: their specificity and evolution. J. Plant Res. 117: 495–505.

Schlodder, E. Shubin, V.V., El-Mohsnawy, E.,R€ogner, M. and Karapetyan, N.V. (2007). Steady-state and transient polarized absorption spectroscopy of photosystem I complexes fromthe cyanobacteria Arthrospira platensis and *Thermosynechococcus elongatus*. Biochim. Biophys. Acta. 1767: 732–741.

Schubert, W.D., Klukas, O., Krauß, N., Saenger,W., Fromme, P. and Witt, H.T. (1997). Photosystem I of *Synechococcus elongatus* at 4 A° resolution: comprehensive structure analysis. J. Mol. Biol. 272: 741–769.

Schubert, W.D., Klukas, O., Saenger, W., Witt, H.T., Fromme, P. and Krauß, N. (1998). A common ancestor for oxygenic and anoxygenic photosynthetic systems: a comparison based on the structural model of Photosystem I. J. Mol. Biol. 280: 297–314.

Schwenkert, S., Netz, D.J., Frazzon, J., Pierik, A.J., Bill, E.,Gross, J., Lill, R. and Meurer, J. (2010). Chloroplast HCF101 is a scaffold protein for [4Fe–4S] cluster assembly. Biochem. J. 425: 207–214.

Semenov, A.Y., Vassiliev, I.R., Est Avan Der, Mamedov, M.D., Zybailov, B., Shen, G., Stehlik, D., Diner, B.A., Chitnis, P.R. and Golbeck, J.H. (2000). Recruitment of a foreign quinone into the A1 site of Photosystem I. Altered kinetics of electron transfer in phylloquinone biosynthetic pathway mutants studied by time-resolved optical, EPR, and electrometric techniques. J. Biol. Chem. 275: 23429–23438.

Sener, M., Park, S., Lu, D.Y., Damjanovic´, A., Ritz, T., Fromme, P. and Schulten, K. (2003). Excitation transfer dynamics in monomeric and trimeric forms of cyanobacterial Photosystem I. Biophy. J. 84: 274a–274a.

Shen, G.Z., Antonkine, M.L., Est Avan der, Vassiliev, I.R., Brettel, K., Bittl, R., Zech, S.G., Zhao, J.D., Stehlik, D., Bryant, D.A. and Golbeck, J.H. (2002). Assembly of

Photosystem III. Rubredoxin is required for the in vivo assembly of F-x in *Synechococcussp* PCC 7002 as shown by optical and EPR spectroscopy. J. Biol. Chem. 277: 20355–20366.

Sommer, F., Drepper, F., Haehnel, W. and Hippler, M. (2004). The hydrophobic recognition site formed by residues PsaATrp651 and PsaB-Trp627 of Photosystem I in Chlamydomonas reinhardtii confers distinct selectivity for binding of plastocyanin and cytochrome c6. J. Biol. Chem. 279: 20009–20017.

Stöckel, J. and Oelmüller, R. (2004). A novel protein for photosystem I biogenesis. J. Biol. Chem. 279: 10243–10251.

Sun, J., Xu, Q., Chitnis, V.P., Jin, P. and Chitnis, P.R. (1997). Topography of the Photosystem I core proteins of the cyanobacterium *Synechocystis*sp. PCC 6803. J. Biol. Chem. 272: 21793–21802.

Touraine, B., Boutin, J.P., Marion-Poll, A., Briat, J.F., Peltier, G. and Lobréaux, S. (2004). Nfu2: a scaffold protein required for [4Fe–4S] and ferredoxin iron–sulphur cluster assembly in Arabidopsis chloroplasts. Plant J. 40: 101–111.

Varotto, C., Pesaresi, P., Jahns, P., Lessnick, A., Tizzano, M., Schiavon, F., Salamini, F. and Leister, D. (2002). Single and double knockouts of the genes for photosystem I subunits G, K, and H of Arabidopsis. Effects on photosystem I composition, photosynthetic electron flow, and state transitions. Plant Physiol. 129: 616-624.

Vasiliev, S. and Bruce, D. (2004). Optimization and evolution of light harvesting in photosynthesis: the role of antenna chlorophyll conserved between Photosystem II and Photosystem I. Plant Cell 16: 3059–3068.

Vassiliev, I.R., Yu, J., Jung, Y.S., Schulz, R., Ganago, A.O., McIntosh, L., Golbeck, J.H. (1999). The cysteine-proximal aspartates in the F(x)-binding niche of photosystem. I: effect of alanine and lysine replacements on photoautotrophic growth, electron transfer rates, single-turnover flash efficiency, and EPR spectral properties. J. Biol. Chem. 274: 9993–10001.

Wang, P. and Dalbey, R.E. (2011). Inserting membrane proteins: the YidC/Oxa1/ Alb3 machinery in bacteria, mitochondria, and chloroplasts. Biochim. Biophys. Acta. 1808: 866–875.

Watanabe, M., Kubota, H., Wada, H., Narikawa, R. and Ikeuchi, M. (2011). Novel supercomplex organization of photosystem I in *Anabaena* and *Cyanophora paradoxa*. Plant Cell Physiol. 52: 162–168.

Webber, A.N. and Lubitz, W. (2001). P700: the primary electron donor of photosystem I. Biochim. Biophys. Acta. 1507: 61–79.

Wilde, A., Lünser, K., Ossenbühl, F., Nickelsen, J., Börner, T. (2001). Characterization of the cyanobacterial ycf37: mutation decreases the photosystem I content. Biochem. J.357: 211–216.

Xu, W., Chitnis, P., Valieva, A., Est Avan der, Pushkar, Y.N., Krzystyniak, M., Teutloff, C., Zech, S.G., Bittl, R., Stehlik, D., Zybailov, B., Shen, G. and Golbeck, J.H. (2003).

Electron transfer in cyanobacterial Photosystem I: I. Physiological and spectroscopic characterization of site-directed mutants in a putative electron transfer pathway from A0 through A1 to FX. J. Biol. Chem. 278: 27864–27875.

Yeremenko, N., Jeanjean, R., Prommeenate, P. Jeanjean, R., Prommeenate, P., Krasikovm V., Nixon, P.J., Vermaas, W.F., Havaux, M. and Matthijs, H.C. (2005). Open reading frame ssr2016 is required for antimycin A-sensitive photosystemI-driven cyclic electron flow in the cyanobacterium *Synechocystis* sp. PCC 6803. Plant Cell Physiol. 46: 1433–1436.

Yu, L., Bryant, D.A. and Golbeck, J.H. (1995). Evidence for a mixedligand [4Fe-4S] cluster in the C14D mutant of PsaC. Altered reduction potentials and EPR spectral properties of the FA and FB clusters on rebinding to the P700-FX core. Biochemistry 34: 7861–7868.

Zhao, J., Li, N., Warren, P., Golbeck, J. and Bryant, D. (1992).Sitedirected conversion of a cysteine to aspartate leads to the assembly of a [3Fe-4S] cluster in PsaC of Photosystem I – the photoreduction of FA is independent of FB. Biochemistry 31: 5093–5099.

Zhao, J., Snyder, W.B., Mu¨hlenhoff, U., Rhiel, E., Warren, P.V., Golbeck, J.H. and Bryant, D.A. (1993) Cloning and characterization of the psaE gene of the cyanobacterium *Synechococcus sp.* PCC 7002: characterization of a psaE mutant and overproduction of the protein in Escherichia coli. Mol. Microbiol. 9: 183–194.

Zheng, L., Li, Y., Li, X., Zhong, Q., Li, N., Zhang, K., Zhang, Y., Chu, H., Ma, C., Li, G., Zhao, J. and Gao, N. (2019). Structural and functional insights into the tetrameric photosystem I from heterocystforming cyanobacteria. Nature Plant 5: 1087–1097.

Zouni, A., Witt, H.T., Kern, J., Fromme, P., Krauß, N., Saenger, W. and Orth, P. (2001). Crystal structure of Photosystem II from Synechococcus elongatus at 3.8 A ° resolution. Nature 409: 739–743.

Chapter 27

Biofuel Production from Lignocellulosic Biomass: A Green Technology

Hrudayanath Thatoi, Sonali Mohapatra and Manish Paul

[1]*Department of Biotechnology, Maharaja Sriram Chandra Bhanja Deo University, Baripada, Odisha-757003, India.*
[2]*Department of Biotechnology, Odisha University of Technology and Research, Biju Patnaik University of Technology, Kalinga Nagar, Ghatikia, Bhubaneswar, Odisha 751003, India.*

ABSTRACT

"Second generation" bioethanol, with lignocellulose material as feedstock, is a promising alternative for first generation bioethanol. This paper provides an overview of the current status and reveals the bottlenecks that hamper its implementation. The current literature specifies a conversion of biomass to bioethanol of 30 to ~50 per cent only. Novel processes increase the conversion yield to about 92 per cent of the theoretical yield. New combined processes reduce both the number of operational steps and the production of inhibitors. Recent advances in genetically engineered microorganisms are promising for higher alcohol tolerance and conversion efficiency. By combining advanced systems and by intensive additional research to eliminate current bottlenecks, second generation bioethanol could surpass the traditional first generation processes.

Introduction

Petroleum (or crude oil), a form of fossil fuel, is a nonrenewable energy. That means once petroleum is used, it cannot be regenerated at a pace to sustain its current consumption rate. Crude oil has taken 50–300 million years to form, but half of all global oil reserves have been consumed in approximately 125 years (Zou *et al.,* 2016). The excessive use of petroleum results in severe environmental issues, such as greenhouse gas emissions, air pollution, and subsequent climate problems (Jun *et al.,* 2001). According to the information reported by the World Energy Council it is presumed that biomass can provide about 16 per cent of the expected energy consumption of 17208 Mtoe in 2020 (World Energy Outlook, 2012).

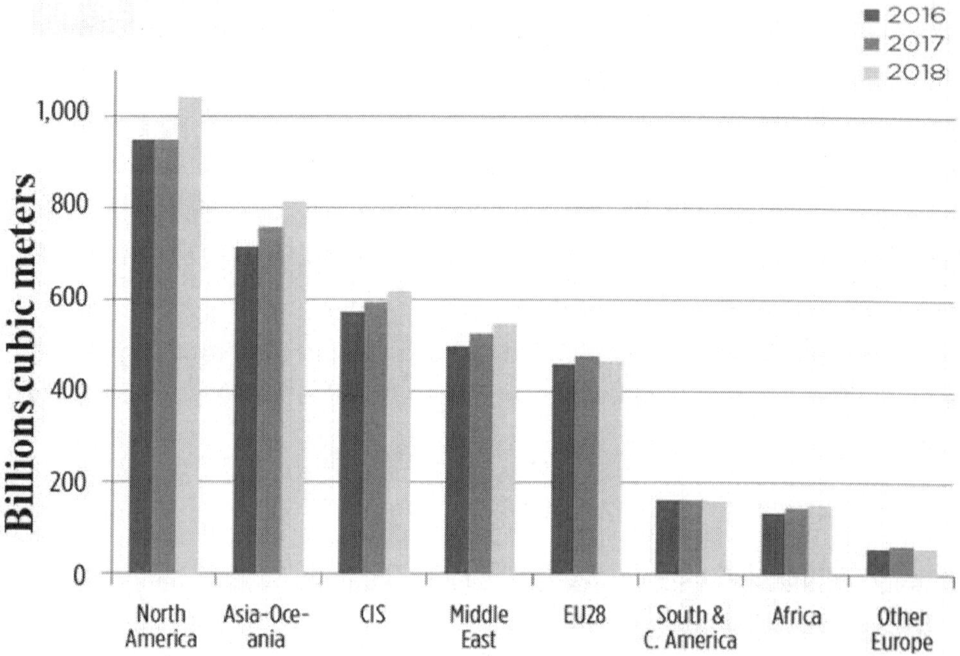

Figure 27.1: Consumption of Natural Gas Worldwide in different Countries.

Lignocellulose is a generic term for describing the main constituents in most plants, namely cellulose, hemicelluloses, and lignin. Lignocellulose is the most abundant renewable biomass. The yield of lignocellulose can reach approximately 200 billon metric tons worldwide per year (Zhang *et al.,* 2007). Lignocellulose is a complex matrix, comprising many different polysaccharides, phenolic polymers and proteins. Cellulose, the major component of cell walls of land plants, is a glucan polysaccharide containing large reservoirs of energy that provide real potential for conversion into biofuels. Lignocellulosic biomass consists of a variety of materials with distinctive physical and chemical characteristics. It is the non-starch based fibrous part of plant material. First-generation biofuels (produced primarily from food crops such as grains, sugar beet and oil seeds) are limited in their ability to achieve targets for oil-product substitution, climate change mitigation, and economic growth. Their sustainable production is under scanner, as is the possibility of creating undue competition for land and water used for food and fibre production.

The cumulative impacts of these concerns have increased the interest in developing biofuels produced from non-food biomass. Feedstocks from ligno-cellulosic materials include cereal straw, bagasse, forest residues, and purpose-grown energy crops such as vegetative grasses and short rotation forests. These second-generation biofuels could avoid many of the concerns facing first-generation biofuels and potentially offer greater cost reduction potential in the longer term.

The largest potential feedstock for biofuels is lignocellulosic biomass, which includes materials such as agricultural residues (corn stover, crop straws and bagasse), herbaceous crops (alfalfa, switchgrass), short rotation woody crops, forestry residues, waste paper and other wastes (municipal and industrial). Bioethanol production from these feedstocks could be an attractive alternative for disposal of these residues. Importantly, lignocellulosic feedstocks do not interfere with food security. Moreover, bioethanol is very important for both rural and urban areas in terms of energy security reason, environmental concern, employment opportunities, agricultural development, foreign exchange saving, socioeconomic issues *etc.* Lignocellulose is a feedstock with a low production cost. Lignocellulose can be found everywhere and is available as waste biomass, including agricultural residues (wheat straw, sugarcane bagasse, and corn stover), energy crops (switch grass), and municipal solid waste (paper and paperboard products) (Hamelinck *et al.,* 2005).

Lignocellulosic biomass consists mainly of lignin and the polysaccharides cellulose and hemicellulose. Compared with the production of ethanol from first-generation feedstocks, the use of lignocellulosic biomass is more complicated because the polysaccharides are more stable and the pentose sugars are not readily fermentable by *Saccharomyces cerevisiae.*

In order to convert lignocellulosic biomass to biofuels the polysaccharides must first be hydrolysed, or broken down, into simple sugars using either acid or enzymes. Several biotechnology-based approaches are being used to overcome such problems, including the development of strains of *Saccharomyces cerevisiae* that can ferment pentose sugars, the use of alternative yeast species that naturally ferment pentose sugars, and the engineering of enzymes that are able to break down cellulose and hemicellulose into simple sugars.

Lignocellulosic processing pilot plants have been established in the EU, in Denmark, Spain and Sweden. The world's largest demonstration facility of lignocellulose ethanol (from wheat, barley straw and corn stover), with a capacity of 2.5 Ml, was first established by Iogen Corporation in Ottawa, Canada. Many other processing facilities are now in operation or planning throughout the world.

Economically, lignocellulosic biomass has an advantage over other agriculturally important biofuels feedstocks such as corn starch, soybeans, and sugar cane, because it can be produced quickly and at significantly lower cost than food crops. Lignocellulosic biomass is an important component of the major food crops; it is the non-edible portion of the plant, which is currently underutilized, but could be used for biofuel production. In short, lignocellulosic biomass holds the key to supplying society's basic needs for sustainable production of liquid transportation fuels without impacting the nation's food supply. Bioethanol is one form of renewable energy. It is renewable because it is made from bioresources like crops and agricultural residues, which can be grown in a short period of time. Using lignocellulose to produce ethanol avoids competition with the food industry, as opposed to first generation biofuels, which are directly converted from sugar and starch crops found in arable areas (Hamelinck *et al.,* 2005).

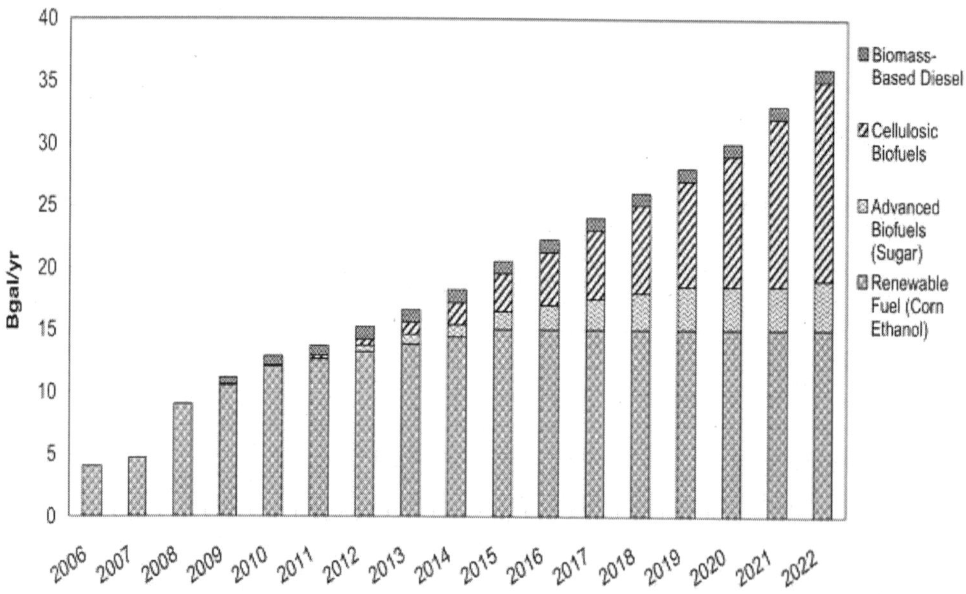

Figure 27.2: Production of Biofuels (Billions gallons/years) over the Years.

Ethanol is burned more completely than other fuels, which results in less emission of greenhouse gases, such as nitrogen oxides and carbon dioxide (Gupta, 2009). On the other hand, the energy content of ethanol is approximately 33 per cent lower than gasoline, which results in vehicles getting less mileage from ethanol. Adopting bioethanol is relatively easy because it can be integrated into the existing road transport fuel system. In the United States, ethanol can be blended with gasoline at rates of up to 10 per cent. Cellulosic ethanol has a larger net energy gain than corn ethanol. The net energy of biofuel is defined as the difference between the consumable energy of biofuel produced and the energy expended during biofuel production (Farrell *et al.*, 2006). The process of producing cellulosic ethanol is generally divided into four steps: 1) pretreatment, 2) enzyme hydrolysis, 3) fermentation, and 4) distillation.

Physico-chemical Characteristics of Lignocellulosic Biomass

All plant materials are mostly composed of three major units *i.e.*, cellulose, hemicellulose and lignin. Lignocellulosic materials including agricultural wastes, forestry residues, grasses and woody materials have great potential for bio-fuel production. Typically, most of the agricultural lignocellulosic biomass is comprised of about 10–25 per cent lignin, 20–30 per cent hemicellulose, and 40–50 per cent cellulose (Iqbal *et al.*, 2011; Kumar *et al.*, 2009; Malherbe and Cloete, 2002). Cellulose is a major structural component of plant cell walls, which is responsible for mechanical strength while, hemicellulose macromolecules are often repeated polymers of pentoses and hexoses. Lignin contains three aromatic alcohols (coniferyl alcohol, sinapyl alcohol and p-coumaryl alcohol) produced through a biosynthetic process and forms a protective seal around the other two components *i.e.*, cellulose and hemicelluloses (Calvo-Flores and Dobado, 2010; Jiang *et al.*, 2010; Menon and Rao, 2012). In general the composition of lignocellulose highly depends on its source whether it is derived from the hardwood, softwood, or grasses.

Physical and Structural Properties of Cellulose

Cellulose is a highly stable polymer consisting of glucose and attached with linear chains up to 12,000 residues. It is majorly composed of (1,4)-d-glucopyranose units, which are attached by β-1,4 linkages with an average molecular weight of around 100,000 (Himmel *et al.*, 2007). Plant biomass contain 40–50 per cent of cellulose molecules which are held together by intermolecular hydrogen bonds in native state, but they have a strong tendency to form intra-molecular and intermolecular hydrogen bonds and this tendency increases the rigidity of cellulose and make highly insoluble and highly resistant to most organic solvents. Naturally cellulose molecules are exists as bundles which aggregated together in the form of micro-fibrils order *i.e.*, crystalline and amorphous regions (Iqbal *et al.*, 2011; Taherzadeh and Karimi, 2008).

Physical and Structural Properties of Hemicelluloses

Hemicellulose is the second most abundant heterogeneous polymers that mainly consist of glucuronoxylan, glucomannan and trace amounts of other polysaccharides. Grasses and straws contain arabinan, galactan and xylan, while mannan is a component of hardwood and softwood hemicellulose (Brigham *et al.*, 1996). They are catalogued with sugar as a backbone, *i.e.*, xylans, mannans and glucans, with xylans and mannans being the most common (Wyman *et al.*, 2005). Galactans, arabinans and arabinogalactans are included in the hemicellulose group; however, they do not share the equatorial β-1,4 linked backbone structure.

In hardwoods, glucuronoxylan (O-acetyl-4-O-methyl-glucurono-β-d-xylan) is the predominant component. Xylospyranose is the backbone of the polymer and connected with β-1,4 linkages. Hemicellulosic biomass contains 25–35 per cent of hemicellulose, with an average molecular weight of <30,000. Cellulose and hemicellulose binds tightly with non-covalent attractions to the surface of each cellulose micro-fibril. Hemicelluloses were originally believed to be intermediates in the biosynthesis of cellulose (Vercoe *et al.,* 2005).

Physical and Structural Properties of Lignin

Lignin is generally the most complex and smallest fraction, representing about 10–25 per cent of the biomass by weight. It has a long-chain, heterogeneous polymer composed largely of phenyl-propane units most commonly linked by ether bonds. Lignin acts like a glue by filling the gap between and around the cellulose and hemicellulose complexion with the polymers. It is present in all plant biomass; therefore, it is considered byproduct or as a residue in bio-ethanol production process. Lignin is comprised of complex and large polymer of phenyl-propane, methoxy groups and non-carbohydrate poly phenolic substance, which bind cell walls component together (Hamelinck *et al.,* 2005). Phenyl-propanes (3 carbons attached with 6 carbon atom rings) are main block of lignin. These phenyl-propanes denoted as 0, I, II methoxyl groups attached to rings give special structure I, II and III. These groups depend on the plant source which they are obtained. Structure I exist in plants (grasses) and structure II found in the wood (conifers) while structure III present in deciduous wood.

Biotechnological Importance of Lignocellulosic Biomass

From the biotechnological point of view a wide variety of lignocellulosic biomass resources are available as potential candidate that are also convert able into high value bio-products like bio-ethanol/bio-fuels. From the last several years a considerable improvement from the green biotechnology related to lignocellulose biomass has appeared. The ever increasing costs of fossil fuels and their greenhouse effects are creating a core demand to explore alternative cheaper and eco-friendly bio-fuels resources as a strategy for reducing global warming (Asgher *et al.,* 2013; Iqbal *et al.,* 2013). Environmental pollution, global warming, and the future of oil production are among major causes of public and private interests in natural bio based resources as an alternative or substitute for fossil fuel oil. One potential method for the low-cost production of bio-ethanol is to utilize the lignocellulosic or agro-industrial biomass because they contain carbohydrates that must be first converted into simple sugars (glucose) and then fermented into ethanol (Alonso *et al.,* 2008; Balat and Balat, 2009; Lin and Tanaka, 2006). Given this reality, nations around the world are investing in alternative sources of energy, including bio-ethanol. The conversion of lignocellulosic biomass into higher value added products like fine chemicals or bio-fuel production normally requires a multi-step processing that include (i) pre-treatment (mechanical, chemical, or biological *etc.*) (ii) enzymatic hydrolysis (iii) fermentation process (Wyman, 1999; Xiao *et al.*, 2012).

Role of Pre-treatment

Pre-treatment is an important step for the recovery of cellulosic content from lignin based biomass as compare to the starchy materials. While dealing with lignocellulosic biomasses, pre-treatment is also required to break down the lignin barrier to recover cellulose, which is further subjected to enzymatic hydrolysis to convert into fermentable sugars. During the past few decades, several pre-treatment approaches have been developed for generating cost-effective fermentable sugar from most of the agricultural cellulose and hemicellulose containing lignocellulosic materials (Yang and Wyman, 2008). In this background, there are a number of reports on pre-treatment technologies for a variety of feed-stocks. Some of the most promising pre-treatment categories have already been commercialized for the productions of bio-energy are summarized in the Table 2. An effective pre-treatment is characterized by several criteria: preserving hemicellulose fractions, to yield maximum fermentable sugar contents, limiting the loss of carbohydrate, to minimize the formation of inhibitors due to degradation products, minimizing energy input, and the process is economically efficient as well as cost-effective. While, comparing various pre-treatment options, all of the above mentioned criteria should be comprehensively considered as a basis to achieve maximal end product of interest. Hydrolysis of biomass can be done by different ways mainly including physico-chemically, chemically or biologically (Asgher *et al.*, 2013; Hamzeh *et al.*, 2013; Rohowsky *et al.*, 2013; Yang *et al.*, 2011; Yang and Wyman, 2008).

Chemical Pre-treatments

To date chemical pre-treatment is the most studied technique among various pre-treatment categories that was originally developed and therefore has extensively been used for delignification of cellulosic materials. Chemical hydrolysis is an important treatment method for recovery of sugar monomers from cellulose and hemicellulose polymers from lignocellulosic biomass by optimizing chemical reagents. The most commonly used chemical pre-treatments include: acid and alkali based hydrolysis approaches.

Acid Based Hydrolysis

Chemical treatment of cellulosic biomass with concentrated hydrochloric acid or sulphuric acid is conventional procedure. The entire process of pre-treatment can be operating at very low temperature as compared to dilute-acid pre-treatment. On the other hand one of the possible drawbacks of this process is that it is required in higher concentration (30–70 per cent), therefore cause high level of corrosive reaction. In this background the whole process needs further expenditure in the form of specialized non-metallic or non corrosive material such as ceramic or carbon-brick lining. In comparison to the other pre-treatment procedures particularly dilute-acid hydrolysis the environmental hazards and high operating cost involved in concentrated-acid hydrolysis reduce the interest on industrial scale (Katzen *et al.,* 1995; Wyman, 1999). Dilute-acid pre-treatment has some advantages over concentrated-acid hydrolysis to solve the issues like acid recovery, toxicity, acid and special maintenance against corrosion materials (Sivers and Zacchi, 1995; Sun and Chen, 2007). Acid pre-treatment has been applied on several biomass feed-stocks like herbaceous material (grass), hardwoods and agricultural wastes. Most of the substrates give better results by solubilizing the hemicellulose (Liao *et al.,* 2007; Wyman *et al.,* 2005). Other two factors including temperature and incubation time had also important impact on alteration the structure of biomass. Major disadvantage of this process is the formation of secondary products which can lower the yield of sugars due to conversion of products in to furfural and hydroxyl-methyl furfural compounds and these compounds interfere in bio-ethanol fermentation process.

Alkali Based Hydrolysis

Alkali based pre-treatment involves the use of bases, such as sodium and ammonium hydroxide, for the pre-treatment of agricultural lignocellulosic feed-stocks. Alkaline hydrolysis causes various structural alterations inside the lignocellulosic material during treatment process such as the depletion of lignin barrier, cellulose swelling, and partial decrystallization and solvation of cellulose and hemicelluloses, respectively (Cheng *et al.,* 2010; Ibrahim *et al.,* 2011; Sills and Gossett, 2011; Zhu *et al.,* 2010). Lignocellulosic feed-stocks that have been shown to benefit from the method of alkaline pre-treatment are corn stover, switch-grass, bagasse, wheat, and rice straw (Hu *et al.,* 2008; Zhao *et al.,* 2008; Zhu *et al.,* 2012). Zhao *et al.* (2008) had showed the effectiveness of sodium hydroxide pre-treatment for hardwoods, wheat straw, switch-grass, and soft-woods with less than 26 per cent lignin content.

Biological Pre-treatment

Biological pre-treatment employs wood degrading microorganisms, including white rot fungi (WRF), brown or soft-rot fungi, and bacteria to modify the chemical composition and/or structure of the lignocellulosic biomass. Bio-delignification is useful for pre-treatment purposes because it replaces or supplements the chemical-based pre-treatments, which include mechanical treatment with acid, alkali, and steam explosion (Iqbal *et al.,* 2013). In spite of this biological pre-treatments are more effective, economical, eco-friendly and less health hazardous as compare to the physico-chemical or chemical-based pre-treatment approaches. Therefore, from the last few years research scientists are directing their interests towards biological delignification. Recent advances in the characterization of ligninolytic enzymes involving the degradation of lignin have given new impetus to the research in this area, which has now become amenable to biotechnological exploitation (Asgher *et al.,* 2012). Bio-conversions of lignocellulosic materials to useful products normally require multi-step processes that include pre-treatment, enzymatic hydrolysis, and fermentation (Xiao *et al.,* 2012), so that the modified or pre-treated biomass is more amenable to enzyme digestion. Increasing understanding of termites and fungal systems has provided insights for developing more effective pre-treatment technologies to realize the above mentioned advantages or benefits of biological pre-treatment over some others. However, biological pre-treatment is a very slow process that also requires careful control of growth conditions and large amount of space to perform treatment. In addition to this most of the lignolytic microorganisms solubilize/consume not only lignin but also hemicellulose and cellulose. Because of these drawbacks/limitations the biological pre-treatment faces techno-economic barriers and therefore is less attractive commercially (Eggeman and Elander, 2005).

Enzymatic Hydrolysis

Enzymatic hydrolysis is an effective and economical method to achieve fermentable sugars under mild and eco-friendly reaction conditions from the pre-treated cellulosic biomass (Wyman *et al.,* 2005). The entire process of enzymatic hydrolysis critically depends on variety of factors *viz.,* pH, time, temperature substrates and enzyme activities, *etc.* Enzymatic saccharification is done separately from fermentation known as separate hydrolysis and fermentation (SHF). When cellulose hydrolysis and fermentation are carried out simultaneously the phenomenon is known as simultaneous saccharification and fermentation (SSF). Now a days this process of simultaneous saccharification of both cellulose and hemicellulose is achieved by co-fermentation of both hexoses and pentoses sugars (SSCF) with the help of genetically engineered microbes that ferment xylose and glucose in the same medium where both enzymes for cellulose and hemicelluloses are available. The major advantage of this technology is that SSF and SSCF can be performed in the same tank which makes the entire process cheap, feasible and cost-effective (reduce the capital and operational investment). Biological, physical and chemical methods have been employed for detoxification (removal of inhibitory compounds in fermentation) of lignocellulosic hydrolyzates (Olsson and Hahn-Hagerdal, 1996). Lignocellulosic materials have different degree of inhibition and tolerance levels vary according to different microbial strains. Degree of tolerance varies with different strains of Saccharomyces cerevisiae, so inhibitory compounds are detoxified by changing the substrate concentration and altering the pH of media (Linden *et al.,* 1992; Palmqvist *et al.,* 1998).

Fermentation Strategies

Ethanol production from biomass is mainly categorized into three steps process (1) achieve a fermentable sugars (2) conversion of fermentable sugars into ethanol and (3) ethanol separation and purification through distillation (Asgher *et al.,* 2014; Demirbas, 2005). Difference between lignocelluloses or starch ethanol production is the step for obtaining sugars before fermentation. Sugar crops or starchy crops need milling and grinding for recovery of sugars by extraction and fermentation becoming a relatively simple process that requires no hydrolysis or pre-treatment steps for obtaining sugars and transformation into ethanol (Icoz *et al.,*

2009). Bio-ethanol production is mainly done by fed-batch process and low ethanol produce by multi-stage continuous fermentation. Basic steps for the conversion of lignocellulosic biomass are: (1) pre-treatment process which can reduce the lignin content and render cellulose and hemicellulose content for enzymatic hydrolysis (2) steps to convert enzymatic hydrolysis to break down polysaccharide to simple sugars (3) conversions of sugars (hexoses and pentoses) for ethanol production through microorganisms (4) production of ethanol from pentose sugars. Several fungal species belonging with genera *Fusarium, Rhizopus, Monilia, Neurospora* and *Paecilomyces* have been found potential for fermenting glucose as well as xylose (Singh *et al.,* 1992). Production of bio-ethanol from cellulose is mostly conducted by using fermentative organism, but the conversion rate is very low due to byproducts formation. Filamentous fungus *Fusarium oxysporum* is also known for the production of bio-ethanol through SSF by direct utilizing the cellulose, but their conversion rate is low due to production of acetic acid as a byproduct (Panagiotou *et al.,* 2005).

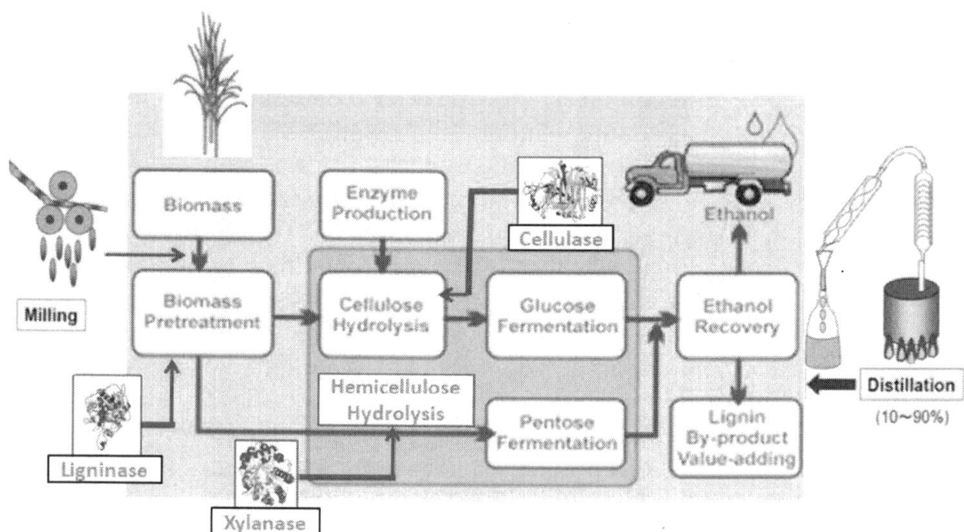

Figure 27.3: Schematics of Bioethanol Production from Biomass.

Enzymes Production

To date, the production of various ligninolytic enzymes including LiP, MnP, versatile peroxidise (VP), and laccases and other lignocellulolytic mainly endoglucanases (EC 3.2.1.4), cellobiohydrolases (EC 3.2.1.91) and β-glucosidases (EC 3.2.1.21) have been widely studied in submerged and solid culture processes in the laboratory, ranging from flask shake to large scale (Elisashvili *et al.,* 2006; Moldes *et al.,* 2007; Xia and Len, 1999). There are large numbers of microorganisms capable for degrading cellulose (Jungebloud *et al.,* 2007). *Trichoderma, Aspergillus,*

Penicillium and *Fusarium* species are commonly used for cellulases production (Iqbal *et al.,* 2011). Selections of desired fungal strains depend on several factors and selection of substrate for optimizing the cellulase producing conditions (Shazia *et al.,* 2007). Ligninolytic, cellulases and hemicellulases are important industrial enzymes having numerous applications and biotechnological potential for various industries including chemicals, fuel, food, brewery and wine, animal feed, textile and laundry, pulp and paper and agriculture (Asgher *et al.,* 2011; Asgher and Iqbal, 2011; Eun *et al.,* 2006; Iqbal and Asgher, 2013; Iqbal *et al.,* 2011; Irshad *et al.,* 2013; Oberoi *et al.,* 2010; Papinutti and Forchiassin, 2007; Papinutti and Lechner, 2008; Stoilova *et al.,* 2010; Yoon *et al.,* 2012). A range of different lignocellulosic materials that has successfully been adopted for the production of different enzymes having industrial importance.

Conclusion and Future Outlook

The energy and environmental crises which the modern world is experiencing is forcing to re-evaluate the efficient utilization or finding alternative uses for natural, renewable resources, using clean technologies. In this regard, lignocellulosic biomass holds considerable potential to meet the current energy demand of the modern world. This is also essential in order to overcome the excessive dependence on petroleum for liquid fuels. Further advanced biotechnologies are crucial for discovery, characterization of new enzymes, and production in homologous or heterologous systems and ultimately lead to low-cost conversion of lignocellulosic biomasses into bio-fuels and bio-chemicals. In current scenario future trends are being directed to lignocellulose biotechnology and genetic engineering for improved processes and products. To overcome the current energy problems it is envisaged that lignocellulosic biomass in addition of green biotechnology will be the main focus of the future research.

REFERENCES

Alonso, A., Pe´rez, P., Morcuende, R., Martinez-Carrasco, R. (2008). Future CO_2 concentrations, though not warmer temperatures, enhance wheat photosynthesis temperature responses. Physiologia Plantarum, 132, 102e112.

Asgher, M., Ahmed, N., Iqbal, H.M.N. (2011). Hyperproductivity of extracellular enzymes from indigenous white rot fungi (P. chrysosporium IBL-03) by utilizing agro-wastes. BioResources, 6(4), 4454e4467.

Asgher, M., Iqbal, H.M.N. (2011). Characterization of a novel manganese peroxidase purified from solid state culture of Trametes versicolor IBL-04. BioResources, 6, 4302e4315.

Asgher, M., Ahmad, Z., Iqbal, H.M.N. (2013). Alkali and enzymatic delignification of sugarcane bagasse to expose cellulose polymers for saccharification and bio-ethanol production. Industrial Crops and Products, 44, 488e495.

Asgher, M., Iqbal, H.M.N., Irshad, M. (2012). Characterization of purified and xerogel immobilized novel lignin peroxidase produced from Trametes versicolor IBL-04 using solid state medium of corncobs. BMC Biotechnology, 12(1), 46.

Asgher, M., Shahid, M., Kamal, S., Iqbal, H.M.N. (2014). Recent trends and valorization of immobilization strategies and ligninolytic enzymes by industrial biotechnology. Journal of Molecular Catalysis B: Enzymatic, 101, 56e66.

Balat, M., Balat, H. (2009). Recent trends in global production and utilization of bio-ethanol fuel. Applied Energy, 86(11), 2273e2282.

Brigham, J.S., Adney, W.S., Himmel, M.E. (1996). Hemicellulases: diversity and applications. In C. E. Wyman (Ed.), Bioethanol: Production and utilization (pp. 119e141).

Calvo-Flores, F. G., Dobado, J. A. (2010). Lignin as renewable raw material. ChemSusChem, 3(11), 1227e1235.

Cheng, Y.S., Zheng, Y., Yu, C.W., Dooley, T.M., Jenkins, B.M., VanderGheynst, J.S. (2010). Evaluation of high solids alkaline pretreatment of rice straw. Applied Biochemistry and Biotechnology, 162, 1768e1784.

Demirbas, A. (2005). Bio-ethanol from cellulosic materials: a renewable motor fuel frombiomass. Energy Sources, 27, 327e337.

Eggeman, T., Elander, R.T. (2005). Process economic analysis of pretreatment technologies. Bioresource Technology, 96, 2019e2025.

Elisashvili, V., Penninckx, M., Kachlishvili, E., Asatiani, M., Kvestiadze, G. (2006). Use of Pleurotus dryinus for lignocellulolytic enzymes production in submerged fermentation of mandarin peels and tree leaves. Enzyme and Microbial Technology, 38, 998e1004.

Eun, J.S., Beauchemin, K.A., Hong, S.H., Bauer, M.W. (2006). Exogenous enzymes added to untreated or ammoniated rice straw: effects on in vitro fermentation characteristics and degradability. Animal Feed Science and Technology, 131, 87e102.

Farrell, A.E., Plevin, R.J., Turner, B.T., Jones, A.D., O'Hare, M., Kammen, D.M. (2006). Ethanol Can Contribute to Energy and Environmental Goals. Science, 311(5760): 506–8.

Gupta, G.K. (2009). Characterization and Pretreatment of Cotton Gin Waste. Master's Thesis. National Institute of Technology, Rourkela.

Hamelinck, C.N., Geertje, V.H., Faaij, A.P.C. (2005). Ethanol from Lignocellulosic Biomass: Techno-Economic Performance in Short-, Middle- and Long-Term. Biomass and Bioenergy 28(4): 384–410.

Hamzeh, Y., Ashori, A., Khorasani, Z., Abdulkhani, A., and Abyaz, A. (2013). Pre-extraction of hemicelluloses from bagasse fibers: effects of dry-strength additives on paper properties. Industrial Crops and Products, 43, 365e371.

Himmel, M.E., Ding, S.Y., Johnson, D.K., Adney, W.S., Nimlos, M.R., Brady, J.W., *et al.* (2007). Biomass recalcitrance: engineering plants and enzymes for biofuel production. Science, 315, 804e807.

Hu, Z., Wang, Y., Wen, Z. (2008). Alkali (NaOH) pretreatment of switchgrass by radio frequency-based dielectric heating. Applied Biochemistry and Biotechnology, 148, 71e81.

Ibrahim, M.M., El-Zawawy, W.K., Abdel-Fattah, Y.R., Soliman, N.A., Agblevor, F.A. (2011). Comparison of alkaline pulping with steam explosion for glucose production from rice straw. Carbohydrate Polymers, 83, 720e726.

Icoz, E., Tugrul, M.K., Saral, A., Icoz, E. (2009). Research on ethanol production and use from sugar beet in Turkey. Biomass and Bioenergy, 33, 1e7.

Iqbal, H.M.N., Ahmed, I., Zia, M.A., Irfan, M. (2011). Purification and characterization of the kinetic parameters of cellulase produced from wheat straw by Trichoderma viride under SSF and its detergent compatibility. Advances in Bioscience and Biotechnology, 2(3), 149e156.

Iqbal, H.M.N., Asgher, M. (2013). Characterization and decolorization applicability of xerogel matrix immobilizedmmanganese peroxidase produced from Trametes versicolor IBL-04. Protein and Peptide Letters, 5, 591e600.

Iqbal, H.M.N., Asgher, M., Bhatti, H.N. (2011). Optimization of physical and nutritional factors for synthesis of lignin degrading enzymes by a novel strain of Trametes versicolor. BioResources, 6, 1273e1278.

Iqbal, H.M.N., Kyazze, G., Keshavarz, T. (2013). Advances in valorization of lignocellulosic materials by bio-technology: an overview. BioResources, 8(2), 3157e3176.

Irshad, M., Anwar, Z., But, H.I., Afroz, A., Ikram, N., Rashid, U. (2013). The industrial applicability of purified cellulase complex indigenously produced by Trichoderma viride through solid-state bio-processing of agro-industrial and municipal paper wastes. BioResources, 8(1), 145e157.

Jiang, G., Nowakowski, D.J., Bridgwater, A.V. (2010). A systematic study of the kinetics of lignin pyrolysis. Thermochimica Acta, 498(1), 61e66.

Jun, P., M. Gillenwater, W. Barbour. (2001). CO_2, CH_4 and N_2O Emissions from Transportation-Water-Borne Navigation. Good Practice Guidance and Uncertainty Management in National Greenhouse Gas Inventories. Accessed March 20, 2020.

Jungebloud, A., Bohle, K., Go¨ cke, Y., Cordes, C., Horn, H., Hempel, D.C. (2007). Quantification of product-specific gene expression in biopellets of Aspergillus niger with real-time PCR. Enzyme and Microbial Technology, 40(4), 653e660.

Katzen, R., Madson, P.W., Monceaux, D.A. (1995). Use of cellulosic feedstocks for alcohol production. In T. P. Lyons, J. E. Murtagh, D. R. Kelsall (Eds.), The alcohols textbook (pp. 37e46).

Kumar, P., Barrett, D.M., Delwiche, M.J., Stroeve, P. (2009). Methods for pretreatment of lignocellulosic biomass for efficient hydrolysis and biofuel production. Industrial and Engineering Chemistry Research, 48, 3713e3729.

Liao, W., Liu, Y., Wen, Z., Frear, C., and Chen, S. (2007). Studying the effects of reaction conditions on components of dairy manure and cellulose accumulation using dilute acid treatment. Bioresource Technology, 98, 1992e1999.

Lin, Y., Tanaka, S. (2006). Ethanol fermentation from biomass resources: current state and prospects. Applied Microbiology and Biotechnology, 69, 627e642.

Linden, T., Peetre, J., Hahn-Hagerdal, B. (1992). Isolation and characterization of acetic acid tolerant galactose-fermenting strains of Saccharomyces cerevisiae from a spent sulfite liquor plant. Applied and Environmental Microbiology, 58, 1661e1669.

Malherbe, S., Cloete, T.E. (2002). Lignocellulose biodegradation: fundamentals and applications. Reviews in Environmental Science and Biotechnology, 1, 105e114.

Menon, V., and Rao, M. (2012). Trends in bioconversion of lignocellulose: biofuels, platform chemicals and biorefinery concept. Progress in Energy and Combustion Science, 38(4), 522e550.

Moldes, A.B., Bustos, G., Torrado, A., Dominguez, J.M. (2007). Comparison between different hydrolysis processes of vinetrimming waste to obtain hemicellulosic sugars for further lactic acid conversion. Applied Biochemistry and Biotechnology, 143, 244e256.

Oberoi, H.S., Chavan, Y., Bansal, S., Dhillon, G.S. (2010). Production of cellulases through solid state fermentation using kinnow pulp as a major substrate. Food and Bioprocess Technology, 3, 528e536.

Olsson, L., Hahn-Hagerdal, B. (1996). Fermentation of lignocellulosic hydrolysates for ethanol production. Enzyme and Microbial Technology, 18, 312e331.

Palmqvist, E., Galbe, M., Hahn-Hagerdal, B. (1998). Evaluation of cell recycling in continuous fermentation of enzymatic hydrolysates of spruce with Saccharomyces cerevisiae and online monitoring of glucose and ethanol. Applied Biochemistry and Biotechnology, 50, 545e551.

Panagiotou, G., Villas-Boas, S.G., Christakopoulos, P., Nielsen, J., Olsson, L. (2005). Intracellular metabolite profiling of Fusarium oxysporum converting glucose to ethanol. Journal of Biotechnology, 115, 425e434.

Papinutti, L., Lechner, B. (2008). Influence of the carbon source on the growth and lignocellulolytic enzyme production by Morchella esculenta strains. Journal of Industrial Microbiology and Biotechnology, 35, 1715e1721.

Rohowsky, B., Ha¨ ßler, T., Gladis, A., Remmele, E., Schieder, D., Faulstich, M. (2013). Feasibility of simultaneous saccharification and juice co-fermentation on hydrothermal pretreated sweet sorghum bagasse for ethanol production. Applied Energy, 102, 211e219.

Shazia, S., Bajwa, R., Shafique, S. (2007). Cellulase production potential of selected strains of Aspergilli. Pakistan Journal of Phytopathology, 19(2), 196e206.

Sills, D.L., Gossett, J.M. (2011). Assessment of commercial hemicellulases for saccharification of alkaline pretreated perennial biomass. Bioresource Technology, 102, 1389e1398.

Singh, A., Kumar, P.K.R., Schugerl, K. (1992). Bioconversion of cellulosic materials to ethanol by filamentous fungi. Advances in Biochemical Engineering Biotechnology, 45, 29e55.

Sivers, M.V., Zacchi, G. (1995). A techno-economical comparison of three processes for the production of ethanol from pine. Bioresource Technology, 51, 43e52.

Stoilova, I., Krastanov, A., Stanchev, V. (2010). Properties of crude laccase from Trametes versicolor produced by solid substrate fermentation. Advances in Bioscience and Biotechnology, 1, 208e215.

Sun, F.B., Chen, H.Z. (2007). Evaluation of enzymatic hydrolysis of wheat straw pretreated by atmospheric glycerol autocatalysis. Journal of Chemical Technology and Biotechnology, 82(11), 1039e1044.

Taherzadeh, M.J., Karimi, K. (2008). Pretreatmentof lignocellulosic wastes to improve ethanol and biogas production: a review. International Journal of Molecular Sciences, 9, 1621e1651.

Vercoe, D., Stack, K., Blackman, A., Richardson, D. (2005). A study of the interactions leading to wood pitch deposition. In 59th Appita annual conference and exhibition: Incorporating the 13th ISWFPC (international symposium on wood, fibre and pulping chemistry) (p. 123). Auckland, New Zealand, 16e19 May 2005: Proceedings.

World Energy Outlook 2012, OECD/IEA, 2012.

Wyman, C.E. (1999). Biomass ethanol: technical progress, opportunities and commercial challenges. Annual Review of Energy and the Environment, 24, 189e226.

Wyman, C.E., Dale, B.E., Elander, R.T., Holtzapple, M., Ladisch, M.R., Lee, Y.Y. (2005). Coordinated development of leading biomass pretreatment technologies. Bioresource Technology, 96, 1959e1966.

Xia, L., Len, P. (1999). Cellulose production by solid-state fermentation on lignocellulosic waste from the xylose industry. Process Biochemistry, 34, 909e912.

Xiao, W., Wang, Y., Xia, S., Ma, P. (2012). The study of factors affecting the enzymatic hydrolysis of cellulose after ionic liquid pretreatment. Carbohydrate Polymers, 87, 2019e2023.

Yang, B., Wyman, C. (2008). Pre-treatment: the key to unlocking low-cost cellulosic ethanol. Biofuels, Bioproducts and Biorefining, 2, 26e40.

Yang, B., Dai, Z., Ding, S.-Y., Wyman, C.E. (2011). Enzymatic hydrolysis of cellulosic biomass. Biofuels, 2, 421e450.

Yoon, L.W., Ngoh, G.C., Chua, A.S.M. (2012). Simultaneous production of cellulase and reducing sugar from alkali pretreated sugarcane bagasse via solid state fermentation. BioResources, 7, 5319e5332.

Zhang, Y.H.P., Ding, S.Y., Mielenz, J.R., Cui, J.B., Elander, R.T., Laser, M., Himmel, M.E., McMillan, J.R., Lynd, L.R. (2007). Fractionating Recalcitrant Lignocellulose at Modest Reaction Condition. Biotechnology and Bioengineering 97(2): 214–23.

Zhao, Y., Wang, Y., Zhu, J.Y., Ragauskas, A., Deng, Y. (2008). Enhanced enzymatic hydrolysis of spruce by alkaline pretreatment at low temperature. Biotechnology and Bioengineering, 99, 1320e1328.

Zhu, J., Wan, C., Li, Y. (2010). Enhanced solid-state anaerobic digestion of corn stover by alkaline pretreatment. Bioresource Technology, 101, 7523e7528.

Zhu, H., Sheng, K., Yan, E., Qiao, J., Lv, F. (2012). Extraction, purification and antibacterial activities of a polysaccharide from spent mushroom substrate. International Journal of Biological Macromolecules, 50(3), 840e843.

Zou, C., Zhao, Q., Zhang, G., Xiong, B. (2016). Energy revolution: From a fossil energy era to a new energy era. Natural Gas Industry B, 3(1): 1-11.

Chapter 28

Physicochemical Characteristics of Water Samples and Analysis of Heavy Metals in Atharbanki Creek, East Coast of India

Dibyajyoti Sr Pattanayak and Samarendranath Naik

Department of Botany, College of Basic Science and Humanities, OUAT, Bhubaneswar – 751 003, Odisha
e-mail: dibyajyotisrpatt@gmail.com/samarendranaik09@gmail.com

ABSTRACT

Atharbanki waterways (lat 20°15'01.34"N, long 86°38'21.20"E; to lat 20°17'26.13"N, long 86° 42'16.23"E) is stretched 13.5 km, adjoining the Paradeep Township skirting Paradeep phosphate Limited Township, Paradip Phosphate Limited (PPL), Indian Farmers Fertilizer Co- operative Limited, Essar steel, including industrial pipelines which runs along the waterways. Various anthropogenic activities leads to its pollution, including municipal sewage discharge from Paradeep Township and industrial effluents. In the present study, an attempt has been made to monitor the physiochemical properties and estimation of heavy metal in the Atharbanki creek, the East Coast coastal belt of India. The result revealed a remarkable variation in physicochemical properties such as (pH, TDS, Resistivity, Electrical conductivity, Salinity) and presence of dissolved heavy metals such as (Fe, Mn, Zn and Cr) in the surface water of the creek which are attributed to the runoff water getting discharge in Atharbanki creek, the east coast belt of India. Samples were collected from three Stations along the course of the creek. It is being recommended to treat waste water and industrial effluents before discharging them to the aquatic environment.

Keywords: *Physicochemical, Water, Pollution, Atharbanki creek.*

Introduction

Paradeep is the port town of Odisha and is the only public – sector port operating in the state. It is a major industrial sector with a number of public and private undertaking companies operating in the region. (Paradeep Phosphate Limited, Indian Farmers Fertilizer Co-operative Limited, Essar steel, Indian Oil Corporation Limited). These industries lack waste disposal unit and discharges the chemical effluents in the nearby waterbodies (including Atharbanki creek, Kaudia River, Sahara Machyadiha). Atharbanki (20.2912° N, 86.6552° E) is a prominent place of Paradeep and a major trading centre for fisheries in Jagatsinghpur district. Atharbanki waterway covers 13.5 km skirting industrial units of Paradeep phosphate Limited, Essar steel and further opens into the mangrove fringed estuarine environment of Mahanadi. Municipal sewage of Paradeep Township, industrial effluents, agricultural run-off and fishing harbour activities are discharged into the waterways. Local community of artisans, fishermen's, farmers residing along the banks of waterway are prone to the exposure of these pollutants. In this context, the author have made a water sampling test of the Atharbanki creek. The present study aim to find out physicochemical parameters and heavy metal analysis of Atharbanki creek.

Materials and Methods

Location and Duration of Study

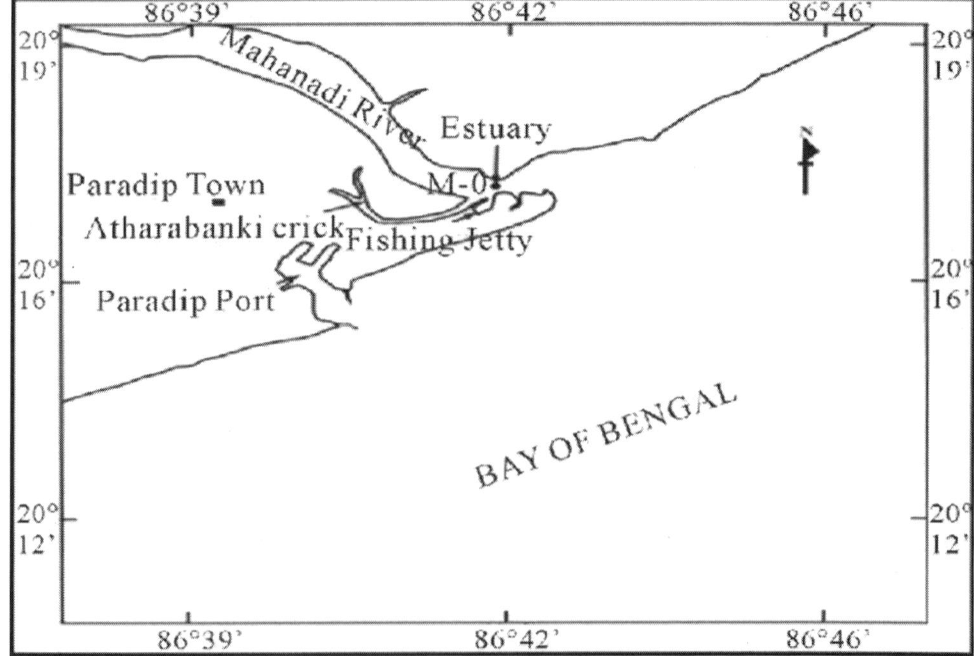

Figure 28.1: Atharbanki Creek

Figure 28.2: Google Map Showing Locations of Stations in Atharbanki Creek.

The Atharbanki waterway under study is located within the state of Odisha (20°17'18.9"- 20°17'54.4"N Latitude and 86°42'08.7"- 86°38'32.2"E Longitude) on the east coast of India. During high tide it is flooded with water from Mahanadi estuarine system and exposes intertidal mudflats at low tide. The climatic conditions are tropical with four distinct seasons namely the cool season (December to February), hot (March to May), monsoon (June to September) and post monsoon season (October to November). Atharbanki waterways originate near Kansaripatia (Lat 20°15'01.34"N and Long 86°38'21.20"E), covers a distance of 13.5 km skirting Paradeep Phosphate Limited Township, Paradeep Phosphate Limited (PPL), Indian Farmers Fertilizer Co-operative Ltd. (IFFCO) along its path, flowing adjacent to the fishing harbour (Lat 20°17'26.13"N and Long 86°42'16.23"E) and further merges into Mahanadi which eventually enters Bay of Bengal. Other major industrial unit located in the vicinity of its course are Essar Steel, Deepak Fertilizer, Sea Food Processing, Petroleum Coke, Cargil oil and Indian Oil Corporation Limited. Mangrove vegetation occurring abundantly have decreased to greater extent in these areas. Water samples has been collected from three stations along the waterways having coordinates as follows: Station 1(20°17'18.9"N, 86°42'08.7"E), Station 2(20°17'31.7"N, 86°40'16.9"E), Station 3(20°17'54.4"N, 86°38'32.2"E). Water samples were collected in clean sterilized reagent bottles. While collecting samples, the surface temperature and pH was measured simultaneously. Under laboratory conditions physicochemical parameters were measured and the sample was acid digested (using Aquaregia) prior Heavy metal detection using Atomic Absorption Spectrum (AAS).

Results and Discussion

Hydrography

Water samples were collected in the month of December. The variations in the hydrological parameters are mentioned in Table 28.1.

Table 28.1: Hydrological Parameters, Atharbanki Creek

Sampling Stations	*pH*	*Electrical Conductivity (in mS/cm)*	*Salinity (in PSU)*	*Surface Temp (°C)*	*Resistivity (in Ω)*	*Total Dissolved Solid (in ppt)*
1	6.85	15.24	18.33	36.6	33.63	14.75
2	6.54	6.947	7.693	34.7	74.56	6.681
3	7.82	0.204	0.192	34.6	2481	0.009
Min	6.85	0.204	0.192	34.6	33.36	0.009
Max	7.82	15.24	18.33	36.6	2481	14.75

Temperature

The seasonal water temperature varied from 34.6 to 36.6°C along the creek.

pH

The pH of the water across the channel fluctuated between 6.85 to 7.82, higher pH values were observed at stations fringed with mangroves, this could be due to the emergence of creek's water with the estuarine water and mangrove photosynthetic activity, which utilized CO_2 thereby shifting the equilibrium towards highly alkaline (Ruttner 1953). Similar readings were also obtained from Pitchavaram mangrove regions (Ramanathan, 1999).

Salinity

Salinity varied from 0.19 PSU to 18.33 PSU. Salinity towards the station 1 was marked higher due to tidal influx whereas in the rest of the waterway it remained relatively low.

Total Dissolved Solid

Total dissolved solid (TDS) value varied from 0.009 ppt to 14.75 ppt. Dissolved organic and inorganic matter were maximum towards station 1 which gradually decreased from station 1 to 3.

Electrical Conductivity

Electrical conductivity (EC) is the measure of the ability of water to pass an electrical current. Conductivity increases from 0.204 mS/cm to 15.24 mS/cm as we move from station 3 to 1 towards the estuarine mouth.

Resistivity

Resistivity indicates the resistance offered to the flow of electric current. Higher salt concentration towards the estuarine mouth leads to the higher electric conductivity and lower resistivity. Thus resistivity and station 1 is minimal to that of station 2 and 3.

Concentration of Heavy Metals

Atharbanki creek water analysis predicts the following concentrations of the detected metals:

Table 28.2: Concentration of Heavy Metals (in ppm)

Sampling Station	*Fe (in ppm)*	*Mn (in ppm)*	*Cr (in ppm)*	*Zn(in ppm)*
1	0.608	0.157	0.159	0.058
2	0.667	0.388	0.108	0.056
Maximum permissible limit	0.10	0.05	0.05	0.05

Above result indicates the concentrations of Iron were within the range 0.608 to 0.667ppm, the concentration levels of Manganese were within the range 0.157 to 0.388. The observed values for Chromium and Zinc varied from 0.159 to 0.108

ppm and 0.058 to 0.056 ppm respectively. The levels of heavy metal though not present in higher concentration but are greater than the maximum permissible limit except zinc thus making it unusable.

The study of physicochemical characteristics and heavy metal analysis of the Atharbanki banki water gave us following remarks. Atharbanki creek is heavily polluted from the municipal sewage waste from Paradeep Township and effluents from industries such as PPL and passes it to Mahanadi estuary, which is due to the influx of solid waste from the phosphatic fertilizer plants situated in the estuarine region. Proper sewage treatment plants should be set before discharging the sewage into the creek, otherwise the situation may be alarming to the inhabitants of the study area. Thus adequate attempts should be made to treat the wastewater before discharging it to the water bodies to keep the aquatic environment healthy.

References

Dixit, Pravat Ranjan, Kar, Biswabandita, Chattopadhyay, Partha and Panda, Chitta Ranjan (2013). Seasonal Variation of the Physicochemical Properties of Water Samples in Mahanadi Estuary, East Coast of India. Journal of Environmental Protection, 2013, 4, 843-848.

Nayak, Aswini, Raut, Dipti and Patnaik, Lipika (2014). Icthyofauna of Atharbanki waterway, Mahanadi estuary, East coast of India. Vol. 5, Issue, 10, pp.1871-1878.

Report on Environmental Pollution Monitoring at Paradeep Port, Sponsored by Paradeep Port Trust, Quarterly Report.

Sahu, S.K., Sarangi, D. and Pradhan, K.C. (2006). Water Pollution in Orrisa, Orissa Review.

Chapter 29

Understanding Life: A Journey from Genes to Omics

Jnyan Ranjan Mishra

Ganeshayan, Mahadev Nagar, Jharpara,
Bhubaneswar – 751 006, Odisha
e-mail: mishrajnyan@gmail.com

ABSTRACT

Prior to discovery of principles of inheritance, our ancestors were concerned with the application of genetics to practical matters like improving crop plants and domestic animals by selective breeding. After the establishment of genetic principles, DNA was proved to be the genetic material by several scientists. Then in the first half of DNA revolution, its structure and function could be established. The second half of DNA revolution involved the development of recombinant DNA technology followed by DNA sequencing which lead to the birth of the industry of biotechnology. The next achievement in the study of genetics was the sequencing of the entire genome of human beings. This extraordinary technological achievement made possible for opening of the new field called OMICS and analysis of multi-omics data has become the ultimate approach for investigation of diseases.

Introduction

The art of breeding plant and animal is much older than the science of genetics. In fact, prior to Bateson's coining the term genetics for science of inheritance, our ancestors were concerned with the application of genetics to limited practical matters like improving crop plants and domestic animals by generations of careful selection breeding during 18th century. However, Gregor Mendel is considered as the Father of Genetics for establishing the principles behind inheritance of characters in 1865. Simultaneous to his experiments on hybridization of pea plants, great naturalist Charles Darwin working on hereditary and variations published his famous book "Origin of Species" to explain organic evolution. Work of Darwin inspired his cousin and friend Francis Galton to start a social and genetic work which he named Eugenics to describe the application of agricultural breeding to human beings. The consequences were to sterilize 2, 25,000 Jews to whom Nazis Considered inferior and categorised them as 'useless eaters'. Ultimately 75,000 of them were mass murdered in the gas chamber by the order of Hitler, whom they placed under the group 'not worth living'. Subsequently the branch Eugenics had lost its credibility in the scientific community and ultimately proved as a disaster for the emerging science of genetics. Most of the members including T. H. Morgan left the 'Eugenic Club' established by Galton and further activities in this line was stopped.

Classical Genetics

Research and discovery of different phenomena in the field of genetics progressed very fast after the rediscovery of Mendelian principles in 1900, with a gradual change in experimental materials from Mendel's pea plant to viruses on the basis of their generation period. Some of the works in the field of genetics are worth mentioning here. With the help of improvised microscope, Walter Sutton (studying the chromosome behaviour in grasshoppers) and Theodor Boveri independently discovered the correlation between mysterious Mendelian factors and chromosomes in 1902. Working on fruit fly *Drosophila*, Morgan and his boys (1910-12) effectively proved the Sutton-Boveri theory and proposed the chromosomal theory of inheritance.

The DNA Revolution

Discovery of transformation phenomenon by Griffith (1928) on *Pneumococcus,* One-gene-one-polypeptide hypothesis by Beadle and Tatum on *Neurospora* and with the development of more effective technologies during 1930s and 1940s, scientists became curious to know about the greatest biological mystery- 'What is the chemical nature of gene?' Experimental evidences by Avery *et al.,* on *Pneumococcus* and Harshey and Chase on bacteriophase established the DNA as genetics material.

There was a big jump from experiments on organisms to the DNA molecule and the discovery of the structure of double helix, 'the blue print of life', by Watson and Crick in 1953 brought a great change in the science of genetics. Contributions of Maurice Wilkins, John Kendrew, Linus Pauling, Luria, Rosalind Franklin, Erwin Chargaff and many others in this connection are praiseworthy. The first half of DNA revolution that is from the discovery of its structure up to the discovery of genetic code by Crick, Jacob and Monod, Nerenberg and Khorana, *etc.* during 1960s, involved around 3000 scientists, many of them being awarded with Nobel Prize.

The second half of DNA revolution started after 1972 with the development of Recombinant DNA technology followed by DNA sequencing, which led to the birth of new industry-Biotechnology. Development of RDT during 1960s-1970s with the discovery of DNA ligase by Gilbert and Lehman; restriction enzymes by W. Arber and the method to induce *E.coli* cells to import plasmids from outside by Stanley Cohen resulted in customization of DNA molecule. H. Boyer, the master of restriction enzymes and S. Cohen, the first among plasmid pioneers are called the world's first genetic engineers, who provided a chance to 'Play God'. Development of Biotechnology during 1980s resulted in commercialization of the genome. Technology to produce marketable proteins, cloned human insulin and interferon and its application in silk, dairy, pharmaceuticals and enzyme industries made biology a big-money game (DNA, Dollars and Drugs). The founders of biotech companies were typically university professor and the funding companies. Basing on the DNA sequencing methodology developed simultaneously by Sanger and Gilbert, Alec Jeffrey developed the genetic finger printing technology in 1986 which has become the main technique for criminal investigation and establishment of parentage (DNA's day in the court).

Human Genome Project (HGP)

The next target, as well as one of the most significant achievements in genetics was sequencing of entire genome of human beings. It was initiated by International Human Genome Sequencing Consortium (IHGS). Universities and research institutes of US, UK, France, Germany, Japan and China participated in the Human Genome Project and submitted the final draft in 2003. HGP revealed the presence of 3.1 to 3.2 billion base pairs with 30,000 to 35,000 genes in the entire genome. But the protein coding genes are only 20,000-25,000 which are less than 2 per cent of the total genome. Further, 99.9 per cent of base pairs are same in all people and the vast variation among them is due to single base pair difference in about 50,000 locations. This led to use of DNA markers as a new strategy for identifying genes that cause genetic diseases. Most important applications of the strategy are Gene Hunting (tracking the genes that cause diseases, *e.g.*, Duchenne Muscular Dystrophy- a progressive muscular- wasting disease) and defying genetic disorders. The most commonly known example of defying genetic disorders is the case of David Vetter who suffered from severe combined immunodeficiency disorders (SCID). To remain away from even slightest infection, he spent the entire life period in a sterile world. The boy in the plastic bubble was known as the 'bubble boy'. However, tracking a disease causing gene and making arrangements to allow the patient survive is not a solution for cure of the disease.

The Omics Revolution

The Human Genome Project is an extraordinary technological achievement which led to open a new field called Omics. It is the molecular underpinning of development in an egg-to-adult journey. Different Omics in a multi-omics approach are Genomics, Epigenomics, Transcriptomics, Proteomics, Metabolomics and Microbiomics.

Genomics is the study of entire genome as opposed to single gene (*i.e.*, genetics). It involves recombinant DNA technology, entire genome sequencing and application of bioinformatics to analyse the sequences. It focuses on interaction between alleles and genes and identification of variants associated with diseases.

Epigenomics is the study of modification of DNA by methylation and associated proteins by histone acetylation. These modifications are major regulators of gene transcription and are influenced by both genetic and environmental factors. Epigenetic maps or epigenetic signatures in multiple human tissues are available in International Human Epigenomics Consortium.

Transcriptomics is the analysis of entire collection of mRNA sequences to determine when and where each gene is 'turned on' or 'turned off' in the cell and tissue. While 80 per cent of genome is transcribed only 3 per cent of genome encodes protein. This protein coding transcriptome system is extremely complex as thousands of novel isomers of transcriptomes are available. The non-coding transcriptomes are also very important, which include long non-coding RNAs, short non-coding RNAs and circular non-coding RNAs. Short non-coding type includes micro RNA, piwi-interacting RNA and small nuclear RNA. Dysregulation of these non-coding RNAs leads to various diseases like diabetes, cancer, heart diseases, *etc.* Some of the transcriptome database resources are National Human Genome Research Institute (NHGRI), National Cancer Institute (NCI), Genotype Tissues Expression Project (GTEx) and Encyclopaedia of DNA Elements (ENCODE). An example of Database uses is, if an unknown gene is expressed in fat tissue but not in other tissue, then it may be responsible for obesity.

Proteome is a set of proteins produced in an organism or a system and proteomics is the large-scale study of proteomes. Proteomics deals with protein abundance, modifications and interactions between proteins. Different database resources of proteomes are UniprotKB, IntAct, Reactome, PRIDE, *etc.*

Metabolomics is the large-scale study of small molecules like amino acids, fatty acids, carbohydrates and other products of cellular metabolism. Concentration of metabolites and their relative ratios directly reflect the biochemical activities of the cells and beyond normal range are indicative of diseases. Further, novel genetic loci have been discovered which regulate small molecules. Hence, Metabolomics is influenced by both genetic and environmental factors.

Microbiomics includes all the microorganisms of a given community investigated together. The bacteria, viruses and fungi colonized in human skin, gut and mucosa surface are collectively known as microbiota and their gene constituents as microbiome. Human microbiome is very complex. The gut contains 100 trillion bacteria from 1000 species and differences of microbiomes between individuals depend on age, environment, diet, drugs used, *etc.*

Under omics studies, diseases are categorised as simple diseases which arise by simple gene mutants and complex diseases which involve both genetic and environmental factors. Integrative analysis of multiple omics data which is done by statistical analysis of different omics layers and network modelling is an appropriate step towards investigation of diseases.

REFERENCES

Hasin Y., Seldin M. And Lusis A. (2017) Multi-omics approaches to disease. Genome Biology, 18: 83.

Watson, J. D. and Berry A. (2003) DNA - The secrete of life. Knopf A.A., Random House, Inc, New York.

Chapter 30

Arabidopsis Decapping Protein Factor Mobilizes Ubiquitin 5 into the Processing Bodies

*Gagan Kumar Panigrahi and Kunja Bihari Satapathy**

School of Applied Sciences,
Centurion University of Technology and Management, Odisha
**e-mail: kbs_bot@rediffmailmail.com*

ABSTRACT

In eukaryotes, the expulsion of methylated-7-guanosine-di-phosphate (m7GDP) present at the 5' end of the messenger RNA (mRNA) results into the turnover of mRNA. This significant event is referred as decapping process. Primarily, decapping of mRNA modulates plant developmental processes. The decapping process is executed by several decapping complex protein factors, of which DCP1, DCP2 and DCP5 are indispensable. In plants, DCP5 is well involved in regulating the functioning of processing bodies (P bodies). P bodies are the cytoplasmic foci whereby the anomalous mRNAs are accumulated for decay. The molecular mechanisms fundamental to the DCP5 mediated processing of mRNAs are alluring. The results of the present study showed that the DCP5 interacts with Ubiquitin 5. The ubiquitin is a 76-amino acid polypeptide that acts as a covalent modifier of numerous proteins, resulting in cellular homeostasis. Ubiquitin protein initiates the process of ubiquitination resulting in paving path for the proteins to decay. Further studies may reveal the plausible role of ubiquitin proteins in the process of mRNA decay.

Keywords: *mRNA decapping, P bodies, Ubiquitin, mRNA decay.*

Introduction

In eukaryotes, mRNA turnover primarily involves the elimination of m^7GDP from the 52 end. The 5' monophosphate is a substrate for the 5' exonuclease XRN1 (Poole and Audrey, 1997) resulting in the quick decay of the mRNA. Primarily, a protein complex mediates the decapping reaction. The decapping protein complex is well characterized in yeast and human but not in plants. In *Arabidopsis thaliana*, DCP1, DCP2, DCP5 are the central decapping protein factors. The m^7GDP cap is specifically removed by the DCP2 which posses the pyrophosphatase activity (Xu *et al.,* 2006; Iwasaki *et al.,* 2007). DCP1, DCP2 and DCP5 localize in *Arabidopsis* processing bodies (P bodies), depicting their role in mRNA decay (Xu *et al.,* 2006; Xu *et al.,* 2009). mRNA decapping event occurring inside P bodies is tightly regulated, chiefly an irreversible process. In eukaryotes, messenger ribo-nucleoproteins (mRNPs) assemble together into the P bodies (Sheth and Parker, 2003). Most importantly, P bodies function includes mRNA storage, mRNA decapping and translational repression (Yu *et al.,* 2019). Regulation of mRNA translatability is associated with the P-body assembly. But, the translational arrest may be reversed, enabling the reengagement of some mRNAs into the translational machinery to resume translation (Brengues *et al.,* 2005). The mechanisms regulating the mRNA decapping process are still unknown.

Ubiquitination, a unique post-translational modification event in eukaryotes is vital for maintaining the protein homeostasis (Hershko and Ciechanover, 1998; Khoury *et al.,* 2011; Walsh, 2006). Mainly, ubiquitination involves covalent add-on of a small ubiquitin protein into the target protein through a series of reactions, essentially facilitated by three classes of enzymes, E1, E2 and E3. E1 is the ubiquitin-activating enzyme (UBA), E2 is the ubiquitin-conjugating enzyme (UBC) and E3 is the ubiquitin ligase (Callis, 2014). Ubiquitin, a highly conserved protein contains 76-amino acid residues (Callis and Vierstra, 1989). In eukaryotes, multiple genes encode ubiquitin resulting into two dissimilar types of translational fusions referred to as homomeric fusions and heteromeric fusions. Homomeric fusions comprise of multimers of ubiquitin coding regions (polyubiquitin) whereas, heteromeric fusions, the 76-amino acid residue ubiquitin is followed by a different protein (ubiquitin-extension proteins), mostly small ribosomal proteins or else, a ubiquitin-like protein, RUB (Burke *et al.,* 1988; Callis and Vierstra, 1989; Callis *et al.,* 1990; Callis *et al.,* 1995). Till date, ubiquitin gene count in *Arabidopsis thaliana* is twelve (Callis and Vierstra, 1989; Callis *et al.,* 1990; Sun and Callis, 1993; Callis *et al.,* 1995). The present investigation shall try to show the physical interaction between *Arabidopsis thaliana* UBQ5 and DCP5. It is intriguing to reveal the basis of this interaction as it would facilitate in one hand to understand the role of DCP5 beyond its canonical activity and in other hand, the role of ubiquitin in maintaining the protein homeostasis.

Materials and Methods

Plant Materials and Growth Conditions

Nicotiana benthamiana plants for transient expression of target proteins, particularly for confocal experiments were grown in soil. Temperature was maintained at 22° C with 70–80 per cent relative humidity under long day (12/12 h light/dark) conditions.

Plasmids

DCP5 and UBQ5 full-length cDNAs were obtained by RT-PCR using Col-0 RNA. The PCR products were cloned into pDONR207 and sequenced. Wild type (WT) coding sequences of DCP5 and UBQ5 were cloned into expression vectors (GFP, RFP and BiFC vectors).

Agro-Infiltration and Confocal Laser Scanning Microscopy (CLSM)

The binary clones were transformed into *Agrobacterium tumefaciens* strain (GV3010), and agro-infiltrated into *Nicotiana benthamiana* leaves. Two days post-agro-infiltration, the infected leaves were analyzed under CLSM.

Results and Discussion

UBQ5 and DCP5 Localizes to Cytoplasm and P Bodies Respectively

One of the central decapping protein factors, DCP5 is comprised of several vital domains including DFDF positioned from 453-489 amino acid residues and FFD/TFG positioned from 512-554 amino acid residues (Figures 30.2a-c). Likewise, UBQ5 contains ubiquitin coding domain fused with the 40S ribosomal protein S27a-3 (Figure 30.1a-c). UBQ5 when over-expressed under *35S* Cauli Mosaic Virus (CMV) promoter localized to the cytoplasm (Figure 30.3). Under similar conditions, over-expressed DCP5 localized to P bodies. Canonically, DCP5 is involved in mRNA decapping, primarily which occurs at P bodies. Moreover, UBQ5 localization points out towards the role of ubiquitin, mainly in the protein homeostasis.

UBQ5 and DCP5 Co-localize and Physically Interact in the P Bodies

UBQ5 re-localized to the P bodies when co-expressed with the DCP5 (Figure 30.4a). To validate whether there is any physical interaction among the DCP5 and UBQ5, we co-infiltrated DCP5-VYNE and UBQ5-CYCE (BiFC fusion proteins). BiFC studies pointed out the DCP5-UBQ5 interaction as the fluorescence signal was seen in the P bodies (Figure 30.4b), suggesting that the DCP5 is involved in re-localization of the UBQ5 into the P bodies by physical interaction. This is quite interesting, since; previously it was known that DCP5 is well involved in targeting mRNA (Xu

1a

At UBQ5 (AT3G62250)

```
  1 MQIFVKTLTGKTITLEVESSDTIDNVKAKIQDKEGIPPDQQRLIFAGKQLEDGRTLADYNIQKESTLHLVLRLRGGAKKRKKKTYTKPKKIKHKHKKVKL
101 AVLQFYKIDGSGKVQRLRKECPNATCGAGTFMASHFDRHYCGKCGLTYVYQKEGVEA
```

1b

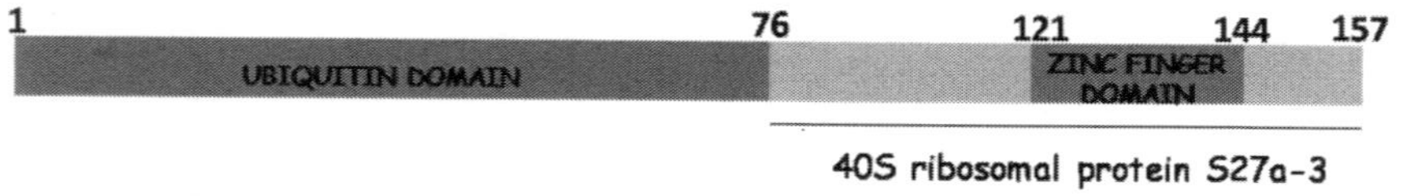

1c

Amino acid residue	Domain/motif/compositional bias
1-76	Ubiquitin
77-157	40S RIBOSOMAL PROTEIN S27a-3
121-144	Zinc Finger Domain
77-79	Lys-rich

Figure 30.1: Amino Acid Sequence and Protein Domain Organization of UBQ5:
a: Sequence of UBQ5 protein is shown using single-letter codes for amino acids;
b: Domain organization of *Arabidopsis* UBQ5: Domain information was generated with UniProt;
c: Amino acid residues of UBQ5 representing domain/motif/compositional bias.

2a **At DCP5 (AT1G26110)**

```
  1 MAADNTGSKSSSAADSYVGSLISLTSKSEIRYEGILYNINTDESSIGLQNVRSFGTEGRKKDGPQVPPSDKVYEYILFRGTDIKDLQVKASPPVQPPAST
101 INNDPAIIQSHYPSPMPTSGSLPSTASGSLPDISSHNGQPGQHGMGFQNAMPLYQPGGNLGSWGASPQPPMYWQGFYTPPPNGLPQLHQQSLIRPPHGLP
201 MPNSLQQPLQYPNFNTPPPPTGSSSLQGSSLPEAPSSLFPFSTSSQMLAPSSLPFPGLPPVTLSSSLQSTLQSAPSPSLASEMAPPLLSNKAPITAPPTL
301 PQDTNLLSFSLSTTRATEASTGLPLSNKPSVVTGPISPPQTTPLTSAPVAGVSSSISQDKPKPLLVTPGQLLQSGSSAVSLSPPSTNADKDVEVVQVSSS
401 AGLEQSVPVTSEAQPPILPLPSSARPTQKPNGHSFPNHNGYRGRGRGRGRGAGRSHQVMKFTEDFDFTAMNEKFNKDEVWGHLGKSTTLDGDEDDDSPTV
501 DEAELPKIEAKPVYNKDDFFDSLSSNTIDRESQNSRPRFSEQRKLDTETFGEFSRFRGGRGGRGGYGRNNGYSRGGYGGRGYGGYGGRGGGGGGYGYGGR
601 GQGRGVSNRTT
```

2b

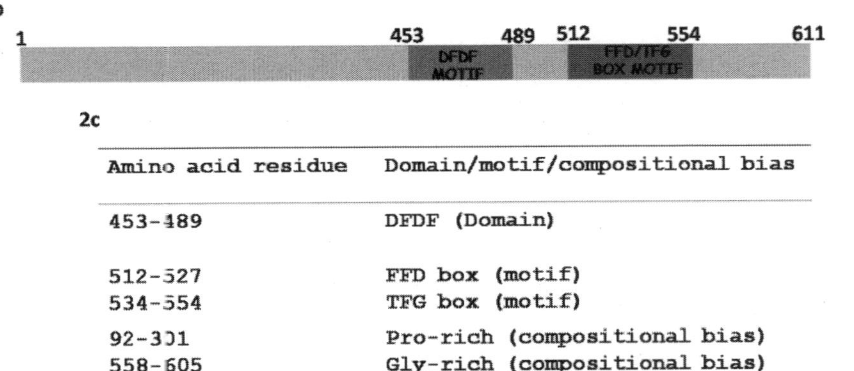

2c

Amino acid residue	Domain/motif/compositional bias
453-489	DFDF (Domain)
512-527	FFD box (motif)
534-554	TFG box (motif)
92-301	Pro-rich (compositional bias)
558-605	Gly-rich (compositional bias)

Figure 30.2: Amino Acid Sequence and Protein Domain Organization of DCP5:
a: Sequence of DCP5 protein is shown using single-letter codes for amino acids;
b: Domain organization of *Arabidopsis* DCP5: Domain information was generated with UniProt;
c: Amino acid residues of DCP5 representing domain/motif/compositional bias.

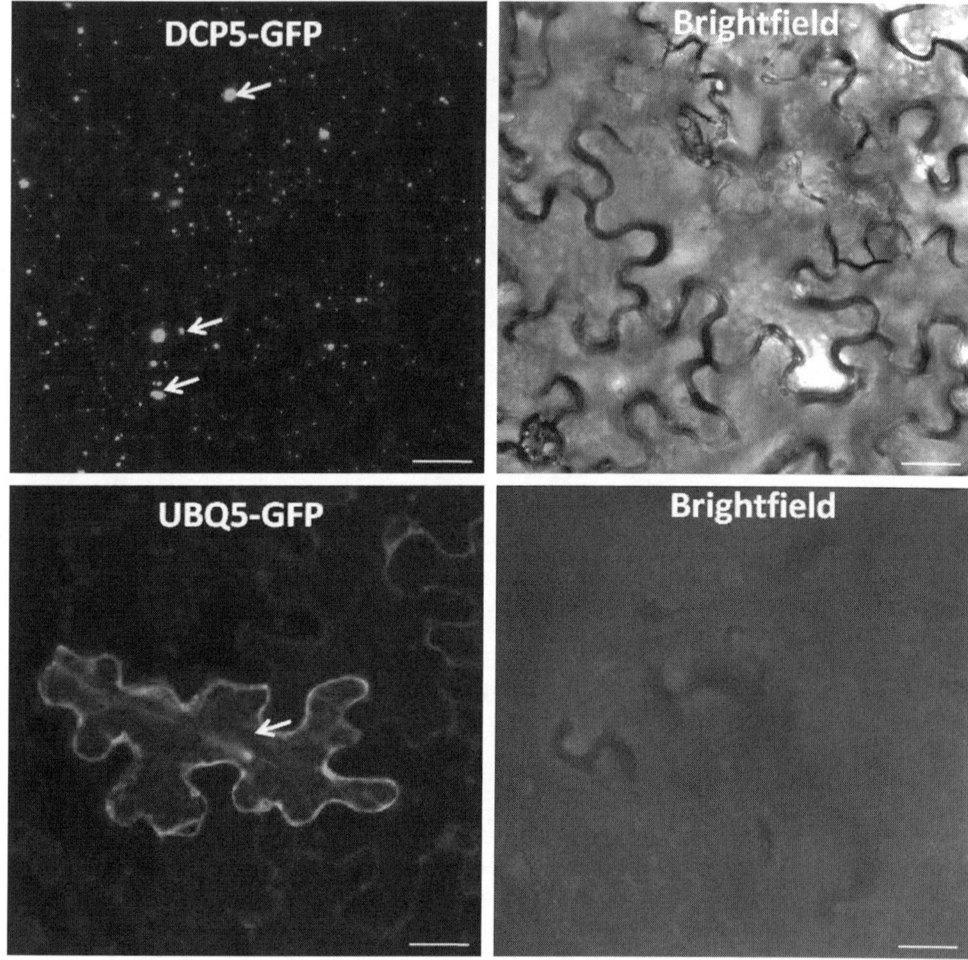

Figure 30.3: Independently, DCP5 and UBQ5 Localizes to P Bodies and Cytoplasm Respectively. DCP5-GFP localizes to P bodies, whereas, UBQ5-GFP localizes to the cytoplasm when transiently expressed in the *Nicotiana benthamiana* as indicated by the arrow marks. (Scale bar: 10μm).

et al., 2009) whereas, UBQ5 performs its canonical role by targeting proteins (Callis, 2014). Interaction among two distantly related protein factors particularly in terms of their targets possibly may provide new directions to understand the underlying molecular mechanisms, mainly the role of mRNA decapping factors in the protein degradation pathway.

Conclusion

Decapping of mRNA is quite vital, primarily for maintaining the homeostasis of functional mRNA. Even though several studies have revealed the underlying

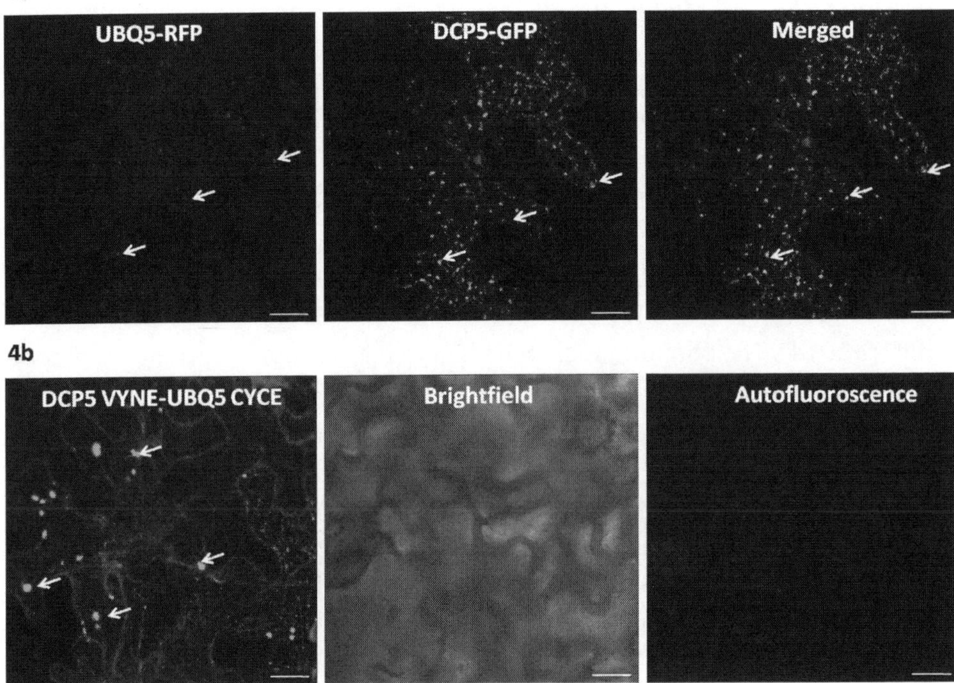

Figure 30.4: UBQ5 Physically Interacts with DCP5: Subcellular localizations are indicated by the arrow marks. (Scale bar: 10μm). a: UBQ5-RFP and DCP5-GFP co-localizes in P bodies; b: DCP5 and UBQ5 physically interact with each other in the P bodies.

principles behind this significant molecular event, still the role of decapping complex proteins are at primitive level. In plants, DCP5 being a key player in maintaining the P body function by actively participating in the mRNA decapping process, its non-canonical function may also be quite intriguing. Ubiquitin, primarily being responsible for targeting specific proteins into the protein degradation pathway interacts with DCP5 and moreover effects the sub-cellular localization of the DCP5, thus suggesting its role in the turnover of mRNA. In future studies, it would be quite interesting to reveal the role of plant ubiquitins beyond their canonical function.

Acknowledgements

Authors are thankful to the administration and management of Centurion University of Technology and Management, Odisha, India for providing necessary facilities to conduct the experiment.

REFERENCES

Brengues, M., Teixeira, D. and Parker, R. (2005). Movement of eukaryotic mRNAs between polysomes and cytoplasmic processing bodies. Science. 310: 486–489.

Burke T., Callis J. and Vierstra R.D. (1988). Characterization of a polyubiquitin gene from *Arabidopsis thaliana.* Mol. Gen. Genet. 213: 435-443.

Callis J., Carpenter T., Sun C.W. and Vierstra R.D. (1995). Structure and evolution of genes encoding polyubiquitin and ubiquitin-like proteins in *Arabidopsis thaliana* ecotype Columbia. Genetics. 139: 921-939.

Callis J., Raasch J.A. and Vierstra R. (1990). Ubiquitin extension proteins of *Arabidopsis thaliana;* structure, localization and expression of their promoters in transgenic tobacco. J. Biol. Chem. 265: 12486-12493.

Callis J. and Vierstra R.D. (1989). Ubiquitin and ubiquitin genes in higher plants. Oxford Surveys in Plant Mol. Biol. 6: 1-30.

Callis, J. (2014). The ubiquitination machinery of the ubiquitin system. Arabidopsis Book, 12, e0174.

Hershko A. and Ciechanover A. (1998). The ubiquitin system. Annu. Rev. Biochem. 67: 425-479.

Iwasaki, S., Takeda, A., Motose, H. and Watanabe, Y. (2007). Characterization of Arabidopsis decapping proteins AtDCP1 and AtDCP2, which are essential for post-embryonic development. FEBS Lett. 581: 2455-2459.

Khoury, G.A., Baliban, R.C. and Floudas, C.A. (2011). Proteome-wide posttranslational modification statistics: frequency analysis and curation of the swissprot database. Sci. Rep. 1: 90.

Poole, Toni L. and Stevens, A. (1997). "Structural Modifications of RNA Influence the 52 Exoribonucleolytic Hydrolysis by XRN1 and HKE1 of Saccharomyces cerevisiae". Biochemical and Biophysical Research Communications. 235 (3): 799-805.

Sheth, U. and Parker, R. (2003). Decapping and decay of messenger RNA occur in cytoplasmic processing bodies. Science. 300: 805-808.

Walsh, C. (2006). Posttranslational modification of proteins: expanding nature's inventory. Englewood, Colo.: Roberts and Co. Publishers.

Xu, J. and Chua, N.H. (2009). Arabidopsis Decapping 5 Is Required for mRNA Decapping, P-Body Formation, and Translational Repression during Postembryonic Development. Plant Cell. 21: 3270-3279.

Xu, J., Yang, J.Y., Niu, Q.W. and Chua, N.H. (2006). Arabidopsis DCP2, DCP1, and VARICOSE form a decapping complex required for postembryonic development. Plant Cell. 18: 3386-3398.

Xiao Yu., Bo Li., Geng-Jen Jang., Shan Jiang., Dao hong Jiang., Jyan-Chyun Jang., Shu-Hsing Wu., Libo S. and Ping He. (2019). Orchestration of Processing Body Dynamics and mRNA Decay in *Arabidopsis* Immunity. Cell Reports. 28: 2194-2205.

Chapter 31

Comparative Study on Performance of Tasar Silkworm, *Antheraea mylitta*, Drury (Daba TV) under Direct Rearing and Chawki Method on Tasar Silkworm Food Plant *Terminalia tomentosa* (Roxb.)

*Manjula Gupta and Dara Singh Gupta**

University Department of Botany, Kolhan University, Chaibasa – 833 202, Jharkhand
**e-mail: darasgupta.jcc@gmail.com*

ABSTRACT

Tasar silkworm Antheraea mylitta, Drury is very important for sericulture industry in India. The present study aimed to compare the direct and chawki reading of Tasar silkworm on the host plant Asan. The chawki rearing was carried out during I and II instars larvae whereas, direct rearing method was from I to V instar on Terminalia tomentosa (Roxb.) in common plot. The results indicated that larval duration took six days lesser wihen reared with chawki than direct method. Mortality rate of silkworm was found less in Chawki method as compared to direct method. Tasar silkworm reared under chawki method had higher larval weight, maximum cocoon weight as compared to direct brushing method. Hence, it was concluded that Tasar silkworm rearing with chawki garden during initial stage improves the rearing performance and quantitative cocoon traits.

***Keywords**: Chawki, Cocoon, Direct, Mortality, Silkworm tasar.*

Introduction

Antheraea mylitta Drury (Lepidoptera, Saturniidae) is a semi-domesticated tropical tasar silkworm, distributed in India between the range of 16-24°N latitude and 80-88° E longitude from West Bengal in the East to Karnataka in the South and in the forest area of Jharkhand, Bihar, Odisha, Chhattisgarh, Madhya Pradesh and Andhra Pardesh (Thangavelu and Sinha, 1993). It is exploited commercially for the production of tasar silk (Dash *et al.,* 1992). In India, it is trivoltine (three crops in a year) at lower altitude and for commercially it is exploited as bivoltine, reared two times in a year in July to August (Ist crop, Rainy cocoon crop) and September to October (IInd crop, Autumn crop). There are forty four ecoraces of tasar silkworm in varied geographical tropical zone in India (Mohanty, 2003). Out of these, only Daba and Sukinda are mainly contributing for country's tasar raw silk (Hansda *et al.,* 2008). Tasar silkworm is a polyphagous insect feeding on a variety of food plant primarily on *Terminalia tomentosa* (Roxb.), *Terminalia arjuna* (Roxb.) Wight and Arn and *Shrea robusta* Gaertn. (Reddy *et al.,* 2010). The host plants, which silkworm normally prefers are known as primary host plants. Other host plants, where the silkworm can sustain its life, but normally do not prefer, are called secondary host plants. Direct (traditional) rearing of tasar silkworm on forest grown trees resulted in 80-90 per cent crop loss due to natural calamities diseases pests, predators (Hari and Seth, 1999). The majority of crop loss is ascribed to pest, squirrel and birds such as crow during young stage larval period called chawki stage (I and II instars stage). Outdoor rearing is conducted in the forest, there is no control over the climatic conditions such as rain, temperature and humidity and thereby these climatic conditions that often lead to viral attacks (Singh *et al.,* 1992). Chawki methods of tasar silkworm rearing, which enables the farmers to reap good harvest of quality cocoons. Young age silkworm rearing comprises of rearing of 1st and 2nd instar larvae under the care with nutritious hygienic and easily digestible leaves. In traditional (direct) method of tasar silkworm rearing, young larvae is directly brushed on host plants without care which results in frequent loss in cocoon yield. Thus, chawki rearing method is best choice to increase the quantity of crops as well as increase the quality and quantity of silk. For this, an investigation was done to compare the performance of tasar silkworm in direct and chawki rearing methods

Materials and Method

The experiment conducted during 2nd rearing crop in the month of September 2018 between Kharsawan block of Seraikella-Kharsawan district Kharsawan Garh is situated at 22.79°N and 23°09'34"N altitude 85°30'4" and 86°15'29" E longitude.

It has an average elevation of 3000 meters ASL. 100 Dfls (disease free laying) were collected from PPC, Kharsawan Dfls were washed properly by 5 per cent Depuratax then dried under eggs dryer machine for 20 mins). Average maximum and minimum temperature during the growing season were 27.0 and 21.5°C, respectively.

Chawki garden was established with Asan plantations (Figure 31.1). In order to maintain chawki garden, four years old economic plantation of Asan plant were selected in a spacing of 4' x 4', comprising 10 plants. Plants were pruned at 3 feet height during first fortnight of March and clipping of upper portion of branches during July for second crop rearing. Cultural operations did after pruning *i.e.* vermicompost (2kg) was applied per plant by deep placement. Fertilizer such as Nitrogen applied in two split doses (each dose of 16 gram Urea/plant in Phosphate applied (46 gram Single Super Phosphate/plant and Potash (12 gram Muriate of Potash/plant as a single dose. 10 plants were selected with newly fresh grown leaves were divided equally. 100 eggs collected from PPC, Kharsawan for Direct and Chawki method of rearing. 3-5 per cent of bleaching powder were applied in and around the field disinfection measure NaOCl (0.02-0.05%) was sprayed before starting of each instar. Before the start of young age rearing, the plants of chawki garden are covered by nylon net with size of 6m X 6m. After second instar those larvae were shifted to separate block plantation for late age rearing. On other hand for direct method traditional practices of tasar silkworm rearing procedure was followed. Figure 31.2 shows the laying of eggs, second instar larvae and fifth instar larvae.

It this experiment larvae parameter such as larvae weight, mortality and lifespan were studied from both types of rearing. The larval weight was measured using electronic balance from first to fifth instar. The larval mortality was calculated

Chawki garden

Forest grown tree

Figure 31.1: Chawki Garden and Forest Garden.

by counting the number of larvae lost due to various reasons such as natural calamities, diseases, pests, predator during each instar. Cocoons parameters such as, cocoons, weight, shell weight, shell ratio and cocoons volume were also recorded. Shell ration is calculated by the formula shell ratio = (Shell weight/cocoon weight) x 100

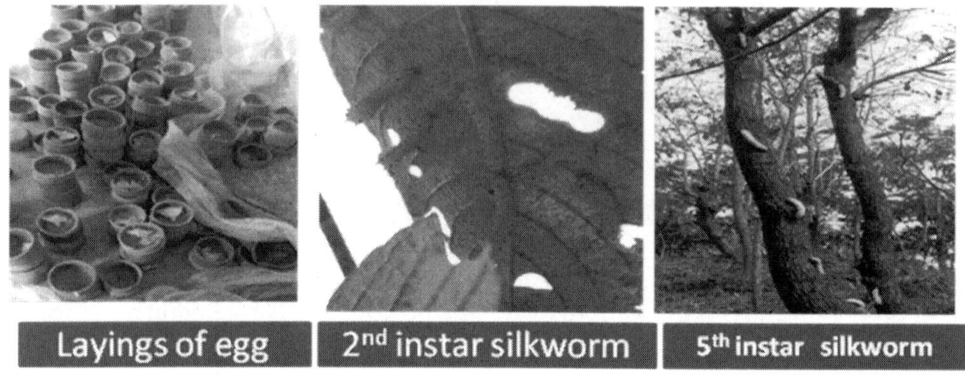

Figure 31.2: Laying of Eggs, Second Instar Silkworm and Fifth Instar Silkworm.

Results and Discussion

Rearing was conducted with 50 dfls each in the same period in direct and chawki method. Table 31.1 shows the significant difference of larval duration from 1st to 4th instar. There was no any variations in 5th instar larvae period. Date of spinning also differed in both the method. Chawki method took six days less to attend spinning stage as compared to direct one. So, the total larval duration

Table 31.1: Rearing Performance of Tasar Silkworms on Direct Rearing Method and Chawki Methods

Parameters	*Direct Method*	*Chawki Method*
1st Instar (duration)	5 days	4 days
2nd Instar (duration)	6 days	5 days
3rd Instar (duration)	8 days	6 days
4th Instar (duration)	9 days	7 days
5th Instar (duration)	15 days	15 days
Date of spinning	11-11-2018	05-11-2018
Total larval period	43 days	37 days
Date of harvesting	21-11-2018	15-11-2018
Total number of harvest cocoons	24	43
Loss of silkworm due to diseases/pest and predator	26	7

of chawki rearing was 37days whereas direct method of silkworm rearing performance was 43 days to from the cocoon. This may be due to nutrition rich and hygiene of leaves available in chawki garden as special care was taken to maintain the chawki garden plants. Sekar (1995) also indicated similar results. In direct method of silkworm rearing was observed more losses of silkworms due to pest, disease and other climatic factors than chawki method. In this study loss of worm for direct method was observed of 26 as compared to 7 in chawki method. Hence, higher number of tasar silkworm cocoons were obtained in chawki method (43) than direct method (24).

Mortality Rate

The mortality by bacteriosis, pebrine, pests and predators of tasar silkworm, during the rearing period are presented in Table 31.2 and Figure 31.3. During the rearing period, losses in direct rearing due to bacteriosis, pebrine, pest,

Table 31.2: Mortality Rate of Tasar Silkwormsm on Direct Rearing Method and Chawki Method

Parameters	*Direct Method*	*Chawki Method*
Bacteriosis	4	0
Pebrine	2	1
Pest and predators	12	4
Malnourishment	5	1
Others	3	1
Total	**26**	**7**

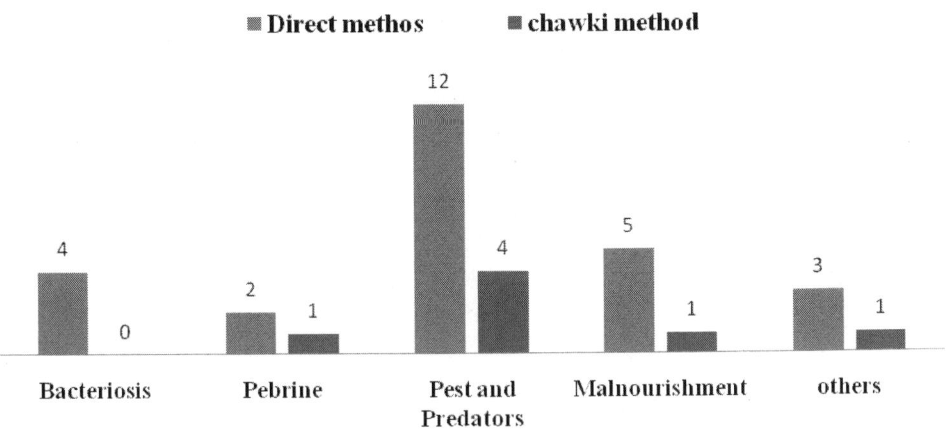

Figure 31.3: Mortality Rate of Tasar Silkworms on Direct Rearing Method and Chawki Method.

predators, malnourishment and others were 4, 2, 12, 5, 3 respectively (total loss 26) while that in the chawki rearing were 0, 1, 4, 1, 1 respectively (total loss 7). The data indicate that loss due to bacterial, pebrine and pest and predators and other factors in direct method rearing much higher in compare to that in chawki rearing method. This might be due to the fresh green hygenic and nutritive leaves provided to the silkworm grown in chawki garden which improved the immune system through which they sustained against biotic and abiotic factor. Mahobia and Yadav (2010) investigated the reasons for low productivity of *Antheraea mylitta*, Tasar silkworm rearing, and found that crop loss due to 15.01 per cent of bacterial incidence followed by 12.08 per cent of virosis, 10.58 per cent of Pebrine and the other accounted for 7.07 per cent at larval, moth and cocoon levels.

Figure 31.4: Cocoons Obtained from Chawki Method and Direct Method.

Larval Weight

Data presented in table 31.3 and Figure 31.5 illustrate that larval weight under chawki method was observed consistently increasing from I instar stage to V instar stage as compared to direct method silkworm rearing method. It was observed that the larvae fed with nutrition rich hygienic leaves in chawki method recorded higher larval weight (0.09, 0.98, 8.03, 14.01, 43.37 g during I, II, III, IV and V instars, respectively) as compared to direct method rearing of silkworm (0.06, 0.75, 6.01, 12.02 and 34.05 g during I, II, III, IV and V instars, respectively). The healthy leaves with supplemented feed increase larvae weight significantly (Etebari and Fazilati, 2003). These finding was also supported by Mahmoud (2013). Larvae were fed with healthy nutritious and hygienic Asan leaves in chawki methods might be possible reason to significant increase in body mass compared to the direct method of silkworm rearing.

Table 31.3: Larva Weight of Tasar Silkworm on Direct Rearing Method and Chawki Rearing Method

Stages of Silkworm	*Larval Weight in chowki Method (g)*	*Larval Weight in Direct Method (g)*
I instar	0.09	0.06
II instar	0.98	0.75
III instar	8.03	6.01
IV instar	14.01	12.02
V instar	43.37	34.05

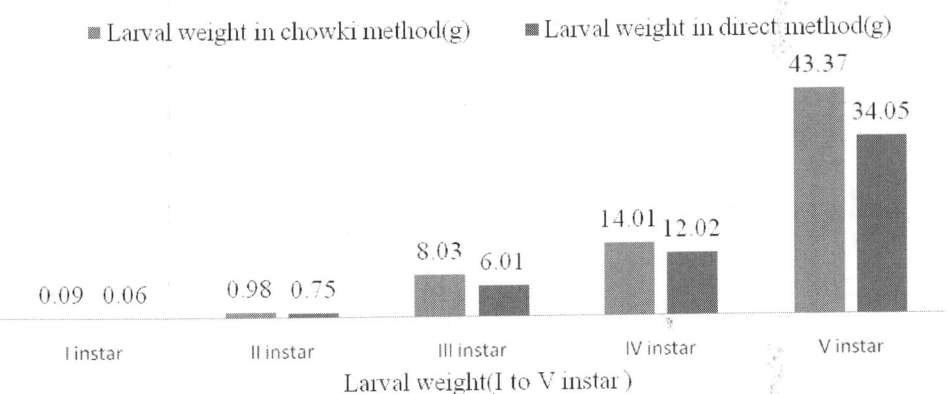

Figure 31.5: Larva Weight of Tasar Silkworm on Direct Rearing Method and Chawki Rearing Method.

Cocoon Characters

Tasar silkworms reared under chawki method showed significant influence on cocoon characters. Table 31.4. Average cocoon weight of tasar silkworm in direct and chawki rearing method were 10.39 and 13.57g, respectively (Figure 31.6). Similarly, larvae reared with chawki method were found higher shell weight (1.6g) than direct method of rearing. The shell ratio percentage was found higher (14.07 per cent) under chawki method than direct rearing method. Similar finding were observed by Subburathinum and Krishna (1992) who have supplied with nutrient rich leaves source that influenced higher shell percentage. Also the higher cocoon volume was observed under chawki rearing method (13cc). Similar finding also observed by Rajegowda (2002) that the higher cocoon and shell weight when silkworm fed nutrient rich hygiene leaves. Similar findings were observed by

Kamaney et al., (2017) that confirmed, chowki methods not only reduce larval. Montality but also improved the cocoon weight, shell weight, shell ratio due to the feeding of nutrient rich hygienic leaves.

Table 31.4: Cocoons Character under Direct Method and Chawki Method

Cocoons Character	*Direct Method*	*Chawki Method*
Cocoons Weight(g)	10.39	13.57
Shell weight(g)	1.2	1.6
Shell ratio(per cent)	10.55	14.07
Cocoons volume(cc)	12	14

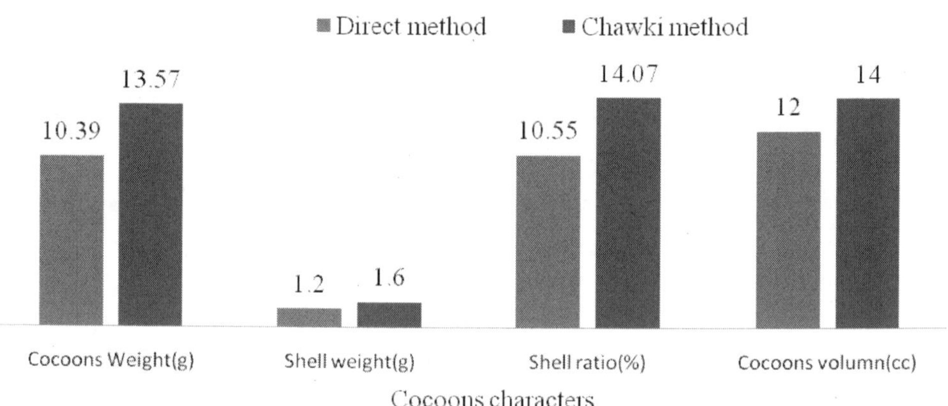

Figure 31.6: Cocoon Characters of Tasar Silkworm on Direct Rearing Method and Chawki Rearing Methods.

Conclusions

From the above observation, it is concluded that young age tasar silkworm reared under chawki garden had less larval duration higher harvestable cocoons and less mortality rate. Also improved cocoon characters like cocoon weight, silk ratio silk weight and volume of tasar silk cocoon.

Acknowledgement

Authors are very thankful to tribal people and farmers of Kharsawan Block and Kolhan University Chaibasa for their cooperation during this work.

REFERENCES

Dash A. K., Nayak B. K. and Dash M. C. (1992). The effect of different food plants on cocoon crop performance in the Indian tasar silkworm Antheraea mylitta Drury (Lepidoptera: Saturniidae). J. Res. Lepidopt., 31, 127-131.

Etebari, K. and Fazilati M. (2003). Effect of feeding on mulberry's supplementary leaves with N, P, and K in some biological and biochemical characteristics of silkworm, *Journal of Science Technology Agriculture and Natural Resources,* 7, 233-244.

Hansda G., Reddy R. M., Sinha M. K., Ojha N. G. and Prakash N. B. V. (2008). Ex-situ stabilization and utility prospects of Jata ecorace of tropical tasar silkworm Antheraea mylitta Drury. Int. J. Industrial Entomol., 17 (2), 169-172.

Hari Om Agrawal and Seth M. K. (1999). Food plants of the Tasar silkworms, *Sericulture in India,* Eds. 761-777

Mohanty P. K. (2003). Tropical wild silk cocoons of India, Book, *Daya Publishing House,* New Delhi.

Muhammad Ashfaq, Sajjad Ali, Muhammad Asif and Hanif, (2009). Bioaccumulation of cobalt in silkworm (*Bombyx mori* L.) in relation to mulberry, soil and wastewater metal concentrations, *Process Biochemistry,* 44, 1179–1184.

Reddy R. M., Sinha A. K., Kumar R. and Prasad B. C. (2010) Parental combination and rearing season compatibility for silk yield in tropical tasar silkworm, *Antheraea mylitta Drury.,* World Applied Sci. J., 9 (8), 855-859.

S. Kamarej, T. Pandiaraj, Immaual Gilwax Prabhu, Nidhi Vishwakarma and A.K. Sinha, (2017). Comparative study on performance of Tasar Silkworm Antheraea Mylitta, Druary chowki garden and direct rearing method. *Int. J. Curr. Microbiol. App. Sci.* 6(4): 2421-2425.

Sekar, K. (1995). Foliar supplementation of soya flour for improving the silk yield of mulberry silkworm, *Bombyx mori* L. M.Sc. Thesis, Tamil Nadu Agricultural University,

Singh R.N., Bajpayee C.M., Jayaswal J. and Thangavelu, V. (1992). Perspective of biological control in tasar culture, *Indian silk, 31(7) 48-50*

Subburathinum, K. M. and M. Krishna (1998). Role of soya bean protein in silkworm nutrition, *Indian silk,* 9 – 10.

Thangavelu K. and Sinha A K. (1993) Population ecology of *Antheraea mylitta Drury* (Lepidoptera: Saturniidae) in Wild Silk Moths'92, edited by H. Akai, Y. Kato, M. Kiuchi and J. Inouchi. International Society for wild silkmoths. National Institute of Sericulture and Entomological Science Tsukuba, Japan. 87-92.

Chapter 32

Prospecting of Bioactive Constituents from Plants using Ethnobotanical Approaches against Metabolic Syndrome: A Link Towards better Health

*Madhabi Madhusmita Bhanjadeo[1] and Nihar Ranjan Singh[2]**

[1]Department of Biochemistry,
[2]Department of Botany and Center of Excellence in Environment and Public Health, School of Life Sciences, Ravenshaw University, Cuttack – 753 014, Odisha*
**e-mail: nihar.singh@gmail.com*

ABSTRACT

In all ancient civilisation, there is a long history of the use of traditional knowledge from medicinal plants for the treatment of various diseases and disorders and for the well-being of the human health. Traditionally, ethnobotanists are known to record and engage in the collection and documentation of knowledge that has been associated with plants that is related to their medicinal benefits. Such a knowledge will give immense data for the researchers in the field of pharmacology and phytochemical studies that may lead to the novel discovery of biomolecules or natural compounds. Many ethnobotanical data generated through ethnobotanical approaches by the modern ethnobotanists are potential source of novel drugs and molecules. Despite the progress in the field of pharmacology, there is still a huge demand in the traditional knowledge among the present generation and it is highly promoted. Ethnobotany is the use of plants by the indigenous people of ethnic group or tribal communities and is normally associated with wide biogeographic and ethnographical diversity. Worldwide many tribal communities have been reported to practice plants with medicinal importance for alleviation of different diseases. India represents the second largest populated country of the world with half of its population dwells in the rural and tribal regions. The traditional knowledge from this population is needed to be preserved and protect from disappearing and thus ethnobotany plays an important role in it. In recent past, many diseases and disorders have been evaluated with plant extract and plant derived compounds and their efficacy are evaluated at various level for its application towards therapeutic needs and betterment of human health. Modern day lifestyle diseases represented by metabolic syndrome is now affecting more than 30 per cent of the Indian urban population with increasing prevalence. It is now considered as the cluster of several life threatening disorders like hypertension, dyslipidemia, central obesity and glucose intolerance that affect the physiology, biochemistry and metabolic profile of the human being. Specifically, metabolic syndrome is associated closely with diabetes, obesity and cardiovascular diseases. This chapter emphasizes on the potential prospection of bioactive compounds from plant sources reported by ethnobotanical studies in different geographical areas against the pathophysiology of metabolic syndrome.

Introduction

1. Ethnobotany: Knowledge of Plants in Sustaining People

Since the beginning of human civilisation, dependency on plants have been evidenced for their crucial necessities and nourishment towards human need. Over several generations, folk cultures as well as civilisations have tried and continued using plants and parts of plants for their day to day liveliness. Our ethos evolved by transition of more sophisticated knowledge of plants and their usefulness from generation to generation and is now studied as ethnobotany. Ethnobotany has a wide definition that deals with the study of interactions between people and plants. While many research groups prefer to define this discipline as the methodical investigation of the relationship between human cultures and plants (1). Although, there are room for study involving plants used in a variety of economic and non-economic societal purposes (as fodder, fuel and construction materials, in sacred rituals, for ornamentation, and in social life), but in this chapter the attention is on the influences of ethnobotanical approaches to modern medicine and pharmacological interventions in traditional societies. This subsection of ethnobotany is referred to as ethnopharmacology (2).

Initially, the branch of ethnobotany has been focussed on the relationships and interactions between plants and restricted to people in traditional tribal cultures only, but now the definition points ethnobotanical studies comprises all culture areas, including modern western communities, who employ plants in conventional popular folk medicine or long-established traditional remedies in contradiction to the use of modern medicine to heal ailments and injuries. Ethnobotany addresses both knowledge of diverse plants widely known by different folk culture as well by selective plant knowledge by specialists or healers within the ethnic group. Since, an extensive range of topics are overlaid to an even wider spectrum of disciplines under the umbrella of ethnobotany making it a composite platform is science, A person having interdisciplinary knowledge in botany and another academic field may find interesting problems for research in ethnobotany. The detailed documentation records, description and explanation regarding multifaceted relationships among the folks and their dealings with the plants as their financial and therapeutic resources are profiled by the ethnobotanists (3). They also record the sacramental and ceremonial utilisation of plants in society, deliberating the part of a specific plant or plant group within a cultural and social framework of a community. Moreover, ethnobotanists also keep the records regarding the parallel practices, ideas, or collective perceptions surrounding a specific plant and are shared among several diverse ethnic groups across a larger geographical area like a continent (4). Hitherto in areas where traditional folk hold different beliefs about the reason of illness, mode of treatment and cure within a single geographic region (e.g., the tribals of the Amazon Basin, hill tribes of North-east India), it is a big challenge to comprehend such information and preserve them for future generation. The history of ethnobotany began long time, may be thousands of years ago with the origin of human beings. Today it is one of the distinguished area of study within botany and is active field of research in many countries. The first scripted documents of plants and plant products used for medicinal purposes appear in ancient Egypt the Ebers Papyrus, copies of which date to 1550 B.C., as well as in numerous other monumental texts, sarcophagi, mural art forms, and ancient Sumerian culture script listing physician's prescriptions, dated to about 3,000 B.C (5). While these early artefacts represent the knowledge held by medical specialists of ancient civilizations, and it was preserved in different forms for future generations. Such traditional practices from Greek, Roman and Islamic foundations are intensified with the colonial expansion and geographic exploration of the known world leads to the creation of ethnobotanical knowledge. In this pursuit of knowledge, the first hallmark work was in the form of Dioscorides' De Materia Medica, whose date of compilation has variously been estimated at 64 through 77 AD. Dioscorides relied on the texts of Crateuas, physician to King Mithrades VI from 120 to 63 B.C. Nevertheless, systematic study of plants is also evidenced in eastern part of the world. The scripted documents are verified by Emperor Shen Nung around 2,700 B.C in ancient china. In the same line, the Rig Vedas and Ayurvedic medicine, brought together in ancient India, included the information on numerous plants used as medicines for treatment of illness and injuries (6,7). Ayurvedic medicine is assumed to be existing since 5,000 – 10,000 years, and the Rig Veda around 2500 B.C (4,7) or earlier probably. Other medicinal traditions that exploits medicinally important plants, established in Buddhism, are documented in palm-leaf scriptures found in the south Asiatic regions of Burma, Cambodia, Vietnam, Thailand and elsewhere in Indonesia (8,9). Tibetan medicine, also rooted in Ayurvedic medicine, evolved from the works of Charaka Samhita around 1,500

B.C., the Susruta Samhita (300 – 400 A.D.) and the Ashtanga Hridyam (around 500 A.D.). Records containing medicinal uses of plants are found in extant Mayan codices, such as the Book of Chilam Balam of Kaua (7,9).

The breakthrough inventions and uses of plants for medicinal purposes are reported with the progression of complex societies and sociopolitical structures but the written texts on medicinal plant knowledge and healing only demonstrate how specialized such knowledge had become. Besides the written scripture there is a tradition of passing such knowledge orally along generations among the progenies from the ancestors. Many studies have unearthed that many traditional individuals possess an intimate knowledge of their environment and are keen observers of subtle changes across the landscape, including changes in weather, where and when medicinal plant resources may be found at different times of year. This extensive oral tradition also includes knowledge of illnesses, remedies, treatments and healing techniques, learned through a training or apprenticeship to the next generation (5,6,7).

The ground-breaking discoveries and interest in exotic and economically important plants from elsewhere was piqued after Marco Polo's discovery of spice islands (10,11). Furthermore, the discovery of the New-World towards the end of the 15th century, and the succeeding the Columbian exchange or the dogmatic expansion, exponentially rose the information including new foods, medicines, heart wood and wooden materials, and commercial items like dyes, tobacco, pearl *etc.* The establishment of botanical gardens and periodicals regarding botanical treatises in Renaissance Europe only initiated during the sixteenth century and spread rapidly (12). Subsequent explorations culminated into travelogues and indices of traditional peoples and their behaviour, provided further attainment including proceedings on medicinally and economically beneficial plants (12,13). Carl Linnaeus, the builder of the Latin binomial system for naming plant and animal species, conducted ethnographic field survey among the Laplanders in 1733 which are now being reviewed by modern ethnobotanists (12,13,14). In the same context, European powers fought for economic supremacy, which often took the form of endeavours to dominate the market in exotic and commercially important plants. Some of the initial ethnobotanical studies include the medicinal effects of *Digitalis purpurea* or foxglove by William Withering in 1785 (15). Another eminent botanist John Harshberger, first coined the term "ethnobotany" in his book and worked on the purposes of ethnobotany in 1896. The eighteenth and nineteenth centuries produced additional explorations, including the collection of medicinally and economically important plants to be brought back to museums and herbaria in Europe (16). Then Increasingly, these collections chronicled insights on behavioural studies of traditional peoples and their management of plants and the relationship between traditional peoples and their surrounding natural environments (8,10,16). Subsequent innovations on economically important plants lead to discovery of such formulations which became the key player revolutionizing the history of mankind like discovery of morphine from opium, and then quinine from cinchona in the 18th century (17,18,19).

Modern studies in ethnobotany however differs from former approaches towards gathering information in few perspectives, which comprises more information about the ethnic groups by elaborate procedures of field work and then connecting the link with anthropological aspects. These perspectives involve social beliefs in regard with the illness, mode of medication, and curing, including the beliefs about the plant and human ecological relationships, along with ceremonial, and other uses of medicinal plants without neglecting the passed down role of the traditional healers. These healers or other ritual specialist who uses medicinal plants to treat patients, whether the ailment is believed to have a natural cause or beliefs involving by spiritual forces, evil forces or some other non-organic cause. In addition, contemporary approaches often search for the evidence of the efficacy of the formulation and the medicinal effects of the plant on the human body (20). Besides, the existence of literature on medicinally important plants used by traditional people since a long time, yet there exists several gap areas on being systematized study to upgrade the knowledge to discover new line of drugs or health supplements involving a better understanding on human body's response to these medicines. The discovery of new drugs depends predominantly on the identification of the active compounds against specific illness. Identification, isolation and characterization and then evaluation of those compounds not only augments the efficacy but also creates new scopes for a livelihood enhancement for the indigenous people by different approaches of prospecting. It is now also evident that the plants and plant products are efficient food and medicinal source from prehistoric time and people used then not only for their nutritional needs but for treatment towards various illness and also in general wellbeing. The bioconstituents present in plants are the major thrust areas in present day and are potential target for identification of novel drug as these constituents contains the key bioactive compounds that incarnate the medicinal properties. In further sections, it is more pertinent to understand metabolic syndrome (MetS) and global knowledge of ethnobotanical traditional system and correlate with medicinally beneficial constituents from plants and their prospecting approaches.

2. Critical Definition of MetS

Metabolic Syndrome (MetS) also known as lifestyle disorders can be defined as a collective state of disordered metabolic integrity related to increased risk of cardiovascular disease, type 2 diabetes mellitus, renal failure, atherosclerosis and cognitive deficits. The disordered condition incorporates hypertension, hyperglycaemia, obesity, abdominal fat, hypertriglyceridemia, low high-density lipoprotein cholesterol (21). The concept of the MetS has been existing since 1920s (22). This ensemble of metabolic disturbances, combination of risk factors for cardiovascular disease, was first described in by a Swedish physician Kylin, where he grouped them together and relate it with a disease state. Following this, in 1947, further augmentation to the concept of metabolic abnormality and its relation to upper visceral adiposity or male-type obesity was clamped and this obesity phenotype along with type 2 diabetes and cardiovascular disease clustered to form a major disorder as mooted by Vague *et al.* (23). However, the past three decades, there is a spectacular intensification in the number of people affected with the MetS across the globe. This upsurge is directly connected with the global rampant of obesity and diabetes, which is very much associated with modern life style and livelihood. Along with the pre-eminent threat of diabetes and correspondingly of coronary heart disease risks following the MetS, there is an urgency to formulate strategies to counteract the burgeoning global epidemic (24). The MetS is also known as syndrome X, the insulin resistance syndrome and the deadly quartet (25). This notorious masquerade includes complications like type II diabetes or impaired fasting glycaemia, insulin resistance, central obesity, dyslipidaemia, and hypertension, which are completely well-established risk factors for cardiovascular disease These complications co-exist in an individual more often and not by chance. When ensemble together, they connect towards increased risk of MetS (26). In the same context the importance of abdominal obesity and hypertriglyceridaemic waist phenotype is suggested to be a central component for prediction of MetS by Lemieux and colleagues (27).

In 1988, the concept of MetS was re-specified by Dr. Reaven with the terminology "syndrome X" with simultaneously grouped risk factors (28). A decade later in 1998, WHO proposed a definition for MetS which was denoted by presence of insulin resistance or fasting glucose greater than 6.1 mmol/L (110 mg/dl), 2 h post-glucose level more than 7.8 mmol (140 mg/dl) with any two or more of the following: 1. HDL cholesterol < 0.9 mmol/L (35 mg/dl) in men, < 1.0 mmol/L (40 mg/dl) in women 2. Triglycerides > 1.7 mmol/L (150 mg/dl) 3. Waist/hip ratio > 0.9 (men) or > 0.85 (women) or BMI > 30 kg/m2 4. Blood pressure > 140/90 mmHg (110). But this complicated anomaly of metabolic disorders is an evolving concept that continues to be data driven and evidence based (29). For different community and populations, differential threshold MetS criteria can be linked to different lifestyle, gene pool drifting, different food habit leading to differential risk factors in atherosclerotic cardiovascular diseases. However, there is still a scope for more evidence-based studies. The limitations can be related to lacuna of data containing smoking habits, comorbidities, and medications. The information of the family and personal history of chronic diseases as well as lifestyle on an optional basis or routinely basis is required from the patients seeking health check. The incomplete and unverified information may lead to misinterpreted data. During the last three decades, different organizations have developed different definitions for this condition. In 2001, the National Cholesterol Education Program Adult Treatment Panel III included in the MetS criteria the fasting blood plasma glucose level. Consequently, the International Diabetes Federation postulated a fresh characterization of the MetS, enlisting ethnic cluster–selective thresholds for waist circumference to describe central adiposity (30). The American Heart Association/National Heart, Lung, and Blood Institute (AHA/NHLBI) established that MetS is detected when three or more of the following criteria are present: waist circumference >102 cm in men, and >88 cm in women, blood pressure ≥130/≥85 mmHg, triglyceride (TG) plasma level ≥150 mg/dL (1.7 mmol/L), HDL cholesterol (31). However, during the past several decades, the pervasiveness of MetS has strikingly amplified globally. The world-wide occurrences of this disorder are reported 14 per cent to 34 per cent which increases with aging in both sexes. The pervasiveness of the MetS in type II diabetes mellitus is high in both genders and increases with age, thus posing a potential high cardiovascular risk in both the genders (32).

2.1 Pathophysiology of MetS and Associated Complexities

It is worth mentioning here that MetS is the next major cause of fatalities and can be placed next to cancer in years to come. Therefore, it is required to understand the underlying necessities to comprehend the disease pathology and find a sustainable method to delay it if not prevent it. According to different research groups different pathophysiological factors that contribute for MetS are listed as (i) insulin resistance and atherosclerosis (ii) increased tissue and plasma levels of proinflammatory cytokines; (iii) increased plasma levels of neurotrophins such as nerve growth factor (NGF) and brain-derived neurotrophic factor (BDNF); (iv) vegetos dystonia or disbalance of neurotransmitters; (v) hypothalamus-pituitary-adrenal axis activation or hormonal disbalance and (vi) genetic identity

(i) Insulin resistance (IR) precedes to elevated levels of serum insulin and glucose and advances the onset of diabetes mellitus. Furthermore, insulin resistance also culminates to hyperinsulinemia resulting in excessive renal sodium retention and elevations in blood pressure. Obese cases are observed to have increased concentrations of serum insulin, which is entailed to uphold glucose and fatty acid metabolism. Traditional insulin-sensitive tissues such as skeletal muscle, liver, and adipose tissue often have insulin resistance and sometimes non-traditional insulin-sensitive tissues such as heart, aorta, and kidney also show insulin resistance. Therefore, emphasis of some movements of insulin metabolic influences and corresponding resistance to erstwhile actions, initiates diverse clinical manifestations related with the MetS (33,34).

(ii) Inflammatory distress is not a clear factor that causes MetS, but it is known that a chronic, subclinical inflammation is associated with MetS. Diligent investigation in search for intercessor molecules of the inflammatory process has thus become, one of the most hot-topic explorative areas casting a light on the cellular and molecular mechanisms of the pathogenesis of these socially significant diseases. Adipose tissue secrete proinflammatory cytokines, such as tumour necrosis factor alpha (TNFα), IL-6, IL-8 and MCP-1 and this proinflammatory cytokines cause serine phosphorylation of insulin receptor substrate-1 via nuclear factor κB (NFκB) and c-Jun N-terminal kinase signalling inducing inflammatory response (35). Adipocyte hypertrophy causes hypoxia and thus can induce cell necrosis with the production of TNFα, IL-6, plasminogen activator inhibitor-1 (PAI-1), and macrophage infiltration. In turn, TNFα enhances inflammatory response in adipose cells. Non-Alcoholic Fatty Liver Disease (NAFLD) is often found in patients with MetS (36). Both MetS and NAFLD are also related to variation in the gut micro-flora and associated pathways. Changes in intestinal inflammation, gut permeability, energy provision, anaerobic fermentation and inflammatory response have been reported (37). It is also believed that the proinflammatory state is associated with MetS through C-reactive protein (CRP) whose elevated level impart major risk to coronary system (38).

(iii) Neurotransmitter disorder communicate between the altered plasma corticosteroid concentrations along with pro-inflammatory cytokines that rise in chronic stress resulting from the hypothalamic neurotransmitters. Traumatic stress leads to obesity, other components of MetS and alters circadian rhythm through activation of neuropeptide Y and gluco-corticosteroid system (39, 40). This combinatorial effect in which adipokines, cytokines along with immune function confirms the association between neurosystem and insulin resistance leading to the progress of MetS.

(iv) Disbalance in hypothalamo-pituitary-adrenal axis that associated with the psychopathology, human behaviour, sleep and sleep disturbances, energy balance and immunity also substantially supports the succession of MetS. Elevated proinflammatory cytokines and neurotrophins trigger variation that leads to hormonal disbalance in MetS consisting in hypercortisolaemia, hyperinsulinemia, hyperprolactinemia, hypothyroidism, hypogonadism, or hyposomatotropism (41).

(v) Genetic of an individual plays vital role in deciding the ethnicity, diet and anthropological details that strongly influence the pathogenesis of the MetS on case to case basis. This complexity of genetic framework of an individual makes extremely difficult in identification of replicable genetic associations with MetS that might one day form the basis of individual organism specific basis of treatment (42).

2.1.1. Current Scenario of Global Epidemic and Associated Symptomatic Complexities

Individuals with the MetS are effectively on two-fold more risk for cardiovascular disease when compared with those without the syndrome. Additionally, the syndrome promotes the risk for type 2 diabetes by about 5 times. A relatively high prevalence of the MetS is a rampant episode and the phenomenon appears to be rising due to a corresponding increase in the pervasiveness of obesity. The probability of a further increase in the MetS can be foreseen due to projections of a greater predominance of obesity in the future. However, it must be noted that determining the prevalence of the MetS in different regions depends on defining criteria. Most of the scientific reports have used the National Cholesterol Education Program (NCEP) definition to define the MetS. International Diabetic

Federation (IDF) uses waist circumference for the threshold of MetS. One of the major challenges for defining the syndrome is that of the appropriate waist circumference. The primary difference between NCEP and IDF definitions is that the waist-circumference cut points for Whites, Blacks, and Hispanics is higher in NCEP than in IDF. In the below subsections, major symptomatic of MetS have been briefly described to understand that why these factors are detrimental to human health and in this fast moving society why there is an urgent need to keep this factor under check. Also, prevalence of any of this health indicator can lead to the genesis of MetS (130).

I. *Abdominal Obesity*

It is characterized by a redistribution of adipose tissue from peripheral to central sites of the body, mainly in the trunk region and visceral depots and this central obesity is one of the main components of the MetS. Cortisol appears to play a role in adiposity in MetS. Indeed, an increase in urinary free cortisol excretion was reported in patients with MetS. The positive association of cortisol excess with intraabdominal fat was also shown the correlation between HPA axis activity and MetS. Some researchers reported no relationship between cortisol and waist circumference. Higher cortisol concentrations were related to a reduced insulin secretion, a finding consistent with in vivo and in vitro data showing that glucocorticoids regulate insulin secretion (43, 44).

II. *High TG and Low HDL-C Levels*

Hypertriglyceridemia and low HDL-C concentration are a common finding in Cushing syndrome and MetS. Higher cortisol (both urinary free and serum overnight) levels have been associated with high TG and low HDL-C levels. Genetic alterations in the glucocorticoid receptors (GR) correspondingly influences the normal functioning of the HPA axis and lipid metabolism that also connect to MetS (44, 45, 46).

II. *Hypertension*

Hypertension is one of the most distinctive features of MetS because it is present in about 80 per cent of adult patients and in almost 50 per cent of children and adolescent patients (47). Many studies reported an association between cortisol and systolic and diastolic BP levels. This correlation might be attributed to the effect of stress, which is associated with the activation of the HPA axis and sympathetic nervous system (48). Indeed, patients with MetS and hypertension appear to have higher urine levels of both cortisol and catecholamine metabolites than healthy individuals. Obesity is also associated with hypertension. The possible underlying mechanisms include volume expansion, increased cardiac output and systemic vascular resistance, increased sodium reabsorption, increased activity of the sympathetic nervous system and the renin-angiotensin-aldosterone system, high leptin levels and concurrent leptin resistance, as well as increased ET-1 and decreased NO (49).

IV. Hyperglycemia

MetS frequently has increased blood glucose levels. In patients with MetS, serum cortisol levels are significantly associated with fasting glucose concentration. The connection between fasting hyperglycemia and cortisol is because of the glucocorticoid effects on hepatic gluconeogenesis and insulin secretion (50).

V. Endothelial Dysfunction and Hypercoagulability

Since it is already mentioned before, MetS is networked with endothelial dysfunction that remarkably affects to an increased risk for CVD. Endothelial dysfunction induces pathogenesis of atherosclerosis and also increases the risk for insulin resistance and type 2 diabetes. It is also believed that the endothelial dysfunction involves host immune responses and also generated reactive oxygen species. Further, in such cases impaired fibrinolysis due to the high levels of plasminogen activator inhibitor-1 observed and that contributes to hypercoagulability (51, 52)

VI. Serum Uric Acid and Gout

Higher serum uric acid (SUA) levels are shared by MetS and regarded as a predicting symptom of MetS. High SUA levels are considered detrimental to cardiac and uprise overall mortality in patients with stroke (53).

From the above discussion on MetS, it is evident that many risk factors are involved and is prevalent among the present generation population due to modern and sedentary lifestyle along with individual genetic makeup. A general scheme is given in the form of Figure 32.1. The major goal of treatment is to reduce the risk of getting into MetS and one of the suitable remedial answer other than chemical pills is to look for ethnopharmacological aspects and search for suitable phytoconstituents. The present day knowledge of plants and plant based compounds towards alienating MetS is subjected to scientific investigation and many studies shows a positive link towards reducing MetS and also towards better health. In the next part, major phytoconstituents from plant sources is discussed that can possibly be pool of warehouse for novel drugs.

3. Phytochemical Constituents: Prospective Pool of New Drug Molecules

Ethnopharmacological approaches perform a crucial job for the discovery of new drug and bioformulations from native plants having medicinal importance as well as offer potential for alternative sustainable pharmaceuticals (54). There are immense probabilities for novel drug discoveries because of the presence of unmatched availability of chemical diversity in the plant extracts. Due to unanticipated side effects of synthetic drugs, the herbal formulations from plant extracts contribute advantage over synthetic drugs combination of medicinal constituents coupled with minerals and vitamins. In this context, the traditional branch of ethnobotany has been drawing attention from many research groups

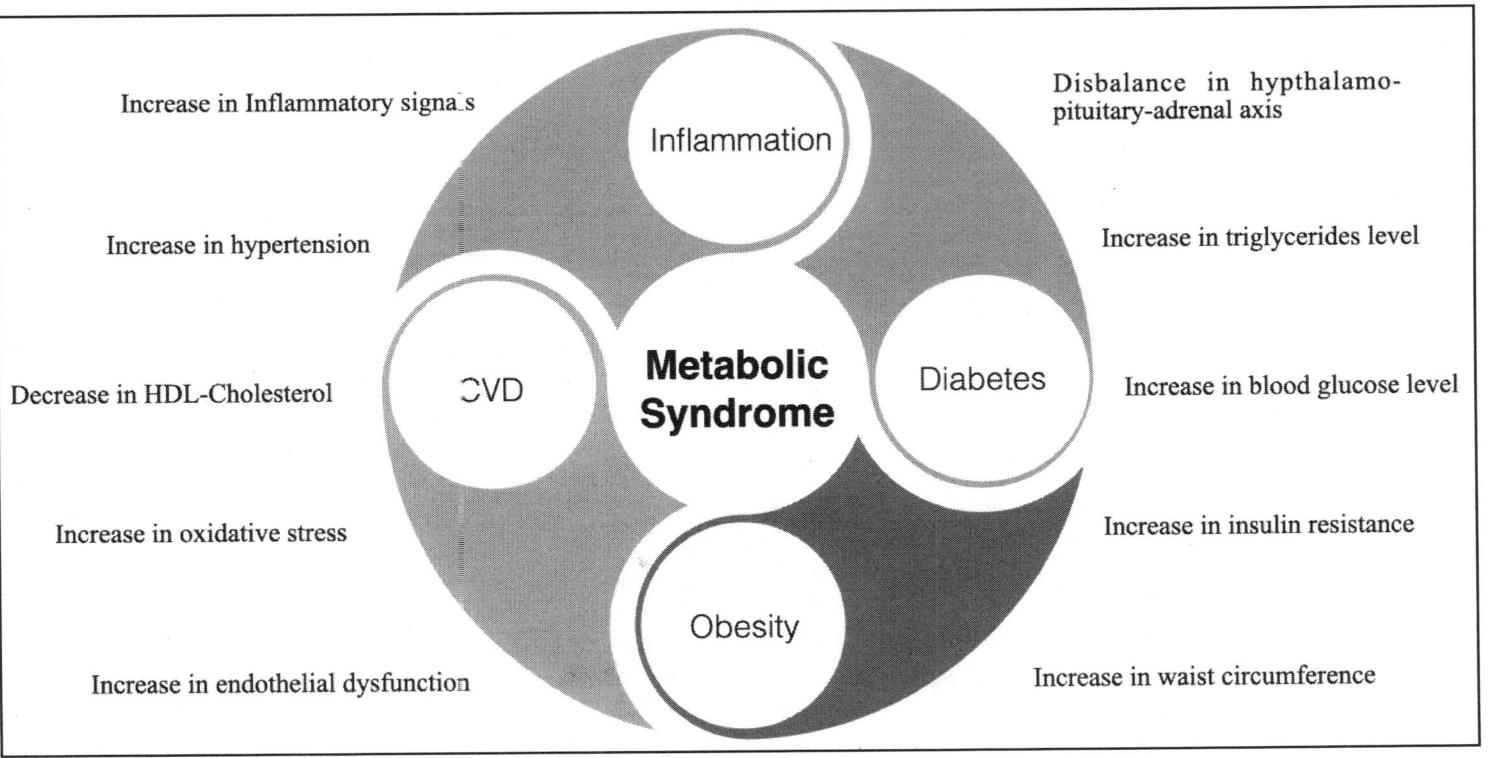

Figure 32.1: Metabolic Syndrome and Associated Risk Factors.

for novel medicines for the revival of knowledge in customary health practices throughout the world specifically alleviating lifestyle related diseases like MetS, cancer *etc.* Several contemporary drugs like taxol, serpentine, vincristine, vinblastine have been discovered lately as the ethnobiologists from all over the world started to gave priority and paying a distinct importance to the documentation of the traditional information of medicinal plants. Traditional knowledge are now organised in digital library, portals and other databases like PLANTS, PRELUDE, KEW, NMPB serve the storage of information regarding cited plants of a specific geographical area and the ethnic group or country to restrict biopiracy and assist in patenting of a plant based traditional knowledge. From different medicinal plants, various natural bio-active phytochemical constituents and their derived molecules are being used as a promising alternative treatment for many diseases including infectious diseases. However, the present chapter emphasizes on the natural bioactive compounds that are effective against the complications of MetS (55,56).

There are different classifications regarding plant-based bioactive compounds. Generally, these compounds are categorised as the secondary metabolites as these compounds are produced by plants towards their defence mechanism against different biotic and abiotic stress such as herbivores, pests, diseases, ultraviolet radiations, and free radical species (57). Sometimes they are secreted to invade the intra-specific competition. Coherent experimental confirmations from epidemiological, *in vitro, in vivo,* and clinical studies have validated about the efficacy of the food regime enriched with herbal nutrients, which can alleviate the risk management of several diseases including diabetes, obesity, cardiovascular complications, and cancer. Alongside the progression of advanced analytical techniques for instance liquid and gas chromatography, sophisticated mass spectrometry and nuclear magnetic resonance spectrometry, the identification and structure resolution of solo compound from plant extracts solutions have become possible (58, 59). A plethora of biologically active compounds including polyphenols, alkaloids, terpenes, saponins, have been extensively employed as dietary ingredients to provide a health benefit beyond basic nutrition. Furthermore, many research groups also suggest that the potent functions could be due to the synergistic influences of these phytochemicals and nutrients present in the plant foods rather than the purified chemicals themselves. This subsection will focus on the different class of compounds cited for alleviation in different diseases and play vital role in the well-being of the human health (60).

Polyphenols

Polyphenols, a group of molecules containing one or more hydroxyl groups (-OH) attached directly to the aromatic phenyl ring structure (61). These are the most profuse secondary metabolites distributed in all vascular plants playing a major role in giving flavour, aroma, pigment and dietary properties of plant-based foods, such as fruits, whole grains, coffee, tea, legumes, cocoa, and wine. Owing to excessiveness in the human diet, an expansive investigation has been performed to analyse the influence of dietary polyphenols, including simple phenolics, coumarins, lignans, flavonoids, anthocyanins, tannins, quinines, and stilbenes continuously (62). Increasing evidence indicates that polyphenols have several benefits on humans such as being a potent antioxidant compound, prevention of various diseases induced by oxidative stress and prevention and delaying of particularly cardiovascular diseases, neurodegenerative diseases, and cancers. However, the health benefits of polyphenols depend on the type of polyphenols, the amount consumed, and their bioavailability (63).

Alkaloids

Alkaloids are group of compounds with heterocylic nitrogen rings biosynthesized as secondary metabolites from precursor amino acids. These compounds can be categorized depending on different criteria like that of their biological function, biosynthetic pathways, precursor molecules and principal carbon skeleton (64). There is a large diversity in these groups of compounds which gives a characteristic bitter flavour and is secreted also by many members of kingdom fungi and bacteria. According to the position of nitrogen atom in the carbon ring, the compounds vary in different plants and other organisms. Most widely accepted classification includes their division into five major groups depending on their biosynthetic amino acid precursor (33, 34): tropane, pyrrolidine, and pyrrolizide alkaloids (ornithine); benzylisoquinoline (tyrosine); indolequinoline (tryptophane); pyridine (pyridine); and quinolizidine and piperidine alkaloids (lysine). After the first medically useful alkaloid, morphine, was isolated in 1805 from *Papaver somniferum* (opium poppy), more plant-derived alkaloids have been discovered from over 300 different plant species (65,66). Nevertheless, presently over and above 10,000 naturally occurring alkaloids are reported to possess pharmacologically active values. With the employment of diverse purification strategies and structure evaluation methods, the structure and biological activity of plant derived alkaloids are gaining attention due to their application in pharmaceuticals, narcotics and toxicity values (67). The group of indole alkaloids is considered as the largest class of alkaloids, which contains more than 1,500 natural products and are reported to be used in traditional medicines. Chemically, indole alkaloids contain a structural moiety of indole and many of them also have an isoprene groups. The amino acid tryptophan is the biochemical precursor of all indole alkaloid (68,69). Pyrrolidine alkaloid is a class of alkaloids comprising around 80 natural compounds derived from ornithine and lysine by adding acetate/ malonate units. The basic unit is the five-membered carbon rings containing the nitrogen atom. The simple pyrrolidine alkaloid structures can be exemplified by hygrine and cuscohygrine isolated from members of family Solanaceae. Tropane alkaloid is a group of alkaloids that contains a tropane ring and is often esterified with a variety of acids, such as acetic acid, atropic acid isobutyric acid, isovaleric acid, propanoic acid, tiglic acid, tropic acid, and 2-methylbutyric acid. Most of them occur naturally in species of Solanaceae, Convolvulaceae, and Erythroxylaceae (70).

Terpenoids

The terpenes or isoprenoids isolated are the secondary metabolites associated with different stages of plant growth (71). Terpenes are considered as the most distinguished class of metabolites while there had been more than 30,000 types of terpenes discovered. Chemically, isoprene has the molecular formula C_5H_8 which is the basic structural formula for terpenes, and the basic molecular formulae of terpenes can be represented as $(C_5H_8)_n$ where n is the number of linked isoprene units. Since the isoprene units may form linear head to tail chains by getting linked together or form rings, terpenes contain diverse compounds with variable structures. Therefore, the general classification of terpenes is in accordance with their molecular formula, such as monoterpenes, C10; sesquiterpenes, C15; diterpenes, C20; sesterterpenes, C25; triterpenes, C30; and so on (72). In the food and pharmaceutical industry, the most popular terpenes belong to the group of monoterpenes, sesquiterpenes, and tetraterpenes. For example, the monoterpene from thyme oil, parthenolide, lenalool, valerenic acid from valerian root lutein, and lycopene are considered as bioactive compounds in different products (73). Thymol (2-isopropyl-5-methylphenol) belongs to the monoterpene group and is biosynthetically derived from cymene, isomeric with carvacrol. It is generally extracted from *Thymus vulgaris* (common thyme) plant source. The purified thymol is a white crystalline substance with strong antiseptic properties and pleasant aromatic odor along with this in foods, thyme is used as a flavouring agent (74,75).

Saponins

Saponins are specifically plant derived bioactive compounds but found in some insects and marine organisms also. These compounds are usually found as polycyclic triterpenes or steroids with glycosides. Owing to their chemical properties, saponins can disturb the permeability of cell membranes as well as can decrease the surface tension of aqueous solutions. Chemically, saponin can be divided into two groups, the triterpenoid saponins and the steroidal saponins. Steroidal saponins contain 27 carbon atoms forming their core structures, such as spirostan. Among n(25R)-spirostan derivatives or "isosaponins" and (25S)-spirostan derivatives "real" saponins or neosaponins are the most commonly occurring saponins, which have the A/B/C/D ring structures linked together in the order of cis-trans-trans (5β derivatives) or trans-trans-trans (5α derivatives). Triterpenoidal saponins or their nor-derivatives contain 30 carbon atoms which form their core structures, such as tetracyclic dammarans and pentacyclic oleanans. There are also other triterpenoidal saponins, *i.e.*, lupan, oleanan, ursan, and hopan, oleanan and lycopene (76, 77).

A representative set of plant based phytochemicals and their phytoconstituents are listed in Table 32.1. In view of these plant based secondary metabolites that can be a potential molecule of interest against a MetS, it is interesting to evaluate the role of global ethnobotanical traditional system and their contribution towards the treatment of MetS.

Table 32.1: A List of different Classes of Pharmacologically Active Compounds from different Plant Sources

Sl. No.	*Class of Bioactive compound and their Biosynthesis Pathway*	*Categories*	*Compound Examples*	*Plant Sources*
(1)	Poly phenols -(Styrene pathway/ Shikimate pathway)	Simple phenolics	Gallic acid	Green tea, grapes, mulberry
			Proto catechuic acid	Dates, goose berry, blue berry
			Hydroxy Cinnamic acid	Woody plant biomass precursor of lignin synthesis
		Coumarins	Furano coumarins (Psoralene)	*Psoralia coryfolia* commonly known as babchi (essential oil from seeds)
			Pyrano coumarins (Seselin)	*Clausena anisata* Essential oils of Leaves
		Lignans	Secoisolariciresinol	Flax seed oil
			Lariciresinol	Sesame seed oil
			Pinoresinol	Sesame seed oil, olive oil and crucifer vegetables
		Flavonoids and Iso-flavonoids	Flavones	Leutolin in red purple fruits
			Flavonols	Catechin of *mimosa catechu*
			Flavonols	Gallocatechol in green tea
			Flavanones	Hesperidin in citrus fruits
			Isoflavones	Soybean and soy derived food
			Rotenoids	Members of family Nyctaginaceae
		Quinones	Emodin	Rhizome of *Rheum palmatum*
		Anthocyanins	Cyanidin	Berry extracts, berry fruits
			Delphinidin	
		Tannins	Proanthocyanidins	Pomegranate and purple coloured fruits
			Gallotannins	
			Elagitannins	

Sl. No.	Class of Bioactive compound and their Biosynthesis Pathway	Categories	Compound Examples	Plant Sources
2	Alkaloids -Amino acid derived secondary metabolites	Pyrrolidine	Hygrine	Members of family solanaceae
		Piperidine,	Lobeline	*Lobelia inflata*
		Amino	Ephedrine	*Ephedra sinica*
		Steroidal	Solanidine	Members of solanaceae
		Purine	Caffeine	*Coffee arabica*
		Diazocin	Lupanine	*Lupinus exaltatus seeds*
		Imidazole	Pilocarpine	*Pilocarpus microphyllus*
		Indole	Ergolines	*Turbine corymbosa*
		Aporphine	Nupharin	*Nymphaea speciosum*
		Isoquinoline	Morphine	*Papaver somnifera*
		Quinoline	Quinine	*Cinchona officinalis*
		Tropane	Atropine	*Atropa belladonna, Datura innoxa*
		Pyrrolizidine	Senecionine	*Jacobaea vulgaris*
3	Terpenoids -Mevalonic acid pathway	Mono terpene	Limonene,	Citrus fruits
		Sequiterpene	Humulene	*Panax ginseng Cannabis sativa*
		Sester terpene	Dioyl-butanoate	*Oryza sativa*
		Diterpene	Phytol	Almost every flowering plant
		Triterpene	Ursolane	*Salvia Rosmarinus* Rosemary essential oil *Thymus vulgaris* thyme essential oil
		Tetraterpene	β-carotene, lycopene	Yellow orange vegetables carrots *Daucus carrota*, pumpkin *Cucurbita maxima,* Lycopene is found in tomatos *Lycopersicum esculenta*
4	Saponins -Saponin are derivatives of steroids, terpenoids, with glycones and with enzyme glucosyltransferases	Dammaranes	Protopanaxadiol, Panaxatriol sapogenins	Ginseng *Panaxa ginseng* ginsengosine
		Tirucallanes	Protolimonoids, norlimonoids	Stem bark or *Toona ciliate*
		Lupanes	Schekwan-glupaside A and B	*Schefflera kwangsiensis*
		Hopanes	Glinusides F, G, I	*Glinus lotoide*

Sl. No.	*Class of Bioactive compound and their Biosynthesis Pathway*	*Categories*	*Compound Examples*	*Plant Sources*
		Oleananes	Longispinogenin 3-O-beta-D-glucurono-pyranoside	*Zymnema sylvestris*
		Taraxasterane	Taraxasterol glucopyranoside	*Taraxacum officinalis*
		Ursanes	3-epi-ursolic acid 3-O-palmitoyl 28-O-a-L-arabino furanosine	*Ajuga multiflorea*
		Cycloartanes	Epoxy-9beta,19-cyclolanostane-3 beta,6 alpha,16 beta,25-tetrol Rhamno pyranosyl glucopyranoside	*Astragalus peregrinus*
		Lanostanes	4,4,14 α-Trimethyl-cholestane	*Ganoderma lusidum*
		Cucurbitanes	Cucurbitacine	Members of family cucurbitaceae
		Steroids derived saponins	Digitoxin	*Digitalis purpurea*

4. Review of Traditional Systems Reported for Treating MetS Across the Globe

In accordance with the above discussion it can be stated that traditional medicinal and healing practices represent an important source of knowledge about the use of biodiversity and some of the practices, formulations have been experimentally proven to effect positively in complexities related to MetS. However, the distribution of different plant resources in different part of globe and employability by different communities are different. Though there are several reports regarding usage of medicinal plants and their phytochemical constituents for alleviation in MetS and associated complexities but there are very less information regarding an overall picture of the plants reported worldwide by different communities. Therefore, keeping this requirement in mind an approach was taken to bring maximum of the data under a single umbrella and tried to cluster the information continent wise. There is a wide distribution of distinct traditional medicinal plants due to great migration and drifting of population and large geographical barriers. The largest continent asian region has ethnic diversities, uses most diverse traditional medicinal plant-based formulations which include, Chinese herbal medicines, Korean system, Mongolian, and Japanese medicinal system in the east Asian region whereas Ayurveda, Siddha, Srilankan, Vietnams, and Thai practices in south east Asia. In middle eastern part, Iranian, Arabic, Byzantine practices are known for their traditional healing methods. In European continent with a comparatively lesser rich biodiversity, the Greek, Roman, Spanish and mostly Mediterranean region have diversities in herbal medicines. Besides that Russian and Ukraine also possess their own specific traditional medicinal practices. In African continent owing to its rich biodiversity, it has seen a variety of ethnic culture based traditional medicinal formulations. Egypt, Nigeria, Ghana, Togo, Kenya, Namibia, Madagascar islands, and some regions of South Africa specific tribes like Ibago, Muti, Hausa *etc.* have traditional healers or shamans, these folk medicinal healers have specific knowledge of plants medicinal properties. In the same context in Northern America, regions of Alaska and major parts of Mexico, the Aztec and Maya codices are famous for traditional herbal medicines. In Southern American continent, Curandero, Kallanwaya and Mapuche are known traditional healers. These groups are a traditional native healer/shaman found both in Latin America as well as Southern Europe. Some curanderos, such as Don Pedrito, the Healer of Los Olmos, make use of simple herbs, waters, and even mud to effect their cures. Others add catholic elements, such as holy water and pictures of saints. Currently, Chinese herbal formulations for several disease are leading the global market following the Mexican herbal remedies, India's Ayurveda and Ghana's traditional medicinal remedies constitute the top four members for herbal medicines. In next subsections plants reported for MetS by different ethnic groups are discussed. Use of herbal medicine in Asia denotes an extended history of human communication with the environment. Scripted accounts prove the purposefulness of Himalayan herbs for medicinal purposes. The earliest such use was mentioned in the Vedas in about 4500 B.C. This represents the oldest repository of human knowledge and comprises 67 plant species. Ayurveda provides further details for the therapeutic use of as many as 290 herbal drugs. In China, Sheng-Nongs Herbal book is known to be one of the earliest sources of folk knowledge using plants; it comprises 365 plants, animals, and minerals useful as medication purpose from the period of Sheng-Nong (3000 B.C.). Current investigations have shown that the earliest Tibetan medicine was in the eighth century A.D. The Unani Medicine in South and West Asia also has a very long tradition and is being practiced in many countries of the region. According to the World Health Organization (WHO), around 1200 plants are used as traditional formulations for MetS and more than 80 per cent of the world's population relies on traditional herbal medicine for their primary health care needs (78, 79). These precious plant-based traditions discovered in third world countries have been considered an important component of the ethnic heritage of the world. Nevertheless, over the last three decades, environmental and cultural transitions, and the swift shifting of the economies from sustenance to commercial exploitation have twisted the impact on all aspects of traditional medical systems by influencing traditional medicine's resource base and environment. Overharvesting of some medicinal plants and animal species have culminated in resource degradation, disturbance in biodiversity, and the loss of indigenous medical knowledge (80).

Searching databases like MEDLINE, Science Direct, Scopus, PUBMED, Web of Science and Google Scholar were used in the present repertoire. Traditional medicinal plants in prevention and treatment of MetS from last twenty years (2000-2020) were surveyed. The reference lists of searched articles were also reviewed for additional relevant revisions. The search terms were MetS, pharmacologically active molecules, ethnomedicinal approaches, traditional and herbal medicine. The criteria included were randomized control trials studies assessing efficacy of medicinal plants on MetS in diabetic subjects (animal models) with the outcome of changes in pathophysiological factors like, Fasting Blood Glucose (FBG), Postprandial Blood Glucose (PBG) and glycosylated haemoglobin HbA1c. the present study is restricted to those published in the English language only. However, excluded criteria are examination involving in vitro studies and unpublished data. The overlapped plant species were eliminated which were being repeated in different geographical area.

4.1 Traditional Medicinal Plants for MetS in Asian Continent

Asiatic traditional medicines have three major names for traditional healing. Chinese herbal medicines are the leaders in the global market of traditional herbal therapies whereas Indian traditional medicinal systems like Ayurveda, Siddha, Unani practices, then Thai and Vietnamese practices in the south-eastern part of Asia and byzantine ethnic medicines influenced from Islamic traditions of herbal knowledge in the middle-east countries like Iran, Jordan and Turkey.

Traditional medicinal plants in Chinese province are extensively distributed in across the whole country, but the species richness and abundance are higher in southern than in northern China. Traditional herbal medicine in china is featured by holistic regulation, in which the organism and not the organ system is considered as a single physiological unit (81). According to the guidelines with references to the human body, practitioners of chinese herbal healing practices pay more attention to the diseased patients instead of the suffered diseases itself unlike the Western medicine which always has a strong effect on a specific disease with focusing on a specific physiological target. Another important feature is diagnosis and treatment based on an overall analysis of the illness and practitioners and differential formulae according to the varying syndromes. Thus, it is a personalized treatment for different patients even with the same disease. Although the efficacy of these herbal healing solutions still needs more improvement scopes than chemicals on a pathological process of MetS and associated disease risks. From the Table 32.2, it is observe that most of the herbal formulations for MetS are collected from the aqueous extracts of leaves, or barks and the bioactive compounds are poly-phenolic compounds containing hydroxy groups which acts as the electron accepter in redox reactions and blocking signalling pathways.

In the same line,Ayurvedic practices are more prominent in northern and central part of India. Particularly, Himalayan belt has rich biodiversity and rich in medicinal plants which are practiced with different formulation in Ayurveda, where as southern part specifically Tamilnadu and Kerala are known to occupy with siddha herbal therapy. The medicinal plants used in siddha practices belong mostly to the western-ghat biodiversity complex and now commercial plantation of some medicinal plants are being done by farmers. Another major source of traditional therapy is that of north-eastern Unani and other allied therapy. They also include plants from the eastern Himalayan belt biodiversity hotspot. These systems also differ from modern medicine in concepts and epistemologies. The approaches are majorly focused on the patient and not just the effected physiological system. The holistic concepts of Ayurveda give emphasis to health promotion, disease prevention, early diagnosis and personalized treatment. However, it is important to state that many plants and their formulations are overlapped for use of different complexities caused by MetS. *Emblica officinalis* or Indian goose berry native to india but also found in southern part of China, Nepal, Vietnam, Cambodia, Pakistan and Bangladesh. The plants description and uses are scripted in Sushrutasanhita in ayurvedic recipes for preparation of decoctions as well as in Chinese herbal formulations as Yu Gan Gzi. Glibenclamide is a nonselective K^+ ATP channel blocker which is used in type 2 diabetes. Methanolic extracts of fruits of *Emblica officinalis* given orally to streptozotocin-induced type 2 diabetic rats decreased fasting blood glucose from 9.2 to 5.4 mmol/L (normal 4.3 mmol/L; glibenclamide at 5 mg/kg/day: 5.5 mmol/L) and increased serum insulin from 41.2 to 89.4 pmol/L. The extract also increased β-cell area from 6522 to 10,525 μm^2 in diabetic rodents (82). Aqueous extract of bark of *Garuga pinnata* Roxb. which is commonly used in Chinese herbal solutions as well as Indian traditional healing systems also act against the metS. Wistar rats given a single oral dose of 500 mg/kg lowered plasma glucose concentration from 110 to 70 mg/dl from and shifted maximal glycemia from 30 minutes to 1 hour suggesting delayed absorption of glucose. This extract when supplemented at a dose of 500 mg/kg per day for 15 days to streptozotocin-nicotinamide induced type 2 diabetic Wistar rats lowered fasting glycemia from 242.5 to 110.1 mg/dL (131)

4.2 Traditional Medicinal Systems of Europe

In Europe there is a very extensive account of customary ethnic medicine that has a proper historical and methodical background,besides many European citizens still trust in this system for resolution of many diseases. The classic Mediterranean Materia Medica also influenced the structure, the literary form and the content of the middle and northern European herbals. During the Middle-ages Benedictine monks were directed by monastic rule to duplicate herbals and to tend medicinal herb gardens such as those of the Monasteries of Montecassino (Italy) and St. Gall (Switzerland). During the Renaissance, the classic Greco-Roman herbals and medical texts were critically analysed, commented upon and translated into modern languages receiving important print runs. However, Renaissance herbals also contained the consensus on central and northern European medical folk knowledge and were accompanied by woodcut illustrations imposing the plant identifications of the commentators. In Czech Republic and Central-East Europe, strong herbal and folk traditions persisted throughout history, and liberalisation of health care systems has led to a surge of interest in complementary therapies (83). Decoctions, ointments and herbal steam bath from medicinal plants are often a first choice for many illnesses, and local knowledge of specific medicinal herbs is high. In Europe, several medicinal plants are traditionally used to treat gout and rheumatic diseases. Gout, arthritis are some of the most common metabolic disorders affecting humans and plant like *Andrographis paniculata L, Artemisia herba-alba, Artemisia dracunculus, Rosmarinus officinalis* have compounds that can delay the disease. The bioactive compounds from secondary metabolites are responsible for the inhibition of α amylase and glucosidase and repair of β cells in pancreas.

4.3 Traditional Systems of Medicine in Africa

Africa is blessed with enormous biodiversity of resources, but plagued with several diseases, including those of metabolic disorders such as diabetes. Diabetes is a major health problem, with its frequency increasing every day in most countries. However, biodiversity and traditional rituals wise Egyptian, traditional medicine system is quite similar to the Byzantine traditional healing system of Turkey probably due to the great migrations from southern part of Europe due to outbreak of disease. The South Africa, Nigeria, Kenya share another set of biodiversity and reported different tribal communities like Iboga, Muti, Nagangan are some documented traditional healing formulations.

In Africa, *Sclerocarya birrea* (A. Rich.) Hochst. tree commonly found in savanna regions, and its geographical distribution expanses from the Gambia in the west across to Nigeria and Cameroon, in Central Africa, and to Ethiopia and Sudan in East. This plant is highly valued in Southern Africa, and numerous studies have delineated its antidiabetic potential. Several studies have reported the hypoglycaemic effect of aqueous stem bark extract of the plant in normal and STZ-treated diabetic rats (84). *Catha edulis* belonging to Celastraceae family found in South Africa concoction prepared from leaves rich in terpenoids, saponins, flavonols, flavones, flavonoids, chalcones, tannins, free anthraquinones, and bound anthraquinones are reported to decrease fasting blood glucose and show antidiabetic properties. *Vernonia amygdalina* belonging to family Asteraceae found in Nigeria and South Africa shows bioactive components in leaves containing Saponins and alkaloids, terpenes, steroids, coumarins are used by healers to patients of cardiovascular diseases and type II diabetes (85).

4.4 Traditional Systems of Australia

The practice of traditional aboriginal medicine within Australia is at risk of being lost due to the impact of colonisation. Displacement of people from traditional lands as well as changes in family structures affecting passing on of cultural knowledge are two major examples of this impact. Prior to colonisation traditional forms of healing, such as the use of traditional healers, healing songs and bush medicines were the only source of primary health care. It is unclear to what extent traditional medical practice remains in Australia. Most of the identified and examined traditional medicine practices within aboriginal Australia today faces a lack of academic literature and research on this subject. The literature found suggests that traditional medicine practice in aboriginal Australia still remains and the extent to which it is practiced varies widely amongst ethnic communities across Australia. This variation was found to depend on association with culture and beliefs about disease causation, type of illness presenting, success of biomedical treatment, and accessibility to traditional healers and bush medicines (86).

4.5 Traditional Medicines from American Continents

The two American continents with rich biodiversity or the gifts new world discovered after Columbian era after 1492. The rich biodiversity has gifted mankind many new world crops as well as herbal medicines like cinchona. The traditional medicinal systems include the Aztec and Maya systems of traditional healing as well as the migrated European systems like Curandero, and Kallanwaya. The north American continent contains traditional medicinal practices mostly restricted to Mexico and some parts of Alaska. In Mexico, most of the traditional knowledge that is about medicinal plants comes from pre-Hispanic times and different ethnic groups currently retain it. The therapeutic effects of plants are due to the content of different secondary metabolites such as essential oils, tannins, phenolic acids, sesquiterpenlactones, ketones and flavonoids, among others. Mexico has a rich tradition in the use of medicinal plants among its several popular healing practices. The first book of Aztec medicinal herbs and one of the most important bibliographical historical sources in America in medicine, named "Book of the medicinal herbs of the Indians" that describes more than 150 Mexico native plants that have medicinal use. It is estimated that Mexican medicinal flora contains between 3,000 and 5,000 plants that have therapeutic potential used by various ethnic groups. Therefore, more clinical and ethnobotany research, is needed to determine the medicinal benefit of these plants. Although the reports indicate a large number of medicinal plants used in this country includes cudweed (*Gnaphalium* sp.), blue gum tree (Eucalyptus globulus), peppermint, chamomile (*Matricaria chamomilla*), indian fig (ficus-indica), mexican arnica (Heterotheca sp.), trumpet tree or guarumo (Cecropia obtusifolia), vervain (*Verbena carolina*), aloe vera (*Aloe vera*), rue (*Ruta graveolens*), rosemary (*Rosmarinus officinalis*), among others (87,88).

In the same line in Latin American continents Brazilian system is the place with a large potential for biodiversity and has a wealth of traditional knowledge accumulated by local people who have direct access to nature and the products of biodiversity. Traditional knowledge related to medicinal plants is the basis of folk medicine in Brazil, which is derived from a mixture of Brazilian indigenous cultures and European and west African influences. Traditional medicine is still widely practiced in Peruvian communities, as a health resource of its own, even for people living in urban settings Albán. In Peru, most of the published studies have taken place in the oriental upper and lowland amazon basin, and in a much lesser degree in the western costal area of Peru. Despite the Andean area represent an important bio-geographical area, with the third of the overall peruvian population living in this mountainous ecosystem in small peasant villages or in bigger cities (Cajamarca, Huancayo, Tarma, la Oroya, Cusco, Arequipa, Abancay, Andahuaylas, Ayacucho, *etc.*). Callejón de Huaylas is the narrow valley of the Río Santa, running down south to north in the central Andes, marking the separation between the summits of the cordillera blanca and the cordillera negra. It stretches over 180 km in length, starting from the main city of Huaraz, ending in the lower and much smaller Caraz is characterized by a rich flora with 900 species. There is remains of archaeological evidence of a very ancient human presence, dating back to 10000 BC (89). High altitude crops are mainly tubers: numerous varieties of potatoes (*Solanum tuberosum* L.), are cultivated together with oca (*Oxalis tuberosa* Molina), arracacha (*Arracacia xanthorrhiza* Bancr.), mashua (*Tropaeolum tuberosum* Ruiz and Pav.), papalisa (*Ullucus tuberosus* Caldas), maca (*Lepidium meyenii* Walp.). Staple food is also based on cereals and andean seeds *i.e.* quinua (*Chenopodium quinoa* Willd.), cebada (*Hordeum vulgare* L.), yuyo (*Amaranthus deflexus* L.), kiwicha (*Amaranthus caudatus* L.,) and chocho (*Lupinus mutabilis* Sweet.). In the sub-tropical lower part of the valley, cash crops varieties are cultivated, i.e palta (Persea americana Miller), melocotón (*Prunus persica* (L.) Stokes), One of the main important features of plants classification in the Andes, is their division in 2 groups: "cold" and "hot" plants. These "hot" and "cold" qualities apply to any kind of plant (medicinal or not), are also the base of the Andean body physiology and, consequently, disease aetiology. Ethnobotanical studies are recognized as effective methods of finding locally important plants for discovery of crude drugs. Traditional medicinal system is prevailed in different parts of globe gaining recognition as alternative medicine among the indigenous communities for their primary healthcare needs. But keeping in mind that individualized intervention with less severe adverse effects makes the traditional healing method a better candidate and will continue to make great contributions to the health of MetS patients associated with complications like cardiovascular disease, and gout. Examples of online databases in this field are: The Phytochemical DB-list, the Madaus DB2 Database which provide basic information on medicinal plants found in Europe, Database of Central China Medicinal Plants, or the Raintree Tropical Plant Database. There have as well been created databases of medicinal plants used in particular diseases, like diabetes or asthma (90, 91, 92).

5. Plant Sources against MetS: Present Status and Emerging Solutions

First-line treatment of MetS consists of lifestyle changes, including diet and physical activity. The Mediterranean diet and other dietary regimens rich in vegetable-derived products were shown to prevent metabolic syndrome. The advantageous influences of these diets are thought to be related to their content of bioactive substances, monounsaturated and polyunsaturated fatty acids. Although low-density lipoprotein cholesterol (LDL-C) does not exclusively belong to MetS, subjects with MetS have a high concentration of small dense LDL particles and a low LDL-C/HDL-C ratio, and both of them are powerful atherogenic risk factors. When lifestyle changes alone are unsuccessful, phytochemicals can contribute to improve lipid profile. Bioactive compounds are plant secondary metabolites found in foods able to modulate metabolic processes. Research on phytochemicals suggests their possibility as therapeutic and preventive agents against diseases including inflammation, oxidative stress, dyslipidaemia and IR. This article addresses current knowledge about the phytochemicals in the treatment or reducing the effects of MetS components such as tolerating insulin resistance and high cholesterol levels and how ethnobotanical practices can enhance the prospecting of such nutraceutical components (93, 94). Several clinical studies have investigated the use of curcumin in the components of MetS. Yang *et al.,* in a randomized double blind controlled trial evaluated the effect of 1950 mg daily of curcumin extract (95 per cent curcuminoids) on weight, glucose and lipid profile in patients with MetS. After 12 weeks of treatment, the levels of HDL-C were significantly increased, while the levels of LDL-C were decreased (95). Several foods and beverages containing polyphenols have been found to be useful in the MetS features. Green tea, nuts, grape seeds, berries and dark chocolate, which contain both flavanol monomeric forms (catechins) and their oligomers (pro-anthocyanidins), are the principal resources of polyphenolic contents. Green tea, that is sourced from the fresh leaves of the *Camellia sinensis* L. plant, contains high levels of flavanols (catechins) and the alkaloid caffeine, both virtually able to produce metabolic effects. A report by Nagao *et al.,* suggested significant reduction of body weight, BMI, waist circumference and body fat mass in male patient group treated for 12 weeks with 690 mg of catechins per day (96). Also, caffeine has been found to influence the energy balance by increasing energy expenditure and decreasing energy intake. According to Basu *et al.,* when 35 MetS subjects divided in three groups treated with control, green tea beverage and green tea extract for 8 weeks significant loss of body weight and BMI was observed in both in green tea beverage group (–2.5±0.7 kg and –0. was 9±0.3 kg/m, 2 respectively) and in green tea extract group (–1.9±0.6 kg and –0.7 ±0.2 kg/m, 2 respectively) compared to controls (97). Suliburska and coworkers, observed that the 12 weeks of supplementation with 378 mg of green tea containing 208 mg of epigallocatechin gallate per day significantly decreased BMI and waist circumference in obesity subjects (98). The effects of the tea on body weight governed by mainly on the effect of caffeine and catechins on the adrenergic system. Caffeine (1,3,7 trimethylxantine) is a purine alkaloid inhibiting the phosphodiesterase enzyme that hydrolyzes cyclic adenosine monophosphate (cAMP) to AMP. cAMP signal is activated by beta-adrenergic stimulation, inducing some adrenergic effects including appetite reduction and increase of energy expenditure and lipolysis (99).

Moreover, caffeine enhances the expressions of the uncoupled proteins that increase thermogenesis through enabling phosphodiesterase inhibition and protein kinase A activating by cAMP. The catechins of green tea interfere also with catecholamine inactivation, inhibiting the catechol-O-methyl transferase (COMT) enzyme and inducing an adrenergic overstimulation. Like the caffeine-induced inhibition of phosphodiesterase, the COMT inactivation increases the energy expenditure, fat oxidation and lipolysis (100). On the other hand, the studies that evaluated the effects of different doses of the green tea catechins on the energy expenditure and on body mass index (BMI) and waist circumference showed no significant effect (101). Anthocyanins are another group of water-soluble polyphenolic compounds widely found in various vegetables and fruits, including grapes, plums, cherries, cranberries, strawberries, blueberries, blackberries, elderberries, currants, beetroot, red cabbage and red onions. These substances have antioxidant property due to their ability to transfer electrons or to donate the hydrogen atoms from various hydroxyl groups to free radicals. Animal studies showed that anthocyanins are mainly absorbed in the intact glycosidic form and they are found in the systemic circulation within 0.25–2 hrs. Their plasma concentration declines rapidly within 2 hrs upon a single oral administration in rat models (102). Blueberries (*Vaccinium myrtillus L.*) are fruits rich in anthocyanidins, chlorogenic acid, flavonoids and stilbenes, such as pterostilbene and resveratrol. Studies on blueberries showed several preventive and therapeutic properties, including the reduction of oxidative stress and inflammatory responses, protecting against cardiovascular disorders, hypertension and diabetes (103). Blueberry-enriched diet evidenced a protective effect on the MetS-related pro-inflammatory status in the obese Zucker rat by increasing adiponectin expression and suppressing liver expression of NF-kB (104). An *in vivo* study showed that blueberry reduces TGs, body weight gain, liver weight, abdominal fat mass, and improved adipose and skeletal muscle peroxisome proliferator activated receptor activities involved in glucose uptake and in glucose and fat oxidation in obese animal models. The plasma levels of interferon gamma (IFN-γ) and urinary level of 8-isoprostane were decreased. However, all parameters related to the glucose and lipid metabolisms were found to be unchanged after intake of the beverage made of Acai (Euterepe oleracea Mart.) (105). Pomegranate (*Punica granatum Linn*) is a fruit cultivated since ancient times throughout the Mediterranean region. Pomegranate is rich in polyphenols such as ellagitannins and anthocyanin and therefore it has been proposed in the prevention and treatment of MetS. *In vivo* studies evidenced that pomegranate reduces blood glucose, increases insulin sensitivity, inhibits α-glucosidase and enhances the glucose transporter type 4 activity. Moreover, pomegranate exerts anti-inflammatory effects, acting through the modulation of peroxisome proliferator-activated receptor pathways and a recent trial evidenced that this fruit reduces blood pressure in subjects with MetS (106). It has been seen that aspalathin from green Rooibos, were effective at reducing blood glucose levels of nondiabetic mice following ingestion of glucose, sucrose, and starch solutions, while suppressing the activities of α-glucosidase and α-amylase, key enzymes involved in carbohydrate hydrolysis.

Also under further investigated the molecular mechanisms associated with the ameliorative effect of pure aspalathin and an aspalathin-rich green rooibos extract on palmitate-induced insulin resistance in C2C12 myotubes and 3T3-L1 adipocytes. It has been observed that the aspalathin enhancing glucose uptake by promoting AMPK phosphorylation and GLUT4 translocation in L6 myotubes showed that aspalathin promotes insulin sensitivity in cardiomyocytes from young and aged rats (107). Studies *in vitro* and in animal models showed that polyphenols can induce an improvement of MetS features through several mechanisms including inhibition of digestive enzymes and intestinal glucose transporters, stimulation of incretin response, improvement of insulin signaling and modulation of gut microbiota. Cocoa is a rich source of polyphenols such as catechins, anthocyanins and proanthocyanidins. Epidemiological data and short-term human intervention studies evidenced that the cocoa polyphenols exert cardiovascular benefits of in humans. Cocoa polyphenols exert antioxidant and anti-inflammatory effects acting on some important signaling pathways such as toll like receptor 4/NFκB signal transduction and activation of transcription. Cocoa polyphenols induce release of NO through the activation of endothelial NO synthase which, in turn, accounts for vasodilation and cardioprotective effects (108). It is evident from many studies that plant based natural compounds carry a significant potential in curbing the MetS and help in bringing the well being of human health but there is still lots of ground to cover to validate their efficacy through well defined scientific reports. A general scheme is given in Figure 32.2 that links major groups of phytochemicals in shielding MetS.

Terpenes	*Polyphenols*	*Alkaloids*
	Improves Liver Activity	
	Improves Metabolic Activity	
	Maintains Blood Sugar Level	
	Improves Gut Microbiota	
	Reduces Endothelial Dysfunction	
	Decreases ROS generation	
	Improves lipid profiling	
	Reduces pro-inflammatory signals	
	Maintains BMI	

Figure 32.2: Major Groups of Phytochemical from different Plant Sources Helps in Shielding from Metabolic Syndrome by Reducing the Metabolic Risk Factors.

6. Potential Approaches for Prospecting Bioactive Compounds

To address the demand of new line of drug formulation of plant origin, identification of bioactive molecules involves extensive screening methodologies, tedious and time-consuming extraction procedures and further difficult scaling up the isolated compound for efficient dose composition. A classic example can be given that of the discovery of morphine from opium poppy. The opium plant and the sedative properties were long known to the mankind, but the discovery of drug morphine only took place in the eighteenth century. Friedrich Wilhelm Sertürner succeeded in isolating a crystalline substance from opium then known as morphium. He named the new substance 'morphine' after Morpheus, the ancient Greek god of sleep and dreams. He isolated morphine, the first pure opioid available for premeditated pain therapy. Moreover, he was the founder of a new class of pharmaceutical drugs, the alkaloids (109,110). Current scenario for screening of active compounds involves two general approaches of random selection succeeding chemical screening, also known as "bioassay-guided fractionation" and "random-bioassay fractionation" (111). In comparison with the random collection and screening methods, the classical ethnobotanical approaches like field surveys and preparation of analytical questionnaire to the traditional healers or practitioners can amend the lengthy and cumbersome processes by shortening and selective collection (112). The *in silico* approaches along with the multiomics approaches such as phylogenetic analysis and selected molecular markers can further facilitate the screening and identification of bio-active compounds. The structural activity relation can also be explored with the aid of molecular simulation and docking platforms by computational methods (113). The specific therapeutic properties of bio active compounds are governed by their chemical structures including binding sites, number, positions, and types of substitutions which can be traced by *in silico* methods. Few recent reports can be taken into consideration where a catechol group in ring B and a 3-hydroxy group in ring C were essential for the antioxidant activity in flavonoids of quercetin, rutin, kaempferol, and luteolin (114). Therefore, the study of structure driven activity relations is a prerequisite when attempting to discover novel pharmacologically active compounds. The subsequent investigations are in regard with characterization and isolation of the active constituents using different separation techniques and spectroscopic investigation (115). Succeeding evaluation of their medical efficacy in disease models *in vitro* and *in vivo* systems and then preclinical and clinical trials in specific patient groups culminates in the development of new drugs. Schematic representation of different methodical approaches for prospecting of therapeutically important ingredients is given in Figure 32.3. Moreover, this chapter addresses three principal aspects that are ethnobotanical practices, bioactive compounds from plants and their intervention in a very complicated health disorder MetS which has become global threat among noncommunicable diseases. Although the first two aspects have been discussed in the previous sections, the next subsection addresses basic pathophysiological etiology and current scenario or status. It is also worth mentioning here that the rationale behind choosing MetS specifically is to discover a new line of drug candidates by ethnomedicinal practices with natural ingredients that can be considered therapeutically by the population who are undergoing lifestyle disorders.

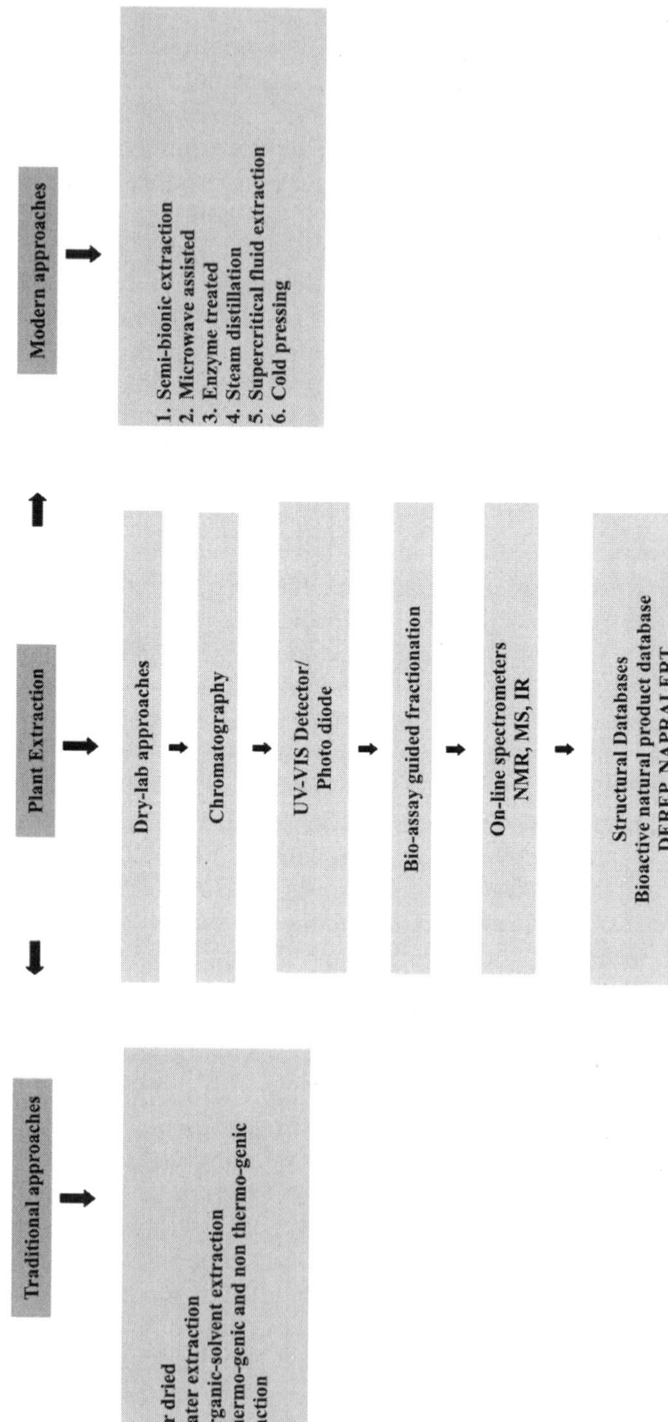

Figure 32.3: Modern Approaches taken by Investigators to Assess the Ethnomedicinal Possibilities to Identify Plant Based Components.

7. Intervention of Ethnobotanical Approaches for Mining of Bioactive Compounds

Intervention of "Modern ethnobotany" in prospecting bioactive compounds can be a sustainable option since a stage of ethnobotanical progression that focuses the research of inter-relationships between ethnic community and plants using advanced technologies. It can provide more systematic data in sustenance of traditional plant practices and augment the concepts and methodology. Traditional medicines have always been one of the principal investigation interests of ethnobotany in contribution to the theoretical advancement of the subject. Phytochemical constituents have been the source of many active ingredients in medicines. The discovery of these pharmacologically active constituents was usually based on ethnobotanical information, and many drugs used today were developed from medicinal plants employed in indigenous societies (116). The results indicated that contributions of ethnomedicinal uses have higher probability in finding the active molecules in drug discovery screening. There were numerous additional motivations corroborating the inventions of pharmacologically functional components from medicinal plants employingmodern ethnobotany. Ethnobotanical investigation could offer ample materials for wet-laboratory experiments as well as dry-lab based platform for generation digital data repository. Ethnobotanical questionnaires provide a short route for chemists for selective screening of plants collection for phytochemical investigation and further pharmacological activities screening (117). Ethnobotanical analysis is a substantial foundation of novel active components as well as formulations. Even though, standard data of traditional medicine is important to social well-being, countless crucial unestablished folk practices or aboriginal customary medicines and formulations could possibly be intermittent and could easily vanish with rapid urbanization and technological progressions. Since indigenous comprehensions about medicinal practices is not based on scientific research, there is a possibility to find novel active constituents and remedies which are not explored or established earlier. Nevertheless, ethnobotanical surveys commonly follow the narration of a rather limited set of well-documented medicinal plants (118,119). Ethnobotanical research can also amalgamate cultural ethnic diversity in miscellaneous traditional healing remedies. Visualizing globally, it can be stated that the northern hemisphere, ethnic folkloric cluster became widely diversified succeeding an extended account of migrations and resettlements but moreover stayed culturally significant and genetically interlinked (120). These groups were influenced by remarkably parallel medicinal flora. This is indicative of the fact that the selection of medicinal plants and straight understanding with their bioactivity are influenced by cultural differences. In addition, periodicals and publications that communicate a multiplicity of cultural beliefs rather than biochemical evaluations also affect the medicinal plant choices (121). Therefore, organized or semi-structured interviews and individual interactions deliver prospects to integrate the cultural diversity of medicinal plant usage. Founded on the motives above, ethnobotany serves as a competent, resourceful approach for the innovation in therapeutically effective elements originations. A very successful case study can be stated here that of *Catharanthus roseus* of family Apocynaceae by ethnic culture native to southeastern Madagascar as an oral supplement for hypoglycemia. Subsequent laboratory research unearthed about 75 active alkaloids including vinblastine and vincristine now established as anticancer drugs for childhood leukemia and Hodgkin's disease (122,123). Numerous ethnobotanical studies aimed at identifying new pharmaceutical products have been initiated in recent times. Drugs derived from plants along with their uses are presented in Table 32.2.

7.1 Harvesting of Medicinal Plants

The harvesting of plants, meaning removal of plants or their parts for medicinal purposes, was considered as most important to healers to ensure long-term availability of the plants and plant parts. Very often the exploitation of wild harvested resources (medicinal plants) is seen degrading the richness of biodiversity. Documenting the eroding plants and associated indigenous knowledge can be used as a basis for developing management plans for conservation and sustainable use of medicinal plants in the area. The principal threatening factors reported were deforestation, agricultural expansion and overgrazing. Learning of correct methods of harvesting was considered an essential part of local conservation efforts by healers. Most healers have certain beliefs like that no plant part should be removed until the plant is strong enough to withstand the loss (124). In general, it was felt that a mature plant should be chosen for removal of the required parts. The right method of harvesting varies according to the growth of the plant and nature of the cure. The time at which plant parts are harvested is also important. For instance, many healers, believe that Saturday night is an auspicious time for harvesting. In addition, knowledge of special care or practices related to safer harvesting is also considered important. Most of the healers have revealed that their shadow should not fall on the plants while harvesting. During such prayers, they appreciate the effectiveness of the medicinal plants and express their gratitude. Knowledge of planetary movement is also a part of their learning of ceremonial rituals by healers. Traditional harvesting practices were taken into consideration while developing sustainable harvesting practices of medicinal plants. Keeping the importance of harvesting on availability of plant we have standardized sustainable harvesting practices of some important medicinal plants *e.g., Andrographis paniculate, Rauvolfia serpentina, Gymnema sylvestre, Tinospora cordifolia, Celastrus paniculatus, Terminalia arjuna, Litsea glutinosa, Saraca asoka, Holarrhena antidysentrica, Bauhinia veriegata, Embelia tsjerium-cottam, Phyllanthus emblica*. However, there is a need for increased efforts to develop technologies to sustain their extraction. By adopting non-destructive harvesting practices these important valuable resources can be conserved. Income from collection and sale of medicinal plants is thought to be marginalized by lack of awareness regarding local and overseas market requirement, local shopkeepers, agents and medicine man. The consumers obtain supplies from individuals who have little experience in

Table 32.2: List of Selected Medicinally Important Plants with their Active Components that Play Crucial Role against Metabolic Syndrome and their Mechanism of Action

Sl. No.	Plant Parts and Scientific Names	Formulations and Active Compounds	Actions	Targets
1	*Aloe vera* leaves Chinese traditional systems	Pulp extracts, ethanolic extract	FBG, TC, TG, FFA, hepatic transaminases, phospholipids ↓ Insulin ↑	Hydroperoxides, GSH ↓
2	*Cinnamomum zeylanicum* Chinese and Indian herbal formulations	Bark extract, aqueous extract	FBG, insulin, protein-bound sugars, TC, TG ↓ HDL ↑	α glycosidase ↓ Glucose uptake, insulin signalling In skeletal Muscle ↑
3	*Crataegus cuneats* Sieb. Et Zucc Chinese herbal	Alcoholic extract of fruit, aqueous extract of leaf	TG, LDL, VLDL, lipid deposits in liver and aorta, FBG ↑	
4	*Cornus officinalis* Sieb. et Zucc Chinese herbal	Alcohol extract, Anthocyanin and ursolic acidoleanolic acid	Lipid accumulation in liver, TG, body weight, FBG ↓ Insulin, C-peptide ↑	Proliferation of islet, GLUT4 ↑
5	*Ganoderma lucidum* (Leyss.ex Fr.) Karst. Chinese, Tibetan and Indian herbal	Polysaccharides, ganoderan B	TC, TG, LDL, LPO, fecal cholate, FBG ↓ Insulin, HDL ↑	Ca2+ inflow to the pancreatic β cellsinsulin-releasing, hepatic glucokinase, phosphofructokinase and glucose6-phosphate dehydrogenase ↑ HMG-CoA reductase, hepatic glucose-6-phosphate, glycogen synthetase ↓ Antioxidation

Sl. No.	Plant Parts and Scientific Names	Formulations and Active Compounds	Actions	Targets
6	*Radix Notoginsen Panax notoginseng* (Burk.) F.H.Chen Chinese herbal	n-Butanol extract, pulverized root, ginsenoside Rg1	TC, TG, LDL, FBG ↓	FXR/LXR α agonist Accumulation of abnormal lipid ↓
7	*Rhizoma Polygonati* Chinese herbal	Methanol extract	FBG ↓	Hepatic glucose output, GLUT2 ↓
8	*Salix matsudana Koidz.* Chinese herbal	Leaves extract, Polyphenol fractions, apigenin-7-O-dglucoside, luteolin-7-O-dglucoside and chrysoeriol-7-Od-glucoside	Body weight, parametrical adipose tissue weights, TC ↓	Amylase and palmitic acid uptake into small intestine ↓ Norepinephrine-induced lipolysis ↑
9	*Thea chinensis* and *Camellia sinensis* Chinese herbal	Leaves Black tea, green tea, water extracts, crude tea polysaccharides, tea polysaccharide fraction, oolong tea	FBG, phosphate, body weight, GSP, TC, TG, FFA, LDL, systolic blood pressure, LPO ↓ GSH, vitamin, glucose tolerance, adiponectin, LDL particle size, insulin, HDL ↑	Insulin response enzyme activities of carbohydrate metabolism and antioxidant defences, PPAR α, PPAR γ ↑, Weight of the intestine ↓
10	*Ficus religiose* Indian Ayurveda, Sidhha	Bark decoction Sitosterol-d-glucoside leucocyanidin 3-O-beta-d-galactosyl cellobioside, leucopelargonidin-3-O-alpha-L rhamnoside furanocoumarin bergapten and bergaptol	FBG, TG, TC, insulin-stimulated glucose uptake in skeletal muscle ↓ muscle and hepatic glycogen ↓	Enzymes of antioxidant defence system ↑ Oxidative stress↓ Glutathione ↑ Malondialdehyde ↓
11	Seeds, leaves, fruits, and bark of Black plum (*Eugenia jambolana*) Indian, Sri Lanka, Vietnam, Cambodia	Anthocyanins, glucoside, ellagic acid, isoquercetin, kaempferol, myricetin, and hydrolysable tannins (1-0-galloyl castalagin and casuarinin). The seeds also contain alkaloid jambosine and glycoside jamboline	Insulin from the pancreas β cells ↓ cholesterol, triglycerides ↓ and free fatty acids ↓	3-HMG Co-A reductase ↓ Glycosuria ↓ Blood uria ↓

Sl. No.	Plant Parts and Scientific Names	Formulations and Active Compounds	Actions	Targets
12	*Momordica charantia* Indian and Sri Lanka traditional system	Fruits, seeds, vicine, charantin, and triterpenoids	Repairs Pancreatic β cells epididymal and retroperitoneal white adipose tissue ↓ GLUT 4 ↑	α Glucosidase ↓ α Amylase ↓
13	*Berberis aristate* Indian Ayurveda Iranian traditional medicine	Roots, stem bark, berberine, mangoflorin, berberamine, polyphenols and flavonoids	SIRT3 in mitochondria ↑ Impaired oxidative phosphorylation ↓ CD36 in hepatocytes↑ Hepatocellular lipid accumulation ↓	Insulin sensitivity ↑ GLUT1↑
14	*Tinospora cordifolia* Indian Ayurveda	Diterpenoid glycosides, steroids, sesquiterpenoid, phenolics and polysaccharides	FBG ↓ Plasma triglycerides ↓ Malondialdehyde ↓	Protein carbonylation ↓ GSH, SOD ↓ Oxidative stress ↓
15	Roots alcoholic extracts of *Rubia cordifolia* Indian Ayurveda	Carbomethoxy prenyl Naphthoquinone, Alizarin	MAPKK/MAPK pathway ↓ Adipokines, cytokines ↓ Blood urea ↓	Restores anti-oxidant level ↑
16	*Ocimum sanctum* Indian Ayurveda	Eugenol, triterpenoids, rosmarinic acid, apigenin, isothymusin, isothymonin, cirsimaritin, orientin, and viceninursolic acid, carvacrol, linalool, limatrol, and caryophyllene	Glycoxidation of haemoglobin ↓ FBG ↓ Blood Urea ↓ Cholesterol, triglyceride, and LDL	Repairs dyslipidaemia by restoring endogenous antioxidant levels ↑

Sl. No.	Plant Parts and Scientific Names	Formulations and Active Compounds	Actions	Targets
17	*Pterocarpus marsupium* Indian Ayurveda	β-sitosterol, lupenol, aurone glycosides, epicatechins, and iso-flavonoids	Regeneration of pancreatic beta cellsrepairs glycosylated haemoglobin	Serum protein, insulin, alkaline and acid phosphatase, and albumin levels ↑
18	*Trigonella foenum-graecum* Indian ayurveda	Saponins4-hydroxyisoleucine trigonelline	Somatostatin ↓ Plasma glucagon levels ↓ Normalizes creatinine kinase in liver, skeletal muscles, and heart	Insulin sensitivity ↓ Renal toxicity ↓
19	*Gymnema sylvestres* Indian ayurveda	Gymnemic acids and gymnemasaponins	Intracellular calcium and insulin secretion in β cells adrenal hormones ↑ Utilization of glucose ↑ decreasing mobilization of fat	chronic inflammation ↓ body weight, plasma proteins ↓
20	*Allium sativum* Indian, Chinese Thailand traditional system	Allicin, allixin, ajoene,Diallyl trisulfideN-Acetylcysteine	FBG ↓ Glucose tolerance ↑ Oxidative stress Serum cholesterol Glycation-derived free radicals and AGEs ↓	Antioxidant enzymes superoxide dismutase, catalase, and glutathione peroxidase TBARS and GSH ↓
21	*Adiantum flabellulatum* L Vietnam and Thai traditional therapeutic system Decoction	Aquous extracts of whole plant	FBG ↓ Glucose sensitivity ↑ Oxidative stress ↓	Levelling up antioxidant enzymes

Sl. No.	Plant Parts and Scientific Names	Formulations and Active Compounds	Actions	Targets
22	*Garcinia cambogia* Thai herbal formulation	Aquous extracts alcoholic extracts	Fatty acid synthase (FAS) mRNA expression ↓ Peroxisome proliferator-activated receptor α (PPARα) ↑ carnitine palmitoyltransferase ↑ β-oxidation ↑	Visceral adiposity ↓ Body fat accumulation glucose intolerance ↓
23	*Polyscias fruticosa* (*L.*) Indian system	3-*O*-[β-d-glucopyranosyl-(1→4)-β-d-glucuronopyranosyl] oleanolic acid 23-*O*-β-d-glucopyranosyl ester saponins	3-HMG Co-A Reductase ↓ Glycosuria ↓	α Glucosidase ↓ α Amylase ↓
24	*Emblica officianlis* L. China, India, Vietnam, Thailand, Srilanka	EtOAc extract of fruits anthocyanins, glucoside, ellagic acid, isoquercetin, kaempferol, myricetin, and hydrolysable tannins	TAG ↓ Total Cholesterol ↓ Hepatic SREBP-1	COX-2, NF-kB, bcl-2 ↓
25	*Paederia foetida* Indian and Chinese traditional medicine	Methanol extract of *P. foetida* L. leaves β-sitosterol	Regeneration of pancreatic beta cells Repairs glycosylated haemoglobin	Insulin resistance Impaired glucose intake
26	*Urtica dioica* Iranian traditional medicines	Lectins, phenols, sterols, and lignans	TC, TG ↓ TNF-α ↓ Interlukin (IL-6) ↓ C Reactive protein (CRP) ↓	Inflammatory distress insulin sensitivity

Sl. No.	Plant Parts and Scientific Names	Formulations and Active Compounds	Actions	Targets
27	*Silybum marianum* Iranian and europian medicines	Lignans and tannins poly phenols, Sylmarin	Liver level of TG $\downarrow$ HDL$\uparrow$ Modified lipoprotein profile in plasma $\downarrow$ VLDLC lipid levels in plasma $\downarrow$	Liver phosphatidate phosphohydrolase activity, Effect on apoAI
28	*Citrullus colocynthis* Iranian as well as European traditional medicine	Fruits Oleic and linoleic acid	FBG $\downarrow$ Impaired glucose uptake Glucose sensitivity $\uparrow$	α amylase $\downarrow$
29	*Rheum ribes* Iranian, Indian and European systems	Tannins, terpenoids, alkaloids, flavonoids, phenols, steroids, glycosides, saponins	FBG $\downarrow$ Impaired glucose uptake $\downarrow$	Insulin resistance $\downarrow$
30	*Plantago ovata* Indian, Iranian and Mediterranean medicinal systems	Psyllium methylcellulose	TNF-α and adiponectin $\downarrow$	Insulin sensitivity Hyper-insulinama Regeneration of beta cells
31	*Alisma plantago-aquatica L.* Russian and Chinese traditional medicines	Roots extracts Triterpenes Protostane type saponins	Peroxisome proliferator-activated receptor $\alpha\uparrow$	Dyslipidaemia Free fatty acid accumulation $\downarrow$
32	*Galega officinalis* Mediterranean medicine system	Metformin	3-HMG Co-Areductase $\downarrow$ Glycosuria $\downarrow$ Peroxisome proliferator-activated receptor $\alpha\uparrow$	α Glucosidase $\downarrow$ α Amylase $\downarrow$ Insulin sensivity

medicinal herb preparations or in understanding of its value. As a result, valuable economic and medicinal plants are becoming rare and some are at the verge of local extinction. Therefore, efforts should be to conserve these valuable plants for future generations.

7.2. Processing of Plants

The useful plant parts are either given as medicine in raw form or are processed in a variety of forms. Processing of raw drugs is a key factor which influences the efficacy of drug. The most common processed forms reported are: crushed drugs, pastes, fresh juice, (dried) powders, decoctions, ash, porridges, powder or paste in oil base for external application) and smoke for inhalation. Local healers have indicated that processing is an important skill which indirectly helps in conservation because proper processing of plant parts consumed less quantity of the plants and helped to ensure the whole plant was not uprooted. Processing also becomes critical when preparing the herbal formulations from toxic plants. Some concoctions are given fresh while others are fermented. The fermented alcoholic extracts give rise to the incomplete hydrolysis of the macromolecules and generation of bio active compounds. Ethnobotany has once again become a recognized tool in the search for new pharmaceuticals. Initiatives by governmental agencies and the private sector have helped spark this renewal. Many of these projects are interdisciplinary, efforts involving scientists from the fields of anthropology, botany, medicine, pharmacology and chemistry. The concept of the ethno-biomedical reserve has been introduced. It provides an opportunity for pharmaceutical and herbal industries to contribute to the conservation effort.

7.3. Status of Traditional Knowledge

Many of the traditional methods and general knowledge of medicinal flora is being lost to time. As healers and tribal elders age and die, their knowledge is dying with them. Scientists are searching for ways to preserve this knowledge and to test them against contemporary diseases. Fewer than 5 per cent of tropical forest plant species have been examined for their chemical compounds and medicinal value. This left great potential for even more discovery, but also the potential for great loss as forests are felled around the globe and unstudied species are lost to extinction. There is a profound and growing knowledge gap between old and younger generations. People of more than 50-65 years age know a lot about wild plant products as compared to younger generation (125).

7.4. Benefit Sharing and Protection of Traditional Knowledge

In the context of progressing intellectual property law, defining ownership of traditional knowledge (TK) can be perplexing when claims of origin are conflicting and requires accepting parameters of how uniqueness is defined, and patent law is applied to protect this information. Transforming laws and policies are globally influencing the way traditional knowledge is evaluated and protected. Within this context, suitable types of benefit sharing can only be achieved if appropriate mechanisms are in place to prevent exploitation. Evolving mechanisms and laws are beginning to impact on how TK can be utilized to achieve appropriate benefits for those who are custodians of these types of information. This is not an easy task since issues related to ownership and ways to ensure appropriate protection, oversight and disbursement of the revenues are always likely to arise. The premise of biodiversity prospecting is that appropriate policies and institutions are needed to ensure that the commercial value obtained from genetic and biochemical resources are a positive force for development and conservation (126). Much depends upon the application of current national and international laws and the acceptance within and among nations to work towards logical solutions. Crucial to this evolving process is the acknowledgement that a significant deal of valuable TK from different parts of the globe is being lost not only by its unsuitable dissemination but because a fitting global repository does not exist to safeguard this knowledge while workable options to address its worth are being examined.

7.5. Education Opportunities

There is more interest in ethnobotany today, than at any time in the discipline's history. Ethnobotany, however, suffers from many deficiencies, especially the lack of research support, educational opportunities, and a theoretical basis. Ethnobotanists should expand the definition of ethnobotany to include all plant-people interactions, not just those of traditional societies. They also must integrate more effectively with colleagues in related disciplines and promote ethno-botany's relevance to introductory botany and other courses. Ethnobotany and ethno-biology are natural links to conservation biology, resource management, and environmental education. Expanding ethno-botany's scope to include all plant and human interactions greatly increases the funding, research, and job opportunities for the discipline. With the intention of taking students back to old-school remedies and to the wonders of traditional medicine, the National Medicinal Plants Board (NMPB), Govt. of India New Delhi is introducing the concept of herbal gardens in various schools across the country from early 2010. As per the NMPB, the herbal gardens project will initially cover 1,000 schools in 50 districts across the country. Evidence-based medicine works of famous scientist Archie Cochrane on efficacy and effectiveness and meta-analysis as a method of summarizing the results of randomized trials have led to a powerful research and analysis tool in the form of 'systematic reviews', which empowered clinicians and researcher's decision making (127). These efforts finally led to the evolution of evidence-based medicine as a new approach to bring more rational and analytical evidence for research-backed practice of medicine practice quality and quality of scientific evidence to develop evidence-based practice.

7.6 Development of Evidence-Based Practice

There are plenty of issues needed to be addressed on traditional knowledge to move towards acceptable evidence based. Concerns related to protocols, problems and potential of various traditional knowledge in context to evidence-based conditions have been presently discussed in many research groups. Few critiques have opined that basic concepts of traditional knowledge should not be distorted to suit convenience or availability of biomedical research models. Arguably, prevailing pre-clinical methods and clinical models like Randomized Control Trials may not be suitable to validate traditional medicine. However, the onus of developing suitable models to build necessary evidence must be voluntarily accepted by the traditional knowledge sector. Some efforts in the direction to conduct the whole system clinical trials are already in progress. Traditional medicine is severely deficient in scientific evidence related to clinical practice and scientific research. This sector needs to go beyond mere scholarly recitals, reviews and defensive interpretations, which are abundant in current literature. Ayurvedic medicine needs more rigorous scientific research for evaluating safety, quality and efficacy. Many examples from past may guide our quest for evidence-based Ayurveda in the future. The innovation of the drug BGR34 Blood Glucose Regulator-34 sets a classic example which is an Ayurvedic-derived product from six different traditional medicinal herbs that is being retailed in India for the management of type 2 diabetes. It was developed in 2015 by two government owned laboratories and launched commercially in 2016. It has been tested in only one, modest-sized, human trial. Besides all the trial experiments the drug formulation has been heavily criticized because of ample phases of clinical trials. The efficacy of the drug remains in question. The manufacturers have refused to acknowledge the claims of inefficacy and other concerns. It may be worthwhile to learn from collaborative research networks like IN-CAM from Canada and CAMbrella from Europe.

7.7. Glimpse of Indian Initiatives on Plant Prospecting

Various government agencies like Department of Biotechnology (DBT), Council of Scientific and Industrial Research (CSIR) and Department of Ayurveda, Unani, Siddha and Homeopathy (AYUSH), Ministry of Health and Family Welfare have commenced exertion on bioprospecting. DBT initiated the network programme on 'Bioprospecting of biological wealth using biotechnological tools' during the 9th plan involving 13 institutions. The objectives of the DBT programme were characterization of biodiversity in different agro-ecological regions, bioresources mapping, discovery and monitoring of biological diversity, characterization and conservation of Himalayan endangered species, including medicinal and aromatic plants, and bioprospecting of molecules and genes for product development. The leads obtained from the first phase of bioprospecting have been taken charge for detailed investigation, with a focus on product and process development and commercialization. CSIR has initiated a coordinated programme on drug discovery with a network of 19 CSIR laboratories and other R and D institutions working in the field of traditional medicines as well as universities. The programme was initiated in 1996, and aims at discovering new bioactive molecules from plants, fungi, microbes, insects, *etc.* using new technologies. The Planning Commission sponsored the New Millennium Indian Technology Leadership Initiative (NMITLI), one of the most innovative bioprospecting programmes. NMITLI started a major herbal drug development programme for developing effective herbal remedies for diabetes, arthritis and hepatic disorders, which has shown highly encouraging results within a short period of time. The Ministry of Health and Family Welfare, Govt of India initiated two important task-force programmes relating to creation of Traditional Knowledge Digital Library and designing a Traditional Knowledge Resource Classification. Traditional knowledge resource center has information on around 5000 subgroups and the structure of TKRC is compatible with the International Patent Classification (127).

8. Perspectives and Possibilities: Mass Bioprospecting and Constructing Library for Discovery of New Drug Candidates

Based on the number of known species (300,000–500,000), plants represent the second largest source of biodiversity (15 per cent). Estimates have been made that between 20,000 and 55,000 species of plants have been used medicinally, of which only a small portion has been investigated for drug purposes (128). Among those that have been investigated are plant species that produce important drugs. Overall, only 15–20 per cent of terrestrial plants have been evaluated for pharmaceutical potential. Consequently, plants, including unexplored. The effort should be taken in searching for agents that cures MetS from the biodiversity, in the present context, of plants, is one example of mass bioprospecting. In this effort, field investigations are largely guided by the "random" collection approach, with a larger role of ethnobotanical or ethnopharmacological information. A wide variety of compound classes were isolated and characterized in past few decades that may have clinically.

8.1. Incorporating *In-silico* Approaches for better Screening of Bioactive Compounds

Screening and analysing bioactive components have always been a deadlock in the investigation of natural medicines. Many methods and technologies have been used in the screening and analysis of bioactive components. In the drug discovery industry, the pharmacological effects and activities of Western medicines are related to their ADMET (Absorption, Distribution, Metabolism, Excretion and Toxicity) properties *in vivo* in addition to interaction with their targets. ADME properties are crucial determinants of the ultimate clinical success of a drug. The early introduction of ADMET screening has been realized during the drug discovery process. Some studies have shown that traditional medicines are like western medicines, and their activity has also been related to ADMET properties. Studies of ADMET properties of TCMs have been developed and applied to the screening of natural medicines. including their ADMET properties and synergetic effects of the active ingredients, targets, interaction sites of the active ingredients. It is necessary to establish a series of new screening and analysis methods and pharmacological models for the study of traditional medicines. Biological finger printing high throughput analysis based on biomolecule recognition or chromatography or metabolism analysis has been developed and utilized for screening and analysis of bioactive components in because it is well suited to the characteristics of herbal medicines, such as their multi-component, holistic adjustment and multi-target features (129). In addition, these techniques can be used to simulate and evaluate the actions of drugs in vivo during screening and analysis of bioactive compounds. It can achieve the purpose of evaluating pharmacological parameters of various components of TCMs, screening and analysis of active ingredients and metabolites and the discovery of pharmacological pathways.

8.2. Plant Derived Bioactive Compounds

Plant-derived bioactive peptides inhibit the enzymes like alpha-glucosidase, alpha-amylase, dipeptidyl peptidase-IV and glucose transporter systems involved in type 2 diabetes. *In vivo* studies of a certain plant peptides fraction showed insulin-mimetic action in animal models. Plant-derived bioactive peptides are not explored to their full potential in comparison to other classes of natural products. Few ethnobotanical approaches reveal that the preparation of herbal remedies by incomplete fermentation of preparation of decoction by thermal treatment led to incomplete hydrolysis of certain proteins and leads to formation of bio active peptides of therapeutic importance. But the investigated and evidence-based proofs are lacking in that area. The reason might be the lack of suitable instrumental techniques to identify, purify and characterize the peptides from plant source. However, the recent advancement in LCMS, LC-NMR, and X-ray crystallography has bridged the gap and this would be the right time to embark on plant peptides. In future, systematic studies including dereplication for the early identification of known peptides, target identification followed by *in vivo* studies would help to speed up the plant peptides research.

9 Concluding Remarks

A clear vision arises from the above discussion that natural plant-based nutraceutical bio-active compounds discovered from medicinal plants and their analogues have provided numerous clinically useful drugs for MetS management. Despite the fact that various challenges encountered in the medicinal plant-based drug discovery, natural products isolated from plants will still remain an essential component in the search for new medicines. The fact that only about one-tenth of the flowering species occurring globally are investigated for their pharmaceutical potential, can be the obvious advantage to begin with plant/ medicinal plant-based drug discovery programmes. The diverse genetic resources and associated rich traditional knowledge available across the globe form the strong basis for bioprospecting and need proper network for storage and passing sharing of knowledge without hampering the biodiversity. Proper utilization of these resources and tools in bioprospecting will certainly help in discovering novel lead molecules for complicated disorder like MetS and find a route towards better health management for the complexities arose due to MetS without risking the adverse effects from candidate plants by employing modern drug discovery techniques and the coordinated efforts of various disciplines.

Acknowledgements

The authors like to acknowledge the facilitates given to them by the Department of Botany and Department of Biochemistry of Ravenshaw University. Authors also acknowledgethe financial assistance given to the Department for being the Center of Excellence in Environment and Public Health by Higher Education Department, Government of Odisha under OHEPEE (HE-PTC-WB-02017).

REFERENCES

1) Pandey, Ashok K., and Y. C. Tripathi. "Ethnobotany and its relevance in contemporary research." Journal of medicinal plants studies 5.3 (2017): 123-129.

2) Eldeen, Ibrahim MS, Mohd AW Effendy, and Tengku S. Tengku-Muhammad. "Ethnobotany: Challenges and future perspectives." Research Journal of Medicinal Plants 10.6-7 (2016): 382-387. medicinal plants studies 5.3 (2017): 123-129.

3) Sohani, Smruti, and Sumeet Dwivedi. "Ethnobotany and its connection with Medical sciences and Public health." International Journal of Pharmacy and Life Sciences 10 (2019).

4) McClatchey, Will C., *et al.,* "Ethnobotany as a pharmacological research tool and recent developments in CNS-active natural products from ethnobotanical sources." Pharmacology and therapeutics 123.2 (2009): 239-254.

5) Cunningham, Anthony B. Applied ethnobotany: people, wild plant use and conservation. Earthscan, 2001.

6) Sheng-Ji, Pei. "Ethnobotanical approaches of traditional medicine studies: some experiences from Asia." Pharmaceutical biology 39.sup1 (2001): 74-79.

7) Ozturk, Munir, and Khalid Rehman Hakeem, eds. Plant and Human Health, Volume 1: Ethnobotany and Physiology. Vol. 1. Springer, 2018.

8) Rahman, Inayat Ur, *et al.,* "Historical perspectives of ethnobotany." Clinics in dermatology 37.4 (2019): 382-388.

9) Addoun, N., *et al.,* "Ethnic Medicine and Ethnobotany Concept to Identify and Characterize New Polysaccharide-Based Drug from Arid and Semi-Arid Lands." Adv Appl Chem Biochem 1 (2018): 37-39.

10) Jamshidi-Kia, Fatemeh, Zahra Lorigooini, and Hossein Amini-Khoei. "Medicinal plants: Past history and future perspective." Journal of herbmed pharmacology 7.1 (2018).

11) Jones, Margaret, and Chandani Liyanage. "Traditional Medicine and Primary Health Care in Sri Lanka: Policy, Perceptions, and Practice." Asian Review of World Histories 6.1 (2018): 157-184.

12) Heinrich, Michael, *et al.,* "Ethnobotany and ethnopharmacology—Interdisciplinary links with the historical sciences." Journal of ethnopharmacology 107.2 (2006): 157-160.

13) Schmidt, Barbara M., and Diana M. Klaser Cheng, eds. Ethnobotany: A Phytochemical Perspective. John Wiley and Sons, 2017.

14) Ghorbani, Abdolbaset, Farzaneh Naghibi, and M. MOSADEGH. "Ethnobotany, ethnopharmacology and drug discovery." (2006): 109-118.

15) Leonti, Marco. "The future is written: Impact of scripts on the cognition, selection, knowledge and transmission of medicinal plant use and its implications for ethnobotany and ethnopharmacology." Journal of Ethnopharmacology 134.3 (2011): 542-555.

16) Throat, Malignant Putrid Sore. "William Withering (1741–1799): a Birmingham lunatic." Proc R Coll Physicians Edinb. Vol. 31. 2001.

17) Petrovska, Biljana Bauer. "Historical review of medicinal plants' usage." Pharmacognosy reviews 6.11 (2012): 1.

18) Srivastava, Akhileshwar Kumar. "Significance of medicinal plants in human life." Synthesis of Medicinal Agents from Plants. Elsevier, 2018. 1-24.

19) Sewell, Robert DE, and Mahmoud Rafieian-Kopaei. "The history and ups and downs of herbal medicines usage." Journal of HerbMed pharmacology 3 (2014).

20) Reddy, Jayarama. "Important Medicinal Plant Families and Plant Based Drugs: A Review." Proceeding 5th International Conference on Civil, Architecture, Environment and Waste Management (CAEWM-17) Singapore March. 2017.

21) Tune, Johnathan D *et al.,* "Cardiovascular consequences of metabolic syndrome." Translational research : the journal of laboratory and clinical medicine vol. 183 (2017): 57-70.

22) Sarafidis, Panteleimon A, and Peter M Nilsson. "The metabolic syndrome: a glance at its history." Journal of hypertension vol. 24,4 (2006): 621-6.

23) Oladejo, A O. "Overview of the metabolic syndrome; an emerging pandemic of public health significance." Annals of Ibadan postgraduate medicine vol. 9,2 (2011): 78-82.

24) Saklayen, Mohammad G. "The Global Epidemic of the Metabolic Syndrome." Current hypertension reports vol. 20,2 12. 26 Feb. 2018

25) Lee, Lana, and Renata Arrington Sanders. "Metabolic syndrome." Pediatrics in review vol. 33,10 (2012): 459-66; quiz 467-8.

26) Petrie, John R *et al.,* "Diabetes, Hypertension, and Cardiovascular Disease: Clinical Insights and Vascular Mechanisms." The Canadian journal of cardiology vol. 34,5 (2018): 575-584.

27) Lemieux, I *et al.,* "Hypertriglyceridemic waist: A marker of the atherogenic metabolic triad (hyperinsulinemia; hyperapolipoprotein B; small, dense LDL) in men?." Circulation vol. 102,2 (2000): 179-84.

28) Reaven, G M. "Banting lecture 1988. Role of insulin resistance in human disease." Diabetes vol. 37,12 (1988): 1595-607.

29) Kassi, Eva *et al.,* "Metabolic syndrome: definitions and controversies." BMC medicine vol. 9 48. 5 May. 2011

30) Rezaianzadeh, Abbas *et al.,* "National Cholesterol Education Program Adult Treatment Panel III Versus International Diabetic Federation Definition of Metabolic Syndrome, Which One is Associated with Diabetes Mellitus and Coronary Artery Disease?." International journal of preventive medicine vol. 3,8 (2012): 552-8.

31) Tsai, Sung-Sheng *et al.,* "A comparison of different definitions of metabolic syndrome for the risks of atherosclerosis and diabetes." Diabetology and metabolic syndrome vol. 10 56. 16 Jul. 2018

32) Ogbera, Anthonia O. "Prevalence and gender distribution of the metabolic syndrome." Diabetology and metabolic syndrome vol. 2 1. 12 Jan. 2010

33) Wilcox, Gisela. "Insulin and insulin resistance." The Clinical biochemist. Reviews vol. 26,2 (2005): 19-39.

34) Hema S. Bagry, Sreekrishna Raghavendran, Franco Carli; Metabolic Syndrome and Insulin Resistance: Perioperative Considerations. Anesthesiology 2008;108(3):506-523

35) de Luca, Carl, and Jerrold M Olefsky. "Inflammation and insulin resistance." FEBS letters vol. 582,1 (2008): 97-105.

36) Makki, Kassem *et al.,* "Adipose tissue in obesity-related inflammation and insulin resistance: cells, cytokines, and chemokines." ISRN inflammation vol. 2013 139239. 22 Dec. 2013

37) Lau, Louis H S, and Sunny H Wong. "Microbiota, Obesity and NAFLD." Advances in experimental medicine and biology vol. 1061 (2018): 111-125.

38) Nash, David T. "Relationship of C-reactive protein, metabolic syndrome and diabetes mellitus: potential role of statins." Journal of the National Medical Association vol. 97,12 (2005): 1600-7.

39) Maury, Eleonore *et al.,* "Circadian rhythms and metabolic syndrome: from experimental genetics to human disease." Circulation research vol. 106,3 (2010): 447-62.

40) Shpakov, Alexander O *et al.,* "Brain signaling systems in the Type 2 diabetes and metabolic syndrome: promising target to treat and prevent these diseases." Future science OA vol. 1,3 FSO25. 1 Nov. 2015

41) Pasquali, Renato *et al.,* "The hypothalamic-pituitary-adrenal axis activity in obesity and the metabolic syndrome." Annals of the New York Academy of Sciences vol. 1083 (2006): 111-28.

42) Abou Ziki, Maen D, and Arya Mani. "Metabolic syndrome: genetic insights into disease pathogenesis." Current opinion in lipidology vol. 27,2 (2016): 162-71.

43) Shuster, A *et al.,* "The clinical importance of visceral adiposity: a critical review of methods for visceral adipose tissue analysis." The British journal of radiology vol. 85,1009 (2012): 1-10.

44) Jeong, In-Kyung. "The role of cortisol in the pathogenesis of the metabolic syndrome." Diabetes and metabolism journal vol. 36,3 (2012): 207-10.

45) Ferraù, Francesco, and Márta Korbonits. "Metabolic Syndrome in Cushing's Syndrome Patients." Frontiers of hormone research vol. 49 (2018): 85-103.

46) Akalestou, Elina *et al.,* "Glucocorticoid Metabolism in Obesity and Following Weight Loss." Frontiers in endocrinology vol. 11 59. 20 Feb. 2020

47) Mendizábal, Yolanda *et al.,* "Hypertension in metabolic syndrome: vascular pathophysiology." International journal of hypertension vol. 2013 (2013): 230868.

48) Ljung, T *et al.,* "The activity of the hypothalamic-pituitary-adrenal axis and the sympathetic nervous system in relation to waist/hip circumference ratio in men." Obesity research vol. 8,7 (2000): 487-95

49) Hall, John E *et al.,* "Obesity-induced hypertension: interaction of neurohumoral and renal mechanisms." Circulation research vol. 116,6 (2015): 991-1006.

50) Alexander, Charles M *et al.,* "Metabolic syndrome and hyperglycemia: congruence and divergence." The American journal of cardiology vol. 98,7 (2006): 982-5.

51) Nieuwdorp, Max *et al.,* "Hypercoagulability in the metabolic syndrome." Current opinion in pharmacology vol. 5,2 (2005): 155-9.

52) Tziomalos, K *et al.,* "Endothelial dysfunction in metabolic syndrome: prevalence, pathogenesis and management." Nutrition, metabolism, and cardiovascular diseases : NMCD vol. 20,2 (2010): 140-6.

53) King, Christopher *et al.,* "Uric Acid as a Cause of the Metabolic Syndrome." Contributions to nephrology vol. 192 (2018): 88-102.

54) Verma, Sheetal, and S. P. Singh. "Current and future status of herbal medicines." Veterinary world 1.11 (2008): 347.

55) Rates, Stela Maria Kuze. "Plants as source of drugs." Toxicon 39.5 (2001): 603-613.

56) Singh, Mahesh Prasad, and Himadri Panda. "Medicinal herbs with their formulations". Daya Books, 2005.

57) Pagare, Saurabh, *et al.,* "Secondary metabolites of plants and their role: Overview." Current Trends in Biotechnology and Pharmacy 9.3 (2015): 293-304.

58) Ceylan-Isik, Asli F., *et al.,* "Herbal and traditional Chinese medicine for the treatment of cardiovascular complications in diabetes mellitus." Current Diabetes Reviews 4.4 (2008): 320-328.

59) Devalaraja, Samir, Shalini Jain, and Hariom Yadav. "Exotic fruits as therapeutic complements for diabetes, obesity and metabolic syndrome." Food Research International 44.7 (2011): 1856-1865.

60) Wink, Michael. "Modes of action of herbal medicines and plant secondary metabolites." Medicines 2.3 (2015): 251-286.

61) Stevanovic, Tatjana, Papa N. Diouf, and Martha Estrella Garcia-Perez. "Bioactive polyphenols from healthy diets and forest biomass." Current Nutrition and Food Science 5.4 (2009): 264-295.

62) Zhao, Yueliang, Y. Wu, and Mingfu Wang. "Bioactive Substances of Plant Origin 30." Handbook of food chemistry 967 (2015).

63) Joven, J., *et al.,* "Multifunctional targets of dietary polyphenols in disease: a case for the chemokine network and energy metabolism." Food and chemical toxicology 51 (2013): 267-279.

64) Masci, Valentina Laghezza, *et al.,* "Medicinal Plants as a Source of Alkaloids." Medically Important Plant Biomes: Source of Secondary Metabolites. Springer, Singapore, 2019. 85-113.

65) Roy, Arpita. "A review on the alkaloids an important therapeutic compound from plants." IJPB 3.2 (2017): 1-9.

66) Yang, Liuqing, and Joachim Stöckigt. "Trends for diverse production strategies of plant medicinal alkaloids." Natural product reports 27.10 (2010): 1469-1479.

67) Kawasaki, Tomomi, and Kazuhiro Higuchi. "Simple indole alkaloids and those with a nonrearranged monoterpenoid unit." Natural product reports 22.6 (2005): 761-793.

68) Singh, Thokchom P., and Okram Mukherjee Singh. "Recent progress in biological activities of indole and indole alkaloids." Mini reviews in medicinal chemistry 18.1 (2018): 9-25.

69) Scott, Alastair Ian. "Biosynthesis of the indole alkaloids." Accounts of Chemical Research 3.5 (1970): 151-157.

70) Jirschitzka, Jan, *et al.,* "Plant tropane alkaloid biosynthesis evolved independently in the Solanaceae and Erythroxylaceae." Proceedings of the National Academy of Sciences 109.26 (2012): 10304-10309.

71) Ashour, Mohamed, Michael Wink, and Jonathan Gershenzon. "Biochemistry of terpenoids: monoterpenes, sesquiterpenes and diterpenes." Annual Plant Reviews online (2018): 258-303.

72) Zwenger, Sam, and Chhandak Basu. "Plant terpenoids: applications and future potentials." Biotechnology and Molecular Biology Reviews 3.1 (2008): 1.

73) Ludwiczuk, A., K. Skalicka-WoŸniak, and M. I. Georgiev. "Terpenoids." Pharmacognosy. Academic Press, 2017. 233-266.

74) Abu-Lafi, Saleh, *et al.,* "Thymol and carvacrol production from leaves of wild Palestinian Majorana syriaca." Bioresource technology 99.9 (2008): 3914-3918.

75) Ludwiczuk, A., K. Skalicka-WoŸniak, and M. I. Georgiev. "Terpenoids." Pharmacognosy. Academic Press, 2017. 233-266.

76) Thakur, Mayank, *et al.,* "Chemistry and pharmacology of saponins: special focus on cytotoxic properties." Botanics: Targets and Therapy 1 (2011): 19-29

77) Elekofehinti, Olusola Olalekan. "Saponins: Anti-diabetic principles from medicinal plants–A review." Pathophysiology 22.2 (2015): 95-103.

78) Yin, Jun, Hanjie Zhang, and Jianping Ye. "Traditional chinese medicine in treatment of metabolic syndrome." Endocrine, Metabolic and Immune Disorders-Drug Targets (Formerly Current Drug Targets-Immune, Endocrine and Metabolic Disorders) 8.2 (2008): 99-111.

79) Rajaei, Peyman, and Neda Mohamadi. "Ethnobotanical study of medicinal plants of Hezar mountain allocated in south east of Iran." Iranian journal of pharmaceutical research: IJPR 11.4 (2012): 1153.

80) Sen, Tuhinadri, and Samir Kumar Samanta. "Medicinal plants, human health and biodiversity: a broad review." Biotechnological applications of biodiversity. Springer, Berlin, Heidelberg, 2014. 59-110.

81) Shea, Jeanne L. "Applying evidence-based medicine to traditional Chinese medicine: debate and strategy." Journal of Alternative and Complementary Medicine 12.3 (2006): 255-263.

82) Kanthe, Pallavi Shantkumar, *et al.,* "Protective effects of ethanolic extract of Emblica officinalis (amla) on cardiovascular pathophysiology of rats, fed with high fat diet." Journal of clinical and diagnostic research: JCDR 11.9 (2017): CC05.

83) Cornish, Caroline, and MARK NESBITT. "Historical perspectives on Western ethnobotanical collections." Curating biocultural collections: a handbook. Richmond: Kew Publishing (2014): 271-285.

84) Gouwakinnou, Gerard N *et al.,* "Local knowledge, pattern and diversity of use of Sclerocarya birrea." Journal of ethnobiology and ethnomedicine vol. 7 8. 1 Feb. 2011

85) J Afolayan, Anthony, and Taofik O Sunmonu. "In vivo Studies on Antidiabetic Plants Used in South African Herbal Medicine." Journal of clinical biochemistry and nutrition vol. 47,2 (2010): 98-106.

86) Oliver, Stefanie J. "The role of traditional medicine practice in primary health care within Aboriginal Australia: a review of the literature." Journal of ethnobiology and ethnomedicine vol. 9 46. 2 Jul. 2013

87) Cámara-Leret, Rodrigo *et al.,* "Ethnobotanical knowledge is vastly under-documented in northwestern South America." PloS one vol. 9,1 e85794. 9 Jan. 2014

88) VanPool, Christine. "Ancient medicinal plants of South America." Proceedings of the National Academy of Sciences of the United States of America vol. 116,23 (2019): 11087-11089.

89) Cruz, Lia Mendes, and Natalia Hanazaki. "The Ethnobotany of an Afro-Brazilian Community at Sertão do Valongo, Santa Catarina, Brazil." Functional Ecosystems and Communities 2 (2008): 59-65.

90) Bussmann, Rainer W., and Douglas Sharon. "Traditional medicinal plant use in Northern Peru: tracking two thousand years of healing culture." Journal of Ethnobiology and Ethnomedicine 2.1 (2006): 47.

91) Antal, Diana Simona. "Medicinal plants with antioxidant properties from Banat region (Romania): a rich pool for the discovery of multi-target phytochemicals active in free-radical related disorders." Analele Universitatii din Oradea-Fascicula Biologie 17 (2010): 1.

92) Zilio, Daniel C., *et al.,* "DB2 design advisor: integrated automatic physical database design." Proceedings of the Thirtieth international conference on Very large data bases-Volume 30. 2004.

93) Hosseini, Azar, *et al.,* "Effects of a standardized extract of Rheum turkestanicum Janischew root on diabetic changes in the kidney, liver and heart of streptozotocin-induced diabetic rats." Biomedicine and Pharmacotherapy 86 (2017): 605-611.

94) Shahat, Abdelaaty A., *et al.,* "Regulation of Obesity and Lipid Disorders by Foeniculum vulgare Extracts and Plantago ovata in High-fat Diet-induced Obese Rats." Am J of Food Tech 7 (2012): 622-632.

95) Gupta, Subash C *et al.,* "Therapeutic roles of curcumin: lessons learned from clinical trials." The AAPS journal vol. 15,1 (2013): 195-218.

96) Krotkiewski, Marcin. "Formulation for treating obesity and associated metabolic syndrome." U.S. Patent Application No. 10/585,546.

97) Basu, Arpita *et al.,* "Green tea supplementation affects body weight, lipids, and lipid peroxidation in obese subjects with metabolic syndrome." Journal of the American College of Nutrition vol. 29,1 (2010): 31-40.

98) Suliburska, Joanna *et al.*, "Effects of green tea supplementation on elements, total antioxidants, lipids, and glucose values in the serum of obese patients." Biological trace element research vol. 149,3 (2012): 315-22

99) El-Merahbi, Rabih *et al.*, "The adrenergic-induced ERK3 pathway drives lipolysis and suppresses energy dissipation." Genes and development vol. 34,7-8 (2020): 495-510.

100) Landis-Piwowar, Kristin *et al.*, "Inhibition of catechol-Omicron-methyltransferase activity in human breast cancer cells enhances the biological effect of the green tea polyphenol (-)-EGCG." Oncology reports vol. 24,2 (2010): 563-9.

101) Suzuki, Takuji *et al.*, "Beneficial Effects of Tea and the Green Tea Catechin Epigallocatechin-3-gallate on Obesity." Molecules (Basel, Switzerland) vol. 21,10 1305. 29 Sep. 2016

102) Naseri, Rozita *et al.*, "Anthocyanins in the Management of Metabolic Syndrome: A Pharmacological and Biopharmaceutical Review." Frontiers in pharmacology vol. 9 1310. 4 Dec. 2018

103) Basu, Arpita *et al.*, "Berries: emerging impact on cardiovascular health." Nutrition reviews vol. 68,3 (2010): 168-77.

104) Achari, Arunkumar E, and Sushil K Jain. "Adiponectin, a Therapeutic Target for Obesity, Diabetes, and Endothelial Dysfunction." International journal of molecular sciences vol. 18,6 1321. 21 Jun. 2017

105) Kim, Hyemee *et al.*, "Açaí (Euterpe oleracea Mart.) beverage consumption improves biomarkers for inflammation but not glucose- or lipid-metabolism in individuals with metabolic syndrome in a randomized, double-blinded, placebo-controlled clinical trial." Food and function vol. 9,6 (2018): 3097-3103.

106) Medjakovic, Svjetlana, and Alois Jungbauer. "Pomegranate: a fruit that ameliorates metabolic syndrome." Food and function vol. 4,1 (2013): 19-39.

107) Johnson, Rabia *et al.*, "Aspalathin from Rooibos (Aspalathus linearis): A Bioactive C-glucosyl Dihydrochalcone with Potential to Target the Metabolic Syndrome." Planta medica vol. 84,9-10 (2018): 568-583.

108) Jaramillo Flores, Maria Eugenia. "Cocoa Flavanols: Natural Agents with Attenuating Effects on Metabolic Syndrome Risk Factors." Nutrients vol. 11,4 751. 30 Mar. 2019

109) Huxtable, Ryan J., and Stephan KW Schwarz. "The isolation of morphine—first principles in science and ethics." Molecular interventions 1.4 (2001): 189.

110) Norn, S., P. R. Kruse, and E. Kruse. "History of opium poppy and morphine." Dansk medicinhistorisk arbog 33 (2005): 171-184.

111) Eloff, Jacobus Nicolaas. "Quantifying the bioactivity of plant extracts during screening and bioassay-guided fractionation." (2004).

112) Pieters, Luc, and Arnold J. Vlietinck. "Bioguided isolation of pharmacologically active plant components, still a valuable strategy for the finding of new lead compounds?" Journal of ethnopharmacology 100.1-2 (2005): 57-60.

113) Medema, Marnix H., and Michael A. Fischbach. "Computational approaches to natural product discovery." Nature chemical biology 11.9 (2015): 639.

114) Rinschen, Markus M., *et al.,* "Identification of bioactive metabolites using activity metabolomics." Nature Reviews Molecular Cell Biology 20.6 (2019): 353-367.

115) Yuliana, Nancy Dewi, *et al.,* "Comprehensive extraction method integrated with NMR metabolomics: a new bioactivity screening method for plants, adenosine A1 receptor binding compounds in Orthosiphon stamineus Benth." Analytical chemistry 83.17 (2011):

116) Farombi, E. Olatunde. "African indigenous plants with chemotherapeutic potentials and biotechnological approach to the production of bioactive prophylactic agents." African Journal of biotechnology 2.12 (2003): 662-671.

117) Gyllenhaal, C., *et al.,* "Ethnobotanical approach versus random approach in the search for new bioactive compounds: support of a hypothesis." Pharmaceutical biology 50.1 (2012): 30-41.

118) Sharma, Vivekanand, and Indra Neil Sarkar. "Bioinformatics opportunities for identification and study of medicinal plants." Briefings in bioinformatics 14.2 (2013): 238-250.

119) Lagunin, Alexey A., *et al.,* "Chemo-and bioinformatics resources for in silico drug discovery from medicinal plants beyond their traditional use: a critical review." Natural product reports 31.11 (2014): 1585-1611.

120) Leonti, Marco. "Traditional medicines and globalization: current and future perspectives in ethnopharmacology." Frontiers in pharmacology 4 (2013): 92.

121) Balick, Michael J., and Paul Alan Cox. Plants, people, and culture: the science of ethnobotany. Scientific American Library, 1996.

122) Retna, M. A., and P. Ethalsa. "A review of the taxonomy, ethnobotany, chemistry and pharmacology of Catharanthus roseus (Apocyanaceae)." International Journal of Engineering Research and Technology 2.10 (2013): 3899-3912.

123) Rout, S., *et al.,* "Characterization of Periwinkle-A Review." International Journal for Agriculture and Research 1.8 (2015).

124) Popoviæ, Zorica *et al.,* "Ethnobotany and herbal medicine in modern complementary and alternative medicine: An overview of publications in the field of I and C medicine 2001-2013." Journal of ethnopharmacology vol. 181 (2016): 182-92.

125) Kelly, Michael, *et al.,* "Evidence based public health: A review of the experience of the National Institute of Health and Clinical Excellence (NICE) of developing

public health guidance in England." Social science and medicine 71.6 (2010): 1056-1062.

126) Laird, Sarah A., and Rachel Wynberg. "Institutional policies for biodiversity research." Biodiversity and Traditional Knowledge, Equitable Partnerships in Practice, Londres, Earthscan Publications Ltd (2002): 39-76.

127) Jachak, Sanjay M., and Arvind Saklani. "Challenges and opportunities in drug discovery from plants." Current science (2007): 1251-1257.

128) Tan, G., C. Gyllenhaal, and D. D. Soejarto. "Biodiversity as a source of anticancer drugs." Current drug targets 7.3 (2006): 265-277.

129) Yang, Haixing, et al., "A novel systems pharmacology model for herbal medicine injection: a case using reduning injection." BMC complementary and alternative medicine 14.1 (2014): 430.

130) Cornier, M. A., Dabelea, D., Hernandez, T. L., Lindstrom, R. C., Steig, A. J., Stob, N. R., Van Pelt, R. E., Wang, H., & Eckel, R. H. (2008). The metabolic syndrome. Endocrine reviews, 29(7), 777–822. https://doi.org/10.1210/er.2008-0024

131) Shirwaikar, A., Rajendran, K., & Barik, R. (2007). Effect of Garuga pinnata. Roxb Aqueous Bark Extract on Tissue Antioxidant Activity in Streptozotocin-Nicotinamide–Induced Type II Diabetic Rats. Pharmaceutical Biology, 45(3), 205-209.

Chapter 33

Phytoremediation as a Green Technology for Mitigation of Fluoride Pollution: A Review

*Sunil Kumar Gochhi, Sudeshna Khatua and Surjendu Kumar Dey**

Post-Graduate Department of Environmental Science,
Fakir Mohan University, Vyasa Vihar, Balasore – 756 089, Odisha
**e-mail: surjendufmu@rediffmail.com*

ABSTRACT

Fluoride pollution in the environment has now emerged as a matter of serious concern throughout the world. Fluorides are not biodegradable and from contaminated water and soil, they enter the food chain causing many adverse effects on plants, animals and human health. Decontamination of sites polluted with fluorides is, therefore, highly essential. Using the conventional methods of wastewater treatments fluoride removal is possible, but the results are not that much encouraging and again for contaminated soils these methods cannot be applied. In this context the innovation of plant based technology for defluoridation has got much attention. Therefore, in this review a thorough discussion on the different aspects of fluoride pollution, conventional treatment methods along with the prospects and limitations of phytoremediation approach in removing fluoride from polluted sites have been made giving much emphasis on the remediation of contaminated water.

Keywords: *Aquatic macrophytes, Fluoride accumulator plants, Germinated seeds, Land plant species, Wastewater treatment methods.*

Introduction

Fluorine is the 13th most abundant element which is widely distributed throughout the earth's crust. It is a pale yellow gas at standard temperature and pressure. Fluorine is considered as the most reactive element of the halogen family. Fluoride is the ionic form of the element Fluorine. Fluoride compounds are formed when the element fluorine combines with other chemical elements. Fluoride does not occur in a free state in nature; it occurs only in combination with other elements. Most of the fluoride associated with monovalent cations such as NaF, KF which are water soluble. Fluoride is a common environmental pollutant. The fluoride rich soil is a potential source of its contamination in groundwater. From contaminated ground water and soil fluoride enters into the food chain and finally also into the human body. Since fluorides are not biodegradable and accumulate in the environment, they cause considerable damage to the environment (Marier and Rose, 1977). Fluorides are poorly detoxified by plants (Abdallah *et al.,* 2006; Gupta *et al.,* 2009) and animals and this is why they have many adverse effects on plants, animals and human health while moving through the food chains (Stevens *et al.,* 2000). Therefore, thorough studies on the sources, routes, and toxic effects of the fluoride along with different remediation strategies for fluoride decontamination are highly essential.

Occurrence and Sources of Fluoride

Fluoride occurrence in groundwater is a natural phenomenon, which is influenced by the local and regional hydro-geological condition of the region. Fluoride occurs abundantly in the earth's crust as a component of rock and minerals. Naturally, fluoride is present in the form of fluorite, fluorspar or calcium fluoride (CaF_2), apatite or rock phosphate [$Ca_3F(PO)_3$], cyolite (Na_3AlF_6), magnesium fluoride (MgF_2) (Majhi, 2016) and released into the groundwater by slow dissolution of such rocks and minerals (Goswami and Purkait, 2011). Chemically, fluoride and OH ions are negatively charged and also have almost the same ionic sizes. Hence, during the chemical reaction, fluoride ion can easily replace OH ions present in rocks and thereby enriched its concentration in rocks and minerals. Whenever carbonate and bicarbonate rich water passed through such types of rocks, fluoride ion is replaced due to the chemical reaction (Reactions 1 and 2) and percolates to the groundwater and increases its concentration (Saxsena and Ahmed, 2001).

$$CaF_2 + Na_2CO_3 \rightarrow CaCO_3 + 2F^- + 2Na^+ \quad (1)$$

$$CaF_2 + 2NaHCO_3 \rightarrow CaCO_3 + 2Na^+ + 2F^- + H_2O + CO_2 \quad (2)$$

The dissolution of fluoride in groundwater occurs through the rainwater and repeated irrigation of agricultural lands. As rainwater percolates through the soil, it comes in contact with rock and minerals in the aquifer. The fluoride content in groundwater becomes higher in summer due to the drop in water level. The higher concentration of fluoride in summer season due to the presence of air in the minerals/rocks cavities, which indicates that oxygen in the cavities of the mineral, catalyzes the fluoride dissolution process.

Industries are also a very important source of groundwater contamination. The effluents and other by-products of industries are the sources of groundwater pollution. Industries that release significant quantities of fluoride in process of waste generation include fluorosilicone, gasoline production and manufactures of coke, ceramic, cement, enamel, fiberglass, electronics, pesticides, fertilizers and wood preservatives (Narsimha and Sudarshan, 2018).

World Scenario

The excess amount of fluoride contamination in drinking water is often reported in many countries all over the world. Around 25 developed and developing countries are suffering from fluorosis. These include India, China, Srilanka, Australia, New Zealand, Japan, Canada, Iran, Iraq, Syria, Kenya, Egypt, Tanzania, South Africa, Persian Gulf, Turkey, Argentina, Morocco, Algeria, and Libya. One of the most badly affected nations is China, where groundwater, as well as air, is polluted with fluoride (Chinoy and Narayana, 1994).

Indian Scenario

In 1937, for the first time fluoride was detected in the Nellore district of Andhra Pradesh. Since then many research works have been done in different states of India to identify the fluoride contaminated water sources and their impacts on living organisms. Now it has been reported that high level of fluoride in water as well as in soil is dominant in around 230 districts of 22 States and in one Union Territory (UT) of India (Mukherjee and Singh, 2018; Singh *et al.*, 2018; Singh *et al.*, 2020). The states with a high rate of fluoride prevalence in India are Rajasthan, Telengana, Karnataka, Assam, Manipur, Andhra Pradesh, Bihar, Chhattisgarh, Gujarat, Haryana, Jharkhand, Jammu and Kashmir (at present reorganized into two Union Territories, *i.e.*, Jammu and Kashmir; and Ladakh), Kerala, Madhya Pradesh, Maharashtra, Odisha, Punjab, Tamil Nadu, West Bengal, Uttar Pradesh, Uttarakhand and Himachal Pradesh; along with the UT of Delhi.

Odisha Scenario

In Odisha, drinking water in 20 districts is heavily contaminated with fluoride. These districts are; Puri, Khurda, Nuapada, Bargarh, Kalahandi, Jajpur, Bolangir, Dhenkanal, Deogarh, Jharsuguda, Kandhamal, Kendrapara, Angul, Balasore, Cuttack, Ganjam, Jagatsinghpur, Nabarangpur, Nayagarh, and Subarnapur (Das *et al.*, 2000). In 2018, potable water sources in the 54 schools and anganwadi premises of Nuapada district were found contaminated with high level of fluoride. Even the fluoride content up to 5.25 mg/L in water from a tube well in the school was found in a laboratory tests certified by the National Rural Drinking Water Programme (Barik, 2018), while consumption of water having fluoride content more than 1.5 mg/L is considered dangerous for health.

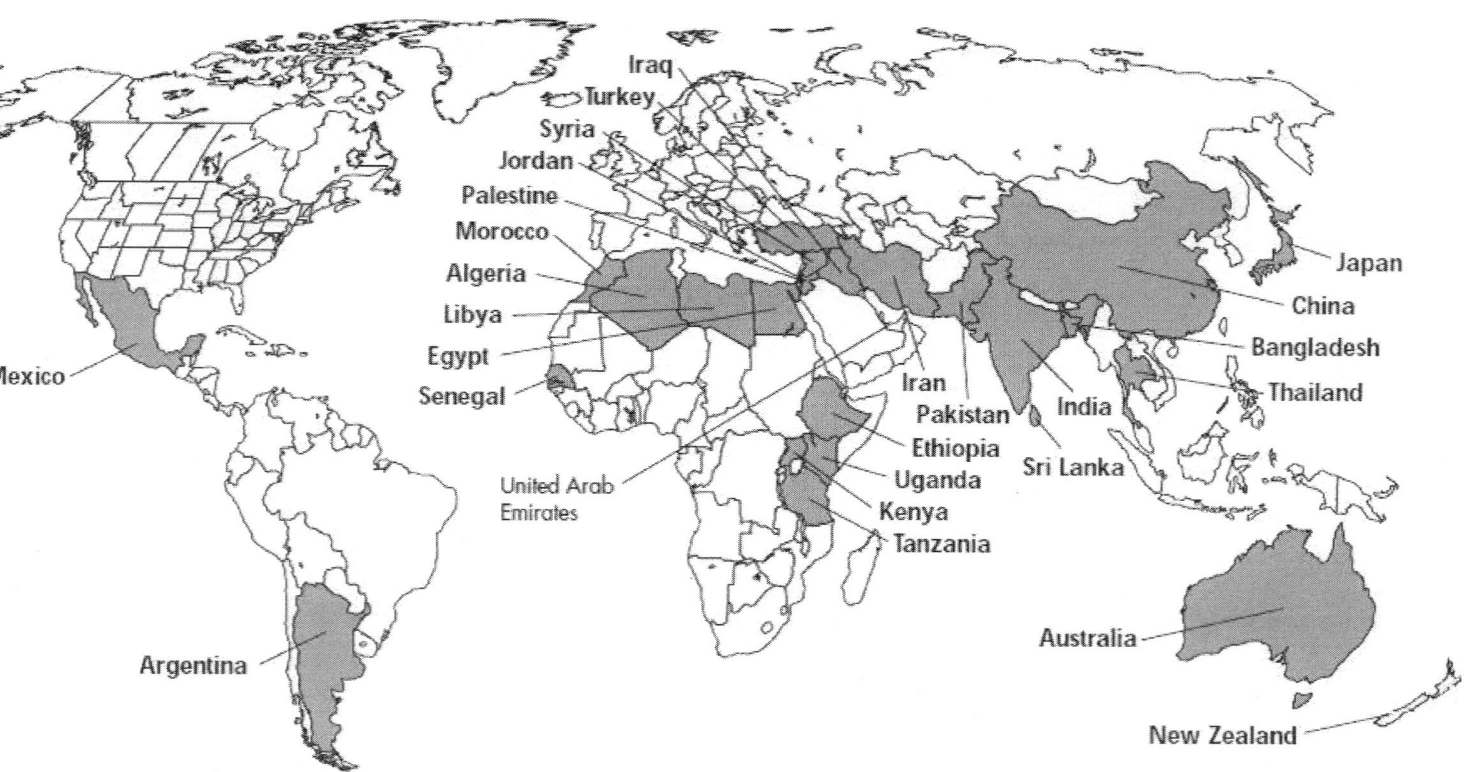

Figure 33.1: Countries with Endemic Fluorosis Due to Excess Fluoride in Drinking Water (Burken and Schnoor, 1996).

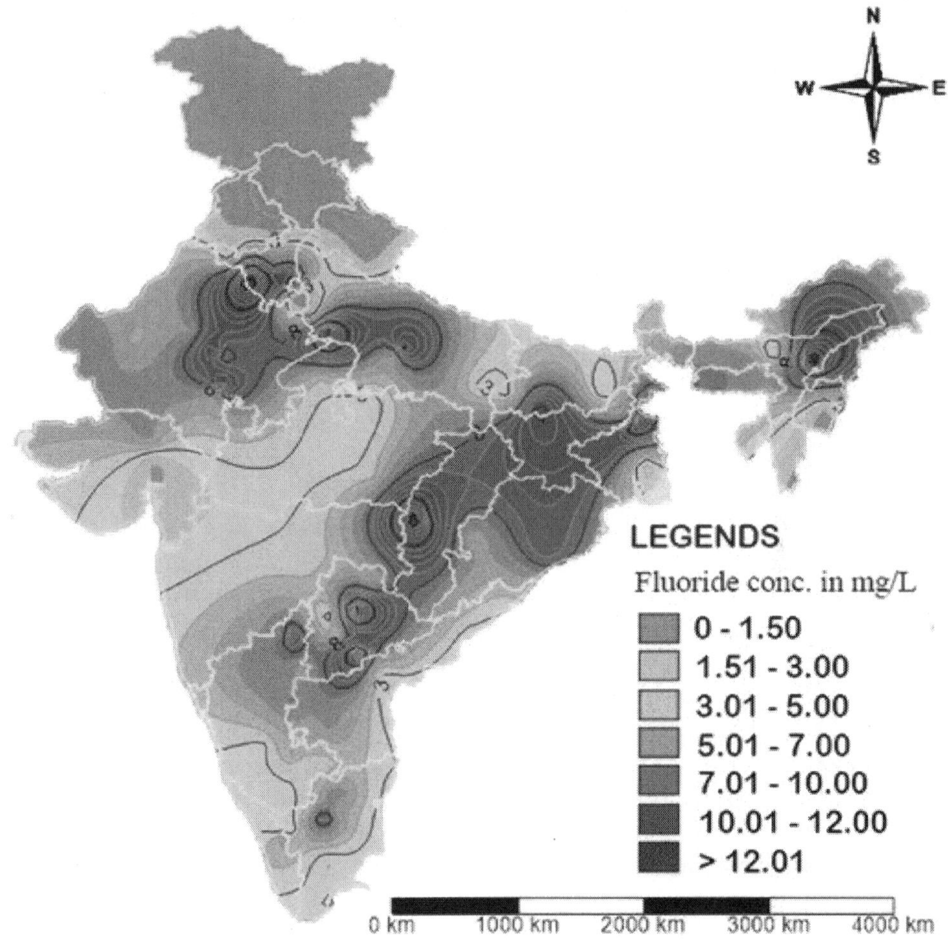

Figure 33.2: Spatial Distribution of Fluoride in India (Mukherjee and Singh, 2018).

Health Effects and Permissible Limits of Fluoride

The dreadful fate of fluorosis is due to the long term intake of excessive fluoride. So far two main kinds of fluorosis, namely dental fluorosis and skeletal fluorosis have been identified. Fluorosis occurs due to the presence of fluoride in the concentration >1.5 mg/L in drinking water. Prolonged consumption of water contaminated with fluoride cause dental and skeletal fluorosis. The disease is manifested with common symptoms like mottled teeth and crippled backbone and limbs. Dental fluorosis is a qualitative defect of enamel. This results from an increase in fluoride concentration within the micro-environment of ameloblasts during the process of enamel formation (Aoba and Fejerskov, 2002). Enamel mineralization is very sensitive to free fluoride ions. When the fluoride ions are incorporated into enamel crystals, the solubility of the mineral is reduced which affects the subsequent mineralization process (Aoba and Fejerskov, 2002). In dental fluorosis teeth mottling, which is characterized initially by opaque white patches on the teeth followed by pitting of teeth surface in advanced stages, occurs. In general, enamel fluorosis is a fluoride-related alteration in enamel that occurs during enamel development and become more severe with increase in fluoride intake and time of exposure (DenBesten and Li, 2011). Dental fluorosis is mostly found in children up to the age of 12 years. Skeletal fluorosis is a bone disease which is caused by consumption of fluoride more than 3 mg/L (Teotia *et al.*, 1998). Skeletal fluorosis results in unbearable pain as well as severe damage to bones and joints (Ares, 1978). The severity of fluorosis depends on the concentration of fluoride in drinking water, daily intake, duration of exposure and climate conditions. As fluoride has dual significance on human health, World Health Organization recommends that water containing a minimum of 0.5 mg/L fluoride and a maximum of 1.5 mg/L fluoride is considered safe for drinking purposes (WHO, 1993).

Besides fluorosis, fluoride has several other mechanisms of toxicity (Li, 2003; Kanbur *et al.*, 2009; Tsai, 2009). The symptoms of fluoride toxicity have been elucidated by very well described mechanisms (Whitford, 1989). Initially it acts locally on the intestinal mucosa, and then it forms hydrofluoric acid in stomach leading to gastrointestinal irritation. The symptoms like nausea, vomiting, diarrhoea and abdominal pain are very common under such situation (Gessner *et al.*, 1994). Once absorbed, fluoride binds with calcium ions leading to hypocalcemia. Fluoride directly interferes with a number of enzyme systems and thereby disrupts oxidative phosphorylation, glycolysis, and neurotransmission (Reddy *et al.*, 2014). The fluoride toxicity also causes the decrease in the levels of CD4, NK cells and IgG1 coupled with significant increase in the lipid peroxidation level and impairment of antioxidative defense system (Reddy *et al.*, 2014).

Common Fluoride Treatment Methods

The most common methods used to remove the toxic metals from wastewater include electro dialysis, reverse osmosis, ion exchange, chemical precipitation and adsorption (Fernandes and Lona, 2005). Except reverse osmosis, all other methods are also used conventionally for fluoride treatment. The methods are outlined in Figure 33.3. The selection of method for the removal of toxic metals depends on factors such as capital and operational cost, effect on the environment and effectiveness of removal of toxic metals.

Electro Dialysis

In this method ion exchange membranes are used along with an electrical potential difference for removing ionic species from aqueous solutions (Altin *et al.*, 2017). It consists of a series of cation and anion exchange membranes which are arranged between the cathode and the anode. The cations move to the cathode while anions to the anode. The limitation of this method is that it is costly and has low efficiency in removing the trace concentration from the aqueous solution.

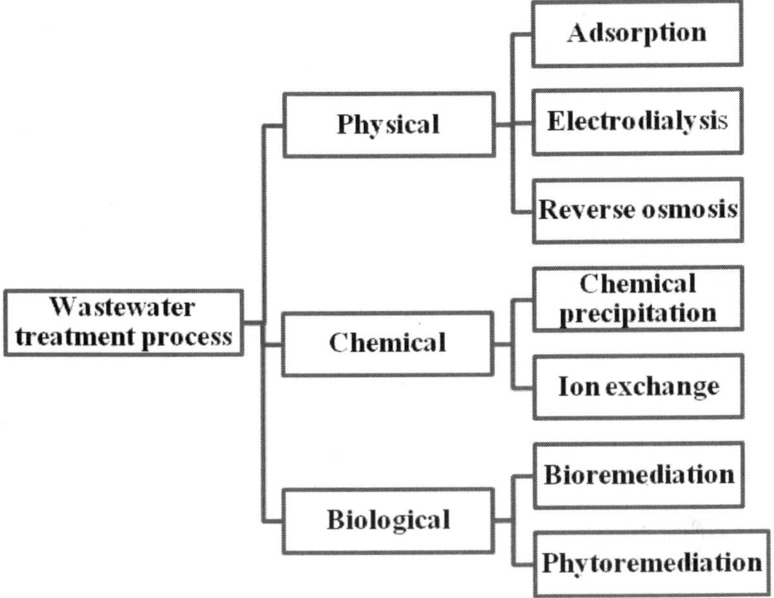

Figure 33.3: Flow Chart of the Wastewater Treatment Process.

Reverse Osmosis

Reverse osmosis (RO) is a water purification technology that uses a partially permeable membrane to remove ions, unwanted molecules and larger particles from the water with the application of osmotic pressure. The process is reverse of natural osmosis as a result of the applied pressure to the concentrated side of the membrane. RO membrane rejects ions based on size and electrical charge and gives extremely high purity of water (Schneiter and Middlebrookes, 1983). This process is very costly and requires regular maintenance. Only fluoride removal from water through this technology is not possible and hence remediation of fluoride contaminated water is not advisable using RO.

Ion Exchange

Ion exchange is a process in which the solid material takes up charged ions from a solution and release an equivalent amount of other ions into the solution (Sahu *et al.,* 2009). The ability to exchange ions is dependent on the properties of the structure of the materials. The exchanger consists of a matrix, with positive or negative excess charge. The removal of anionic or cationic constituents present in water takes place by exchange with ions of the resin. When the resin bed becomes saturated, they are regenerated using acid or alkali. Ion-exchange resins are available selectively for certain ions. However, in the presence of large quantities of competing mono and divalent ions such as sodium and calcium, the efficiency of ion- exchange process decreases.

Chemical Precipitation

The chemical precipitation method is the most common method of removal of metals from water. This is a two step process in which precipitation occurs by adding lime in the first step. Then it is followed by a second step where alum is added to cause coagulation. When alum is added to water, essentially two reactions occur. At first, alum reacts with some of the alkalinity to produce insoluble aluminum hydroxide [$Al(OH)_3$] and then alum reacts with metals present in water. The best fluoride removal is accomplished at a pH range of 5.5-7.5 (Ravindra and Garg, 2007).

Adsorption

Adsorption is a process in which a single or a group of ions/compounds get accumulated on the surface of another solid or liquid. The substance on which the adsorption takes place is known as adsorbent and the substance, which gets adsorbed, is called adsorbate. To reduce the cost of adsorption process people have worked on finding the low-cost adsorbents, as has been done by Sharma and Forster (1994) for Cr (VI). Based on the extent of attraction between the adsorbent and adsorbate, the adsorption process is classified into two types: physical adsorption; and chemisorptions. Adsorption, which is based on the principle of the Vander Waal's force of interaction, is known as physical adsorption or Vander Wall's adsorption. The physical adsorption process is reversible and establishes rapidly. In chemisorption, the chemical interaction/electrostatic force of attraction occur between the adsorbent surface and adsorbate molecules (Kumar *et al.,* 2011).

For fluoride removal from water, adsorption has been proved to be the superior among all the techniques mentioned above. This is because of its low initial cost, flexibility, simplicity of design and ease of operation and maintenance in comparison to others. For the purpose of fluoride removal, a large number of adsorbents including activated charcoal, alumina, calcite, flyash, amberlite resin and layered double hydroxides have been studied (Fan *et al.,* 2003; Abe *et al.,* 2004; Ghorai and Pant, 2005; Meenakshi *et al.,* 2008; Shao-Xiang *et al.,* 2009). Among all these, the activated alumina has been most widely accepted for fluoride removal due to its cost and efficiency. The development and utilization of polymeric composite materials for defluoridation of water is another trend of research in adsorption technology (Islam *et al.,* 2011).

Disadvantages of the Conventional Method of Water Treatment

Most of the conventional methods are affected by high capital and operational cost, high cost of reagent, limited tolerance to pH change and incomplete metal removal. These require technically skilled manpower for operation and also require a large area of land. The treatment efficiency is low and energy requirement is high.

Treatment of Contaminated Sites using Phytoremediation Techniques

Phytoremediation is a technique that utilizes plants and associated rhizosphere micro-organisms to remove or to transform toxic chemicals located in water, soils, sediments, and even the atmosphere. Phytoremediation is now used for remediation of sites contaminated with heavy metals, petroleum hydrocarbons, chlorinated solvents, pesticides, explosives and radio-nuclides and also the landfill leachates. It is reported that approximately 80 per cent of polluted groundwater is within the initial 20 meters of ground level. This suggests that water pollution removal can be carried out using low-cost phytoremediation approaches (Al-Musharafi *et al.*, 2013). The different phytoremediation processes include physical and chemical properties modification, releasing root exudates, improving aeration by releasing oxygen to root zone, intercepting and retarding the movement of chemicals, effecting co-metabolic microbial and plant enzymes; and decreasing vertical and lateral migration of pollutants (Chang *et al.*, 2002). When contaminants like heavy metals are at low concentration then phytoremediation is the most suitable pollution removal technique in terms of cost and effectiveness. This is why the phytoremediation is a long term and commercially viable solution (Wilson *et al.*, 1996). There are many mechanisms by which the plants can remediate the contamination from water and soil. During phytoremediation, the efficiency of the process is affected by physical and chemical properties of the contaminants. These include water solubility, vapor pressure, and molecular weight. The environmental parameters like temperature, pH and organic matter in hydroponics have also important roles. There are many mechanisms by which the plants can facilitate remediation. These are phytoextraction, rhizofiltration, phytostabilization, and phytovolatization.

Phytoextraction

Plants have the ability to phytoextract essential metals like Cu, Mg, Mo, K, Fe, Mn, Ni, P, and Zn from the soil or from the aqueous medium required for growth. At the same time plants also absorb some non-essential metals like Se, B, Cd, Co, Cr, Ag, Pb, and Hg. The non-essential metals are considered to be toxic to plants if present even in very low concentration and essential metals have also become toxic if present in more than the required quantity. In phytoextraction, plants uptake metal contaminants in the soil through roots and translocate the same to other parts. This is also called as phytoaccumulation. A hyperaccumulator is a plant species that is capable of accumulating metals 100 times more than a common non-accumulating plant. The main disadvantage of phytoextraction is that it is significant to only those sites which have low to medium amount of metal contamination since the highly polluted sites are proved to be toxic for the plant growth and development (Padmavathiamma and Li, 2007). There are many plants which have been reported to have the potential to extract different metals. For example *Pteris vittata* and *Chenopodium album* have been reported for the phytoextraction of As and Pb respectively (Silva Gonzaga *et al.,* 2006). *Impatiens glandulifera* and *Noccaea caerulescens* have been reported as Cd hyperaccumulator (Coakley *et al.,* 2019; Robinson *et al.,* 1998). Similarly many plant species are known today to have the potential as accumulator or hyperaccumulator of different metals like Cr, Hg, Al, Cu, Mn, Fe *etc.* To increase the bioavailability of metals, some chelating ligands like ethylene diamine-N,N'-disuccinic acid (EDDS), ethylene diamine tetraacetic acid (EDTA), succinic acid, citric acid, *etc.* are also added to contaminated sites in order to increase the accumulation efficiency of the plants (Padmavathiamma and Li, 2011).

Rhizofiltration

Rhizofiltration mechanism is applied to remove heavy metals as well as radioactive metals like Cd, Cu, Ni, Pb, Cr, As, U and Sr from aqueous solutions using plant root systems. Rhizofiltration is similar to phytoextraction, but here the plants are used to remediate the contaminated water rather than soil. Suitability of the plants for rhizofiltration depends upon the root system as roots filter metals from aqueous solution. Generally the plants having longer and hairy root systems with considerable surface area are identified for use in this technology. Indian mustard (*Brassica juncea*) and sunflower (*Helianthus annuus*) are found to be suitable for rhizofiltration of many metals. *Brassica* can effectively remediate Pb, Cd, Cu, Cr, Ni, Zn, and *Helianthus annuus* rhizofilters Ra and U (Vera Tome *et al.,* 2009).

Phytostabilization

Phytostabilization refers to the process in which plants immobilize metals in their source and transform metallic toxic state to a less toxic state. It is also otherwise known as in-place inactivation. This process decreases the mobility of the contaminant and prevents their migration into the groundwater and also reduces the bioavailability of the metal in the food chain (Eapen and D'Souza, 2005). The plants whose roots can develop into contamination zone and help in immobilization of metal in soils either by root adsorption or by metal precipitation or complexation or reduction are suitable for phytostabilization process. Through phytostabilization process the highly toxic Cr (VI) gets transformed into Cr (III), which is a less soluble and immobile and less toxic form. Phytostabilization is found to be more effective in the case of fine soils and the soil with high organic matter content (Raskin *et al.,* 1997).

Phytovolatization

In phytovolatization technique plants use to take up contaminants from the soil, transform them into volatile forms and then transpire them into the atmosphere. Before reaching the leaves the contaminants get diffused through the stems or other parts of the plant. Some organic pollutants and heavy metals such as arsenic, mercury, and selenium get volatilized by plants. Tritium (the isotope of hydrogen) is stabilized as helium through phytovolatization (Dushenkov, 2003). Selenium found in the soil, is volatized as $(CH_3)_2Se$. This form of Se is 600 times less toxic than elemental Se (De Souza, 2002). This technology has its both advantage and disadvantage. For example through phytovolatization the contaminant like mercuric ion may be transformed into a less toxic substance. But again the mercury released into the atmosphere is likely to be recycled by precipitation and then redeposited back into lakes and oceans, repeating the production of methyl mercury by anaerobic bacteria, which is in fact the disadvantage of the process.

Phytodegradation

Phytodegradation is also otherwise called as phytotransformation. In this case, the complex organic molecules get degraded to simple molecules and then transport into the plant tissue (Zhuang *et al.,* 2005). In phytodegradation process, the contaminants are broken down to simple molecules after they have been taken up by the plant. This method is very effective in remediating some organic contaminants, such as chlorinated solvents and herbicides present in the different components of the environment like soil, sediment or groundwater.

Rhizodegradation

Rhizodegradation refers to the breakdown of contaminants within the plant root zone *i.e.*, the rhizosphere by the help of the bacteria or other microorganisms. It is also known as phytostimulation. Plants are known to exude different metabolites like sugars, amino acids, enzymes and other compounds through their roots to the soil that can stimulate the growth of bacteria and other microorganisms (Burken and Schnoor, 1996). So there is a widespread growth of microorganisms in the rhizosphere zone. The roots also provide additional surface area for microbes to grow on and a pathway for oxygen transfer from the environment. The rhizosphere environment favours the microorganisms to degrade many soil contaminants. Mostly the organic chemicals like petroleum hydrocarbons, polycyclic aromatic hydrocarbons (PAHs), chlorinated solvents, pesticides, polychlorinated biphenyls (PCBs), benzene, toluene, ethylbenzene, and xylenes are degraded through rhizodegradation process (Zhuang *et al.,* 2005).

Prospects and Limitations of Phytoremediation Techniques

Each and every technology has its prospects and limitations; and so is the case of phytoremediation technique. The most important prospect of this technique is that it is environment friendly, cost effective and aesthetically pleasing. The entry of the contaminants into the environment can be reduced by preventing their leakage into the ground water. The metals so absorbed by the plants can also be extracted from the harvested biomass. This is now an expanding technology in which many higher plants are used to clean up the contaminated environments. It has been estimated that the conventional cost of remediation may cost around 10 to 1000 US $ per cubic meter whereas the cost of phytoextraction is as low as 0.05 US $ per cubic meter (Cunningham *et al.,* 1997; McCutcheon and Schnoor, 2003). A wide range of both organic and inorganic pollutants present in different components of environment are cleaned up through phytoremediation technique. Variety of pollutants including metals (Ghosh and Singh, 2005; Pilon-Smits and Pilon,

2002), antibiotics (Gujarathi *et al.*, 2005; Boonsaner and Hawker, 2010), pesticides (Henderson *et al.*, 2006; Hussain *et al.*, 2009), and aromatic compounds (Rentz *et al.*, 2003; Rivera-Cruz *et al.*, 2004) can be removed from the contaminated sites by different plant species. Since the naturally occurring organisms, *i.e.*, the plants are used in this technique to clean up the contaminated sites the environment is preserved in a more natural state.

This technique has also many limitations. The most important among them is that it is a time consuming process and some time it may take several growing seasons to clean up a site, especially the polluted land. Phytoremediation is also limited by the growth rate of the plants which is again dependent on the soil type and other climatic conditions. Many times it has been found that the intermediates formed from the organic and inorganic contaminants are cytotoxic to the plants (Mwegoha, 2008). The conventional methods like excavation, disposal or incineration usually takes some weeks or months to complete the process of remediation whereas phytoremediation takes more time and in some case several years for the same purpose. Therefore, for the sites with acute risks for human and other ecological receptors phytoremediation may not be a recommended approach (US EPA, 2000; Mwegoha, 2008). Since this is a plant-based system of remediation, complete prevention of leaching of contaminants into the ground water is not possible. Phytoremediation is, thus, best suited for the remote areas with limited human contact and where the soil contamination does not require immediate attention (Salido *et al.*, 2003).

Phytoremediation of Fluoride Contaminated Water

The potentials of different plant species in remediating the sites contaminated with diverse types of pollutants including heavy metals have already been established. In case of polluted water, the treatment approaches with conventional techniques, as outlined above, do not give satisfactory results. These methods are time consuming, costly and environmentally destructive. On the other hand, the waste water treatment with aquatic macrophytes has shown encouraging results in many studies and has emerged as an effective phytoremediation approach (Ali *et al.*, 2020 and references therein). Plants have also been reported to have potential in removing fluoride from contaminated water and accumulating ionic fluoride in their tissues. Since the last two decades, people from the different parts of the world are working on this aspect and many encouraging results are also coming. Through a hydroponic culture Santos-Diaz *et al.* (2010) have screened 17 species for their fluoride removal potential and have found some fluoride tolerant and highly tolerant species. They have found that out of 17 species only three species, *viz.*, *Camellia japonica*, *Pittosporum tobira* and *Saccharum officinarum* were able to remove fluoride from water with some efficiency. All these three species could tolerate a maximum concentration of 5 mg F/L and *Saccharum officinarum* was found to remove maximum F (*i.e.*, 40 per cent) from the water, in comparison to other species. Accordingly they have recommended this species as suitable one for phytoremediation of F from contaminated water.

The potential of tree species in removing F from water and soil has also been reported. In their work in hydroponic culture system with eight tree species Baunthiyal and Sharma (2012) have found that three plants, *viz.*, *Acacia tortilis*, *Prosopis juliflora* and *Cassia fistula* accumulated significant amount of F and deposited it in different organs and their sub-cellular fractions. The maximum deposition was found in roots and *P. juliflora* was found most efficient in F removal among all the species they have studied. The aquatic macrophytes like *Hydrilla verticillata*, *Chara carollina* and *Cladophora glomerata* have promising prospect with regard to their use in phytoremediation programme for removal of F from contaminated water and *Chara carollina* showed better removal efficiency among these three (Baunthiyal and Sharma, 2017). Similarly the free floating aquatic macrophytes like *Pistia stratiotes*, *Eichhornia crassipes*, and *Spirodela polyrhiza* have also been studied for their efficiency in F removal by Karmakar *et al.* (2016). They found that the plants could tolerate the F concentration up to 3 mg/L and *Pistia stratiotes* could remove maximum F (*i.e.*, 19.87 per cent) from the water among the three. During several other course of investigations, many aquatic as well as land plants have been examined and some have been found to have their efficiency to accumulate F from contaminated water (Baunthiyal and Ranghar, 2015). The germinating seeds have also shown good results in removing F from water. Four types of germinating seeds, *viz.*, *Eleusine coracana*, *Pennisetum glaucum*, *Setaria italica* and *Sorghum bicolor* were exposed to industrial effluent and artificially made polluted water and were found to have efficiency in removing F (Gandhi, *et al.*, 2016). Among these seeds, good results were obtained with *Pennisetum glaucum*.

Conclusion

Presence of fluoride in drinking water and cultivable soils in high concentrations is a matter of serious concern throughout the world. Since it has many adverse effects on human health, remediation of sites contaminated with fluoride is highly essential. Although many conventional methods like Nalgonda technique (in which alum, lime and bleaching powder are used to precipitate fluoride from water), reverse osmosis, ion exchange, electrolysis *etc.* are available to remove fluoride from water, these methods are not suitable for remediating contaminated soil. With regard to their applicability and efficiency, these methods have their limitations. Under these circumstances phytoremediation has emerged as a green technology to mitigate fluoride from the polluted sites. During the last two decades, a large number of people have worked on it and many plant species, both aquatic and terrestrial, have been identified as fluoride accumulator and hyperaccumulator. Since this is a nature-based technology, it has no damaging effect on the environment. The growth of the plants, which are used in phytoremediation programme, is regulated by the soil type, physico-chemical properties of water and other agro-climatic conditions of the region. Therefore, for wider applicability of the technique to mitigate fluoride, a thorough screening of variety of plant species of different regions for their potential to accumulate fluoride is highly essential. At the same time, methods should have to be formulated to facilitate the better growth of the plants so that their efficiency in mitigating the fluoride from contaminated sites can be enhanced. Although defluoridation through phytoremediation is a new trend of research; it has opened up many windows with vast future opportunities.

REFERENCES

Abdallah, F.B., Elloumi, N., Mezghani, I., Garrec, J.P. and Boukhris, M. (2006). Industrial fluoride pollution of Jerbi grape leaves and the distribution of F, Ca, Mg and P in them. Fluoride 39 (1): 43-48.

Abe, I., Iwasaki, S., Tokimoto, T., Kawasaki, N., Nakamura, T. and Tanada, S. (2004). Adsorption of fluoride ions onto carbonaceous materials. J. Colloid. Interface Sci. 275: 35-39.

Ali, S., Abbas, Z., Rizwan, M., Zaheer, *I.E.*, Yavas, I., Ünay, A., Abdel-DAIM, M.M., Bin-Jumah, M., Hasanuzzaman, M. and Kalderis, D. (2020). Application of floating aquatic plants in phytoremediation of heavy metals polluted water: A review. Sustainability 12: 1927-1959.

Al-Musharafi, S., Mahmoud, I. and Al-Bahry, S. (2013). Heavy metal pollution from treated sewage effluent. APCBEE Procedia 5: 344-348.

Altin, S., Oztekin, E. and Altin, A. (2017). Comparison of electrodialysis and reverse electrodialysis process in the removal of Cu (III) from dilute solution. Kor. J. Chem. Engin. 34 (8): 2218-2224.

Aoba, T. and Fejerskov, O. (2002). Dental fluorosis: Chemistry and biology. Crit. Rev. Oral Biol. Med. 13 (2): 155-170.

Ares, J. (1978). Fluoride cycling near a coastal emission source. J. Air Pol. Contr. Assoc. 28 (4): 344-349.

Barik, S. (2018). Many Odisha schoolchildren hit by fluoride contamination. The Hindu, Bhubaneswar Edn. September 18, 2018; Accessed on 28th March, 2020.

Baunthiyal, M. and Sharma, V. (2012). Phytoremediation of fluoride contaminated water and soil: A search for fluoride hyperaccumulators. J. Agri. Technol. 8 (6): 1965-1978.

Baunthiyal, M. and Ranghar, S. (2015). Accumulation of fluoride by plants: Potential for phytoremediation. Clean Soil Air Water 43 (1): 127-132.

Baunthiyal, M. and Sharma, V. (2017). Pytoremediation potential of selected hydrophytes for fluoride in aquatic environment. World Rev. Sci. Technol. Sustainable Dev. 13 (2): 133-144.

Boonsaner, M. and Hawker, D.W. (2010). Accumulation of oxytetracycline and norfloxacin from saline soil by soybeans. Sci. Total Environ. 408: 1731-1737.

Burken, J. and Schnoor, J. (1996). Phytoremediation: Plant uptake of atrazine and role of root exudates. J. Environ. Engin. 122 (11): 958-963.

Chang, Y., Bae, B., Lee, I. and Corapcioglu, M. (2002). Modelling plant-aided in-situ bioremediation of soils contaminated with a volatile organic compound (VOC). Acta. Biotechnol. 22 (1 and 2): 21-34.

Chinoy, N. and Narayana, M. (1994). In vitro fluoride toxicity in human spermatozoa. Reproductive Toxicol. 8 (2): 155-159.

Coakley, S., Cahill, G., Enright, A., O'Rourke, B. and Petti, C. (2019). Cadmium Hyperaccumulation and Translocation in *Impatiens glandulifera*: From foe to friend? Sustainability 11: 5018-5034.

Cunningham, S.D., Shann, J.R., Crowley, D. and Anderson, T.A. (1997). Phytoremediation of soil and water contaminants. American Chemical Society, Washington DC.

Das, S., Mehta, B.C., Samanta, S.K., Das, P.K. and Srivastava, S.K. (2000). Fluoride hazards in ground water of Orissa, India. Indian J. Environ. Health 1 (1): 40-46.

DenBesten, P. and Li, W. (2011). Chronic fluoride toxicity: Dental fluorosis. Monogr. Oral Sci. 22: 81-96.

De Souza, M. (2002). Selenium assimilation and volatilization from selenocyanate treated Indian mustard and Muskgrass. Plant Physiol. 128 (2): 625-633.

Dushenkov, S. (2003). Trends in phytoremediation of radionuclides. Plant and Soil 249 (1): 167-175.

Eapen, S. and D'Souza, S. (2005). Prospects of genetic engineering of plants for phytoremediation of toxic metals. Biotec. Adv. 23 (2): 97-114.

Fan, X., Parker, D.J. and Smith, M.D. (2003). Adsorption kinetics of fluoride on low cost materials, Water Res. 37: 4929-4937.

Fernandes, F. and Lona, L. (2005). Neural network application in polymerization processes. Braz. J. Chem. Engin. 22 (3): 401-418.

Gandhi, N., Sirisha, D. and Asthana, S. (2016). Phytoremediation of fluoride (F-) from water using germinated seeds. Int. J. Adv. Res. Eng. Appl. Sci. 5 (7): 16-28.

Gessner, B.D., Beller, M., Middaugh, J.P. and Whitford, G.M. (1994). Acute fluoride poisoning from a public water system. New England J. Med. 330: 95-99.

Ghorai, S. and Pant, K.K. (2005). Equilibrium, kinetics and breakthrough studies for adsorption of fluoride on activated alumina. Sep. Purif. Technol. 42: 265-271.

Ghosh, M. and Singh, S.P. (2005). A review on phytoremediation of heavy metals and utilization of its byproducts. Appl. Ecol. Environ. Res. 3: 1-18. Goswami, A. and Purkait, M. (2011). Kinetic and equilibrium study for the fluoride adsorption using pyrophyllite. Sep. Sci. Technol. 46 (11): 1797-1807.

Gujarathi, N.P., Bryan, J., Haney, B.J. and Linden, J.C. (2005). Phytoremediation potential of *Myriophyllum aquaticum* and *Pistia stratiotes* to modify antibiotic growth promoters, tetracycline, and oxytetracycline in aqueous wastewater systems. Int. J. Phytorem. 7: 99-112.

Gupta, S., Banerjee, S. and Mondal, S. (2009). Phytotoxicity of fluoride in the germination of paddy (*Oryza sativa*) and its effect on the physiology and biochemistry of germinated seedlings. Fluoride 42 (2): 142-146.

Henderson, K.L., Belden, J.B., Zhao, S. and Coats, J.R. (2006). Phytoremediation of pesticide wastes in soil. Z. Naturforsch. C: Biosci. 61: 213-221.

Hussain, S., Siddique, T., Arshad, M. and Saleem, M. (2009). Bioremediation and phytoremediation of pesticides: Recent advances. Crit. Rev. Environ. Sci. Technol. 39: 843-907.

Islam, M., Mishra, P.C. and Patel, R. (2011). Fluoride adsorption from aqueous solution by a hybrid thorium phosphate composite. Chem. Eng. J. 166: 978-985.

Kanbur, M., Eraslan, G., Silici, S. and Karabacak, M. (2009). Effects of sodium fluoride exposure on some biochemical parameters in mice: Evaluation of the ameliorative effect of royal jelly applications on these parameters. Food Chem. Toxicol. 47: 1184-1189.

Karmakar, S., Mukherjee, J. and Mukherjee, S. (2016). Removal of fluoride contamination in water by three aquatic plants. Int. J. Phytorem. 18 (3): 222-227.

Kumar, E., Bhatnagar, A., Kumar, U. and Sillanpaa, M. (2011). Defluoridation from aqueous solutions by nano-alumina: Characterization and sorption studies. J. Hazard. Materials 186 (2 and 3): 1042-1049.

Li, L. (2003). The biochemistry and physiology of metallic fluoride: Action, mechanism, and implications. Crit. Rev. Oral Biol. Med. 14: 100-114.

Majhi, A. (2016). Fluoride concentration in ground water around an aluminium smelter plant (a case study) in Angul-Talcher Industrial Belt, Odisha, India. Int. J. Environ. Monit. Anal. 4 (1): 1-5.

Marier, J. and Rose, D. (1977). Environmental Fluoride. National Research Council of Canada. Associate Committee on Scientific Criteria for Environmental Quality. NRCC No. 16081.

McCutcheon, S.C. and Schnoor, J.L. (2003). Phytoremediation, transformation and control of contaminants. John Wiley Interscience, New Jersy.

Meenakshi, S., Sundaram, C.S. and Sukumar, R. (2008). Enhanced fluoride sorption by mechanochemically activated kaolinites. J. Hazard. Materials 153: 164-172.

Mukherjee, I. and Singh, U. (2018). Groundwater fluoride contamination, probable release, and containment mechanisms: a review on the Indian context. Environ. Geochem. Health. 40 (6): 2259-2301.

Mwegoha, W.J.S. (2008). The use of phytoremediation technology for abatement of soil and groundwater pollution in Tanzania: Opportunities and challenges. J. Sustainable Dev. Africa 10 (1): 140-156.

Narsimha, A. and Sudarshan, V. (2018). Drinking water pollution with respective of fluoride in the semi-arid region of Basara, Nirmal District, Telengana State, India. Data in Brief 16: 752-757.

Padmavathiamma, P. and Li, L. (2007). Phytoremediation technology: Hyper-accumulation metals in plants. Water Air Soil Pol. 184 (1-4): 105-126.

Padmavathiamma, P. and Li, L. (2011). Rhizosphere influence and seasonal impact on phytostabilisation of metals- a field study. Water Air Soil Pol. 223 (1): 107-124.

Pilon-Smits, E. and Pilon, M. (2002). Phytoremediation of metals using transgenic plants. Crit. Rev. Plant Sci. 21: 439-456.

Raskin, I., Smith, R. and Salt, D. (1997). Phytoremediation of metals: Using plants to remove pollutants from the environment. Current Opinion Biotech. 8 (2): 221-226.

Ravindra, K. and Garg, V. (2007). Hydro-chemical survey of groundwater of Hisar city and assessment of defluoridation methods used in India. Environ. Monit. Asses. 132 (1-3): 33-43.

Reddy, Y.P., Tiwari, S.K., Shaik, A.P., Alsaeed, A., Sultana, A. and Reddy, P.K. (2014). Effect of sodium fluoride on neuroimmunological parameters, oxidative stress and antioxidative defenses. Toxicol. Mech. Methods 24 (1): 31-36.

Rentz, J.A., Chapman, B., Alvarez, P.J. and Schnoor, J.L. (2003). Stimulation of hybrid poplar growth in petroleum-contaminated soils through oxygen addition and soil nutrient amendments. Int. J. Phytorem. 5 (1): 57-72.

Rivera-Cruz, M.C., Ferrera-Cerrato, R., Sanchez-Garcia, P., Volke-Haller, V., Fernandez-Linares, L. and Rodriguez-Vazquez, R. (2004). Descontaminacion de suelos con petroleo crudo mediante microorganismos autoctonos y pasto aleman (*Echinochloa polystachya*). Agrociencia 38: 1-12. [in Spanish].

Robinson, B.H., Leblanc, M., Petit, D., Brooks, R.R., Kirkman, J.H. and Gregg, P.E.H. (1998). The potential of *Thlaspi caerulescens* for phytoremediation of contaminated soils. Plant Soil 203: 47-56.

Sahu, S., Meshram, P., Pandey, B., Kumar, V. and Mankhand, T. (2009). Removal of chromium (III) by cation exchange resin, Indion 790 for tannery waste treatment. Hydrometallurgy 99 (3 and 4): 170-174.

Salido, A.L., Hasty, K.L., Lim, J.M. and Butcher, D.J. (2003). Phytoremediation of arsenic and lead in contaminated soil using Chinese brake ferns (*Pteris vittata*) and Indian mustard (*Brassica juncea*). Int. J. Phytorem. 5 (2): 89-103.

Santos-Diaz, M.S., Zamora-Pedraza, C. and Potosi, S.L. (2010). Fluoride removal from water by plant species that are tolerant and highly tolerant to hydrogen fluoride. Fluoride 43 (2): 150-156.

Saxena, V. and Ahmed, S. (2001). Dissolution of fluoride in groundwater: A water-rock interaction study. Environ. Geol. 40 (9): 1084-1087.

Schneiter, R. and Middlebrookes, E. (1983). Arsenic and fluoride removal from ground water by reverse osmosis. Environ. Int. 9 (4): 289-291.

Shao-Xiang, T. Shu-Guang, W., Wen-Xin, G., Xian-Wei, L. and Bao-Yu, G. (2009). Removal of fluoride by hydrous manganese oxide-coated alumina: Performance and mechanism. J. Hazard. Materials 168: 1004-1011.

Sharma, D. and Forster, C. (1994). A preliminary examination into the adsorption of hexavalent chromium using low-cost adsorbents. Bioresource Technol. 47 (3): 257-264.

Silva Gonzaga, M., Santos, J. and Ma, L. (2006). Arsenic chemistry in the rhizosphere of *Pteris vittata* L. and *Nephrolepis exaltata* L. Environ. Pol. 143 (2): 254-260.

Singh, G., Kumari, B., Sinam, G., Kriti, Kumar, N. and Mallick, S. (2018). Fluoride distribution and contamination in the water, soil and plants continuum and its remedial technologies, an Indian perspective-a review. Environ. Pol. 239: 95-108.

Singh, G., Sinam, G., Kriti, Pandey, M., Kumari, B. and Kulsoom, M. (2020). Soil pollution by fluoride in India: Distribution, chemistry and analytical methods. In: Shukla, V. and Kumar, N. (eds) Environmental Concerns and Sustainable Development, Springer Nature Singapore Pte Ltd. pp. 307-324.

Stevens, D.P., McLaughlin, M.J., Randall, P.J. and Keerthisinghe, G. (2000). Effects of fluoride supply on fluoride concentrations in five pasture species: Levels required to reach phytotoxic or potentially zootoxic concentrations in plant tissue. Plant and Soil 227: 223-233.

Teotia, M., Teotia, S. and Singh, K (1998). Endemic chronic fluoride toxicity and dietary calcium deficiency interaction syndromes of metabolic bone disease and deformities in India: Year 2000. J. Pediatrics 65 (3): 371-381.

Tsai. W.T. (2009). Environmental hazards and health risk of common liquid perfluoron-alkanes, potent green house gases. Environ. Int. 35 (2): 418-424.

U.S. Environmental Protection Agency. (2000). Introduction to Phytoremediation, National Risk Management Research Laboratory. EPA/600/R-99/107.

Vera Tome, F., Blanco Rodriguez, P. and Lozano, J. (2009). The ability of *Helianthus annuus* L. and *Brassica juncea* to uptake and translocate natural uranium and 226 Ra under different milieu conditions. Chemosphere 74 (2): 293-300.

Wilson, S., Duarte-Davidson, R. and Jones, K. (1996). Screening the environmental fate of organic contaminants in sewage sludges applied to agricultural soils: The potential for downward movement to ground waters. Sc. Total Environ. 185 (2): 45-57.

WHO (1993). Guidelines for drinking-water quality, 2nd ed. Vol. 1. Recommendations. World Health Organization, Geneva.

Whitford, G.M. (1989). The metabolism and toxicity of fluoride. Monographs in Oral Science. Vol. 13, Karger, Basel, Switzerland.

Zhuang, P., Ye, Z., Lan, C., Xie, Z. and Shu, W. (2005). Chemically assisted phytoextraction of heavy metal contaminated soils using three plant species. Plant and Soil 276 (1 and 2): 153-162.

Chapter 34

Integrated Pest Management: An Eco-friendly Approach for Sustaining Rice Productivity and Achieving Food Security

U.S Nayak[1]* *and G. Shial*[2]

[1]*Regional Research and Technology Transfer Station,*
[2]*Krishi Vigyan Kendra, Bhadrak,*
OUAT, Ranital – 756 111, Odisha
**e-mail: usnayak74@gmail.com*

ABSTRACT

A diverse species of phytophagous insect pests infest rice crop at different stages of crop growth causing sizable yield loss every year in different parts of the world. Under the changing climatic scenario the frequency of pest outbreak and magnitude of crop loss has been increasing tremendously that threatens sustainable rice production and global food security. Exclusive reliance on chemical pesticides and their indiscriminate application have many serious adverse effects on human health and surrounding environment. Experiments were carried out in the adopted village of Krishi Vigyan Kendra, Bhadrak, Odisha to assess the efficacy of an eco- friendly integrated pest management (IPM) module against major insect pests of rice during kharif seasons of 2013 and 2014. The results of the experiment reveled that in the IPM plots all the major insect pests remained below the economic threshold level (ETL) and in comparison to non-IPM plots there has been significant reduction in the pest incidence and infestation level. While stem borer incidence was reduced to around 45 per cent, the infestation of foliage feeders (hispa, leaf folder and whorl maggot) was minimized to around 25 per cent in comparison to the non-IPM plots. However, maximum reduction (45-67 per cent) was noticed in the incidence of sucking pests (BPH, WBPH and GLH) in the IPM plots. On the environmental benefit point of view, higher population level of beneficial natural enemies was observed in the IPM plots indicating the positive effect of IPM on insect diversity. Similarly, the plots with IPM registered higher yield (38.05 q/ha) and income (Rs.14,200.00/ha). The reduced cost of plant protection measures and higher benefit cost ratio justified the environmental viability and profitability of IPM.

***Keywords**: Integrated pest management, Rice,*

Introduction

Rice (*Oryza sativa* L.) is an important staple food crop for more than two third of the population of India and more than 65 per cent of the world population (Mathur *et al.,* 1999). India is the largest rice growing country occupying 24 per cent of the gross cropped area and accounts for more than 40 per cent food grain production. This crop is infested by a number of insect pests from nursery to harvesting and out of the 128 insect species reported to attack rice crop, 15-20 insects cause economical damage and substantially reduce the yield and profit (Kalode, 2005). The overall yield loss due to the insect pests varies from 21 to 51 per cent (Singh and Dhaliwal, 1994) and in the year of heavy infestation the yield loss may be as high as 31.5 to 86.0 per cent (Gunathilagaraj and Kumar, 1997) in India. According to Pasalu and Katti, 2006, nearly 300 species of insect pests attack the rice crop at different stages and among them 23 species cause notable damage causing 30-35 per cent yield loss (Prakash *et al.,* 2007). However, in monetary value, losses incurred due to insect pests of rice is to the tune of 5,51,200 lakh rupees, which in turn comes out to 18.6 per cent of total loss (Kumar *et al.,* 2010). Expansion of irrigation favoured year round cultivation of different crops that provides continuous food supply to the phytophagous insects. This has gradually changed the pest situation in crop eco-system and consequently, a shift in pest status and pest population dynamics has been realized both in time and space.

Among the insect pests, Yellow stem borer (*Scirpophaga incertulas* Walker) is one of the most destructive pests of this crop and is widely distributed monophagous pest in the Indian subcontinent (Atwal and Dhaliwal, 2008) with an estimated yield loss of 27-34 per cent every year (Prasad *et al.,* 2007). Similarly, rice leaf folder, *Cnaphalocrocis medinalis* (Guenee), which was earlier a minor pest of rice has gained the major pest status with the extensive cultivation of high yielding varieties, reduced genetic variability in crop ecosystem, multiple cropping patterns, higher dose of nitrogenous fertilisers and prophylactic use of pesticides (De Kraker *et al.,*2000). It can cause leaf infestation up to 61.9 per cent (Chhavi *et al.,* 2016) and can contribute yield loss to the extent of 5 to 25 per cent (Kulgagod *et al.,* 2011). Case worm, *Nymphula depunctalis* Guenee though earlier considered as a minor and localised pest cause extensive damage to the lowland rice where prolonged water stagnation is a problem. In late transplanted lowland rice crop

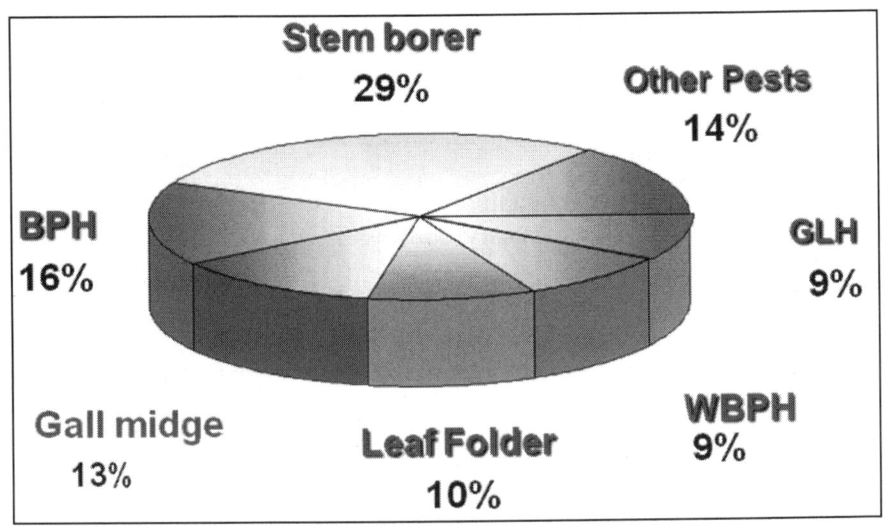

case worm caused up to 18.60 per cent leaf damage (Prasad *et al.,* 2018). Plant hoppers (Brown plant hopper, BPH, *Nilaparvata lugens,* Stal. and White backed plant hopper, WBPH, *Sogatella furcifera* Horváth.) assumed serious significance in the recent years and cause widespread damage to the wet season rice of coastal areas due to favourable ecological niche and injudicious use of chemical insecticides resulting in insecticide resistance and resurgence (Nayak *et al.,* 2019). Generally the yield losses due to hoppers ranges from 10 to 90 per cent but if timely control measures are not taken up, there may be possibility of total crop loss within a very short period (Seni and Naik, 2017). A recent report from DRR estimated that plant hoppers cause losses ranging from 1-2 million tons of rice annually in India. Besides, the direct damage through hopper burn, BPH transmits viral diseases like grassy stunt, ragged stunt and wilted stunt (Bhanu *et al.,* 2014) and therefore often considered as the most destructive insect pest of wet season rice. The white backed plant hopper, *Sogatella furcifera* (Horváth) (Homoptera: Delphacidae),which was earlier recognized as a minor pest of rice is emerging as a serious pest in many rice growing countries (Khan and Saxena, 1986) and the extent of yield losses ranges from 54.4 to 79.8 per cent in different rice varieties (Khan and Kushwaha, 1991).

For reducing the insect damage and minimizing the yield loss, farmers mostly rely on the chemical pesticides and the local pesticide dealers are the major source of information regarding the selection of the chemicals and application procedure. As a result of the excessive and indiscriminate use of various synthetic insecticides alone and in combination there has been a serious adverse impact on the human health and surrounding environment. The farmers even start applying pesticides in the absence of the pest in the field resulting in the widespread killing of the beneficial organisms like earth worm, predators like spiders, mired bugs, water striders *etc.* and parasitoids in the rice eco-system. Similarly excessive application of synthetic pyrethroids and their combination products results in the resurgence of many sucking pests like plant hoppers, leaf hoppers and mites posing a new threat to rice growers. Most of the chemical pesticides are costly and increasing application cost due to repeated application increases the cost of production for the resource poor farming community.

Over reliance of the chemical pesticides coupled with its associated environmental, socio-economic problems and ineffectiveness of any individual method alone to provide satisfactory control of the pest has paved the path of implementing Integrated Pest Management (IPM) as the only viable option for managing insect pests. IPM is the judicious integration of all available pest management strategies to bring the pest population below the ETL without endangering the surrounding environment. It is not only based on the ecological principle but also on the socioeconomic considerations and hence, considered to be effective, economical and eco-friendly. In recent days the application of botanicals, biopestcides, new generation selective insecticides and adoption of mechanical and behavioural control for managing insect pests are gaining popularity as these methods largely avoid and eliminate the problems like pest resistance, resurgence, secondary pest out break, extinction of natural enemies, residual toxicity, environmental pollution *etc.* It is therefore; quite imperative that for successful and profitable rice production serious efforts is to be taken to minimize the pest incidence by adoption of some suitable IPM strategies. Keeping this in view an experiment was conducted to evaluate the performance of an eco-friendly management strategy based on the IPM principle against the insect pest complex of rice by judiciously integrating the cultural, mechanical, behavioural, biological and need based chemical methods of pest control.

Materials and Methods

Field experiments were conducted in the farmers' field of Kshirasahi village of Bhadrak block (adopted village of KVK, Bhadrak, Odisha) during *Kharif* 2013 and 2014 to evaluate the performance of one eco-friendly pest management technique designed in the IPM principle against major insect pest of rice in comparison to the conventional insecticide based management schedule. Both these pest management modules are considered as treatments with thirteen numbers of replications (each farmer represents one replication). The experiments were conducted through a farmer participatory through a farmer participatory research approach where, farmers were actively involved in designing, implementing and evaluating the trial and the scientists simply played the role of external facilitators. The eco-friendly pest management module (IPM) was laid out separately from the conventional insecticide based modules at an isolation distance of 50 m from each other to avoid the negative effect of chemicals on pheromone efficiency and bio-agent activity. Rice variety "Swarna" was sown in the nursery during 3rd week of June and 30 days old seedlings were transplanted in the main field (0.1 ha area each) with a row spacing of 20 x 15 cm. The crop was grown with recommended agronomic package of practices and standard intercultural operations to ensure optimum plant growth. The details of the eco-friendly pest management module and conventional management schedule are mentioned below.

Eco-friendly Pest Management Module (IPM)	*Conventional Management Schedule (Non IPM)*
Ø Nursery treatment with carbofuran 3 G @ 1 kg a.i./ha Ø Bund trimming and destruction of weeds Ø Clipping of the apical portion of the leaves Ø Installation of pheromone traps @ 10 no./ha Ø Release of parasitoids *Trichogramma chilonis* @ 1 lakh/ha at 40, 50, 60 and 70 DAT Ø Installation of bird perches @ 20/ha Ø Spraying of Fipronil 5 SC @ 750 ml/ha at 35 DAT Ø Foliar spraying of neem based formulations @1.5 liter/ha at 50 and 65 DAT Ø Spraying of Buprofezin 20 SC @ 750 ml/ha at 80 DAT based on ETL	Ø Soil application of carbofuran 3 G @ 30 kg/ha at 30 DAT Ø Spraying of Monocrotophos @ 1 l/ha at 40 DAT Ø Spraying of Chloropyriphos + Cypermethrin @ 1.5 l/ha at 55 DAT Ø Spraying of Lambda cyhalothrin @ 750 ml/ha at 65 DAT Ø Spraying of Imidacloprid @ 250 ml/ha at 75 DAT Ø Spraying of Acephate @ 750 g/ha at 82 DAT

The spraying of insecticides and botanicals was done with the help of manually operated Knapsack sprayer using 500 liter of water per hectare. The egg parasitoid *Trichogramma chilonis* Ishii. were procured from the bio-control laboratory of Department of Entomology, Orissa University of Agriculture and Technology (OUAT), Bhubaneswar and released in the rice field during morning hours. Pheromone traps (funnel trap) were installed in the crop field ensuring the lure position 30 cm above the crop canopy and the lure (scirpo lure) was changed at every three weeks interval. Treatment and replication wise observations on the incidence of insect pest and population level of natural enemies were recorded by the farmers in presence of the scientists at weekly interval from 1 m^2 area at 5 randomly selected spots in each plot and the cumulative counts during the crop season were subjected to analysis for comparison. The data on grain yield was also recorded from each plot at harvest and converted to quintal per hectare for comparision and the economics was also worked out in terms of net return and B: C ratio.

Results and Discussion

The efficacy of the IPM module against the major insect pests of rice has been presented in Table 34.1 and it was revealed from Table 34.1 that a significantly lower incidence of all the insect pests was observed in the IPM plots in comparison to the non-IPM plots. The incidence of stem borer expressed as dead heart (per cent), white ear head (per cent) and adult moth population (No/m^2) has been reduced to around 45 per cent in the IPM fields and found to be much lower than the ETL. Similarly, the infestation level of foliage feeders (leaf folder, whorl maggot and hispa) was found to be nearly 25 per cent lower in the IPM treated plots compared to the conventional insecticide based management schedule. The IPM module also retained its superiority in minimizing the incidence of sucking pests like BPH, WBPH and GLH and registered a significantly higher reduction of pest population level (45- 67 per cent).

Table 34.1: Incidence of Major Insect Pests of Rice in the IPM and Non IPM Plots

Insect Pest Incidence	*IPM Plot**			*Non IPM Plots**			*Per cent Reduction from Non IPM*
	2011	*2012*	*Mean*	*2011*	*2012*	*Mean*	
A. Yellow stem borer							
i) Dead heart (per cent)	3.8	3.1	3.45	6.7	5.5	6.10	43.44
ii) White ear head (per cent)	4.6	3.7	4.15	8.1	7.5	7.80	46.79
iii) Adult moth population (No./m^2)	0.7	0.5	0.60	1.3	0.9	1.1	45.45
B. Leaf folder							
i) Infested leaves (No./hill)	1.5	1.8	1.65	2.1	2.3	2.20	25.00
ii) Larval incidence (No./hill)	0.3	0.5	0.40	0.5	0.6	0.55	27.27
C. Whorl maggot (per cent hills infected)	4.3	5.1	4.70	5.1	6.8	5.95	21.01
D. Rice hispa							
i) Insects (Adult/Grub)- No/hill	0.4	0.4	0.40	0.6	0.5	0.55	27.27
ii) Infested leaves (No/hill)	1.1	0.8	0.95	1.4	1.1	1.25	24.00
E. Plant and Leaf hopper							
i) Brown plant hopper (Nos/hill)	5.2	4.1	4.65	15.6	12.9	14.25	67.37
ii) White backed plant hopper (Nos/hill)	4.5	3.9	4.20	13.3	11.8	12.55	66.53
iii) Green Leaf hopper (Nos/hill)	6.1	4.9	5.50	10.7	9.5	10.10	45.54

*Average of 13 replications.

The effect of IPM module on the population level of different predatory fauna in the rice eco-system is depicted in Table 34.2 and the data revealed that IPM modules recorded significantly higher population of predatory spiders (1.55/hill), Staphyllinids (1.30/hill) and Coccinellids (1.60/hill) compared to a minimum 0.35, 0.40 and 0.50 spiders, Staphyllinids and Coccinellids respectively per hill in the non-IPM plots. Similarly the eco-friendly management module also harbored an encouraging number of other predators like damsel flies, dragon flies and Mirid bugs indicating the safety of IPM to the natural enemies.

Table 34.2: Natural Enemies Population of Insect Pests of Rice in the IPM and Non IPM Plots

Insect Pest Incidence	*IPM Plot**			*Non IPM Plots**		
	2011	*2012*	*Mean*	*2011*	*2012*	*Mean*
A. Spiders (Nos/hill)	1.4	1.7	1.55	0.3	0.4	0.35
B. Staphyllinid beetles (Nos/hill)	1.1	1.5	1.30	0.4	0.4	0.40
C. Coccinellid beetles (Nos/hill)	1.5	1.7	1.60	0.6	0.4	0.50
D. Ciccindelid beetles (Nos/hill)	0.3	0.5	0.40	0.2	0.1	0.15
E. Damsel flies (Nos/hill)	0.7	1.1	0.90	0.3	0.5	0.40
F. Dragon flies (Nos/hill)	0.5	0.4	0.45	0.1	0.3	0.20
G. Mirid bugs (Nos/hill)	1.8	2.3	2.05	0.6	0.4	0.50

The effect of IPM module on the yield and economics of rice production has been presented in Table 34.3 and the generated data on this regard revealed that the IPM plot registered higher grain yield of 38.05 q/ha compared to 33.80 q/ha in the non-IPM plots. There has been reduction in the cost of plant protection due to minimum number of application of safer insecticides and botanicals, which contributed higher net return in the IPM treated plots (Rs. 14,200/ha). The economic efficiency of the IPM modules was evidenced with the higher B:C ratio of 1.55 compared to 1.35 in the non-IPM plots.

Table 34.3: Grain Yield and Economics of Rice in the IPM and Non IPM Plots

Yield/Economics	*Non IPM Plots**			*Non IPM Plots**		
	2011	*2012*	*Pooled Mean*	*2011*	*2012*	*Pooled Mean*
A. Grain yield (q/ha)	36.26	39.84	38.05	32.85	34.74	33.80
B. Gross income (Rs./ha)	38073	41832	399530	34493	36477	35485
C. Cost of plant protection (Rs./ha)	3165	3340	3253	3710	3875	3793
D. Gross expenditure (Rs/ha)	25665	25840	25753	26210	26375	26293
E. Net income (Rs./ha)	12408	15992	14200	8283	10102	9192
F. B:C ratio	1.48	1.62	1.55	1.32	1.38	1.35

The results of the experiment was supported by the findings of Chakraborty (2012) who reported that adoption of IPM exhibited significantly lower incidence of YSB (8.55 per cent DH and 5.41 per cent WH) over the non-IPM (11.25 per cent

DH and 7.69 per cent WH). Abundance of predatory Arthropods like spider and coccinellid beetle in the IPM plots indicated the safety of IPM strategy to natural enemies. Besides, IPM plots registered higher grain yield, net return and B:C ratio. Karthikeyan *et al.* (2010) also reported similar results where management module comprising alternate spraying with neem-based formulation coupled with release of egg parasitoids against leaf folder and monitoring of stem borer with sex pheromone traps significantly reduce incidence of dead hearts, white ear heads and damaged leaves. Aggarwal *et al.* (2016) observed that validation of large scale bio-intensive IPM practices strategies in aromatic rice recorded higher population of natural en-emies in the IPM fields in comparison to conventional plots. Alam *et al.*, 2016 opined that a well-designed IPM strategy can be used for effective pest control in rice agro- ecosystems and the principle advantage of this technique is that rice yields increase as the resilience of the cropping system increases.

REFERENCES

Aggarwal N, Sharma S, Jalali SK. 2016. On-farm impact of biocontrol technology against rice stem borer, *Scircophaga incertu*las (Walker) and rice leaf folder *Cnaphalocrocis medinalis* (Guenee) in aromatic rice. *Entomol Gen.* 36: 137–148. https://doi.org/10.1127/entomologia/2016/0135

Alam, M. Z., Alex, R.C., Haque, M., SirajulIslam, M., Hossain, E., Hasan, S.B., Hasan, S.B. and Hossain, M.S. 2016. Effects of Integrated Pest Management on Pest Damage and Yield Components in a Rice Agro-Ecosystem in the Barisal Region of Bangladesh. Frontiers in Environmental Science, 4 (22): 1-10.

Bhanu, K.V., Reddy, P.S., Satyanarayana, J., 2014. Studies on tolerance of rice cultivars against brown plant hopper, *Nilaparvata lugens*. Indian Journal of Plant Protection 42 (3), 199–203.

Chakraborty, K.2012. Evaluation of integrated pest management module against paddy yellow stem borer *Scirpophaga incertulas* Walk. *Karnataka Journal of Agriculture Sciences,* **25** (2): 273-275.

Chhavi, Srivastava, A., Sharma, P.K., 2016. Populations build up of rice leaf folder, *Cnaphalocrocis medinalis* (Guenee) under mid-hill conditions of Himachal Pradesh. Journal of Insect Science 29, 37-42.

De Kraker., Rabbinge, R., van Huis, A., van Lenteren, J.C., Heong, K.L., 2000. Impact of nitrogenous-fertilizations on the population dynamics and natural control of rice leaf folders (Lepidoptera: Pyralidae). International Journal of Pest Management 46: 225-235

Gunathilagaraj K and Kumar MG 1997**.** Extent of damage and pattern of emergence of overwintering larvae of rice stem borer in Punjab. *Indian Journal of Ecology* **23**:104-108.

Kalode, M.B. 2005. Insect pests of rice and their management in Indian perspective edited by S.D. Sharma and B.C. Nayak. Today and Tomorrow printers and publishers, India.819-854 pp.

Karthikeyan K, Jacob S, Beevi P, Purushothaman SM. 2010. Evaluation of different integrated pest manage-ment modules for the management of major pests of rice (*Oryza sativa*). *Indian J Agric Sci.*: **80**: 59–62.

Khan, M.S. and Kushwaha, K.S. 1991.Studies on the population build up of white backed planthopper, *Sogatella furcifera* in relation to abiotic factors on six rice varieties in Haryana. *Ind. J. of Ento.* **53**:93-98.

Khan, Z.R. and Saxena R.C. 1986. Varietal resistance in rice against *Sogatella furcifera* (Horváth). *Crop Protection.* 5 (1): 15-25.

Kulgagod, S.D., Hegade, M., Nayak, G.V., Vastrad, A.S., Hugar, P.S., Basavanagoud, K., 2011. Evaluation of insecticides and biorationales against yellow stem borer and leaf folder in rice crop. Karnataka Journal of Agricultural Sciences 24(2), 244-246.

Kumar V. B., Boomathi, N., Kumaran, N. and Kuttalam, S. 2010. Non Target Effect of Ethiprole + Imidacloprid 80 WG on Predators of Rice Planthoppers. *Madras Agriulture Journal*, **97** (4-6): 153-156.

Mathur, K.C., Reddy, P.R., Rajamali, S. and Moorthy, B.T.S., 1999, Integrated pest management of rice to improve productivity and sustainability. *Oryza*, **36**(3): 195-207.

Nayak, U.S., Das, A. and Shial, G. 2019. Farmer Participatory Assessment of Integrated Pest Management Strategies against the Insect Pest of Lowland Rice in Coastal Odisha. *International Journal of Bio-resource and Stress Management,* 10(4):397-401.

Pasalu, I.C., Katti, G., 2006. Advances in ecofriendly approaches in rice IPM. Journal of Rice Research 1(1), 83-90.

Prakash, A., Rao, J., Singh, O.N., Tyagi, J.P., Singh, S., Rath, P.C.,2007. Rice: The Queen of Cereals, Applied Zoologist Research Association (AZRA), CRRI, Cuttack, Odisha, India pp.202.

Prasad, R., Soren, A., Hembrom,L., Prasad,D., 2018. Status of prevailing insect pest fauna associated with transplanted rice in Jharkhand. Journal of Eco-friendly Agriculture 13(1), 60-64.

Seni, A., Naik,B.S., 2017. Efficacy of some insecticides against major insect pests of rice, *Oryza sativa* L. Journal of Entomology and Zoology Studies 5(4), 1381-1385.

Chapter 35

Estimation of Population, Species Richness and Diversity of Marine Phytoplankton of Paradip Harbour Region, Odisha, India

Ranjan Kumar Mohapatra[1], *Manoranjan Nayak*[2], *Chitta Ranjan Panda*[1] *and Hrudayanath Thatoi*[3]

[1]*Environment and Sustainability Department, CSIR-Institute of Minerals and Materials Technology, Bhubaneswar – 751 013*

[2]*Biorefinery and Bioenergy Research Laboratory, Amity Institute of Biotechnology, Amity University Uttar Pradesh, Noida - 201313, India*

[3]*Department of Biotechnology, Maharaja Sriram Chandra Bhanja Deo University, Sriram Chandra Vihar, Takatpur, Baripada-757003, Odisha, India*

Corresponding Authors: mranjankumar.biotech@gmail.com (R.K. Mohapatra); hn_thatoi@rediffmail.com (H.N. Thatoi)

ABSTRACT

Surface water samples were collected from Paradip port area at eleven different locations during 2014-15 in four seasons such as post monsoon, winter, pre-monsoon and monsoon. The quantitative and qualitative phytoplankton ecology such as total phytoplankton count (TPC), species composition and diversity index (H) was evaluated. Maximum total phytoplankton count and species richness were found during summer and post monsoon. But the highest diversification of the phytoplankton (in the form of Shanon's diversity index) was observed during monsoon. No algal bloom was detected during this study where as some toxic/harmful phytoplanktons are detected in some locations during all the seasons.

Introduction

Phytoplankton population and diversity in marine ecosystem are importance as the primary producers in food webs, which contribute about 95 per cent of the total production in the marine environment and ensuring ecological balance (Nail *et al.,* 2014; Naik *et al.,* 2009; Mishra *et al.,* 2009). Increasing industrialisation and human anthropogenic activity causes insertion of local waste to the marine system through river mouth (Nail *et al.,* 2014; Madhab and Kondalarao, 2004). Phytoplanktons also known as microalgae are autotrophic, buoyant organisms float in the upper part of lakes, rivers and oceans where sunlight penetrates the water surface. The phytoplankton composition in marine system was affected by various physico-chemical factors like pH, temperature, salinity, turbidity, light intensity and nutrients (Naik *et al.,* 2009) Phytoplankton also require inorganic nutrients such as nitrates, phosphates, and sulphur which they convert into proteins, fats, and carbohydrates, which usually results in dense algal blooms under favourable condition (Madhab and Kondalarao, 2004) Besides normal and periodical blooms of phytoplankton, some harmful algal bloom was also occurs exceptionally (Madhab and Kondalarao, 2004). These blooms reduce the penetration of sunlight through subsurface water as a result of which the growth of submerged aquatic vegetation is hampered. This has also impact on dissolved oxygen levels. Extremely dense algal blooms produce high quantity of oxygen levels on sunny days while the algae are photosynthesising (Sarojini and Sarma, 2001). Phytoplankton compositions constitute the most important biological community in aquatic system and change in its dominance and diversity has been used as an indicators of water quality (Sahu *et al.,* 2012; Naik *et al.,* 2009). The marine water of the Paradip Port harbour is affected by various anthropogenic activity and port operations. Hence, in the present study an attempt has been made to study the distribution of phytoplankton community with special emphasis to species richness and diversity index inside paradip port harbour water system in different seasons during 2014-15.

Materials Methods

Sampling Location

Water samples were collected from Paradip Port, a sea port in the east coast of India (Bay of Bengal, Odisha coast), is situated between the latitude 20°15' – 55°44'N and longitude 86°40' – 36°62'E. Inside the port region the study area comprises eleven important locations such as Approach channel (AC), Oil jetty (OJ), Iron ore handling berth (IOHB), ADB-Coal handling berth (ADB), North quay (NQ), General cargo berth (GCB), South quay (SQ), East quay (EQ), Fertilizer berth (FB), Marine site (MS) and Fishing harbour (FH) (Figure 35.1).

Sample Collection and Processing

Surface water from different points of Paradip Port harbour was collected during 2014-2015 in four different seasons like November-2014(Post Monsoon),

Figure 35.1: Sampling Locations in Side Paradip Port.

January-2015 (Winter), April-2015 (Pre-Monsoon/Summer) and August-2015 (Monsoon). Each sample (1L) was collected in polythene bottles. Immediately after collection, 10 ml Lugol's iodine solution was added to the water samples (1L) and the samples were brought to the laboratory and were kept standing undisturbed for at least for 48 hours in a separating funnel. This allowed phytoplanktons present in the sample to be fixed and to settle down to the bottom of the container. The lower layer of water sample containing phytoplankton biomass was removed carefully leaving the upper layer. The samples were then used for phytoplankton estimation under microscope.

Estimation of Phytoplankton

Total phytoplankton count was done for the preserved sample with Lugol's iodine solution under light microscope with the help of a Sedgwick-rafter counting chamber. Each concentrated sample was uniformly mixed and 1 ml of it was loaded in to the phytoplankton counting slide (Sedgwick-Rafter Chamber) with cover slip in position (Sarojini and Sarma, 2001). It was allowed to stand for five minutes and counted under a compound light microscope (40X objective lense). Phytoplanktons present in the sample were identified according to their morphological view by following the identification manual (Hasle *et al.,* 1997; Verlencar *et al.,* 2004). To know the species diversity of the phytoplankton communities, diversity index (H) was calculated by the formulae proposed by Shannon and Weaver (Shannon and Weaver, 1949).

$$H = -\sum_{i=1}^{S} p_i \ln(p_i)$$

where,

i - is an index number for each species present in a sample; S- is the total number of species; ln - is the natural logarithm; pi - is the number of individuals within a species (ni) divided by the total number of individuals (N) present in the entire sample.

Results

Results obtained from the samples collected in the month of November-2014, January-2015 and April-2015, August-2015 have been presented in Figures 35.2–35.4 and the varieties of different genera detected during this study were given in the Tables 35.1 to 35.4. Statistics of phytoplankton analysis data was given in Table 35.5. The dominated species community identified at all the stations during four sampling seasons are *Nitzschia longissma, Coscinodiscus sp., Chaetoceros sp., Melosira sp., Asterionella japonica, Ceratium sp., Rhizosolenia sp., Naviculla sp., Thallasiothrix sp., Thallasionema sp., Pleurosigma sp., Podosira sp., Bacteriastrum sp.* and *Lauderia sp.* Total number of individual phytoplankton species was found between 11 to 38 under the taxonomic class like Bacillariophyceae, Dinophyceae, Cyanophyceae and Chlorophyceae. The phytoplankton genus recorded throughout

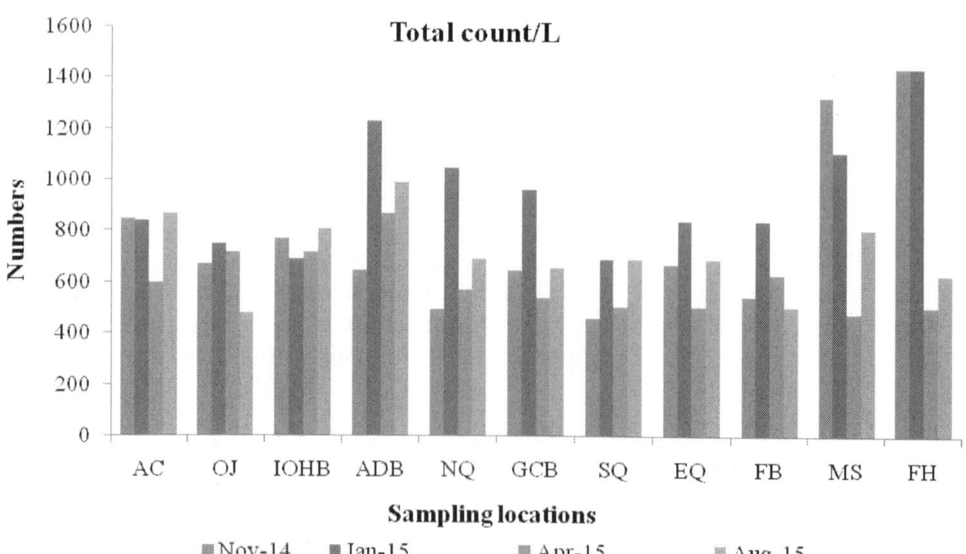

Figure 35.2: Total Phytoplankton Count per Litre.

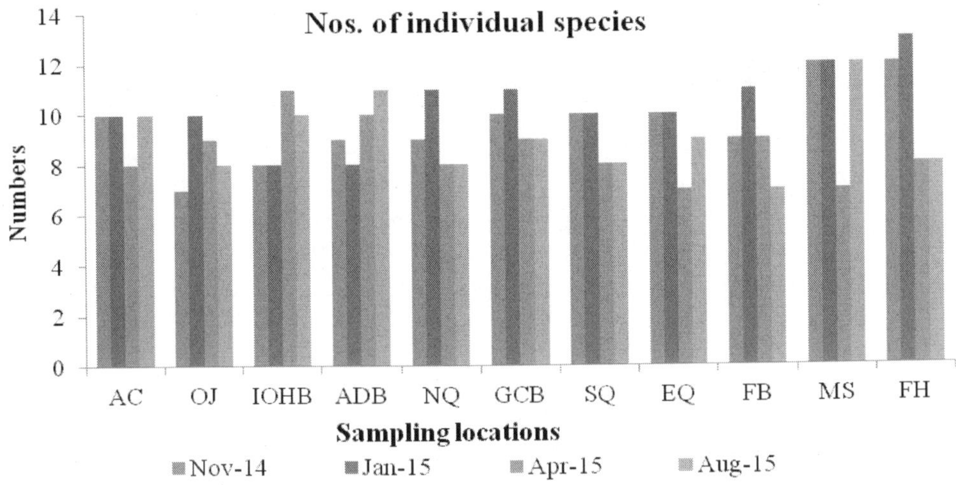

Figure 35.3: Number of Individual Species of Phytoplankton per Litre.

the year in all seasons was dominated by Diatoms (Bacillariophyceae) followed by Dinoflagellates (Dinophyceae). According to percentage occurrence of different phytoplankton specie with respect to total phytoplankton taxa, the decreasing order sequence is Bacillariophyceae (80-85 per cent) > Dinophyceae (6-15 per cent) > Cyanophyceae (3-8 per cent) > Chlorophyceae (1-2 per cent). Similar types of findings were reported earlier by various authors on Dhamra estuary and Bay of Bengal, Odisha coast (Panda *et al.*, 2012; Palleyi *et al.*, 2011; Campbell *et al.*, 2011).

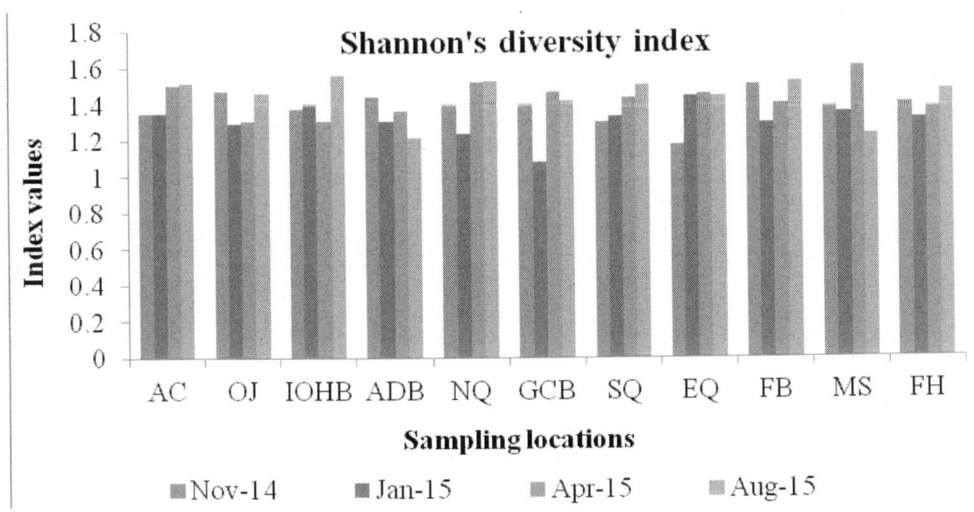

Figure 35.4: Shannon's Diversity Index (H).

Table 35.1: Different Phytoplankton Species Detected during November-2014 (post monsoon)

Station Codes	*Genera of Phytoplankton Found*
AC	*Triceratium sp., Chaetoceros sp., coscinodiscus thorii, Prorocentrum micans,*# *Ceratium sp., Peridinium sp.,*# *Leptocylindrus danicus, coscinodiscus sp.,*@ *Bidulphia haeteroceros, Protoperidinium oceanicum*#.
OJ	*Leptocylindrus danicus, Triceratium sp.,* @ *Streptotheca indicus, Triceratium robertsianum, Planktoneilla sol, Coscinodiscus sp.,* @ *Bellerochea mallecus.*
IOHB	*Coscinodiscus sp.,*@ *Triceratium sp., Hemidiscus hardmanninus, Ceratium furca, Stephanopyxis palmeriana, Prorocentrum micans, Protoperidinium sp.,*# *Pyrophacus steinii.*
ADB	*Coscinodiscus sp.,*@ *Chaetoceros sp., Rhizosolenia sp., Triceratium sp., Pleurosigma elongatum, Ceratium macroceros, Prorocentrum micans, Rhizosolenia stolterfothii, Bidulphia heateroceros.*
NQ	*Coscinodiscus sp.,* @*Protoperidinium sp., Streptotheca indicus, Triceratium sp., Cyclotella sp., Pleurosigma sp., Rhizosolenia sp., Coscinodiscus centralis, Skeletonema costatum.*
GCB	*Rhizosolenia sp.,* @ *Coscinodiscus sp.,* @ *Chaetoceros sp., Coscinodiscus gigas, Thalassiosira subtillis, Cyclotella sp., Naviculla sp., Ceratium furca, Planktoneilla sol, Melosira sp.*
SQ	*Leptrocylindrus danicus, Coscinodiscus sp.,* @ *Triceratium sp., Rhizosolenia styliformis, Coscinodiscus eccentricus, Pleurosigma sp., Hemidiscus hardmanninus, Rhizosolenia sp., Gyrosigma sp., Prorocentrum micans.* #
EQ	*Prorocentrum micans,* # *Pleurosigma sp., Triceratium sp., Coscinodiscus eccentricus, Ceratium inflatum, Coscinodiscus sp.,*@ *Rhizosolenia sp., Bellerochea malleus, Thallasiothrix sp., Nitzschia paradoxa.*
FB	*Coscinodiscus sp.,* @ *Prorocentrum sp., Pyrophacus stenii, Triceratium sp., Prorocentrum micans,* # *Ceratium sp., Peridinium sp., Navicula sp., Rhizosolenia sp.*
MS	*Coscinodiscus sp.,*@ *Triceratium sp., Cyclotella sp., Peridinium sp., Rhizosolenia sp., Rhizosolenia stolterfothii, Prorocentrum micans,* # *Hemidiscus hardmanninus, Chaetoceros sp., Ceratium sp.*
FH	*Ceratium sp., Triceratium sp., Coscinodiscus sp.,*@ *Coscinodiscus gigas, Bacillaria paradoxa, Ceratium inflatum, Hemidiscus hardmanninus, Rhizosolenia sp., Chaetoceros sp., Navicula sp., Gyrosigma sp., Dinophysis caudata*#.

@: Major species, #: Toxic/Pollution indicating phytoplankton.

Discussions

Phytoplankton and Micro organisms are the major indicators of the change in biological environment of sea, which are mainly affected by any alteration of the water quality. Phytoplankton appears very sensitive to climate changes and could be used as a proxy of the variations occurring in the environment plays a major role for indicating pollution in coastal marine ecosystem. Here some

Table 35.2: Different Phytoplankton Species Detected during January -2015 (Winter)

Station Codes	*Genera of Phytoplankton Found*
AC	*Triceratium sp.,@ coscinodiscus sp., Ceratium tripos, Cyclotella sp., Ceratium furca, Chaetoceros currvisetum, Chaetoceros sp., Rhizosolenia stolterforthii, Nitzschia sp, Dinophysis caudate#.*
OJ	*Triceratium sp., coscinodiscus sp.,@ Rhizosolenia stolterforthii, Pleurosigma sp., Prorocentrum micans, Ceratium inflatum, Rhizosolenia styliformis, Chaetoceros sp., Dinophysis caudate#.*
IOHB	*Rhizosolenia sp., Coscinodiscus sp., Chaetoceros sp., Bellerochea malleus, Pleurosigma elongatum,, Nitzschia sp., Ceratium sp., @ Ceratium furca.*
ADB	*Pyrophacus stenii, Chaetoceros sp., Triceratium sp.,@ Rhizosolenia sp., Coscinodiscus sp., Asterionella sp., Prorocentrum micans #, Ceratium sp.*
NQ	*Triceratium sp.,@ Chaetoceros sp., Coscinodiscus sp., Hemidiscus hardmanninus, Pleurosigma sp., Ceratium sp., Biddulphia sp., Dinophysis sp., Chaetoceros Coarctatus, Pyrophacus steinii, Nitzchia sp.*
GCB	*Rhizosolenia sp., Prorocentrum micans, Coscinodiscus sp., Ceratium sp., Dinophysis caudatae#, Ceratocory horrida, Coscinodiscus centralis, Triceratium sp., @ Ceratium sp., Gyrosigma sp., Peridinium sp.*
SQ	*Prorocentrum micans, # Coscinodiscus sp., Triceratium sp., @ Ceratium sp., Cyclotella sp., Thalassiosira subtillis, Rhizosolenia sp., Ceratium furca, Protoperidinium sp., # Rhizosolenia stolterforthii.*
EQ	*Triceratium sp., Rhizosolenia sp., Nitzchia sp., Protoperidinium sp., # Ceratium sp., Coscinodiscus sp., @ Planktoniella sol, Chaetoceros sp., Biddulphia sp., Cyclotella sp.*
FB	*Triceratium sp., Ceratocorys horrida, Rhizosolenia stolterforthii, Prorocentrum micans,# Coscinodiscus sp.,@ Ceratium sp., Ceratium furca, Bacillaria paradoxa, Pyrophacus steinii, Rhizosolenia setigera, Rhizosolenia sp.*
MS	*Coscunidiscus centralis, Chaetoceros sp., Rhizosolenia stolterforthii, Coscinodiscus gigas, Coscinodiscus sp.,@ Triceratium sp., Rhizosolenia sp., Biddulphia sp.,Nitzchia closterium, Dinophysis caudata, # Ceratium sp., Pyrophacus steinii.*
FH	*Coscinodiscus sp.,@ Pleurosigma sp., Rhizosolenia stolterforthii, Biddulphia sp., Skeletonema costatum, Ceratium furca, Bacillaria paradoxa, Hemidiscus hardmanninus, Nitzchia sp., Thalassiothrix frauenfeldii, Gyrosigma sp., Triceratium sp., Coscinodiscus gigas.*

@: Major species, #: Toxic/Pollution indicating phytoplankton.

major biological parameters were discussed below in respect to water quality change during the annual 2014-15. Total phytoplankton population and species richness was found higher during post monsoon and pre monsonn/summer due to availability of more micronutrients and light factor. The obtained results were in accordance with the findings of Sahu *et al.* (2012). But Diversity index value was observed maximum during monsoon may be due to insertion of nutrient rich rain water run-off through river discharges greatly influence the phytoPlankton

Table 35.3: Different Phytoplankton Species Detected during April-2015 (Pre-monsoon/Summer)

Station Codes	*Genera of Phytoplankton Found*
AC	*Coscinodiscus sp.,*@ *Chaetoceros sp., Triceratium sp., Rhizosolenia stolterforthii, Nitzchia sp., Ceratium inflatum, Navicula sp., Gyrosigma sp.*
OJ	*Coscinodiscus sp.,*@ *Prorocentrum micans,*# *Coscinodiscus gigas, Triceratium sp., Chaetoceros sp., Pleurosigma sp., Biddulphia pulchella, Peridinium sp.,*# *Pyrophacus steinii.*
IOHB	*Coscinodiscus sp.,*@ *Chaetoceros sp., Triceratium sp., Coscinodiscus gigas, Rhizosolenia sp., Protoperidinium sp.,*# *Rhizosolenia stolterforthii, Leptocylindrus danicus, Pleurosigma sp., Bacillaria paradox, Prorocentrum micans.*#
ADB	*Triceratium sp.,*@ *Coscinodiscus sp., Ceratium sp., Bellerochea malleus, Chaetoceros sp., Skeletonema costatum, Rhizosolenia sp., Cyclotella sp., Protoperidinium sp, Dinophysis caudate.* #
NQ	*Triceratium sp.,*@ *Coscinodiscus sp., Rhizosolenia sp., Pyrophacus steinii, Streptotheca indicus, Chaetoceros sp., Planktoneilla sol, Navicula sp., Thalassioyhrix frauenfeldii.*
GCB	*Coscinodiscus sp.,*@ *Nitzchia sp., Triceratium sp., Ceratium sp., Cyclotella sp., Rhizosolenia styliformis, Pyrophacus steinii, Protoperidinium sp.,*# *Coscinodiscus radiatus.*
SQ	*Coscinodiscus sp.,* @ *Coscinodiscus gigas, Rhizosolenia sp.,* @ *Pleurosigma sp., Ceratiums sp., Protoperidinium sp.,*# *Prorocentrum micans,*# *Nitzchia longissima*
EQ	*Coscinodiscus sp.,*@ *Pyrophacus steinii, Ceratiums sp., Coscinodiscus eccentricus, Ceratium furca, Chaetoceros sp., Dinophysis caudate.* #
FB	*Coscinodiscus sp.,*@ *Skeletonema costatum, Biddulphia sp., Triceratium sp., Rhizosolenia sp., Ceratium furca, Prorocentrum micans,*# *Pleurosigma sp., Ceratium sp.*
MS	*Coscinodiscus sp*@*., Pyrophacus steinii, Ceratium furca, Triceratium sp., Rhizosolenia sp., Dinophysis caudata,*# *Bacillaria paradoxa.*
FH	*Protoperidinium sp.,*# *Navicula sp., Rhizosolenia sp., Nitzchia sp., Coscinodiscus sp., Coscinodiscus gigas, Cyclotella sp., Ceratium sp.,*@ *Thalassiosira subtillis.*

@: Major species, #: Toxic/Pollution indicating phytoplankton.

diversification (Mishra *et al.,* 2009). Maximum count was observed at MS may due to more nutrients availability and other favourable physico-chemical conditions for their growth and proliferations. During this annual, populations of various encountered harmful pollution indicating phytoplankton like *Porocentrum micans, Prorocentrum sp., Protoperidinium sp.* and *Dinophysis caudatae* were very less in number. So they may have not such effect on the biological quality of water system and other phytoplankton community. As reported earlier, *Porocentrum means*

Table 35.4: Different Phytoplankton Species Detected during August-2015 (Monsoon)

Station Codes	Genera of Phytoplankton Found
AC	*Triceratium sp., Coscinodiscus sp.,*[@] *Prorocentrum micans,*[#] *Ceratium sp., Pyrophacus steinii, Rhizosolenia sp., Coscinodiscus gigas, Navicula sp., Streptotheca sp., Gyrosigma sp.,*
OJ	*Hemidiscus hardmanninus, Coscinodiscus sp.,*[@] *Ceratium sp., Pleurosigma sp., Bellerochea malleus, Chaetoceros sp., Triceratium sp., Rhizosolenia sp.*
IOHB	*Chaetoceros sp., Rhizosolenia sp., Leptocylindrus danicus, Ceratocorys horrida, Coscinodiscus sp., Triceratium sp*[@]*., Bellerochea malleus, Coscinodiscus radiatus, Bacteriastrum hyalinum, Peridinium sp.*[#]
ADB	*Triceratium sp., Skeletonema costatum, Biddulphia haeteroceros, Coscinodiscus sp.,*[@] *Prorocentrum micans,*[#] *Ceratium tipos, Rhizosolenia sp., Gyrosigma sp., Skeletonema sp., Ceratium furca, Cyclotella sp.*
NQ	*Pyrophacus steinni, Coscinodiscus sp., Triceratium sp., Planktoniella sol, Protoperidinium sp.,*[#] *Ceratium sp.,*[@] *Coscinodiscus gigas, Dinophysis caudate*[#] [@]
GCB	*Triceratium sp., Nitzchia sigma, Pyrophacus steinii, Rhizosolenia sp., Coscinodiscus sp.,*[@] *Cyclotella sp., Streptotheca indicus, Biddulphia haeteroceros, Chaetoceros sp.*
SQ	*Cyclotella sp., Thalassiosira subtilis, Pleurosigma sp., Coscinodiscus sp.,* [@] *Ceratium sp., Triceratium sp., Protoperidinium sp.* [#]
EQ	*Triceratium sp., Coscinodiscus sp.,*[@] *Skeletonema costatum Rhizosolenia sp., Biddulphia sp., Bellerochea malleus, Planktoniella sol, Coscinodiscus centralis, trichodesmium erythraeum.*
FB	*Dinophysis caudata*[#]*, Coscinodiscus sp.,*[@] *Pyrophacus steinii,, Triceratium sp., Pleurosigma sp., Navicula sp., Rhizosolenia sp.*
MS	*Rhizosolenia sp.,*[@] *Triceratium sp., Biddulphia haeteroceros, Thalassiothrix sp., Prorocentrum micans,*[#] *Coscinodiscus sp., Pyrophacus steinni, Gyrosigma sp., Bacillaria paradoxa, Pleurosigma elongatum, Chaetoceros sp., Skeletonema sp.*
FH	*Cyclotella sp., Coscinodiscus sp*[@]*., Biddulphia haeteroceros, Chaetoceros sp., Triceratium sp., Pleurosigma sp., Hemidiscus hardmanninus, Rhizosolenia sp.*

@: Major species, #: Toxic/Pollution indicating phytoplankton.

and *proto pyridinium* species were known to be nontoxic, but *Porocentrum* sp. are capable of forming extensive blooms and have been reported to inhibit the growth of other diatoms (Tayler and Seliger, 1979, Anderson *et al.*, 1985, Faust and Steidinger, 1998). *Protoperidinium* blooms can cause discoloration of the water (Balech, 1976; Inoue, 1990). *Dinophysis caudata* produce toxins (Dinophysistoxins and Okadaic acid) that cause diarrheic shelfish poisoning (DSP) and also involved in red tide (Okachi, 1967; Larsen and Moestrup, 1992).

Table 35.5: Statistics for Phytoplankton Analysis

Variables	*November-14 (Post-monsoon)*			*January-15 (Winter)*			*April-15 (Pre-monsoon/Summer)*			*August-15 (Monsoon)*		
	Max	*Min*	*Avg*	*Max*	*Min*	*Avg*	*Max*	*Min*	*Avg*	*Max*	*Min*	*Avg*
TPC*	1440	462	775	1442	690	949	870	480	605	990	480	712
Species*	12	7	10	13	8	10	11	7	9	12	7	9
D.I (H)*	1.499	1.171	1.380	1.442	1.075	1.307	1.597	1.303	1.427	1.559	1.216	1.441

*In numbers, TPC: total phytoplankton count, D.I.: Diversity index.

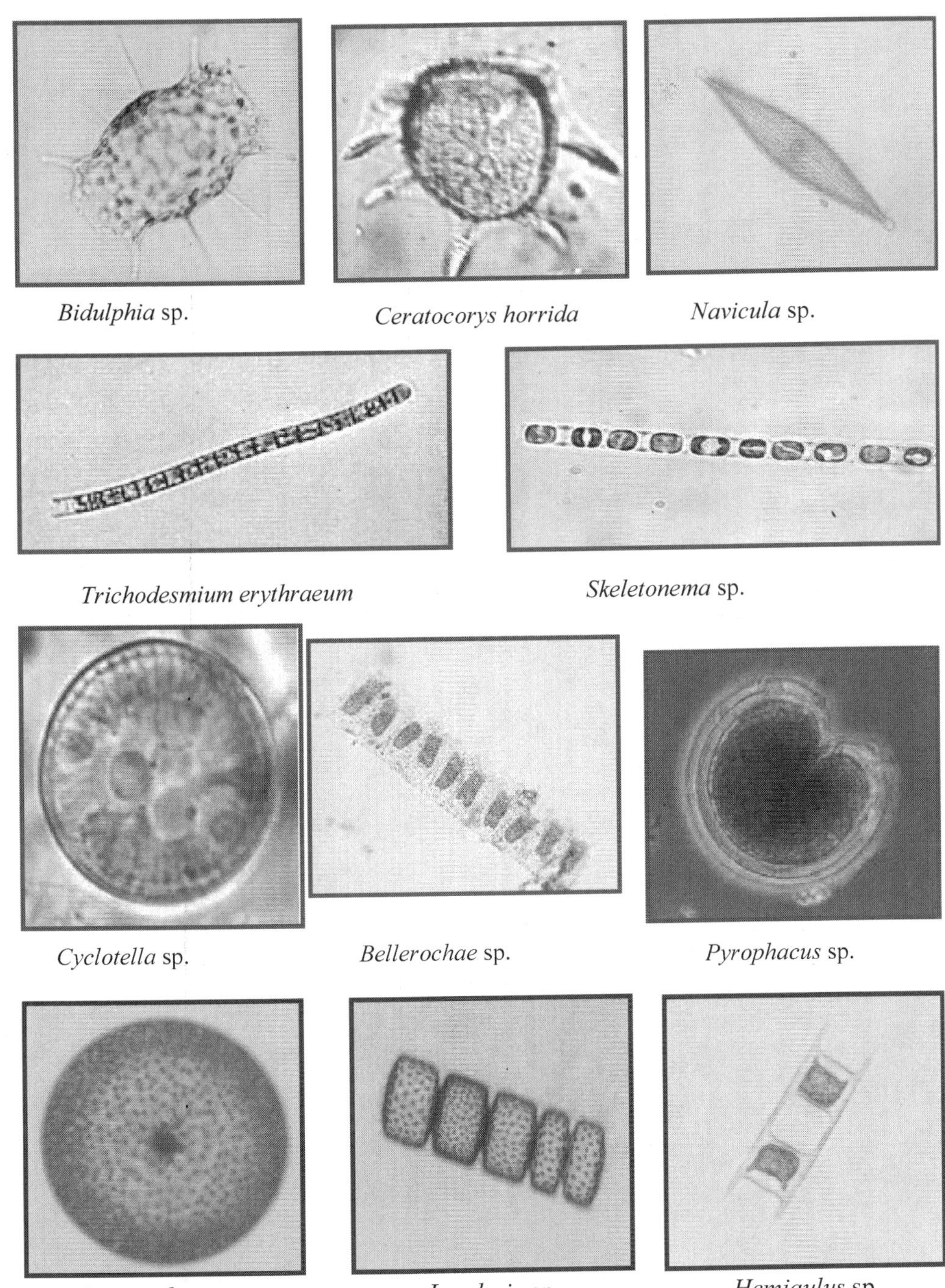

Figure 35.5: Photographs of some Commonly Encountered Phytoplankton Species during November-2014 to August-2015 at Paradip Port Harbour.

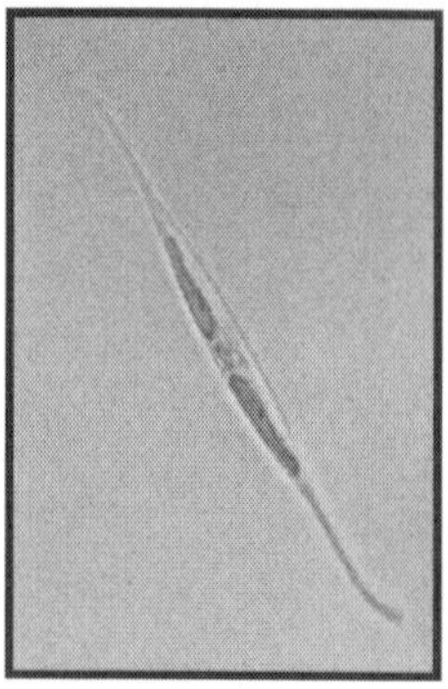

Nitzschia longissma 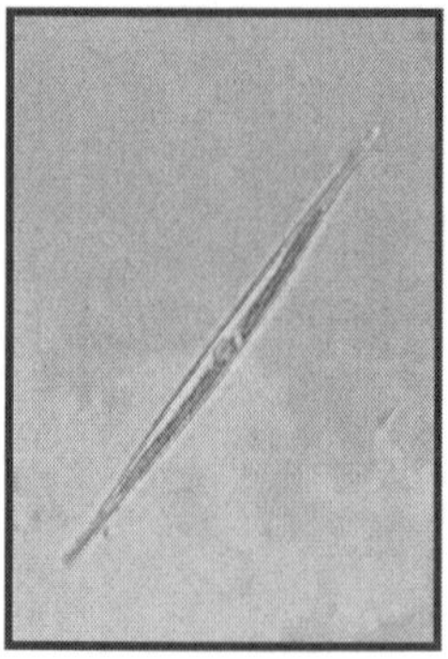 *Nitzschia longissma* 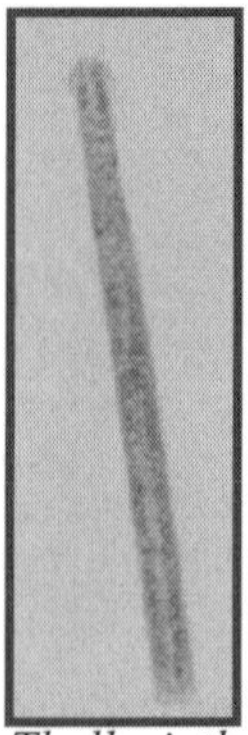 *Thallasiothrix frauenfeldii* 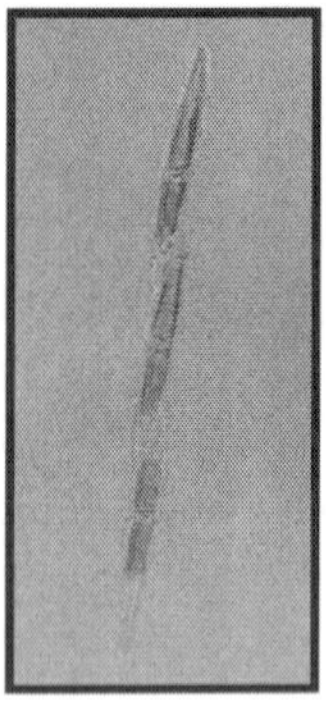 *Pseudo-nitzschia sp/ Nitzschia seriata*

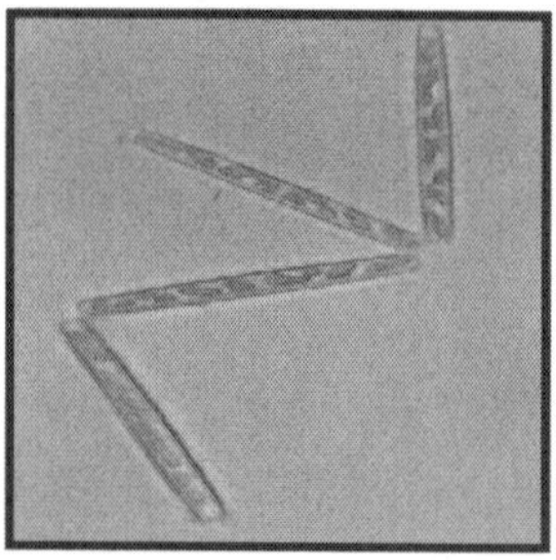

Thalassionema nitzschioides 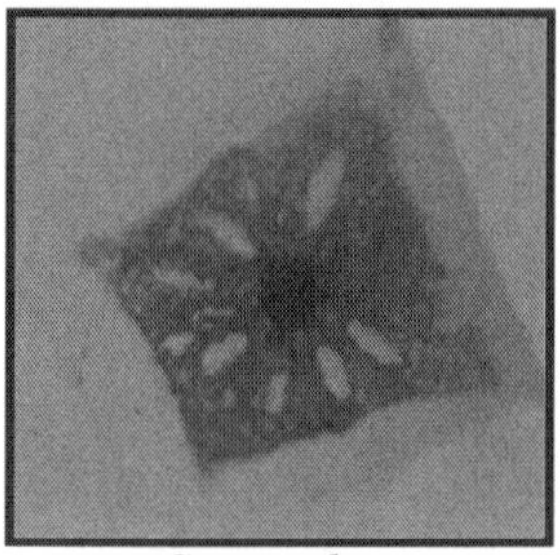 *Streptotheca* sp. 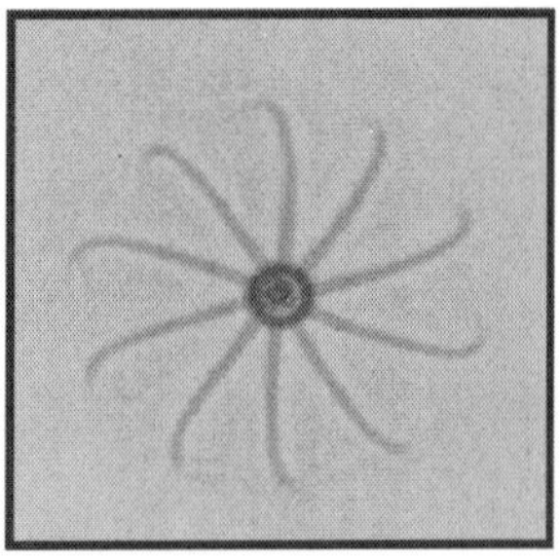 *Bacteriastrum sp*

Figure 35.5: Photographs of some Commonly Encountered Phytoplankton Species during November-2014 to August-2015 at Paradip Port Harbour.

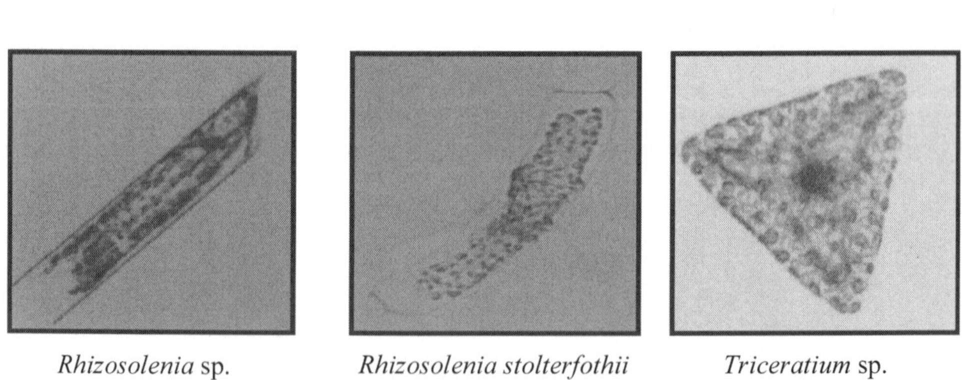

Rhizosolenia sp. *Rhizosolenia stolterfothii* *Triceratium* sp.

Figure 35.6: Some Pollution Indicating Harmful/Pollution Indicating Phytoplankton sps. Encountered at Paradip Port Harbour during November-14 to August-15.

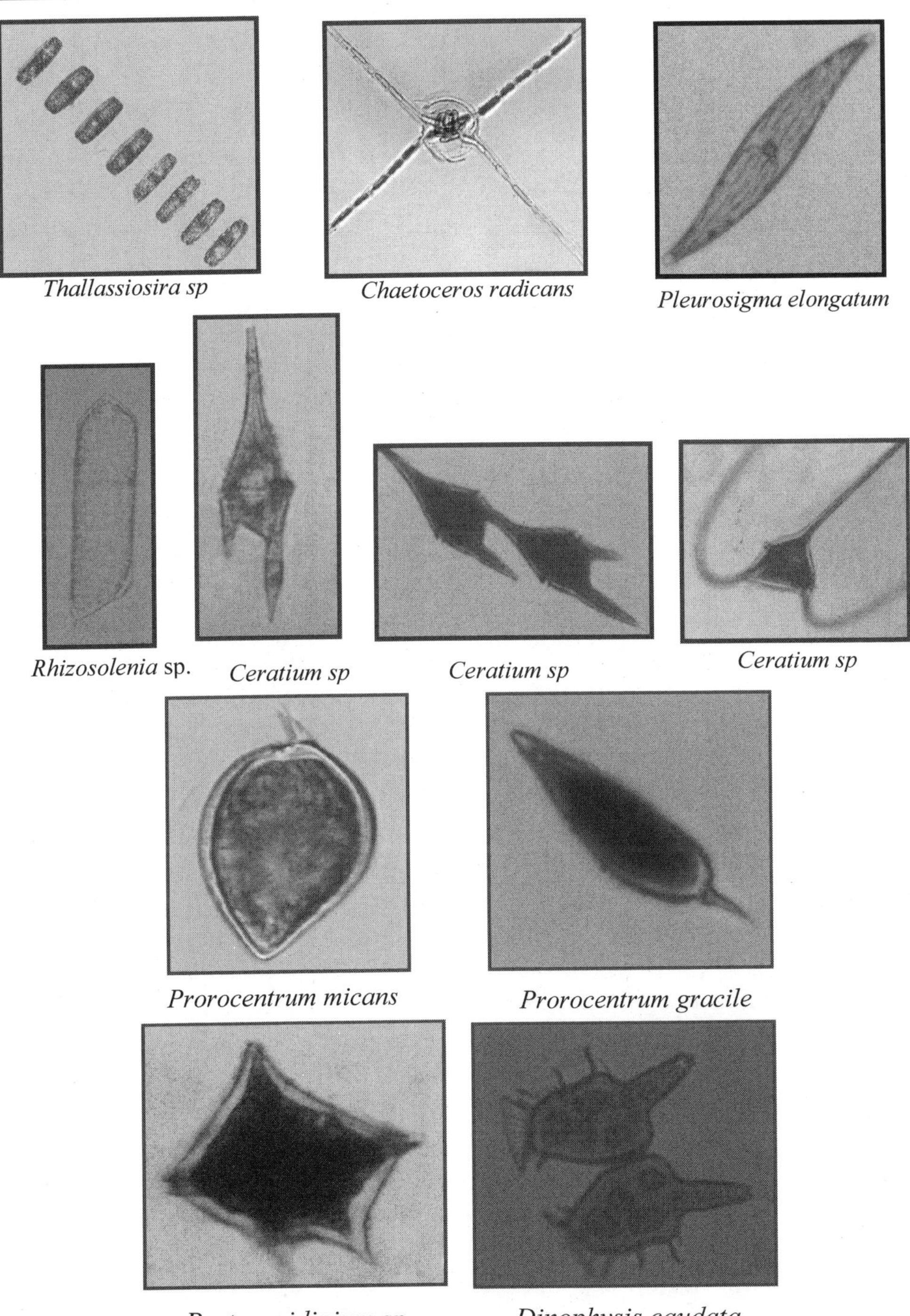

Figure 35.7: Some Pollution Indicating Harmful/Pollution Indicating Phytoplankton sps. Encountered at Paradip Port Harbour during November-14 to August-15.

REFERENCESS

Anderson, D.M., White, A.W. and Baden, 1985. Toxic *Dinoflagellates*, Elsevier, New York. 561.

APHA, *Standard methods for examination of water and wastewater, 19th edn.* (American Public Health Association, AWWA, WEF Inc., Washington, D.C.), 1995

Balech, E. 1976. Sur quelques *Protoperidinium* (Dinoflagellata). Vie et Milieu. 26: 27-46.

Campbell, J.W., Watson, A., Watson, C., Ball, H., Pirkle, R., Escherichia Coli, other coliform, and environmental chemoheterotrophic bacteria in isolated water pools from six caves in northern Alabama and northestern Georgia, Journal of Cave and Karst Studies, 73 (2011) 75-82.

Faust, M.A. and Steidinger, K.A. 1998. *Bysmatrum* gen. nov. (Dinophyceae) and three new combinations for benthic scrippsielloid species. Phycologia 37: 47-52.

Hasle, G.R., Syvertsen, E.E., Steidinger, K.A., Tangen, K., Throndsen, J., Heimdal, B.R., *Identifying marine phytoplankton*, (Academic Press Inc., California, USA) 1997 pp. 1-858.

Inoue, H. 1990. *Dinoflagellates*. In "Red tide organisms in Japan-An illustrated taxonomic guide". Uchida Rokakuho, Tokyo. pp. 150-151

Larsen, J. and Moestrup, O. 1992. Potentially toxic phytoplankton. 2. Genus *Dinophysis* (Dinophyceae). In: J.A. Lindley (ed), ICES Identification Leaflets for Plankton. ICES, Copenhagen, 180: 1-12

Madhav, V. G., and Kondalarao, B. (2004). Distribution of phytoplankton in the coastal waters of east coast of India. Indian journal of Marine sciences, 33(3), 262-268.

Mishra, R. K., Shaw, B. P., Sahu, B. K., Mishra, S., and Senga, Y. (2009). Seasonal appearance of Chlorophyceae phytoplankton bloom by river discharge off Paradeep at Orissa Coast in the Bay of Bengal. Environmental monitoring and assessment, 149(1-4), 261-273.

Naik, S., Acharya, B. C., and Mohapatra, A. (2009). Seasonal variations of phytoplankton in Mahanadi estuary, east coast of India. Indian Journal of Marine Sciences, 38(2), 184.

Naik, S., Mishra, R. K., Mahapatro, D., and Panigrahy, R. C. (2014). Impact of water quality on phytoplankton community and biomass in Dhamara estuary east coast of India. Journal of Environmental Biology, 35(1), 229.

Okaichi, T. 1967. Red tides found in and around the Seto Inland Sea in 1965. Tech. Bull. Fac. Agricul. Kagawa Univ. 15: 181-185.

Palleyi, S., Kar, R.N., Panda, C.R., Influence of Water quality on the biodiversity of phytoplankton in Dhamra River Estuary of Odisha Coast, Bay of Bengal', *Journal of Applied Science Environ. Management,* 15 (2011) 69-74.

Panda, S. S., Dhal, N. K., and Panda, C. R. (2012). Phytoplankton diversity in response to abiotic factors along Orissa coast. Bay of Bengal International Journal of Environmental Sciences, 2(3).

Sahu, G., Satpathy, K. K., Mohanty, A. K., and Sarkar, S. K. (2012). Variations in community structure of phytoplankton in relation to physico-chemical properties of coastal waters, southeast coast of India. Indian Journal of Geo-Marine Sciences, 41(3), 223.

Sarojini, Y., and Sarma, N. S. (2001). Vertical distribution of phytoplankton around Andaman and Nicobar Islands, Bay of Bengal. Indian journal of marine sciences, 30(2), 65-69.

Shannon, C.E. and Weaver, W. (1949). The mathematical theory of communication. The University of Illinois Press Inc., Urbana, pp. 117.

Taylor, D.L. and Seliger, H.H. (1979). Toxic *Dinoflagellate* Blooms. Elsevier, North Holland, New York. 505.

Verlencar, X. N., and Desai, S. (2004). Phytoplankton Identification Manual: National Institute of Oceanography Dona Paula. Goa, India., pp. 1-35.

Index

D

E

F

G

N

O

P

Q

R